Change

Climate change is the single most important global environmental and development issue facing the world today, and has emerged as a major topic in tourism studies. Climate change is already affecting the tourism industry and is anticipated to have profound implications for tourism in the twenty-first century, including consumer holiday choices, the geographical patterns of tourism demand, the competitiveness and sustainability of destinations and the contribution of tourism to international development.

Tourism and Climate Change: impacts, adaptation and mitigation is the first book to provide a comprehensive overview of the theory and practice of climate change and tourism at the tourist, enterprise, destination and global scales. Major themes include the implications of climate change and climate policy for tourism sectors and destinations around the world, tourists' perceptions of climate change impacts, tourism's global contribution to climate change, adaptation and mitigation responses by all major tourism stakeholders, and the integral links between climate change and sustainable tourism. It combines a thorough scientific assessment of the climate–tourism interrelationships with discussion of emerging mitigation and adaptation practices, showcasing international examples throughout the tourism sector as well as actions by other sectors that will have important implications for tourism.

Written by three leading academics in the field, this critical contribution highlights the challenges of climate change within the tourism community and provides a foundation for decision-making for both reducing the risks and taking advantage of the opportunities associated with climate change. This comprehensive discussion of the complexities of climate change and tourism is essential reading for students, academics, business leaders and government policy-makers.

Daniel Scott is a Canada Research Chair in Global Change and Tourism at the University of Waterloo, Canada.

C. Michael Hall is a Professor in the Department of Management, University of Canterbury, New Zealand; Docent in the Department of Geography, University of Oulu; and a Visiting Professor at Sheffield Hallam University, UK and Linneaus University, Sweden.

Stefan Gössling is a Professor at the Department of Service Management, Lund University and the School of Business and Economics, Linnaeus University, both Sweden. He is also research coordinator at the Research Centre for Sustainable Tourism, Western Norway Research Institute.

Contemporary Geographies of Leisure, Tourism and Mobility

Series Editor: C. Michael Hall, Professor at the Department of Management, College of Business and Economics, University of Canterbury, Christchurch, New Zealand

The aim of this series is to explore and communicate the intersections and relationships between leisure, tourism and human mobility within the social sciences.

It will incorporate both traditional and new perspectives on leisure and tourism from contemporary geography, e.g. notions of identity, representation and culture, while also providing for perspectives from cognate areas such as anthropology, cultural studies, gastronomy and food studies, marketing, policy studies and political economy, regional and urban planning, and sociology, within the development of an integrated field of leisure and tourism studies.

Also, increasingly, tourism and leisure are regarded as steps in a continuum of human mobility. Inclusion of mobility in the series offers the prospect to examine the relationship between tourism and migration, the sojourner, educational travel, and second home and retirement travel phenomena.

The series comprises two strands:

Contemporary Geographies of Leisure, Tourism and Mobility aims to address the needs of students and academics, and the titles will be published in hardback and paperback. Titles include:

1. **The Moralisation of Tourism**
   Sun, sand….and saving the world?
   Jim Butcher

2. **The Ethics of Tourism Development**
   Mick Smith and Rosaleen Duffy

3. **Tourism in the Caribbean**
   Trends, development, prospects
   Edited by David Timothy Duval

4. **Qualitative Research in Tourism**
   Ontologies, epistemologies and methodologies
   Edited by Jenny Phillimore and Lisa Goodson

Routledge Studies in Contemporary Geographies of Leisure, Tourism and Mobility
is a forum for innovative new research intended for research students and academics,
and the titles will be available in hardback only. Titles include:

Forthcoming:

Sexuality, Women and Tourism
Susan Frohlick

Gender and Tourism
Social, cultural and spatial perspectives
Cara Atchinson

Backpacker Tourism and Economic Development in the Less Developed World
Mark Hampton

Adventure Tourism
Steve Taylor, Peter Varley and Tony Johnson

Dark Tourism and Place Identity
Elspeth Frew and Leanne White

Scuba Diving Tourism
Kay Dimmcock and Ghazali Musa

Travel, Tourism and Green Growth
Min Jiang, Terry DeLacy and Geoffrey Lipman

Tourism and Climate Change

Impacts, adaptation and mitigation

Daniel Scott, C. Michael Hall
and Stefan Gössling

Routledge
Taylor & Francis Group

LONDON AND NEW YORK

First published 2012
by Routledge
2 Park Square, Milton Park, Abingdon, Oxon OX14 4RN

Simultaneously published in the USA and Canada
by Routledge
711 Third Avenue, New York, NY 10017

Routledge is an imprint of the Taylor & Francis Group, an informa business

© 2012 Daniel Scott, C. Michael Hall and Stefan Gössling

British Library Cataloguing in Publication Data
A catalogue record for this book is available from the British Library

Library of Congress Cataloging-in-Publication Data

Scott, Daniel, 1969-
  Tourism and climate change : impacts, adaptation & mitigation / Daniel Scott,
Michael Hall, Stefan Gossling.
     p. cm.
  Includes bibliographical references and index.
  ISBN 978-0-415-66885-9 (hbk : alk. paper) – ISBN 978-0-415-66886-6
(pbk : alk. paper) 1. Tourism–Environmental aspects. 2. Climatic changes. I. Hall,
Colin Michael, 1961- II. Gössling, Stefan. III. Title.
  G156.5.E58S46 2012
  363.738′74–dc23

ISBN: 978-0-415-66885-9 (hbk)
ISBN: 978-0-415-66886-6 (pbk)
ISBN: 978-0-203-12749-0 (ebk)

Typeset in Times New Roman
by Cenveo Publisher Services

MIX
Paper from
responsible sources
FSC
www.fsc.org   FSC® C004839

Printed and bound in Great Britain by
TJ International Ltd, Padstow, Cornwall

Contents

Figures

Tables

Boxes

Foreword

Climate change has emerged as one of the most significant and controversial areas of academic research and government policy in recent years. This controversy relates not so much to scientific debate over climate change, although differences in scientific understandings certainly exist, but more to the ways in which the possible responses to climate change are interpreted by diverse interests and worldviews. Due to the complexity and contrasting interpretations of climate change, it has been described as a 'wicked problem', one that thus far has defied consensual response by the international community. This, combined with the media's portrayal of climate change research and policy debates, has led to a situation in which tourism students and tourism professionals have been demanding improved and credible resources that can help them navigate the complex dimensions of climate change, as well as assess some of the key areas where knowledge gaps exist and where further research is needed to support tourism-relevant policy, planning and investment decisions.

There is a vast body of literature on climate change that continues to grow rapidly year on year. Given the economic, social and environmental significance of tourism, the coverage of tourism in the climate change literature, and in particular in some key governmental and institutional reports, could be described as patchy to weak at best. In many cases, coverage of tourism is virtually non-existent, or is discussed so generally that the information provided is not helpful for the development of climate change strategies. Although there appears to be consensus that tourism as a sector is particularly vulnerable to climate change, some tourism stakeholders and destinations could realize considerable opportunities. Furthermore, some of the discussion of tourism is based on studies where the assumptions and limitations have not been made clear or are not informed by the substantial body of literature on tourism. There remain limited connections between the increasingly multidisciplinary literature on climate change and tourism, and broader fields of tourism knowledge (particularly tourist behaviour, marketing and sustainable tourism).

Climate change is already affecting tourism destinations and the wider industry, for example with conditions for winter tourism becoming less reliable in some areas, and rising energy prices making travel more expensive. This has led to growing interest from the tourism sector in understanding the potential impacts of climate change on tourism, and in developing adaptation and mitigation responses. The impacts of climate change and climate policy have also attracted much media attention, and tourism stakeholders, including students, are often confronted with information that does not reflect scientific insights, and in some cases has been shown to be outright misinformation. It is very challenging to keep up to date with the multiple facets of the rapidly growing climate change and tourism literature, and sometimes difficult to decipher the credibility of media stories and other assertions about the vulnerability of tourism or its contribution to climate change and emission reduction solutions.

This book offers one response to the educational and professional needs related to climate change. We have endeavoured to provide a comprehensive and critical synthesis of the relevant multidisciplinary work on tourism and climate change under a single cover. The book is written primarily for upper-level undergraduate and postgraduate students to assist them in gaining a better understanding of tourism and climate change, but its critical reflections on the field of scholarship and practice are also designed to be useful to more senior scholars entering this field and professionals seeking to make evidence-based decisions. The book provides introductions to key concepts and terminology in climate change science and tourism studies, as well as a detailed discussion of the significance of weather and climate for tourism. In-depth, critical examinations of the key themes and issues with respect to emissions, mitigation, impacts and adaptation are also provided, along with a comprehensive set of references should readers wish to follow up specific research or policy themes. The book concludes by relating climate change to some of the larger issues of sustainability in general, as well as identifying knowledge gaps and a future research agenda.

As this volume testifies, we do not think tourism will disappear as a result of near-term climate change – far from it. Speculation as to tourism's sudden demise is just that. But make no mistake – climate change will transform tourism. As the book details, all components of the tourism system will need to adapt to the new realities of human-induced climate change and an increasingly carbon-constrained global economy. Tourism, along with much of our world, must change and adapt if we are to keep many of the places that we love and maintain the ability of future generations to travel and experience them. For the sake of our own young children and all future generations, we must genuinely hope that the worst possibilities of climate change are avoided through social change; however, the outcome is largely determined by what *we* do today. We very much hope that this book, even if only a small contribution, will help improve understanding of the challenges climate change poses to society and will foster changes needed to advance the practice of sustainable tourism.

Acknowledgements

Daniel would first like to thank his co-authors for the highly stimulating discussions over the course of this project. Other colleagues also deserve mention for long ago inspiring my interest in global change and tourism, as well as for their special contributions to the knowledge presented in this volume: Geoff McBoyle, Geoff Wall, Chris de Freitas, Allan Perry and Bernard Lane. The support is gratefully acknowledged of the Canada Research Chair programme, without which participation in such international collaborations would not be possible. The late-night research and manuscript preparation efforts of Michelle Rutty and Lindsay Matthews must also be noted. Most of all, I cannot express enough my gratitude to Tonia, Danika and Isabel for all of their love and support.

Stefan wishes to express his gratitude to Meike and Linnea for their patience, love and support during the weekends and holidays that I was busy writing my parts of this book.

Michael would like to gratefully acknowledge the support of his co-authors during the writing of the book, in particular for their support and understanding as a result of the Christchurch earthquakes of 2010–11, in which both house and office were severely damaged and became unusable. Becoming an environmental refugee is certainly one way to gain new insights into crisis, resilience and adaptation, although not one to be recommended. In that light, I cannot emphasize enough my thanks to friends Peter and Jane Firth for providing post-earthquake accommodation that enabled me to carry on lecturing and working at Canterbury University in 2011 (and hence to continue to earn an income). Similarly, I would also like to thank Jim and Nita for welcoming the extended family into their house and for being able to use their garage as my office. Pauline and Keith were also there when needed, as always. For Jim, Nita, Peter and Jane, Pauline and Keith, the thanks for their love and support cannot be expressed enough. The support of numerous colleagues, both in New Zealand and overseas, must also be acknowledged. The help and support of Tim Baird, Melissa Su, Paul Ballantine and Tony Garry at Canterbury, as well as the magnificent office staff, Donna, Irene and Irene, must also be noted.

In addition, a number of colleagues with whom Michael has undertaken research or discussed tourism issues over the years have also contributed to this volume. In particular, thanks to Bill Bramwell, Richard Butler, Tim Coles, Chris Cooper, David Duval, Anna Grundén, David Harrison, James Higham, John Jenkins, Bernard Lane, Dieter Müller, Stephen Page, Jarkko Saarinen, Anna Dóra Sæþórsdóttir, Brian and Delyse Springett, Sandra Wall, Sandra Wilson and Allan Williams for their thoughts, as well as for the stimulation of Jeff Buckley, Nick Cave, Bruce Cockburn, Elvis Costello, Stephen Cummings, Chris Difford and Glenn Tilbrook, Dimmer, Ebba Fosberg, Hoodoo Gurus, Ed Kuepper, Jackson Code, Vinnie Reilly, David Sylvian, Jennifer Warnes, Chris Wilson, and *The Guardian* and BBC – without whom the four walls of many a hotel room would be much more confining. Finally, Michael would like to thank the many people who have supported his work over the years, and especially the Js and the Cs who stay at home, or whatever we now call home.

We would all like to extend our thanks to our editor Emma Travis at Routledge as well as to the rest of the team who have supported us over the project, particularly Carol Barber and Faye Leerink.

Abbreviations and acronyms

3S	sun, sand and sea
A$	Australian dollar
A/C	air conditioning
AGD	Aviation Global Deal
AIC	aviation-induced clouds
AOSIS	Alliance of Small Island States
APD	air passenger duty
APELL	Awareness and Preparedness for Emergencies at the Local Level
AR4	IPPC *Fourth Assessment Report*
ATEC	Australian Tourism Export Council
ATM	air traffic management
ATV	all-terrain vehicle
BRT	bus rapid transit
CARICOM	Caribbean Community
CBDR	Common But Differentiated Responsibilities
CCP	Cities for Climate Protection
CDD	cooling degree days
CDM	clean development mechanism
CEC	Commission of the European Communities
CIE	Coras Iompair Eireann (Irish Public Transport Authority)
COP	Conference of the Parties
COP-3	Third Session of the Conference of the Parties
DMO	destination-marketing organization
DST	decision support tool
EANO	equal advantage for non-ownership
ECA	Emission Control Area
EDEN	European Tourist Destinations of Excellence
EECA	Energy Efficiency and Conservation Authority
EEDI	Energy Efficiency Design Index
EEOI	Energy Efficiency Operational Indicator
ENSO	El Niño Southern Oscillation

EPA	Environmental Protection Agency
ETS	Emissions Trading Scheme
EU 27	European Union with 27 Member States
GATT	General Agreement on Tariffs and Trade
GBEP	global bio-energy partnership
GBR	Great Barrier Reef
GCM	global climate model
GDP	gross domestic product
GDS	global distribution system
GHG	greenhouse gas
GIACC	Group on International Aviation and Climate Change (of the ICAO)
GWP	global warming potential
HDD	heating degree days
IAPAL	International Air Passenger Adaptation Levy
IATA	International Air Travel Association
IBAC	International Business Aviation Council
ICAO	International Civil Aviation Organization
ICLEI	International Council for Local Environmental Initiatives
IEA	International Energy Agency
IMO	International Maritime Organization
IPCC	Intergovernmental Panel on Climate Change
IRU	International Road Transportation Union
LDC	least developed country
LNG	liquefied natural gas
LRET	large-scale renewable energy scheme
LTMS	long-term mitigation scenario
MARPOL	International Convention for the Prevention of Pollution from Ships, 1973 as modified by the Protocol of 1978 (for MARine POLlution)
MEA	Millennium Ecosystem Assessment
NAC	national adaptation capacity
NCB	Nuffield Council on Bioethics
NCCS	national climate change strategy
NECSTour	Network of European Regions for a Sustainable and Competitive Tourism
NGO	non-governmental organization
NOAA	National Oceanic and Atmospheric Administration
NPP	net primary production
O/D	origin–destination (transport)
OECD	Organisation for Economic Co-operation and Development
OUV	outstanding universal value
PATA	Pacific Asia Travel Association
PTLF	Pacific Tourism Leaders' Forum
PV	photovoltaic cell
RCP	representative concentration pathway
REDD	reducing emissions from deforestation and degradation

REFIT	renewable electricity feed in tariff
RET	renewable energy target
RF	radiative forcing
RSB	Roundtable on Sustainable Biofuels
SCCP	Special Climate Change Programme
SECA	SO_x emission control area
SIDS	small island developing states
SME	small to medium-sized enterprise
SRES	small-scale renewable energy scheme
SRES	*Special Report on Emission Scenarios*
STCRC	Sustainable Tourism Cooperative Research Centre
TCI	tourism climatic index
TEEP	Tourism Energy Efficiency Programme
TFO	tourist flight operator
TIA	Tourism Industry Association
TMB	travel money budget
TNA	technology needs assessment
TSA	Tourism Satellite Account
TTB	travel time budget
UCLG	United Cities and Local Governments
UKERC	UK Energy Research Centre
UNCED	United Nations Conference on Environment and Development
UNCLOS	United Nations Convention on the Law of the Sea
UNEP	United Nations Environment Programme
UNFCCC	United Nations Framework Convention on Climate Change
UNWTO	United Nations World Tourism Organization
USCM	US Conference of Mayors
VFR	visiting friends and relations
VRT	vehicle registration tax
WCED	World Commission on Environment and Development
WCI	Western Climate Initiative
WEF	World Economic Forum
WMCCC	World Mayors Council on Climate Change
WMO	World Meteorological Organization
WTTC	World Travel and Tourism Council

CHEMICAL ABBREVIATIONS

C_3	plant species that use the C_3 carbon-fixation pathway (converting CO_2 plus sugars into 3-phosphoglycerate) – they tend to thrive in areas with relatively high CO_2 and water levels
CH_4	methane
CO_2	carbon dioxide
HFC	hydrofluorocarbon
LPG	liquid petroleum gas

N_2O	nitrous oxide
NMVOC	non-methane volatile organic compounds
NO_x	nitrogen oxides
PFC	perfluorcarbon
PM	particulate matter/material
PM_{10}	particulate matter/material with aerodynamic diameter 10 micrometres or less
SO_x	sulphur oxides

UNITS

CO_2-eq	carbon dioxide equivalent
Gt	gigatonne (10^{15} grams), also referred to as a petagram
GWh	gigawatt hour
ha	hectare
m^3	cubic metres
Mt	megatonne
Pg	petagram (10^{15} grams), also referred to as a gigatonne
pkm	passenger kilometre
ppb	parts per billion
ppmv	parts per million by volume
skm	seat kilometre
t	tonne
W/m^2	watts per square metre

1 Introduction

This chapter provides a brief introduction to the subject of tourism and climate change and provides a framework for the rest of the book. It outlines some of the key concepts that help us understand the relationship between tourism and climate change. This includes a brief overview not only of the science of climate change and how it has developed, but also of the notion of a tourism system, and how climate change issues are part of broader concerns about how to make tourism sustainable. The chapter concludes with an overview of the book's structure.

THE CHALLENGE OF CLIMATE CHANGE AND TOURISM

Climate change is arguably one of the most important issues facing the world today. It is also probably one of the most contentious. Not only do many people have an interest in the topic and often some quite firm viewpoints, but also it is a major issue in international diplomatic, business and scientific discussion. When reading a quality newspaper such as *The Guardian*, *The Washington Post* or *Le Monde*, barely a day goes by without an article on some aspect of climate change. The issue also features regularly on the television screen and on talk radio, where everybody seems to be talking about the weather, how strange it is (somewhere), how it might be changing, and why.

Travel and tourism is also a seemingly ubiquitous topic of conversation and media coverage. From television programmes on dream holiday getaways through to regular newspaper coverage on destinations and attractions, the prospect of travel seems never far away. In most countries with satellite or cable television, there are even channels completely dedicated to travel, while the close relationship between tourism and the weather is indicated by coverage of the weather in major international destinations in most national newspapers, television news and weather channels.

From all this media coverage, you would probably think it would be easy to put together a book on tourism and climate change. But actually, no, it isn't. This is because we are often dealing with many taken-for-granted assumptions as to what tourism – and climate change – is. After all, everyone who will pick up this book has experienced both climate and travel. We are not suggesting, of course, that personal knowledge of climate and tourism is of little value, far from it. Indeed, it is a vital part of understanding the puzzle of how tourists experience and perceive weather and climate, and hence incorporate it into their travel decisions. However, in this book we also want to move beyond the individual experience to try and gain an appreciation of the bigger picture of the multiple and complex relationships between tourism and climate, and climate change, at the levels of tourists, destinations, countries, and the planet as a whole. We believe this is important not only for improving understanding of the interrelationships between tourism and climate change, but also to understand better the behaviour of governments, businesses and tourists – our behaviour – in light of a growing awareness of the environmental consequences of travel.

A key role of this book, then, is not only to provide a systematic overview of the interrelationships between tourism and climate change, but also to critique what we know about these interrelationships, and to ask some difficult questions about what the future might look like for the tourism industry, destinations and individual travellers in an environmentally constrained world. Like other areas of climate change research and policy, tourism has seen its share of early speculation and contrasting perspectives, which demand careful, information-based consideration. A central objective of this book is to provide a much needed critical reflection on the first 25 years of climate change and tourism research and practice.

Tourism is currently considered one of the major global economic sectors that are least prepared for climate change. Consequently, this book is designed to inform all tourism stakeholders (government, business, non-governmental organizations, international development agencies and their donors, and academe) about the transformative challenges that climate change poses for tourism. Most importantly, however, this book is aimed principally at the tourism students and young professionals who will experience first-hand the full range of climate change effects on tourism destinations and businesses, and will be the innovators and decision-makers who determine the responses of the tourism community to climate change and its associated risks and opportunities.

As it is a multidisciplinary field of research, a common language for researchers, tourism stakeholders and climate change practitioners is also important. Tourism scholars need to use the terms of climate change science and policy-making correctly. There are recent examples from climate change-related publications in tourism journals and books where the central responses of mitigation and adaptation are not correctly differentiated, or that refer to 'radioactive forcing' instead of 'radiative forcing' (see Box 1.2). Scholars from other fields similarly need to utilize correctly the concepts and terminology of the tourism sector they purport to study. To facilitate improved collaboration going forward, this book defines tourism, climate change and other key terms and concepts within each chapter.

Before moving on to discuss the interrelationships between tourism and climate change, we first provide an introduction to how we understand tourism and its impacts on society and the environment, as well as a brief review of the science of climate change and the international response to this vital development challenge.

TOURISM 101: UNDERSTANDING TOURISM AND ITS IMPACTS

Tourism is a 'slippery', 'fuzzy' concept (Markusen 1999). It is relatively easy to visualize, yet difficult to define with precision because its meaning changes depending on the context of its analysis, purpose and use. Tourism is therefore a concept that, while initially appearing very easy to define, is actually quite complex, with a substantial literature just on the issue of its definition (see Coles *et al.* 2005; Smith 2004; Hall and Lew 2009). Much of the problem with considering the concept of tourism is that most people think of tourism just in terms of leisure travel or being on a holiday or vacation. However, as Figure 1.1 illustrates, the concept is much wider than that.

Three types of tourism are usually recognized with respect to tourism statistics:

- domestic tourism, which includes the activities of resident visitors within their home country or economy, either as part of a domestic or an international trip;
- inbound tourism, which includes the activities of non-resident visitors within the destination country or economy, either as part of a domestic or an international trip (from the perspective of the traveller's country of residence);
- outbound tourism, which includes the activities of resident visitors outside their home country or economy, either as part of a domestic or an international trip.

Confusion over the definition of tourism does not end here. The word 'tourism' is used to describe tourists (people who engage in voluntary return mobility), as well as the tourism

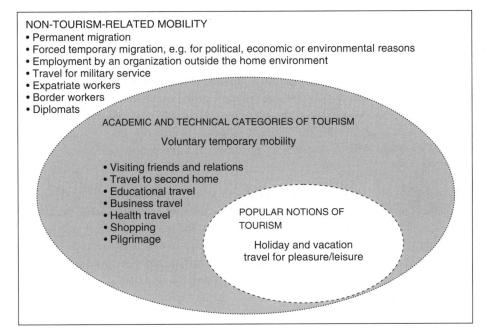

Figure 1.1 *Popular and academic conceptions of tourism*
Source: Hall and Lew (2009)

industry (for-profit businesses, organizations and individuals that enable tourists to travel). And, to complicate matters further, 'tourism' (or the tourism sector) is also used to refer to the whole social and economic phenomenon of tourism, including tourists, the tourism industry, governmental and non-governmental organizations, and the people and places that comprise tourism destinations and landscapes (Hall and Lew 2009).

Tourism products and their consumption present further definitional challenges because many of them serve both tourists and non-tourists (Smith 2004). In particular:

- visitors consume both tourism and non-tourism commodities and services;
- locals (non-visitors) consume both tourism and non-tourism commodities and services;
- tourism industries produce (and often consume) both tourism and non-tourism commodities and services;
- non-tourism industries produce (and often consume) both tourism and non-tourism commodities and services.

In addition to the basic concepts of tourist and tourism, several other seemingly simple terms require surprisingly complicated definitions to understand the positive and negative impacts of tourism on destinations and societies. For example, the concept of the 'home' or 'usual' environment or economy of an individual tourist is an important dimension of tourism definitions and statistics. It refers to the geographical (spatial or jurisdictional) boundaries within which an individual routinely moves in their regular daily life. Tourism exists outside the home environment. The United Nations World Tourism Organization (UNWTO 1994) recommends that an international tourist be defined as:

> a visitor who travels to a country other than that in which he or she has his or her usual residence for at least one night but not more than one year, and whose main purpose of visit is other than the exercise of an activity remunerated from within the country visited.

Although the terms are not used consistently around the world, those who do not stay away from their usual residence for at least one night are called excursionists or day-trippers, or in some cases recreationists. One of the consequences of increased access to fast-transport technology, such as jet aircraft and high-speed trains, is that trips that a number of years ago would have had to be undertaken as an overnight trip can now be done as a day-trip (Hall 2005).

The term trip is also used extensively in tourism studies, and refers to the movement of an individual outside their home environment until they return. The term actually refers to a 'round trip'. The trip concept, and its implications for understanding impacts, can be understood through what is referred to as the tourism system, usually conceptualized as a spatial system (Hall and Page 2006; Hall 2008b). The tourism system includes the various elements that make up a trip: the generation or origin region (or place) of the tourist, the transit region through which the tourist travels, the destination where the tourist is going, and the environment in which these exist (Figure 1.2).

The notion of a tourism system is extremely important when we start to consider the notion of impacts. If we are to evaluate the effects of tourist trips, we need not only to examine where in the system immediate impacts occur (local effects), but also to consider if there

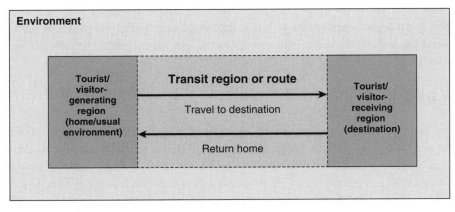

Figure 1.2 *Elements of a geographical tourism system*

are system-wide effects that contribute to change in the broader environment. In the case of maritime ship emissions, for example, nitrogen oxides (NO_x) have both local and global impacts on atmospheric pollution. In the Los Angeles–Long Beach area of California, which is both the United States' busiest port and the country's most polluted area, ocean-going vessels are among the largest sources of NO_x, emitting more NO_x than all power plants and refineries in the region's air basin combined. This is significant because, as well as being a greenhouse gas (GHG) that contributes to global warming and climate change, NO_x reacts with volatile organic compounds in the atmosphere to produce ozone/smog. Particulates from marine vessels also create significant cancer risks, with an estimated more than 700 premature deaths caused in the Los Angeles area annually by these emissions, as of 2008 (testimony of Barry R. Wallerstein, Executive Officer, South Coast Air Quality Management District at Legislative Hearing on the Marine Vessel Emissions Reduction Act of 2007, S. 1499, US Senate, Committee on Environment and Public Works, 14 February 2008, in McCarthy 2009: 1). Because of their often localized nature, the effects of NO_x are felt not just in ports, but also along shipping transit routes. For example, Santa Barbara County in California, 'which has no commercial ports, estimates that by 2020, 67 percent of its NO_x inventory will come from shipping traffic transiting the California coast' (testimony of Bryan Wood-Thomas, US EPA, Office of Transportation and Air Quality, at Legislative Hearing on the Marine Vessel Emissions Reduction Act of 2007, S. 1499, US Senate, Committee on Environment and Public Works, 14 February 2008, in McCarthy 2009: 3). (See chapters 3 and 4 for a more detailed discussion of maritime emissions.)

One historical difficulty in examining the effects of tourism is that studies have tended to examine impacts only in terms of one element of a tourism system, usually the destination, rather than examining the environmental effects of a tourism system as a whole (Hall 2004; Hall and Lew 2009). However, clearly people have to leave from their home environment, travel to the destination, and return. Therefore it is nonsensical to examine the effects of tourism by looking only at destinations. This is not to suggest, of course, that destination impacts are unimportant – they clearly are; rather, it is to highlight that the impacts of tourism go beyond the destination and may even be global in scale, as well as affecting different parts of the tourism system in different ways. As we examine in chapter 5, the impacts associated with climate change exemplify these issues of scale and the need for systems thinking with respect to the assessment of tourism and its role in sustainable development.

CHANGE HAPPENS

The global environment is always changing, although change is never uniform across time and space. Nevertheless, 'all changes are ultimately connected with one another through physical and social processes alike' (Meyer and Turner 1995: 304). However, what is most significant with respect to present environmental change is that it is not just due to natural processes. Rather, the scale and rates of change have increased dramatically as a direct result of human action related to the consumption of natural resources, the creation of new habitat for humans, and the waste products of human consumption and production. Human impacts on the environment can be regarded as having a global character in two ways. First, 'global refers to the spatial scale or functioning of a system' (Turner *et al.* 1990: 15), for example, the climate and the oceans have the characteristic of a global system. Second, global environmental change occurs if a change is cumulative, 'occurs on a worldwide scale, or represents a significant fraction of the total environmental phenomenon or global resource' (Turner *et al.* 1990: 15–16), for example, deforestation and desertification. Tourism is regarded as simultaneously a significant contributor and sensitive to both types of change (Gössling 2002; Gössling and Hall 2006a; Hall and Lew 2009).

Although environmental change occurs at a global scale, regional and even local analyses are essential. Changes have different expressions and different consequences in different regions (Meyer and Turner 1995). Many significant studies of environmental change, including with respect to tourism, are also organized on a regional or national basis so as to reflect geographical, political and even economic and social commonalities and associated responses, as well as national interests. For example, in the past decade a number of studies have been conducted that specifically examine the various dimensions of regional environmental change in the Arctic and its consequences (AHDR 2004; ACIA 2005; Hall and Saarinen 2010a; Forbes 2011).

Thinking of tourism as a system also raises questions about the nature of impacts. An impact is a change in a given state over time as the result of an external stimulus (Hall and Lew 2009). The way the term is used usually implies that tourism has an effect on something, be it a place, person, environment or economy. The term also often suggests that this is a unidirectional or 'one-way' effect. However, within systems thinking it becomes apparent that if, for example, tourists affect one part of the tourism system, then there is potential for feedback to occur, which may affect other parts of the system (Figure 1.3 illustrates these connections). Tourism impacts are very rarely, if ever, a one-way relationship, while it is also essential to be aware of the wider context within which they are situated (see also Abegg *et al.* 1998; Scott 2006a). As Hall and Lew (2009) comment:

> The term tourism impacts is usually used as a kind of shorthand – and a poor one at that – to
> describe changes in the state of something related to tourism over time. A term such as tourism

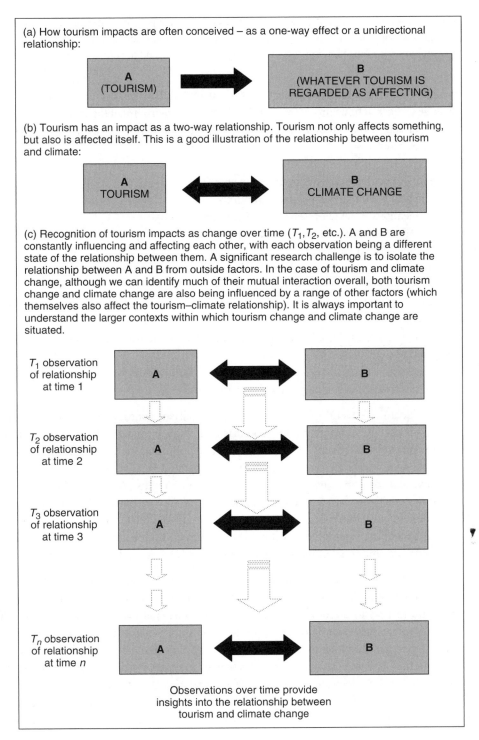

(a) How tourism impacts are often conceived – as a one-way effect or a unidirectional relationship:

A
(TOURISM)

B
(WHATEVER TOURISM IS REGARDED AS AFFECTING)

(b) Tourism has an impact as a two-way relationship. Tourism not only affects something, but also is affected itself. This is a good illustration of the relationship between tourism and climate:

A
TOURISM

B
CLIMATE CHANGE

(c) Recognition of tourism impacts as change over time (T_1, T_2, etc.). A and B are constantly influencing and affecting each other, with each observation being a different state of the relationship between them. A significant research challenge is to isolate the relationship between A and B from outside factors. In the case of tourism and climate change, although we can identify much of their mutual interaction overall, both tourism change and climate change are also being influenced by a range of other factors (which themselves also affect the tourism–climate relationship). It is always important to understand the larger contexts within which tourism change and climate change are situated.

T_1 observation of relationship at time 1

A   B

T_2 observation of relationship at time 2

A   B

T_3 observation of relationship at time 3

A   B

T_n observation of relationship at time n

A   B

Observations over time provide insights into the relationship between tourism and climate change

Figure 1.3 *The nature of impacts*

related change would be a much better way of describing what people mean when they say tourism impacts, but unfortunately people tend to be lazy, and apart from discussions between a few tourism researchers the term impact is the one in common use, and the one we are stuck with!

(Hall and Lew 2009: 3)

Although research on tourism impacts has been undertaken for a number of decades, knowledge of the relationships between tourism and the environment (as well as economic and social effects) is partial and fragmented. This situation exists for several reasons (see Hall and Lew 2009):

- change itself occurs unevenly in time and space;
- different scales of analysis in space and time;
- limited locations at which research has been undertaken;
- use of different and inconsistent research methods;
- lack of longitudinal studies;
- few genuinely comparative studies of different locations;
- lack of baseline data – what existed before tourism commenced;
- paucity of information on the adaptive capacities of ecosystems, communities and economies;
- difficulty of distinguishing between changes induced by tourism and those induced by other human activities;
- concentration of researchers upon particular types of tourism;
- lack of an effective ergodic hypothesis for tourism.

Figure 1.4 identifies several of the ways in which tourism contributes to environmental change, with particular reference to ecotourism in high latitudes (Hall 2010a). These include the more obvious point sources of impact and pollution, such as tourism infrastructure (resorts, roads, attractions); the demands of tourism on local resources, such as water and energy; and the subsequent effects of tourism on habitats. A critical dimension of understanding the impacts of tourism, and one that is core to much of the work in this book, is that some of the most significant impacts, such as GHG emissions, occur as a result of the mobility required to travel between a tourism source area and the destination. (See chapter 4 for more detail on the contribution of transport to global GHG emissions.)

One of the difficulties in assessing the impacts of tourism is the time lag between the initial tourism stimulus and recognition that change has occurred. Change is often quickly recognizable with respect to the development of infrastructure and even human trampling of vegetation, but the introduction of exotic flora, for example, may take some time before it is recognized. Changes in landscape and climate may also take a number of years or decades before they become noticed. Yet such changes may have profound economic, environmental and social effects, and can also affect the relative attractiveness or sustainability of locations as tourist destinations (Hall and Lew 2009). For this reason, one of the main themes in tourism research since the early 1990s has been the search for more sustainable forms of tourism.

Sustainable development is usually defined in terms of the report of the World Commission on Environment and Development (WCED 1987), commonly known as the Brundtland

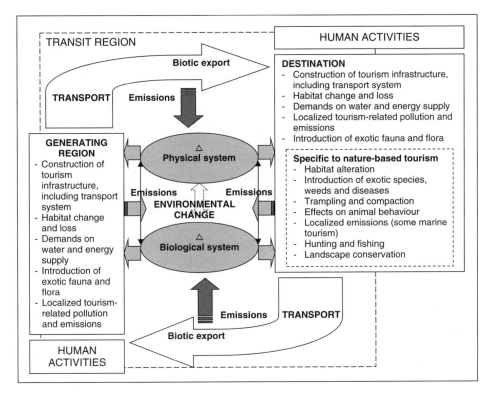

Figure 1.4 *Contribution of tourism to environmental change*

Report, where 'sustainable development is development that meets the needs of the present without compromising the ability of future generations to meet their own needs' (WCED 1987: 49). Five basic principles of sustainability were identified in the report:

- the idea of holistic planning and strategy-making that links economic, environmental and social concerns;
- the importance of preserving essential ecological processes;
- the need to protect both biodiversity and human heritage;
- the need for development to occur in such a way that productivity can be sustained over the long term for future generations (the concept of intergenerational equity);
- the goal of achieving a better balance of fairness and opportunity between nations.

In the 25 years since it was published, the Brundtland Report has been reinforced by consequent international assessments of the state of the planet's environment and statements of intent with respect to sustainability. For example, the Board of the Millennium Ecosystem Assessment (MEA) stated that:

> At the heart of this assessment is a stark warning. Human activity is putting such strain on the natural functions of Earth that the ability of the planet's ecosystems to sustain future generations can no longer be taken for granted … Nearly two thirds of the services provided by nature to humankind are found to be in decline worldwide. In effect, the benefits reaped from our

engineering of the planet have been achieved by running down natural capital assets. In many cases, it is literally a matter of living on borrowed time.

(Board of the Millennium Ecosystem Assessment 2005: 5)

The MEA reported that over the past 50 years, human action had changed the ecosystems on which we depend, 'more rapidly and extensively than in any comparable period of time in human history' (2005: 1) and noted that while some had benefited from such change in material terms, many regions and groups of people had not. According to the MEA (2005: 1), three major problems associated with the management of the world's ecosystems are 'already causing significant harm to some people, particularly the poor, and unless addressed will substantially diminish the long-term benefits' obtained from ecosystems by humankind:

- approximately 60 per cent (15 out of 24) of the ecosystem services examined by the MEA were being degraded or used unsustainably, including fresh water, capture fisheries, air and water purification, and the regulation of climate, natural hazards and pests;
- there was established but incomplete evidence that changes being made in ecosystems were increasing the likelihood of nonlinear ecosystem change (including accelerating, abrupt and potentially irreversible changes) that will have important consequences for human well-being;
- the harmful effects of the degradation of ecosystem services (the persistent decrease in the capacity of an ecosystem to deliver services) were borne disproportionately by the poor, were contributing to growing inequities and disparities, and were sometimes the principal factor causing poverty and a decline in human security.

Tourism cannot be detached from such processes (Gössling and Hall 2006a). Sustainable tourism is a subset of both tourism and sustainable development. Sustainable tourism development is not the same as sustainable development, although the principles of sustainable development, as outlined above, do clearly inform sustainable tourism. The key difference between the two concepts is one of focus or scale. Sustainable tourism refers only to the application of concepts of sustainability at the level of the tourism industry and consequent social, environment and economic effects; whereas sustainable development operates at a broader scale that incorporates all aspects of human interaction with the Earth's environment. The implications of such differentiation of scales of analysis are important because it can be conceived, for example, that a tourism operation may meet criteria of being sustainable if treated in isolation, but if it is placed in a community context, the well-being of that community as a whole may be unsustainable because of the tourism operation, as a result of other development options not being able to be pursued (Hall and Lew 2009).

One of the major ways in which both sustainable development and sustainable tourism have been interpreted is that there needs to be balance between the economic, social and environmental dimensions of development (Hall 2011a). This notion of 'balance' is a thread that runs through much of sustainable tourism policy, industry statements and academic research. For example, according to the then UNWTO Secretary-General Francesco Frangialli, the UNWTO (2007b) is 'committed to seek balanced and equitable policies to encourage both responsible energy related consumption as well as anti-poverty operational patterns. This can and must lead to truly sustainable growth within the framework of the Millennium Development Goals.' UNEP and UNWTO (2005) argue that the concept of sustainable development has evolved since the 1987 Brundtland definition:

> Three dimensions or 'pillars' of sustainable development are now recognized and underlined. These are:
>
> *   Economic sustainability, which means generating prosperity at different levels of society and addressing the cost effectiveness of all economic activity. Crucially, it is about the viability of enterprises and activities and their ability to be maintained in the long term.
> *   Social sustainability, which means respecting human rights and equal opportunities for all in society. It requires an equitable distribution of benefits, with a focus on alleviating poverty. There is an emphasis on local communities, maintaining and strengthening their life support systems, recognizing and respecting different cultures and avoiding any form of exploitation.
> *   Environmental sustainability, which means conserving and managing resources, especially those that are not renewable or are precious in terms of life support. It requires action to minimize pollution of air, land and water, and to conserve biological diversity and natural heritage.
>
> It is important to appreciate that these three pillars are in many ways interdependent and can be both mutually reinforcing or in competition. *Delivering sustainable development means striking a balance between them.*
>
> (UNEP and UNWTO 2005: 9)

This so-called balanced approach, which is arguably the dominant way of thinking with respect to sustainable tourism and sustainable development (Hall 2010e, 2011a), is illustrated in Figure 1.5. The WCED (1987) also argued that there needed to be a change in western lifestyles if global sustainable development was to be achieved. This component of global sustainable development, along with equity issues, has been substantially downplayed in many discussions of sustainable tourism, but remains significant for climate change because of the relationships between lifestyle and carbon emissions.

The widely used WCED (1987) definition is based on the intergenerational equity principle, which stipulates that no avoidable environmental burdens should be inherited by future generations. However, it is also strongly anthropocentric. Sustainable development can also be defined from a more ecocentric perspective: 'Improving the quality of human life, while living within the carrying capacity of supporting ecosystems' (IUCN *et al.* 1991: 10). This approach recognizes that the capacity of the environment to improve living conditions for people is actually limited, and contrasts strongly with perspectives that suggest that economic growth is not environmentally bounded and that there are few limits to both economic growth and natural capital.

Such issues are significant as, although the United Nations Framework Convention on Climate Change (UNFCCC) has as its ultimate objective (Article 2) the stabilization of GHG concentrations at a level that would prevent 'dangerous' interference with the climate system, it specifically requires that this must be achieved within a time frame sufficient 'to ensure that food production is not threatened and to enable economic

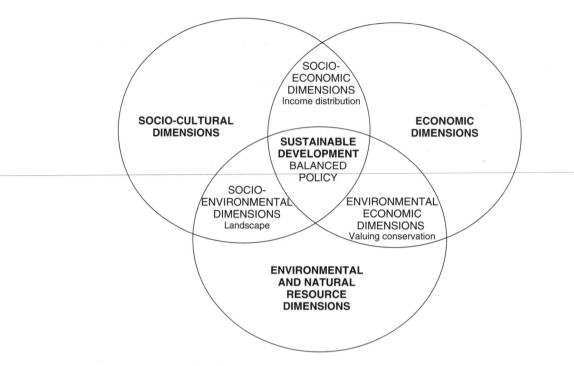

Figure 1.5 *Traditional approaches to the main dimensions of sustainable tourism*

development to proceed in a sustainable manner'. The principles of the convention (Article 3) specifically require that, in achieving this objective, special account should be taken of those countries that are particularly vulnerable to the adverse effects of climate change. The convention's principles also acknowledge each country's right to sustainable development (Martens *et al.* 1997). However, what notion of sustainable development will actually achieve these goals? For example, Hall (2011a) has argued that there are three main formulations of sustainable tourism development: economic sustainability, balanced sustainability, and a third approach that is grounded in ecological economics and a degrowth or steady-state perspective, with the first two approaches being dominant.

In examining sustainable development and sustainable tourism, we are interested primarily in the long-term management of change and judgements as to the acceptability and appropriateness of change. Change refers to the movement from one state or condition to another. Whether such a transition is positive or negative will depend on the original criteria by which change is measured. In the field of tourism studies, investigation and discussion of the impacts of tourism have long been a major research theme (e.g. Mathieson and Wall 1982; Wall and Mathieson 2005; Hall and Lew 2009) as well as a justification for policy interventions that seek to encourage greater sustainability (Hall 2008b).

The physical, socio-cultural and economic environments are always changing, although change is never uniform across time and space. Nevertheless, 'all changes are ultimately connected with one another through physical and social processes alike' (Meyer and

Turner 1995: 304). The scales and rates of change have increased dramatically since the industrial revolution of the nineteenth century because of human actions within which tourism is deeply embedded, particularly the growth of mass mobility and its consequent effects. Concern over the consequences of tourism has grown hand-in-hand with the realization of the changed scale within which these impacts occur. When the effects of tourism were regarded as occurring primarily at a place-specific or destination level, then concern was often expressed only at a local level. However, as impacts have become more commonplace and widespread, so it is that international concern has also grown. Tourism is not alone in this phenomenon. In fact, it characterizes the growth of environmental and social awareness overall (Hall 2008b). Some of the key moments in this change are outlined in the timeline in Figure 1.6. Tourism has therefore become recognized as a contributor to global environmental change in a number of different ways, including such issues as land-use change (often as a result of tourism urbanization), biotic transfer, biodiversity loss, and water and energy demands (Gössling 2002; Gössling and Hall 2006a; Hall 2010b, 2010c; Gössling *et al.* 2010a, 2010c; Hall and James 2011). Nevertheless, perhaps the most significant contribution, especially given its synergy with other forms of change, is tourism's relationship with climate change. But before we study this relationship in detail, we first provide an overview of the science of global climate change and the broad challenges it poses to human development.

THE SCIENCE OF CLIMATE CHANGE: KEY CONCEPTS

> 'Climate change is the pre-eminent geopolitical and economic issue of the 21st century. It rewrites the global equation for development, peace, and prosperity.'
> UN Secretary General Ban Ki-moon (2009)

Weather and climate have a profound influence on natural systems and human societies around the world. Weather can be defined as the state of the atmosphere at a specific moment in time at a specific geographical location, with respect to the simultaneous occurrence of several meteorological variables, including temperature, precipitation, wind, clouds, atmospheric pressure and other variables. As discussed in chapter 2, interpretation of weather as good or bad is subjective to the individual, their activities and livelihood, and what they value.

Climate is commonly considered to be the weather averaged over a period of time, and effectively represents the conditions one would anticipate experiencing at a specific destination and time. The Intergovernmental Panel on Climate Change (IPCC 2001a) more rigorously defines climate as 'the statistical description in terms of the mean and variability of relevant quantities over a period of time ranging from months to thousands or millions of years'. The description of climate is temporally and geographically specific. The World Meteorological Organization (WMO)'s technical regulations specify that 'climatological standard normals' are averages of climate data computed for specific 30-year periods (e.g. 1961–90). However, climate can also describe much longer periods of time,

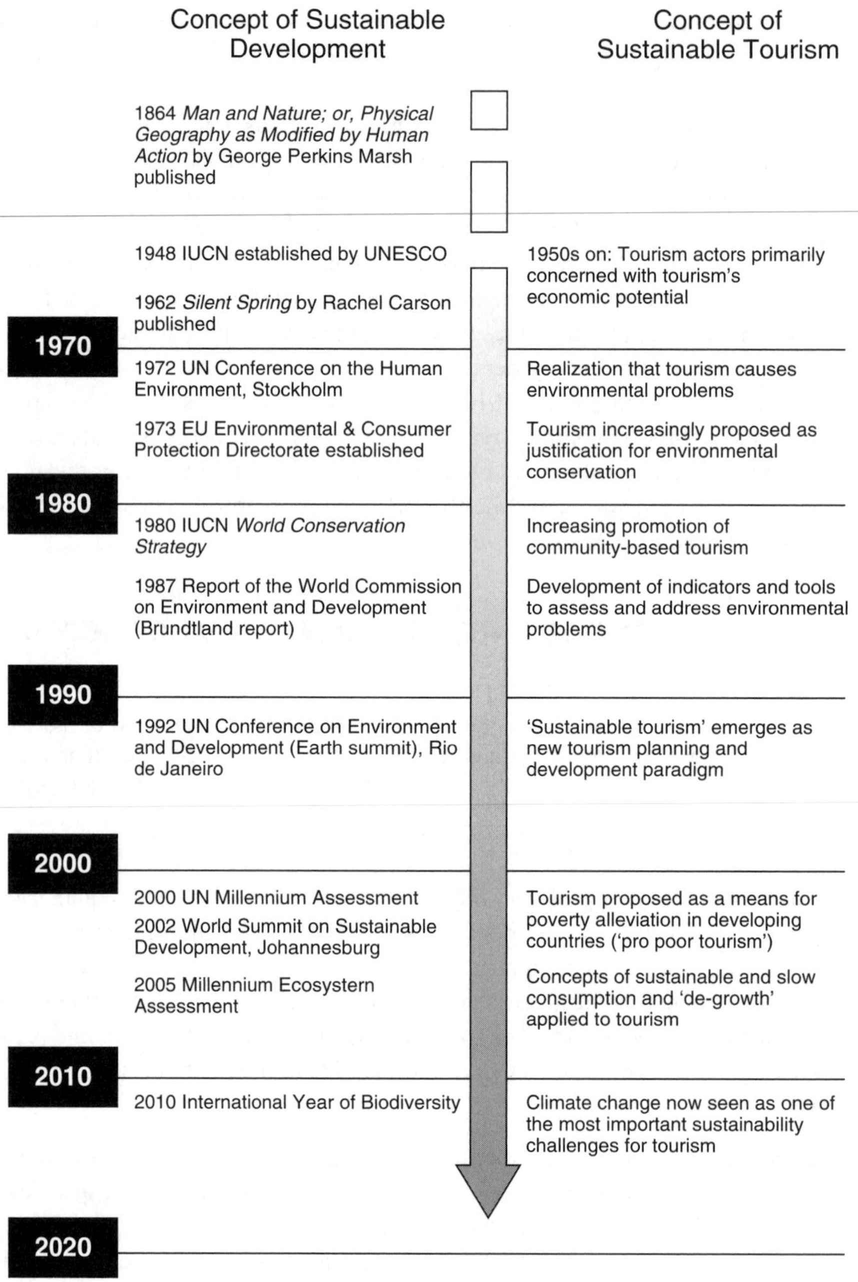

	Concept of Sustainable Development	Concept of Sustainable Tourism
	1864 *Man and Nature; or, Physical Geography as Modified by Human Action* by George Perkins Marsh published	
	1948 IUCN established by UNESCO	1950s on: Tourism actors primarily concerned with tourism's economic potential
	1962 *Silent Spring* by Rachel Carson published	
1970	1972 UN Conference on the Human Environment, Stockholm	Realization that tourism causes environmental problems
	1973 EU Environmental & Consumer Protection Directorate established	Tourism increasingly proposed as justification for environmental conservation
1980	1980 IUCN *World Conservation Strategy*	Increasing promotion of community-based tourism
	1987 Report of the World Commission on Environment and Development (Brundtland report)	Development of indicators and tools to assess and address environmental problems
1990		
	1992 UN Conference on Environment and Development (Earth summit), Rio de Janeiro	'Sustainable tourism' emerges as new tourism planning and development paradigm
2000		
	2000 UN Millennium Assessment	Tourism proposed as a means for poverty alleviation in developing countries ('pro poor tourism')
	2002 World Summit on Sustainable Development, Johannesburg	
	2005 Millennium Ecosystem Assessment	Concepts of sustainable and slow consumption and 'de-growth' applied to tourism
2010	2010 International Year of Biodiversity	Climate change now seen as one of the most important sustainability challenges for tourism
2020		

Figure 1.6 *Timeline of sustainable development, tourism and climate change*

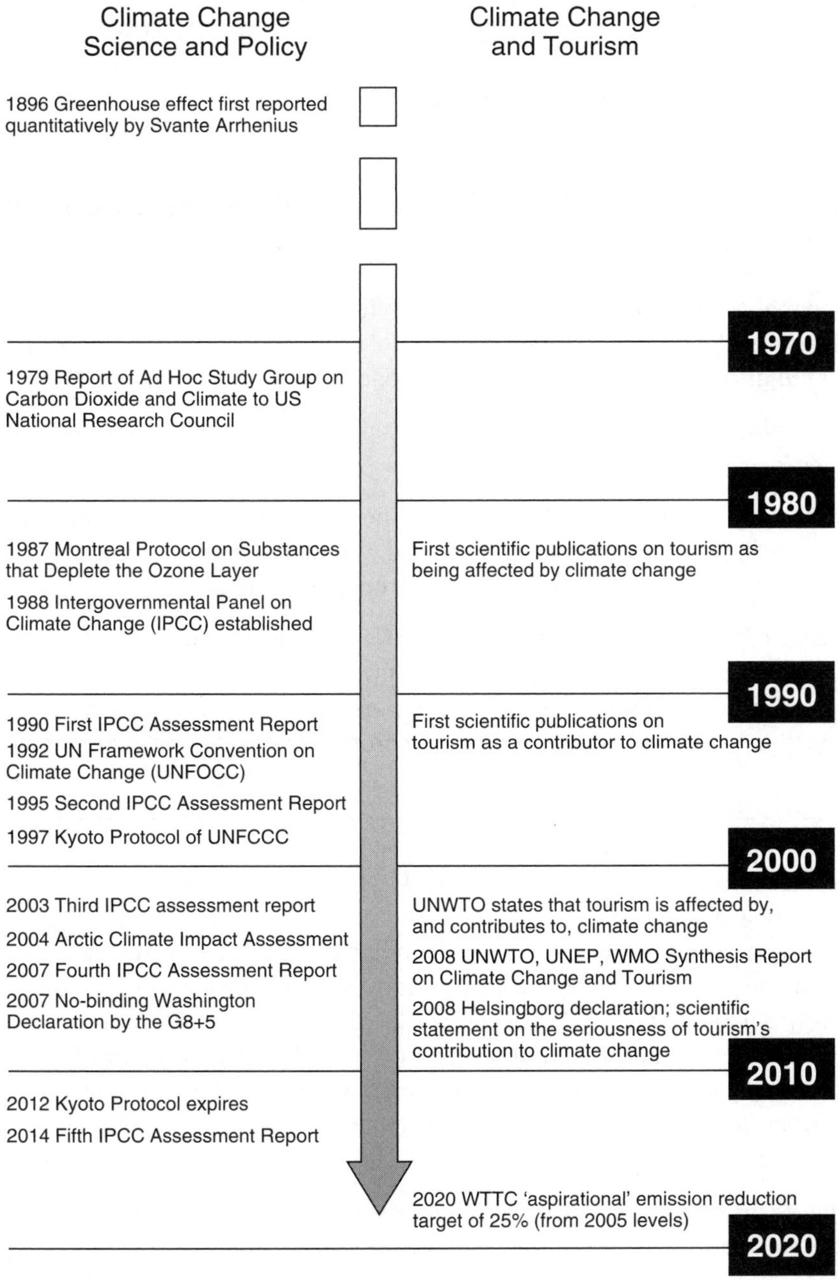

Climate Change Science and Policy

Climate Change and Tourism

1896 Greenhouse effect first reported quantitatively by Svante Arrhenius

1970

1979 Report of Ad Hoc Study Group on Carbon Dioxide and Climate to US National Research Council

1980

1987 Montreal Protocol on Substances that Deplete the Ozone Layer

1988 Intergovernmental Panel on Climate Change (IPCC) established

First scientific publications on tourism as being affected by climate change

1990

1990 First IPCC Assessment Report

1992 UN Framework Convention on Climate Change (UNFOCC)

1995 Second IPCC Assessment Report

1997 Kyoto Protocol of UNFCCC

First scientific publications on tourism as a contributor to climate change

2000

2003 Third IPCC assessment report

2004 Arctic Climate Impact Assessment

2007 Fourth IPCC Assessment Report

2007 No-binding Washington Declaration by the G8+5

UNWTO states that tourism is affected by, and contributes to, climate change

2008 UNWTO, UNEP, WMO Synthesis Report on Climate Change and Tourism

2008 Helsingborg declaration; scientific statement on the seriousness of tourism's contribution to climate change

2010

2012 Kyoto Protocol expires

2014 Fifth IPCC Assessment Report

2020 WTTC 'aspirational' emission reduction target of 25% (from 2005 levels)

2020

Figure 1.6 *Continued*

such as the Cryogenian period (635–850 million years ago), when glacial evidence indicates that global temperatures were substantially cooler than today and much of the Earth's surface was frozen (the so-called 'slushball' or 'ice-house' Earth) (Micheels and Montenari 2008), or the Paleocene–Eocene Thermal Maximum (59–50 million years ago), when global average temperatures may have been as much as 10–12°C higher than today and temperate forests extended to the poles (Katz *et al.* 2001).

BOX 1.1 KEY CONCEPTS: SUBFIELDS OF THE ATMOSPHERIC SCIENCES

Atmospheric sciences is an umbrella term for the study of the atmosphere, its processes, the effects other environmental systems have on the atmosphere, and the effects of the atmosphere on those other environmental systems.

Meteorology is the interdisciplinary scientific study of the atmosphere, with a major focus on weather forecasting.

Climatology is the study of climate, defined scientifically as weather conditions averaged over a period of time, and of climate changes (both long- and short-term) due to both natural and anthropogenic causes.

Applied climatology is the scientific analysis of climatic data in the light of a useful application for an operational purpose in a wide range of economic sectors (e.g. agriculture, transportation, water management, forestry, tourism) or hazard management (e.g. emergency preparedness planning, insurance).

(IPCC 2001a)

Climate conditions can be defined from the local to the global scale. The geographical pattern of contemporary climate is influenced by latitude and by proximity to oceans, other large water bodies and mountains, as well by other factors such as predominant wind and ocean currents, and can be represented globally in classification schemes such as the Köppen system (Peel *et al.* 2007) that contains 24 distinct categories based on temperature, precipitation and dominant vegetation type (Table 1.1). More detailed climatological descriptions at a local to regional scale include the average conditions for multiple atmospheric parameters. Climate also includes descriptions of the variability of conditions, from day-to-day to year-to-year fluctuations.

The global climate system is dynamic and is driven by solar radiation and complex physical, chemical and biological interactions between the atmosphere, hydrosphere (principally the world's oceans), cryosphere, biosphere, land surface and lithosphere. Climate is changeable at varied temporal and geographical scales, and the IPCC (2001a) has defined two central notions in this regard:

- climate variability – variations in the mean state and other statistics (such as standard deviations, the occurrence of extremes, etc.) of the climate on all temporal and spatial scales beyond that of individual weather events;

Table 1.1 Köppen global climate classification system

Classification	Sub-categories		
A	Tropical humid	Af	Tropical wet	No dry season
		Am	Tropical monsoonal	Short dry season; heavy monsoonal rains in other months
		Aw	Tropical savannah	Winter dry season
B	Dry	BWh	Subtropical desert	Low-latitude desert
		BSh	Subtropical steppe	Low-latitude dry
		BWk	Mid-latitude desert	Mid-latitude desert
		BSk	Mid-latitude steppe	Mid-latitude dry
C	Mild mid-latitude	Csa	Mediterranean	Mild with dry, hot summer
		Csb	Mediterranean	Mild with dry, warm summer
		Cfa	Humid subtropical	Mild with no dry season, hot summer
		Cwa	Humid subtropical	Mild with dry winter, hot summer
		Cfb	Marine west coast	Mild with no dry season, warm summer
		Cfc	Marine west coast	Mild with no dry season, cool summer
D	Severe mid-latitude	Dfa	Humid continental	Humid with severe winter, no dry season, hot summer
		Dfb	Humid continental	Humid with severe winter, no dry season, warm summer
		Dwa	Humid continental	Humid with severe, dry winter, hot summer
		Dwb	Humid continental	Humid with severe, dry winter, warm summer
		Dfc	Subarctic	Severe winter, no dry season, cool summer
		Dfd	Subarctic	Severe, very cold winter, no dry season, cool summer
		Dwc	Subarctic	Severe, dry winter, cool summer
		Dwd	Subarctic	Severe, very cold and dry winter, cool summer
E	Polar	ET	Tundra	Polar tundra, no true summer
		EF	Ice cap	Perennial ice
H	Highland			

- climate change – a statistically significant variation either in the mean state of the climate or in its variability, persisting for an extended period (typically decades or longer).

Both climate variability and change may be due to natural internal processes within the climate system, or caused by variations in natural (e.g. fluctuations in solar radiation) or anthropogenic (e.g. changes in the composition of the atmosphere or in land use) forcings. There are a variety of climate change feedback mechanisms that can either amplify or diminish initial climate forcings (e.g. warming temperatures can cause further warming by increasing the amount of water vapour in the atmosphere or melting permafrost and releasing methane – both GHGs).

A brief history of climate change science

'To a patient scientist, the unfolding greenhouse mystery is far more exciting than the plot of the best mystery novel. But it is slow reading, with new clues sometimes not appearing for several years. Impatience increases when one realizes that it is not the fate of some fictional character, but of our planet and species, which hangs in the balance as the great carbon mystery unfolds at a seemingly glacial pace.'

David Schindler (1999)

Our understanding of climate variability and change and its causes has advanced tremendously over the past 50 years, beginning with the period that Lamb (2002) refers to as the 'climate revolution' in the 1960s and 1970s, with the development of a range of new paleoclimate proxy records, including dendrochronology (tree-ring dating), ice cores and palynology (pollen dating), along with vastly improved terrestrial, ocean and space-based climate monitoring systems.

However, the foundations of climate change science trace back well over 100 years. Some credit the medieval Chinese scientist Shen Kuo (1031–1095 AD) as the first to hypothesize that large-scale regional climate change had occurred, after observing petrified bamboo in an arid region (Sivin 1995). In the centuries that followed, others made similar observations that suggested large-scale climate changes had occurred, including glacial features left on the landscapes of Europe, North America and parts of Asia. In the early nineteenth century, scientists endeavoured to understand the causes of past glaciations and climatic changes. In 1824, French scientist Joseph Fourier determined that the Earth would be far colder if it lacked an atmosphere, and was the first to describe what is referred to as the natural greenhouse effect. During the 1850s, John Tyndall's experiments determined that atmospheric gases have varied capacities to absorb or transmit radiant heat (long-wave radiation) – so-called greenhouse gases – and that changes in the concentration of water vapour, carbon dioxide and ozone could result in climatic change.

Using the emerging understanding of atmospheric physics and chemistry, Swedish scientist Svante Arrhenius was the first to hypothesize that human activity could alter the global climate system. In 1896, Arrhenius published the first calculation of global warming from human emissions of CO_2, estimating that a future doubling of atmospheric CO_2 would warm the planet 5–6°C. Given the relatively low rate of CO_2 emissions at the time, which were primarily from coal and wood burning, Arrhenius concluded that it would take 2000–3000 years to have an appreciable impact on the global climate system. The scale of the fossil fuel-based economy that emerged over the next century was not foreseen.

Arrhenius' calculations were erroneously disputed in the early decades of the twentieth century, and scientific investigation into a human-enhanced greenhouse effect did not regain momentum until the 1950s, when Gilbert Plass (a physicist at Johns Hopkins University, USA) completed the second calculation of the impact of a doubling of atmospheric CO_2 on global average temperatures. Plass was prophetic, stating in 1953 that

> In the hungry fires of industry, modern man burns nearly 2 billion tons of coal and oil each year ... his furnaces belch some 6 billion tons of unseen CO_2 ... By conservative estimate, the earth's atmosphere, in the next 127 years, will contain 50% more CO_2 ... At its present rate of increase, the CO_2 in the atmosphere will raise the earth's average temperature 1.5° Fahrenheit (1.1°C) every 100 years.
>
> *(Time Magazine* 1953)

His calculations estimate that a doubling of CO_2 would warm the planet 3.6°C (Plass 1956). By comparison, the IPCC's *Fourth Assessment Report* (Solomon *et al.* 2007) estimated a CO_2 rise of 37 per cent since pre-industrial times and a 1900–2000 warm-up of around 0.7°C. Shortly thereafter, Roger Revelle and Hans Suess (1957) determined that the oceans would not absorb CO_2 from human activities at rates previously expected, and concluded

> ... human beings are now carrying out a large-scale geophysical experiment of a kind that could not have happened in the past nor be reproduced in the future. Within a few centuries we are returning to the atmosphere and oceans the concentrated organic carbon stored in sedimentary rocks over hundreds of millions of years. This experiment, if adequately documented, may yield a far-reaching insight into the processes determining weather and climate.
>
> (Revelle and Suess 1957: 26)

The next year, Charles Keeling established the first significant evidence of annual increases in atmospheric CO_2 concentrations with data from the Mauna Loa Observatory in Hawaii (USA) and Antarctica (the 'Keeling curve').

Three important conferences in the 1970s galvanized scientific opinion around anthropogenic climate change (see Figure 1.6). In 1971, experts from 14 countries met at a conference on 'Man's impact on climate' and concluded that the dangers of humanity's emission of particle pollutants (aerosols) and GHGs could shift the climate dangerously in the next 100 years (Matthews *et al.* 1971). At the end of that decade, the first World Climate Conference (WMO 1979) concluded that

> it appears plausible that an increased amount of CO_2 in the atmosphere can contribute to a gradual warming of the lower atmosphere ... It is possible that some effects on a regional and global scale may be detectable before the end of this century and become significant before the middle of the next century.

The same year, the US National Research Council (1979) published a report that estimated a doubling of atmospheric CO_2 would warm global surface temperatures between 2 and 3.5°C.

The 1980s saw the global cooling trend observed from 1945–75 come to an end; the continued increase of atmospheric CO_2 (the Keeling curve); and the effectiveness of 'clean air' legislation to reduce emissions of aerosol pollution. Concomitant with these important trends was invigorated scientific dialogue on human influence on the global climate system and, towards the end of the decade, calls for political action to mitigate anthropogenic climate change. In 1985, a WMO conference held in Villach, Austria summarized the state of knowledge on GHGs and climate, declaring:

> As a result of the increasing concentrations of greenhouse gases, it is now believed that in the first half of the next century a rise of global mean temperature could occur which is greater than any in man's history. ... the rate and degree of future warming could be profoundly affected by governmental policies on energy conservation, use of fossil fuels, and the emission of some greenhouse gases.
>
> (WMO 1986)

Three years later, the World Conference on the Changing Atmosphere in Toronto, Canada was the first explicitly to recommend political action to reduce CO_2 emissions (20 per cent by 2005). With this recommendation and its significant economic consequences, climate change science has been increasingly politicized ever since.

With the establishment of the IPCC by the WMO and the United Nations Environment Programme (UNEP), 1988 represented a historic turning point in terms of the advancement and international coordination of climate change science (see Figure 1.6). The IPCC is a scientific body and, while its findings are highly policy-relevant, the IPCC is not policy-prescriptive. The IPCC does not conduct its own original research, but rather is tasked with conducting special reports pertinent to key aspects of climate change science (e.g. CO_2 capture and storage, emissions scenarios, aviation and the global atmosphere) and periodic (typically every five to seven years; see Figure 1.6) comprehensive scientific assessments of the existing scientific, technical and socio-economic literature on climate change and its consequences for natural and human systems. Most of the world's leading scientific and government experts on climate change participate in one of the three working groups that make up the current structure of the assessment:

* Working Group 1 – assesses the physical scientific aspects of the climate system and climate change;
* Working Group 2 – assesses the scientific, technical, environmental, economic and social aspects of the vulnerability (sensitivity and adaptability) to climate change of, and the negative and positive consequences for, ecological systems, socio-economic sectors and human health, with an emphasis on regional sectoral and cross-sectoral issues;
* Working Group 3 – focuses on the mitigation of climate change.

The IPCC has broad and geographically balanced participation of experts from relevant fields of knowledge and a rigorous and transparent, multi-stage scientific and government review process. For example, more than 1200 lead and contributing authors and more than 2500 expert reviewers from over 130 countries contributed to the 2007 Fourth Assessment. *All* participating governments endorse the scientific conclusions of the IPCC Assessments before publication, therefore the findings are considered authoritative within the international community. Following the IPCC *Third Assessment Report*, 17 national science academies (from Australia, Belgium, Brazil, Canada, the Caribbean, China, France, Germany, India, Indonesia, Ireland, Italy, Malaysia, New Zealand, Sweden, Turkey and the UK; *Science* 2001: 1261) also issued a joint declaration explicitly acknowledging the IPCC position as representing the scientific consensus on climate change science. In 2007, the IPCC shared the Nobel Peace Prize for the work contained in its *Fourth Assessment Report* (AR4).

Observed changes in the global climate system

The IPCC AR4 (Solomon *et al.* 2007) presented compelling evidence from every continent of the world indicating that the global climate system has changed compared with the pre-industrial era. This has been reinforced further by more recent reports on the state of the global climate system (UNEP 2009b; NOAA 2010b). Notable changes have been observed in the concentrations of atmospheric gases (including GHGs, aerosols and ozone), temperatures (atmospheric, terrestrial and oceanic), and other features of the climate and climate-sensitive natural systems.

Since the onset of the industrial revolution (*c.* 1750), human activity has markedly increased the concentration of carbon dioxide, methane, nitrous oxide and other GHGs (Figure 1.7). Standard measurements of mean atmospheric CO_2 concentration have recorded an increase from 315 parts per million by volume (ppmv) in 1958 to 389 ppmv in 2010. The average annual increase from 2001 to 2010 has been just over 2 ppmv per year. The IPCC noted that the primary source of the increase in CO_2 is fossil fuels, but land-use changes, such as deforestation, also make a contribution (Solomon *et al.* 2007). Comparing the 1990s with 2000–06, the CO_2 emissions growth rate increased from 1.3 to 3.3 per cent per year (McMullen and Jabbour 2009). The efficiency to absorb anthropogenic emissions of the major CO_2 sinks in the oceans and on land has been observed to be declining, further contributing to increased atmospheric concentrations (Canadell *et al.* 2007). The amount of CO_2 in the atmosphere now far exceeds the natural range of the past 650,000 years (180–300 ppm) (Solomon *et al.* 2007).

Although CO_2 is the GHG focused on by most media, business and policy-makers, it is essential to recognize that it is only one of a number of gases and factors that affect climate change. Atmospheric methane is one of the most potent GHGs, with 25 times the impact on temperature of a CO_2 emission of the same mass over the following 100 years (termed global warming potential; see Box 1.2) (Forster *et al.* 2007). Methane concentrations increased sharply during most of the twentieth century, and are now 148 per cent above pre-industrial levels. The amount of methane in the atmosphere in 2005 (1774 parts per billion, ppb) exceeds by far the natural range of the past 650,000 years (320–790 ppb) (Solomon *et al.* 2007). The primary source of the increase in methane is 'very likely'

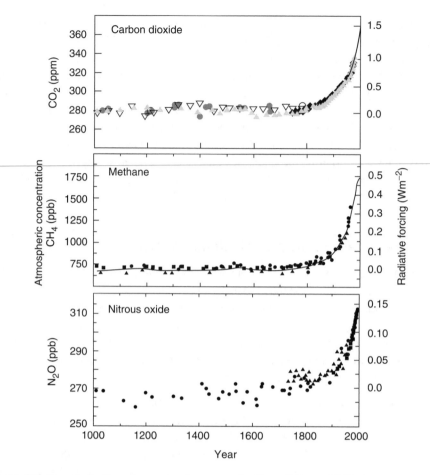

Figure 1.7 *Trends in the atmospheric concentration of three GHGs*
Source: IPCC (2001b)

(see Box 1.3 for IPCC language of uncertainty) to be the combination of human agricultural activities and fossil fuel use; the IPCC noted that relative contributions from other source types are not well determined.

In addition to warming influences on the global climate system through anthropogenic emissions of GHGs, a cooling effect is produced by anthropogenic emissions of aerosols (primarily sulphate, organic carbon, black carbon, nitrate and dust). In effect, human activities have been forcing the global climate system in two directions throughout the twentieth century. Understanding of anthropogenic warming and cooling influences on the global climate has continued to improve, and the IPCC AR4 (Solomon *et al.* 2007) concluded with 'very high confidence' that the global average net effect of human activities since 1750 has been one of warming, with a radiative forcing of +1.6 W/m², and that it is 'likely' that increases in GHG concentrations alone would have caused more warming if not for volcanic activity and anthropogenic aerosols.

BOX 1.2 KEY CONCEPTS: WHAT ARE RADIATIVE FORCING, GLOBAL WARMING POTENTIAL AND EQUIVALENT CO$_2$?

Radiative forcing provides a way to compare the magnitude of different natural and anthropogenic perturbations of the climate system, including cooling (–) and warming (+) influences. When combined, the net radiative forcing indicates the direction and magnitude of influence on the climate.

The IPCC (Forster *et al.* 2007) defines radiative forcing as 'a measure of the influence a factor has in altering the balance of incoming and outgoing energy in the Earth–atmosphere system and ... an index of the importance of the factor as a potential climate change mechanism. In this report radiative forcing values are for changes relative to pre-industrial conditions defined at 1750 and are expressed in watts per square meter (W/m^2).'

The IPCC (Forster *et al.* 2007) defines global warming potential (GWP) as 'An index, describing the radiative characteristics of well-mixed *greenhouse gases*, that represents the combined effect of the differing times these gases remain in the *atmosphere* and their relative effectiveness in absorbing outgoing *infrared radiation*. This index approximates the time-integrated warming effect of a unit mass of a given GHG in today's atmosphere, relative to that of *carbon dioxide*.'

GWP was developed to provide a comparison of the relative ability of GHGs to trap heat in the atmosphere. The higher the GWP, the greater influence each unit of a GHG has on climate change. CO$_2$ has a GWP of 1, as it is the baseline gas to which all other GHGs are compared. The GWP of GHGs varies with time, as a gas with high radiative forcing but a short atmospheric lifetime will have a large GWP on a 20-year scale, but a small GWP on a 100-year scale. Conversely, if a GHG has a longer atmospheric lifetime than CO$_2$, its GWP will increase under longer timescales.

The IPCC (Forster *et al.* 2007) assigns GWP values for selected anthropogenic GHGs:

20 years: methane = 72; nitrous oxide = 289; CFC-12 = 11,000; sulphur hexafluoride = 16,300

100 years: methane = 25; nitrous oxide = 298; CFC-12 = 10,900; sulphur hexafluoride = 22,800

The IPCC (Forster *et al.* 2007) defines equivalent CO$_2$ (CO$_2$-eq) as: 'The concentration of *carbon dioxide* that would cause the same amount of *radiative forcing* as a given mixture of carbon dioxide and other *greenhouse gases*.' Equivalent CO$_2$ is a simple way to include the effects of other GHGs when considering future GHG concentrations of the atmosphere. It can be used to provide a single value, instead of the concentrations of four or more prominent GHGs that contribute to warming. Many stabilization scenarios discussed in policy papers and international negotiations are based on stabilizing total equivalent CO_2 at 450, 550 or 750 ppmv.

BOX 1.3 KEY CONCEPTS: THE IPCC LANGUAGE OF UNCERTAINTY

Assessments of climate change science by the IPCC have always recognized the importance of communicating uncertainties among scientists and to policy-makers. In the AR4, the following terms are used to indicate 'degree of confidence in being correct' and 'likelihood of occurrence', as based on the collective judgement of the authors using the observational evidence, modelling results and theory available to them.

Level of confidence is used to characterize uncertainty as to the correctness of a model, an analysis or a statement:

- very high confidence   at least 9 out of 10 chance of being correct;
- high confidence   about 8 out of 10 chance;
- medium confidence   about 5 out of 10 chance;
- low confidence   about 2 out of 10 chance;
- very low confidence   less than 1 out of 10 chance.

Likelihood refers to a probabilistic assessment of some well defined outcome having occurred or occurring in the future:

- virtually certain   >99 per cent probability of occurrence;
- very likely   >90 per cent probability;
- likely   >66 per cent probability;
- about as likely as not   33–66 per cent probability;
- unlikely   <33 per cent probability;
- very unlikely   <10 per cent probability;
- exceptionally unlikely   <1 per cent probability.

(IPCC 2007a)

Understanding of how climate is changing continues to improve with the extension and broader geographical coverage of numerous observational datasets. In its AR4, the IPCC (2007b: 5) declared that 'warming of the climate system is unequivocal'. The global mean surface temperature has increased approximately 0.76°C between 1850–99 and 2001–05, and the IPCC concluded that most of the observed increase in global average temperatures since the mid-twentieth century is 'very likely' the result of human activities that are increasing GHG concentrations in the atmosphere (Solomon *et al.* 2007). The warming trend over the past 50 years is nearly twice that for the past 100 years, and 15 of the 16 warmest years in the record of global surface temperature since 1850 have occurred between 1995 and 2010. Regional warming trends vary, but were highest in the Arctic region, where average temperatures were observed to have increased at almost twice the global average rate in the past 100 years. Over the past 50 years, the data on extreme temperatures reveal a similar warming signature, with cold days, cold nights and frosts occurring less frequently ('very likely'), while hot days, hot nights ('very likely') and heatwaves occurred more frequently ('likely') over most land areas. Balloon- and satellite-based measurements of lower- and mid-tropospheric temperatures show warming trends that are similar to those of the surface temperature record over the past 30 years (Solomon *et al.* 2007).

The IPCC AR4 also notes that discernible human influences now extend to other aspects of climate, including ocean warming, precipitation patterns, wind patterns and the intensity of extreme storm events. Warming has not been limited to land areas. Observations since the 1960s show that the oceans have absorbed more than 80 per cent of the heat added to the global climate system, and that ocean temperatures have increased to depths of at least 3000 m (Solomon *et al.* 2007). Increased temperatures of upper layers of the oceans have caused thermal expansion for the upper 300–700 m of the ocean, contributing to sea-level rise. Higher surface water temperatures in some regions have also contributed to some of the highest recorded coral bleaching events, resulting in loss of coral cover in some locations and changed coral community structure in others (Donner *et al.* 2007). Over time, the heat already absorbed by the ocean will be released back to the atmosphere, causing additional surface warming, therefore the IPCC warned that some additional atmospheric warming is already 'in the pipeline'.

Change in precipitation patterns in some regions has also been recorded in recent decades, with respect to both amount and intensity. According to the IPCC (Solomon *et al.* 2007), increased precipitation has been observed in eastern parts of North and South America, northern Europe and northern and central Asia, while drying has been observed in the Sahel, the Mediterranean, southern Africa and parts of southern Asia. Long-term trends have not been observed for the other large regions assessed. The frequency of heavy precipitation events has 'likely' increased over most land areas, which is consistent with warming and observed increases in atmospheric water vapour. More intense and longer droughts have been observed over wider areas since the 1970s ('likely'), particularly in the tropics and subtropics (Solomon *et al.* 2007). Changes in sea surface temperatures, wind patterns and decreased snowpack and snow cover have been linked to droughts, but because precipitation is so highly variable spatially and temporally, understanding of the causes of observed precipitation changes remains low.

Varied changes in extreme events have also been recorded. Anthropogenic forcing is 'likely' to have contributed to changes in wind patterns, affecting extra-tropical storm tracks in both hemispheres (Solomon *et al.* 2007), while no systematic changes have been observed in the frequency of tornadoes, thunder days or hail events (IPCC 2001b). There has been an increase in hurricane intensity in the North Atlantic since the 1970s; however, there is no clear trend in the annual number of tropical cyclones (Solomon *et al.* 2007). It was considered more likely than not (>50 per cent) that there has been some human contribution to the increases in hurricane intensity (Solomon *et al.* 2007). Other regions appear to have experienced increased hurricane intensity as well, but data quality precludes a definitive statement for these regions. In addition to being the warmest decade on record, 2001–10 was marked by numerous weather and climate extremes, unique in strength and impact and consistent with impacts anticipated as the Earth's atmosphere warms (WMO 2010) (Table 1.2).

A wealth of evidence about changing climate conditions is available from a range of natural systems. Multiple changes are being recorded in the cryosphere (the area of the Earth's surface where water and/or ground is frozen), with mountain glaciers and polar ice experiencing retreat and volume loss, and contributing to sea-level rise. Evidence of mountain glacier loss has been documented on every continent, and evidence of increasing loss rates is becoming stronger (McMullen and Jabbour 2009). For example, observations of 30 reference glaciers in nine mountain ranges by the World Glacier Monitoring Service reveal the mean loss rate since 2000 has increased to about twice the loss rates observed during the two previous decades (Zemp *et al.* 2009). Snow cover has also declined on average in both Northern and Southern hemispheres (Solomon *et al.* 2007). Lake and river ice cover in the mid- and high latitudes of the Northern hemisphere might have reduced by two weeks over the twentieth century (IPCC 2001b). In the Arctic, annual average sea-ice extent has diminished by 2.7 per cent per decade, with larger decreases in summer months (Solomon *et al.* 2007). In 2007, the sea-ice in the Arctic Ocean shrank to its smallest extent on record, 4.28 million km^2 – 24 per cent less than the previous record set in 2005 and 34 per cent less than the average minimum extent over 1970–2000 (McMullen and Jabbour 2009). Although the 2009 minimum extent is almost 1 million km^2 above the 2007 low point (an increase of around twice the size of Spain), and also larger than that in 2008 (the second-lowest year on record, 4.52 million km^2), there is no indication that the long-term trends are reversing (Schiermeier 2009). The extent of recent warming is such that it has been recognized as the warmest period in the Arctic for the past 2000 years, with four of the five warmest decades in that period occurring in the past 50 years (Kaufman *et al.* 2009). Precipitation in the Arctic has increased at about 1 per cent per decade over the past century, although the trends are spatially highly variable and highly uncertain because of deficiencies in the meteorological record (McBean *et al.* 2005). This rapid seasonal Arctic sea-ice loss; lengthening of the glacial melt season; and the marked retreat, thinning and acceleration of most of Greenland's outlet glaciers south of 70°N have alarmed polar researchers (Forbes 2011). However, Antarctic ice loss remains less well understood.

All meteorological stations on the Antarctic Peninsula show strong and significant warming over the past 50 years, with the peninsula becoming a focus of media attention to global climate change. However, over the wider Antarctic there is considerable variability in temperature trends (Anisimov *et al.* 2007). In contrast to the Arctic, the IPCC indicated that Antarctic sea-ice showed no significant overall trend, but may have increased

Table 1.2 Recent unique climate extremes, 2003–11

Region	Type of extreme	Region	Type of extreme
Globally (2010)	Nineteen countries broke their records for the hottest day ever	Arctic sea-ice (2007)	All-time lowest extent on record in September: surpassed previous record set in 2005 by 23 per cent
South-eastern USA (2011)	Largest tornado outbreak (289 tornado reports in 15 states in single storm system)	Western Europe (2003)	Worst heatwave in 500 years: 21,000 deaths
Caribbean (2005)	Hurricane Wilma becomes the most powerful hurricane on record in the Atlantic Basin	Russia (2010)	Highest temperatures recorded in western Russia contributed to strong drought conditions, wildfires destroyed 25 million acres of land and forest
Caribbean (2005)	Atlantic hurricane season (2005): 27 named storms, most on record	Greater Horn of Africa (2006)	Heavy rain following drought, and worst flooding in 50 years
Mexico (2009)	Worst drought in 70 years, affecting 17 million acres of cropland	India (2009)	Weak summer monsoon created worst drought in 35 years
Ecuador (2008)	Heavy rain and worst flooding in the country's history	China (2009)	Worst drought in 50 years, affected more than 10 million hectares of crops
Brazil (2006)	Worst drought in 60 years, lowest flow on Amazon River in 50 years	Southern Australia (2009)	Exceptional heatwave triggering new temperature records and deadly wildfires
Chile (2008)	Worst drought in 50 years in central and southern parts	Australia (2010)	Wettest spring on record, resulted in worst flooding in 40 years

Source: NOAA Significant Climate Anomalies series 2003–10

(Solomon *et al.* 2007). More recently, researchers have indicated that the ice dynamics in Antarctica are in a state of flux, and estimated that loss of ice (shelves and glaciers) from West Antarctica increased by 60 per cent in the decade to 2006, while ice loss from the Antarctic Peninsula increased by 140 per cent (McMullen and Jabbour 2009). Walsh (2009) suggests that Antarctic changes of recent decades appear to be shaped by ozone depletion and an associated strengthening of the southern annular mode of the atmospheric circulation. Although the signature of greenhouse-driven change in Antarctica is projected to emerge from the natural variability during the present century, the emergence of a statistically significant greenhouse signal may be slower than in other regions (Hall 2010a).

The world's oceans are experiencing two additional major changes. Widespread melt of glaciers and polar ice sheets, combined with warming ocean surface temperatures, have contributed to sea-level rise. The average rate of global mean sea-level rise was approximately 1.7 mm per year between 1900 and 2009, increasing to 3.2 mm per year from 1993 to 2009 (Church and White 2011). Increased concentrations of CO_2 in the atmosphere have also led to CO_2 absorption in the oceans, altering ('acidifying') the chemistry of the top layers. Ongoing ocean acidification may harm a wide range of marine organisms and the food webs that depend on them, eventually degrading entire marine ecosystems (McMullen and Jabbour 2009).

The biological response of ecosystems and individual species consistent with warming climate conditions has been recorded on every continent (Solomon *et al.* 2007). Wide-ranging reviews since the AR4 verify this conclusion (Rosenzweig *et al.* 2008) and raise the prospect of much greater long-term ecosystem change, because many biological systems show significant early inertia in response to climate change (Brovkin *et al.* 2009) (see chapter 5).

As the IPCC (Solomon *et al.* 2007) and UNEP (2009b) clearly document, the impacts of anthropogenic climate change are not consigned to some hypothetical future, but are the reality that scientists, environmental managers and indigenous populations are observing. Furthermore, due to the timescales associated with climate feedbacks, atmospheric warming, sea-level rise and the biological response to climate change would continue for centuries even if atmospheric GHG concentrations were to be stabilized in the near term, a prospect that several recent studies have concluded is remote based on current emission reduction commitments (Meinshausen *et al.* 2009; Parry *et al.* 2009a).

Future climate change

In order to understand the possible consequences of increased GHG concentrations in the atmosphere and other natural and anthropogenic forcings on the global climate system, we rely on complex models referred to as global climate models (GCMs). These are highly complex three-dimensional mathematical representations of the Earth's climate system that are used to simulate past, present or possible future climates under varying conditions, and have been in development for more than 40 years. The complexity of the environmental systems (atmosphere, oceans and land surface components) that GCMs have been able to incorporate has continued to advance as the understanding of the global climate system and its various feedbacks have improved, and as computing technology has progressed. While the capacities of supercomputers have increased dramatically in the past decade, computational power still remains a fundamental constraint on climate modelling, including the ability to parameterize certain physical processes (completeness), and spatial (vertical layers and horizontal grid) and temporal resolution. While incomplete, and not of the high resolution needed to improve regional and local climate projections, GCMs are the best tool available to ask 'what if?' questions regarding the implications of future anthropogenic climate forcings (GHG and aerosol emissions, land use, carbon sink modification, geoengineering) and possible policy alternatives.

Because of the enormous scientific uncertainties involved in assessing the consequences of various human activities (population, economic growth, energy efficiency, energy sources, policy alternatives), GCMs produce climate projections under a range of specified socio-economic conditions or 'scenarios'. Since the IPCC's inception, emission and climate scenarios (see Box 1.4) have been a central component of its work. Scenarios aid better understanding of the uncertainties about human contributions to climate change; the responses of the Earth system to human activities; the impacts of a range of future climates; and the potential implications of different approaches to mitigation (measures to reduce net emissions) and adaptation (actions that facilitate responses to new climate conditions).

Emission scenarios (Box 1.4) attempt to represent the plausible range of future socio-economic, environmental, technological and policy conditions that can affect energy use and sources, and related GHG and aerosol emissions, over decadal timeframes. They serve as foundational inputs to GCMs, but also can be used in climate policy analysis to examine the economic, technological and governance conditions necessary to avoid undesired climate outcomes (to 'prevent dangerous interference with the global climate system' – see the section later in this chapter on the international response to climate change).

The IPCC has led the development of three sets of emission scenarios over the past 20 years. In 1990, the SA90 series of scenarios explored the climate impacts of a 'business as usual' future and three policy scenarios. These were followed by the IS92 series, which began to explore in greater detail the implications of uncertainties in economic growth, population and technology in a number of 'business as usual' energy and economic futures (Leggett *et al.* 1992). The IPCC published a new set of scenarios in 2000 for use in its *Third Assessment Report*. The *Special Report on Emissions Scenarios* (*SRES*) (Nakicenovic *et al.* 2000) also explored a wide range of plausible energy and development (demographic, politico-societal, economic and technological) pathways representative of the literature at the time. The *SRES* scenarios were produced with the involvement of an international and interdisciplinary modelling team (including experts from NGOs and the private sector) and, like the IS92 series, assumed no policy actions to mitigate climate change. The *SRES* scenarios were classified into four families according to their global–regional and development–environmental orientations ('storyline' narratives) (Nakicenovic *et al.* 2000: 4–5).

- The A1 scenario family describes a future world of very rapid economic growth, a global population that peaks mid-century and declines thereafter (to 7.1 billion by 2100), and the rapid introduction of new and more efficient technologies, with a substantial reduction in regional differences in per capita income. The A1 scenario family develops into three groups that describe alternative directions of technological change in the energy system: fossil-intensive (A1FI); non-fossil energy sources (A1T); or a balance across all sources (A1B). Approximate CO_2-eq concentration in 2100: A1T = 700 ppm; A1B = 850 ppm; A1FI = 1550 ppm.

- The A2 scenario family describes a very heterogeneous world. The underlying theme is self-reliance and preservation of local identities. Fertility patterns across regions converge very slowly, which results in continuously increasing global population (15.1 billion in 2100). Economic development is primarily regionally oriented, and per capita economic growth and technological change are more fragmented and slower. Approximate CO_2-eq concentration in 2100 = 1250 ppm.

- The B1 scenario family describes a convergent world with a global population that peaks mid-century and declines thereafter (7.0 billion in 2100), with rapid changes in economic structures toward a service and information economy, reductions in material intensity, and the introduction of resource-efficient technologies. The emphasis is on global solutions to economic, social and environmental sustainability, including improved equity, but without additional climate initiatives. Approximate CO_2-eq concentration in 2100 = 600 ppm.

- The B2 scenario family describes a world in which the emphasis is on local solutions to economic, social and environmental sustainability. It is a world with a continuously increasing global population (10.4 billion in 2100), intermediate levels of economic development, and less rapid and more diverse technological change than in the B1 and A1 storylines. While the scenario is also oriented toward environmental protection and social equity, it focuses on local and regional levels. Approximate CO_2-eq concentration in 2100 = 800 ppm.

Although a total of 40 *SRES* scenarios were developed, each with equal plausibility, a subset of six marker scenarios (A1B, A1FI, A1T, A2, B1 and B2) were used by the climate-modelling community to produce climate projections for the *Third Assessment Report* in 2001 and the AR4 in 2007.

BOX 1.4 KEY CONCEPTS: DISTINGUISHING TYPES OF IPCC SCENARIOS

Scenario (generic) – Is a plausible and often simplified description of how the future may develop, based on a coherent and internally consistent set of assumptions about key driving forces (e.g. rate of technology change, prices and relationships). Scenarios are neither predictions nor forecasts and sometimes may be based on a 'narrative storyline'. Scenarios may be derived from projections, but are often based on additional information from other sources.

Emission scenario – Is a plausible representation of the future development of emissions of substances that are potentially radiatively active (e.g. GHGs, aerosols), based on a coherent and internally consistent set of assumptions about driving forces (such as demographic and socio-economic development, technological change) and their key relationships. Concentration scenarios, derived from emissions scenarios, are used as input into a climate model to compute climate projections.

Climate scenario – Is a plausible and often simplified representation of the future climate, based on an internally consistent set of climatological relationships, that has been constructed for explicit use in investigating the potential consequences of anthropogenic climate change, often serving as input to impact models. Climate projections often serve as the raw material for constructing climate scenarios, but climate scenarios usually require additional information such as about the observed current climate. A 'climate change scenario' is the difference between a climate scenario and the current climate.

(IPCC 2001a)

The IPCC decided not to commission a fourth set of emission scenarios, but instead collaborated with the scientific community to assess the growing scenario literature and establish a new process driven by the research (climate modelling and vulnerability assessment) and policy communities (Moss *et al.* 2010). The new suite of scenarios has been called 'representative concentration pathways' (RCPs). From the 324 scenarios examined in the scientific and policy literature, four were selected. The four RCP scenarios were named after their respective 2100 radiative forcing levels, with two higher forcing scenarios (RCP 8.5 and 6.0) and two lower forcing scenarios (RCP 4.5 and 3-PD) (Table 1.3). Each was developed by a different international modelling group, with differences between the RCPs not directly a result of climate policy, but attributable in part to differences between models and their scenario assumptions (scientific, economic and technological) (Moss *et al.* 2010).

The RCP 3-PD emission pathway is representative of scenarios in the literature leading to very low GHG concentration levels (below the 10th percentile of emissions scenarios in the literature and lower than any of the *SRES* scenarios). It is described as a 'peak-decline' scenario as its influence on the climate peaks by mid-century (at 3.1 W/m²), but then declines by 2100 (to 2.6 W/m²) to levels that are still higher than today (Table 1.3). In order to achieve the radiative forcing levels in this scenario, GHG emissions would need to be reduced substantially and peak before mid-century. Both RCP 4.5 and 6.0 are stabilization scenarios. Radiative forcing is stabilized in RCP 4.5 before 2100, while in RCP 6.0 stabilization occurs after 2100. In both cases, stabilization is well above current forcing levels (Table 1.3). The RCP 8.5 scenario has GHG emissions continuing to increase throughout the twenty-first century, and is representative for scenarios in the literature

Table 1.3 *Representative concentration pathway (RCP) emission scenarios*

RCP scenario	Radiative forcing in 2100 (W/m²)*	CO₂ equivalent concentration (ppm)†	Emissions pathway description in twenty-first century
8.5	>8.5 in 2100	>1370	Steadily rising
6.0	6 at stabilization after 2100	850 at stabilization after 2100	Stabilization (after 2100) without overshoot
4.5	4.5 at stabilization after 2100	650 at stabilization after 2100	Stabilization (before 2100) without overshoot
3-PD	Peak at 3 before 2100 and then declines	Peak at 490 before 2100 and then declines	Peak and decline

*Human activities since 1750 are estimated to have a radiative forcing of +1.6 W/m² (IPCC 2007b)
†2007, CO₂-eq estimated at 375 ppm (IPCC 2007b)
Source: Moss *et al.* (2010)

(approximately the 90th percentile of emission scenarios in the literature and similar to the *SRES* A2 and A1FI scenarios), leading to GHG concentration levels three times higher than current levels.

The IPCC's AR4 assessed the climate projections available from climate models forced with the *SRES* emission scenarios. The projected changes most relevant to tourism are summarized below. The IPCC (Solomon *et al.* 2007) emphasized that observed human-induced climate change had only just begun in the twentieth century, and that the pace of climate change was 'very likely' to accelerate throughout the twenty-first century, with continued GHG emissions at or above current rates. Within the range of *SRES* emission scenarios, warming 0.2°C per decade was projected in the 2010s and 2020s (similar to recent observations), and by 2100 the best estimate was that globally averaged surface temperatures would rise by between 1.8 and 4.0°C (Figure 1.8) (Solomon *et al.* 2007). An important caveat to this best estimate was that the recent emissions trajectory exceeded the most pessimistic of *SRES* scenarios. Importantly, the temperature increase in the twenty-first century would represent only 50 to 90 per cent of the eventual warming that was projected to result from *SRES* emission scenarios (Solomon *et al.* 2007).

The first climate change projections using the new RCP scenarios yield a similar range of additional global warming by 2100: RCP 3.0 = 2.3°C; RCP 4.5 = 3.0°C; RCP 8.5 = 5.6°C (Arora *et al.* 2011). These results are for only one GCM, and so are not directly comparable with the ensemble range from the IPCC AR4.

Future changes in temperature, precipitation, extreme events and other important climate features will manifest themselves differently across the world. The AR4 concluded that

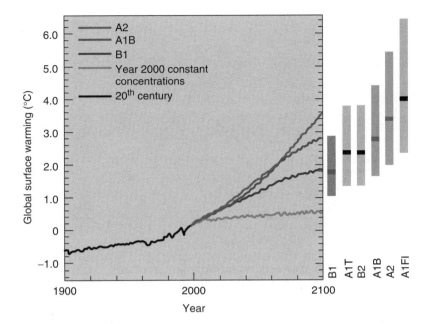

Figure 1.8 *Climate change projections under* Special Report on Emission Scenarios
Source: reproduced with permission of the IPCC

'there is now higher confidence in projected patterns of warming and other regional-scale features, including changes in wind patterns, precipitation and some aspects of extremes and of ice' (IPCC 2007b). Regionally, the greatest warming is expected over land and at high northern latitudes, and least over the Southern Ocean and parts of the North Atlantic Ocean.

According to the IPCC (Solomon *et al*. 2007), it is 'very likely' that extreme high temperatures, heatwaves and heavy precipitation events will continue to become more frequent. Figure 1.9 illustrates the combined impact of warming mean temperatures (a shift in distribution to the right) and greater variability in temperatures (broadening of the distribution) on extreme cold and hot temperatures. The probability of cold temperatures decreases substantially, while the probability of hot temperatures increases far more and the potential for new record high temperatures is introduced. The European heatwave of 2003 was cited as an example of the type of extreme heat event that is likely to become more common in a warmer future climate.

In many regions, in addition to changes in average precipitation amounts, the distribution of precipitation is anticipated to be concentrated into more intense events with longer periods of low precipitation in between (heavy episodic rainfall events with high runoff amounts are interspersed with longer relatively dry periods). Consequently, it was also considered 'likely' that there will be an increase in areas affected by drought, particularly summer drying in mid-latitudes. As average tropical sea surface temperatures increase, it is 'likely' that future tropical cyclones (typhoons and hurricanes) will become

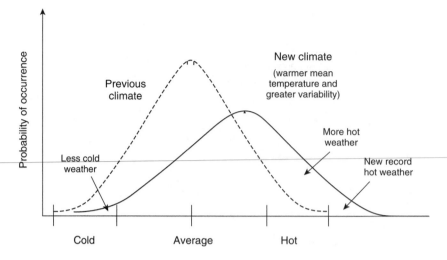

Figure 1.9 *Climate change effects on extreme temperatures*
Source: adapted from IPCC Working Group 1 (Solomon *et al.* 2007)

more intense, with higher peak wind speeds and more heavy precipitation. There is less confidence in projections of a global decrease in numbers of tropical cyclones. While some new evidence suggests that hurricanes may decline in frequency due to increased wind shear, it is not possible to be confident in this, and the possibility of increased frequencies and intensities cannot be ruled out (Knutson *et al.* 2010). Tropical storm tracks in some regions, such as the Caribbean, are projected to expand slightly toward higher latitudes. Extra-tropical storm tracks were also projected to shift poleward. Increases in the amount of precipitation are 'very likely' in high latitudes, while decreases are 'likely' in most subtropical land regions, continuing observed patterns in recent trends.

The IPCC (Solomon *et al.* 2007) also projected important regional changes in the cryosphere, which would also have ramifications for other environmental systems. The observed decreases in snow cover in the Northern hemisphere and increased thawing of permafrost regions were projected to continue. Glacial retreat was also expected in most regions. Sea ice was projected to shrink in both the Arctic and Antarctic. In some projections, Arctic late-summer sea-ice disappears almost entirely by the latter part of the twenty-first century. Continued melting of the cryosphere was expected to contribute to global sea-level rise, together with increased ocean temperatures. The IPCC estimated a sea-level rise of 18–59 cm over the twenty-first century (above the 1990 level), but assumed a near-zero contribution from the Greenland and Antarctic ice sheets on the basis that Antarctica especially was expected to gain mass from an increase in snowfall and insufficient warming (Solomon *et al.* 2007).

Sea-level rise projections have been vigorously discussed since the IPCC AR4, with a number of experts criticizing the IPCC's projections as very conservative (Hansen 2007;

Oppenheimer *et al.* 2007). Recent research suggests that both Greenland and Antarctic ice sheets have been losing mass at an accelerating rate over the past two decades (Rignot *et al.* 2010). Recent studies project that the rate of mean global sea-level rise will increase in the decades ahead, and that the total sea-level rise by the end of the century could reach as much as 1.5–2 m above present levels (Horton *et al.* 2008; Vermeer and Rahmstorf 2009). That these upper estimates would occur is highly unlikely (as is the lower estimated range); however, the central estimates of four of the five most recent studies exceed 1 m sea-level rise in the twenty-first century. Sea-level rise would continue after 2100: even if global temperatures were stabilized, the breakup of parts of the Greenland and West Antarctic ice sheets will continue over centuries.

Climate change scenarios for tourism

When assessing the implications of climate change for tourism (chapter 5), global projections are of relatively little use, as regional manifestations of climate change differ substantially. This is illustrated in Table 1.4, which provides temperature and precipitation projections for some of the world's leading tourism regions.

To assess the potential impact of climate change on tourists' perceptions and decision-making, and to provide tourism decision-makers (tourism operators, marketers, regulators,

Table 1.4 *Projected climate change in major tourism regions in 2100*[*]

Region	Temperature response (°C)		Precipitation response (per cent)	
	Minimum	Maximum	Minimum	Maximum
South Africa	1.9	4.8	–3	+25
Northern Europe	2.3	5.3	0	+16
Southern Europe and Mediterranean	2.2	5.1	–27	–4
East Asia	2.3	4.9	+2	+20
Southeast Asia	1.5	3.7	–2	+15
Western North America	2.1	5.7	–3	+14
Eastern North America	2.3	5.6	–3	+15
Central America	1.8	5.0	–48	+9
North Australia	2.2	4.8	–25	+23
South Australia	2.0	4.1	–28	+12
Caribbean	1.4	3.2	–39	+11
Mediterranean Basin	1.7	4.2	–30	–6

*Ensemble average of 21 for the A1B scenario
Source: Solomon *et al.* (2007)

insurers) with climate information relevant at the destination scale, requires scenarios with greater spatial and temporal resolution than typically provided by GCMs, which have a spatial resolution between 250 and 600 km^2 and output typically provided as seasonal or monthly data.

Several techniques have been developed to produce local-scale climate scenarios and to bridge the gap between the information that the climate-modelling community typically provides and that is required by the impacts research community. These 'downscaling' techniques include regional climate models that have spatial resolutions of as fine as 10 km^2, weather generators and other statistical techniques, and often link GCM scenarios with local climate station data and provide daily temporal resolution. The climate variables typically generated by GCMs are not necessarily useful for tourism decision-making, and translation of climate model outputs to tourism-relevant indicators is often required (e.g. number of rainy days, days with suitable beach-swimming weather, heat emergency days, duration and depth of snowpack, potential snow-making days, forest fire frequency and intensity).

The Day After Tomorrow? *Is abrupt climate change possible?*

In 1996, the IPCC examined the potential for what it called 'climate change surprises' or rapid, nonlinear responses of the climatic system to anthropogenic forcing. The IPCC (1996) and subsequent reviews of abrupt climate change by the US National Academy of Sciences (2002) and the US Climate Change Science Program (Weaver *et al.* 2008) all noted that, although evidence demonstrates that within a decade regional climates can shift substantially into different patterns that can persist for decades to centuries, climate change assessments rarely consider the unexpected, low-probability but high-consequence events associated with abrupt climate change.

In 2008, Lenton *et al.* (2008: 1787) identified nine potential 'tipping points' within the global climate system (outlined in Box 1.5), where exceeding a critical climatic threshold could fundamentally alter the state of an environmental system. They concluded that 'a variety of tipping elements could reach their critical point within this century under anthropogenic climate change'.

The National Academy of Sciences (2002: 1) concluded that while

> available evidence suggests that abrupt climate changes are not only possible but likely in the future, potentially with large impacts on ecosystems and societies ... the concept remains little known and scarcely appreciated in the wider community of scientists, economists, policy-makers, and world political and business leaders.

Studies of the potential ecological and socio-economic impacts of climate change have focused on *gradual* increase in global temperatures, and in some cases the associated changes in weather extremes, but remain typically 'surprise-free' (Weaver *et al.* 2008). No tourism-sector study has examined the implications of abrupt climate change or any of the climate-related tipping points identified by Lenton *et al.* (2008) (Box 1.5), therefore the vulnerability of the sector remains unknown.

BOX 1.5 CASE STUDY: POSSIBLE TIPPING POINTS IN THE GLOBAL CLIMATE SYSTEM

Indian summer monsoon – The regional atmospheric brown cloud is one of the many climate change-related factors that could disrupt the monsoon. Possible timeframe: one year; temperature increase: unknown.

Sahara and West African monsoon – Small changes to the monsoon have triggered abrupt wetting and drying of the Sahara in the past. Some models suggest an abrupt return to wet times. Possible timeframe: 10 years; temperature increase: 3–5°C.

Arctic summer sea-ice – As sea-ice melts, it exposes darker ocean, which absorbs more heat than ice does, causing further warming. Possible timeframe: 10 years; temperature increase: 0.2–2°C.

Amazon rainforest – Losing critical mass of the rainforest is likely to reduce internal hydrological cycling, triggering further dieback. Possible timeframe: 50 years; temperature increase: 3–4°C.

Boreal forests – Longer growing seasons and dry periods increase vulnerability to fires and pests. Possible timeframe: 50 years; temperature increase: 3–5°C.

Atlantic Ocean thermohaline circulation – Regional ice melt will freshen North Atlantic water. This could shut down the ocean circulation system, including the Gulf Stream, which is driven by the sinking of dense saline water in this region. Possible timeframe: 100 years; temperature increase: 3–5°C.

El Niño Southern Oscillation (ENSO) – El Niño already switches on and off regularly. Climate change models suggest ENSO will enter a near-permanent switch-on. Possible timeframe: 100 years; temperature increase: 3–6°C.

Greenland ice sheet – As ice melts, the height of surface ice decreases, so the surface is exposed to warmer temperatures at lower altitudes, which accelerates melting that could lead to ice sheet breakup. Possible timeframe: 300 years; temperature increase: 1–2°C.

West Antarctic ice sheet – The ice sheets are frozen to submarine mountains, so there is high potential for sudden release and collapse as oceans warm. Possible timeframe: 300 years; temperature increase: 3–5°C.

(Lenton *et al.* 2008)

THE INTERNATIONAL RESPONSE TO CLIMATE CHANGE

> 'Climate change is the greatest challenge facing humanity at the start of the 21st century. Failure to meet this challenge raises the spectre of unprecedented reversals in human development.'
>
> UNDP (2007)

This section provides a brief account of the international response to climate change. A fuller discussion of the response at various scales, and by various tourism organizations and sectors, can be found in chapter 3. Their relationship to other aspects of sustainable development and climate change is outlined in Figure 1.6.

The lead international treaty on climate change is the United Nations Framework Convention on Climate Change that was negotiated in part based on the findings of the IPCC's First Assessment in 1990. The UNFCCC was first signed in 1992 and entered into force in March 1994. It now enjoys almost universal membership with 195 Parties (194 countries and one regional economic integration organization), which meet annually at the Conference of the Parties (COP) to consider new scientific findings, discuss experience in implementing climate change policies, and review the implementation of the Convention. The UNFCCC provides an overall framework for international mitigation and adaptation efforts to address the challenges posed by climate change (see Box 1.6). It recognizes the global climate system as a common resource, and that the stability of the climate system can be affected by emissions of GHGs and other human activities.

The overall objective of the Convention (Article 2) is the 'stabilization of greenhouse gas concentrations in the atmosphere at a level that would prevent dangerous anthropogenic interference with the climate system'. The Parties recognize the principle of 'common but differentiated responsibilities' of signatory countries, with developed/industrialized

---

BOX 1.6 KEY CONCEPTS: CLIMATE CHANGE MITIGATION, ADAPTATION AND GEOENGINEERING

Mitigation – An anthropogenic intervention to reduce the sources or enhance the sinks of GHGs (IPCC 2007c).

Adaptation – Adjustment in natural or human systems to a new or changing environment. Adaptation to climate change refers to adjustment in natural or human systems in response to actual or expected climatic stimuli or their effects, which moderates harm or exploits beneficial opportunities (IPCC 2007c).

Geoengineering – The deliberate large-scale manipulation of the planetary environment to counteract anthropogenic climate change (Royal Society 2009).

countries assuming greater responsibility for reducing GHG emissions (mitigation) in the near term and assisting developing countries to adapt to climate change. However, the Convention does not set any mandatory limits on GHG emissions for individual countries and contains no enforcement mechanisms. Instead, the Convention utilizes 'protocols' to establish negotiated emission limits.

The Kyoto Protocol was the first such protocol, signed in Kyoto, Japan in 1997 and entering into force in 2005. As of 2010, a total of 191 countries have both signed and ratified the protocol. The Kyoto Protocol sets binding targets for 37 industrialized countries and the European Union to reduce their collective emissions of a group of GHGs (carbon dioxide, methane, nitrous oxide, sulphur hexafluoride, hydrofluorocarbons and perfluorocarbons) an average of 5.2 per cent against 1990 levels by the period 2008–12.

With the end of the initial commitment period nearing (2012), evaluation by the World Bank (2008a) indicates that the industrialized countries will, as a group, probably meet their Kyoto Protocol targets. Vast differences in performance are noted, however: member nations of the Organisation for Economic Co-operation and Development (OECD) have increased emissions, while members of the 'Economies in Transition' group (largely nations of the former Soviet Union) saw dramatic emission reductions due to economic restructuring in the 1990s. The EU expects that it will meet its collective 8 per cent reduction target for the EU-15 nations. Japan also expects to meet its reduction target. In Canada, Iceland, Australia and New Zealand, emissions have increased by at least 25 per cent compared with 1990, and these and other countries will be able to meet their targets only by purchasing sufficient emissions credits from other countries, a strategy these three countries have pledged not to attempt. Countries that fail to meet their emissions targets by the end of the first commitment period (2012) must make up the difference, plus a penalty of 30 per cent in the second commitment period. This of course assumes there is a second commitment period for the Kyoto Protocol or a successor protocol. Countries not in compliance will also have their ability to sell credits under emissions trading suspended, but given their large need for emission reductions, its is questionable whether any of them will have credits to sell.

International negotiations to establish an emission reduction agreement in which the USA, which did not ratify the Kyoto Protocol and therefore is not bound by it, and major developing countries (China, India, Brazil) would participate have continued since the ratification of the Kyoto Protocol. In 2007, leaders of the G8 announced a goal to reduce global CO_2 emissions 50 per cent by 2050. In 2009, the G8 (2009) nations issued their support for the policy goal of keeping global temperature increases below 2°C (over pre-industrial levels). At the 15th session of the COP in Copenhagen, Denmark in 2009, the delegates took note of the Copenhagen Accord which also recognized the policy goal of keeping global temperature rises below 2°C (over pre-industrial levels), but did not contain commitments for reduced emissions (Table 1.5 provides a selection of individual country commitments to GHG reduction made at Copenhagen). The objective of holding the increase in global average temperature below 2°C above pre-industrial levels was formally endorsed at the 16th COP in Cancun, Mexico in 2010. Because of the lack of progress

Table 1.5 Commitments to greenhouse gas reduction at Copenhagen*

Country	Type of emissions target	Quantitative target for 2020 (per cent)	Base year/ nature of target	Summary of target pledge	Share of global CO$_2$ emissions in 2007 (per cent)
USA	Absolute reduction	–17	2005	Approximately 17 per cent in conformity with anticipated US energy and climate legislation; final target will be reported to the Secretariat in light of enacted legislation	19.91
EU	Absolute reduction	–20 to –30	1990	As part of a global and comprehensive agreement for the period beyond 2012, the EU reiterates its conditional offer to move to a 30 per cent reduction by 2020 compared with 1990 levels, provided that other developed countries commit themselves to comparable emission reductions and that developing countries contribute adequately according to their responsibilities and respective capabilities	14.04
Japan	Absolute reduction	–25	1990	25 per cent reduction is premised on the establishment of a fair and effective international framework in which all major economies participate, and on agreement by those economies on ambitious targets	4.28
Russia	Absolute reduction	–15 to –25	1990	Range of GHG gas emission reductions will depend on appropriate accounting of the potential of Russia's forestry in meeting the obligations of the anthropogenic emissions reduction; undertakings by all major emitters for legally binding obligations to reduce anthropogenic GHG emissions	5.24
Canada	Absolute reduction	–17	2005	17 per cent, to be aligned with the final economy-wide emissions target of the USA in enacted legislation	1.90

Australia	Absolute reduction	−5 to −25	2000	Australia will reduce its GHG emissions by 25 per cent on 2000 levels by 2020 if the world agrees to a global deal capable of stabilizing levels of GHG in the atmosphere at 450 ppm CO_2-eq or lower; Australia will unconditionally reduce emissions by 5 per cent below 2000 levels by 2020, and by up to 15 per cent by 2020 if there is a global agreement that falls short of securing atmospheric stabilization at 450 ppm CO_2-eq and under which major developing economies commit to substantially restrain emissions and advanced economies take on commitments comparable with Australia's	1.28
China	Intensity reduction	−40 to −45	Emissions intensity change 2005–20	China will endeavour to lower its CO_2 emissions per unit of GDP by 40–45 per cent by 2020 compared with 2005 level, increase the share of non-fossil fuels in primary energy consumption to around 15 per cent by 2020, and increase forest coverage by 40 million ha and forest stock volume by 1.3 billion m^3 by 2020 from 2005 levels	22.30
India	Intensity reduction	−20 to −25	Emissions intensity change 2005–20	India will endeavour to reduce the emissions intensity of its GDP by 20–25 per cent by 2020 in comparison with 2005 levels	5.50
Indonesia	Reduction below BAU	−26 to −41	Reduction below BAU at 2020	26 per cent reduction relative to BAU unilaterally, up to 41 per cent reduction with international assistance	1.35
Brazil	Reduction below BAU	−36.1 to −38.9	Reduction below BAU at 2020	Anticipation that reductions in deforestation and other sectors of the economy will lead to reductions relative to projected emissions at 2020	1.26
Mexico	Reduction below BAU	−30	Reduction below BAU at 2020	Aims to reduce its GHG emissions up to 30 per cent with respect to the BAU scenario by 2020, provided there is adequate financial and technological support from developed countries as part of a global agreement	1.61

Continued

Table 1.5 Continued

Country	Type of emissions target	Quantitative target for 2020 (per cent)	Base year/ nature of target	Summary of target pledge	Share of global CO_2 emissions in 2007 (per cent)
	2000	−50	2050	This target is framed as 'aspirational' and is contingent on a multilateral regime that deploys significant financial and technological resources	
South Korea	Reduction below BAU	−30	Reduction below BAU at 2020	Pledge is conditional on a 'fair, ambitious, and effective' international agreement and on international finance, technology and capacity-building support; in addition to the 3.4 per cent reduction by 2020, a 42 per cent reduction is pledged for 2025	1.72
South Africa	Reduction below BAU	−34	Reduction below BAU at 2020	A 34 per cent deviation below the BAU emissions growth trajectory by 2020	1.48
Singapore	Reduction below BAU	−16	Reduction below BAU at 2020	Pledge is 'contingent on a global agreement being reached'	0.18
Costa Rica	Carbon neutral	−100	2021	Unclear whether target pertains to just CO_2 or CO_2-eq	0.03
Maldives	Carbon neutral	−100	2019	Unclear whether target pertains to just CO_2 or CO_2-eq	<0.01

*BAU = business as usual

Sources: Individual countries' submissions and UNFCCC summaries of country submissions, at http://unfccc.int/home/items/5262.php; Fransen (2009); Jotzo (2010); Pew Center on Global Climate Change (2010); UN Statistics Division (2010)

toward an extension of the Kyoto Protocol or some other agreement at the COP-15 (Copenhagen, Denmark) and COP-16 (Cancun, Mexico), and the time required for countries to ratify any agreement that could emerge in 2011, there is very likely to be a gap in any binding GHG emissions between the end of the Kyoto Protocol and any successor agreement.

Critically, recent observations have shown that the current rate of GHG emissions are exceeding the worst-case scenarios in the IPCC *SRES* emission scenarios that have been used for modelling future climate change (Raupach *et al.* 2007). If current emission trends continue, or even if the emission reduction commitments currently made by countries (at the time of writing) are successfully achieved, several studies indicate that temperatures would exceed +2°C average global warming by 2100 (Meinshausen *et al.* 2009; Parry *et al.* 2009a), the level considered by many scientists and the Parties to the UNFCCC to represent 'dangerous interference with the climate system'. Analyses of current GHG emission trajectories and mitigation commitments by the international community have led a number of recent studies to recommend that society should be preparing to adapt to +4°C global warming (Meinshausen *et al.* 2009; Parry *et al.* 2009a; Anderson and Bows 2011).

As a consequence of the very marginal progress to reduce the growth in GHG emissions, let alone realize absolute reductions required to achieve the objectives of the UNFCCC, over the past three or four years the concept of geoengineering (or climate engineering) has shifted from being considered a fringe, science-fiction approach to climate change, to a third response option considered by governments. The Royal Society (2009) reviewed a number of proposed geoengineering schemes and evaluated them in terms of effectiveness, affordability, timeliness and safety. It encouraged the Parties to the UNFCCC to increase their efforts towards mitigating climate change, but concluded that 'research and development of geoengineering options should be undertaken to investigate whether low risk methods can be made available if it becomes necessary to reduce the rate of warming this century' (Royal Society 2009: xi). The US Government Accountability Office (2010) also reviewed the state of geoengineering science, and it likewise highlighted the very limited body of science and tremendous uncertainties and potential dangers of many of the proposed schemes, but also urged the US government to establish a clear strategy for climate-engineering research to ensure preparedness for future climate events. An analysis by a leading economist in the field of costing climate change damages found that the very high impacts associated with low-probability but non-negligible high-temperature scenarios provided a legitimate argument for a geoengineering research programme to develop 'backstop technology' that could be utilized if a global climate emergency is foreseen (Weitzman 2009). Perhaps most significantly from a scientific perspective, the IPCC has now agreed to include geoengineering in its ongoing Fifth Assessment. A growing number of organizations, including the UN Convention on Biodiversity Conference in 2010, have called for a moratorium on climate-engineering field experiments.

As the IPCC (Solomon *et al.* 2007) makes clear, climate change is upon us, and is projected to increase in speed and magnitude in the decades ahead. While the UNFCCC commits all Parties to formulate, implement, publish and update climate change-adaptation measures (see Box 1.6), as well as to cooperate on international adaptation, it has remained a distant second in terms of its salience as a response to climate change, and has been construed by many environmental NGOs and some countries as 'giving up' on climate change mitigation. However, over the past decade, the issue of adaptation has advanced substantially among the scientific community and has moved higher up the UNFCCC negotiating agenda.

While the UNFCCC again reiterated the urgent need for an integrated policy response to climate change in 2010, it has yet to develop institutional structures that facilitate the effective and coherent implementation of adaptation activities. The Cancun Adaptation Framework (established in 2010) once again reinforced the need for enhanced action on adaptation that would reduce vulnerability and build resilience in developing country Parties, through 'long-term, scaled-up, predictable, new and additional finance, technology and capacity to implement adaptation actions, plans, programmes and projects at local, national, subregional and regional levels'. Although new international commitments to fund adaptation have been discussed (US$30 billion between 2010 and 2012, rising to US$100 billion per year by 2020), they fall short of some estimates of annual adaptation financing needs (all in addition to current official development assistance):

- current needs: US$9–41 billion (World Bank 2006) and US$50 billion (Oxfam 2007);
- needs in 2015: US$86 billion (UNDP 2007);
- needs in 2030: US$28–67 billion (UNFCCC 2007a).

Investment by developing nations in adaptation financing also represents a fraction of what is planned for emission reductions in developing countries. The international community will ultimately have to decide what countries and projects to prioritize for adaptation funding, which will be a highly politically contentious process. Furthermore, mobilizing the funds to support adaptation when all developing nation governments are implementing austerity budgets and coping with financial crises will be a major challenge.

It is in this context that the relationship between tourism and climate change is situated. Given its potential for contributing to economic development, tourism is seen as a potentially significant mechanism for encouraging economic growth and employment generation in both developed and less developed countries. Yet, at the same time, tourism is also contributing to GHG emissions and the effects of climate change to which countries are trying to respond. As we shall see throughout this book, this quandary represents the basic environmental economic dilemma that countries and researchers need to address.

> 'Addressing climate change is considered a prerequisite to sustainable development and therefore germane to advancing sustainable tourism.'
>
> Scott (2011: 17)

OUTLINE OF THE BOOK

This chapter provides an introduction to some of the issues surrounding the relationships between tourism and climate change. It argues that responding appropriately to this relationship is one of the most critical issues facing the sustainability of tourism businesses and destinations in the longer term. This chapter also provides an outline of the science of climate change and how this has developed over time. Yet, just as importantly, it also stresses that climate change, including its relationship with tourism, is an area of considerable controversy and debate. This is not just because of the activities of climate change 'sceptics', but also because there are important uncertainties in climate change science and the socio-economic future that inform GHG emission scenarios and the capacity of society to cope with climate change; issues surrounding the assumptions made in global climate and integrated assessment models used to understand the implications of climate change; differences in approach between various disciplines with respect to how the problem of climate change – and hence solutions – are to be defined; and politics of why and how governments, industry organizations, corporations, small and medium-sized enterprises, destinations and individuals choose to act – or not to act, as the case may be – on climate change (see the exchange between Weaver (2011) and Scott (2011) for a tourism-specific discussion of some of these topics).

Before examining the interrelationships between tourism and climate change in detail, the book provides background to the overall role of climate in tourism. Chapter 2 answers the question 'Why is climate important for tourism?' and looks at the effects of climate on tourist behaviour and decision-making, as well on tourism operations. Because of the role of climate and weather in affecting where people travel for leisure and activity-based holidays, many destinations and tourism businesses often seek to position themselves favourably to the consumer. Therefore chapter 2 also looks at the use of weather and climatic information in tourism marketing and in communicating information to tourists. Building on this general discussion of the interface of climate and tourism, the chapter briefly traces the development of research and practice (both government and industry engagement) specific to tourism and climate change, and outlines how this relationship has been approached and has changed over time – from an initial perspective that saw tourism as being affected by climatic and associated environmental change, to recognition of a two-way relationship where tourism is also a significant contributor to GHG emissions and therefore climate change. This brings us to the present, more systemic understanding of the relationship that frames this book.

The contribution of tourism to climate change is detailed in chapter 3. The chapter outlines how growth in tourism and mobility contributes to GHG emissions. This is done several ways at different scales: for the global industry overall; by various sectors (air, car and other transport, accommodation, activities); on a national basis; and by the travel of individuals, including by trip. The chapter concludes with a discussion of the need for a systems perspective in trying to develop a low-carbon tourism system.

Having established tourism's contribution to climate change, chapter 4 discusses how the tourism sector can seek to mitigate its GHG emissions. The chapter first provides a discussion of how policy frameworks for tourism and climate change operate at various scales of governance – which is recognized as an extremely complex response to what is described in policy terms as a 'wicked' problem. The chapter then covers policy frameworks at the international, suprational, national and subnational scales, as well as looking at specific sectoral and organizational (World Travel and Tourism Council, UNWTO) responses before noting some of the issues in implementing 'serious' climate policy in tourism. The chapter then examines the role of carbon management in reducing emissions, emphasizing that while many businesses appear to see emission reductions as a response to regulation and policy, there are sound economic reasons for engaging in carbon management – what is here understood as the strategic reduction of energy use and emissions based on economic considerations.

As noted at the start of this book, tourism both contributes to climate change and is affected by it. Chapter 5 details the impacts of climate change on tourism destinations by looking at the implications of four major types of impact pathway on different types of destination: snow and winter sport tourism, sea-level rise and coastal tourism, changes in biodiversity and landscapes, and recreation seasonality. One of the implications of climate change has been the emergence of what has been described as 'last chance' or 'disappearing destination' tourism – the need to visit a destination or see an attraction before it is 'lost' to climate change. The chapter examines these proposed new forms of tourism.

Having identified the ways in which climate change affects tourism, chapter 6 indicates how tourism organizations, including industry, destination and government, have or plan responses to the impacts of climate change (adaptation). We review the climate change perceptions of tourism professionals; investigate a number of cases of entrepreneurial responses from around the world to assess the variability in business action; and seek to provide explanations for the different responses to climate change.

Chapter 7 examines the response to climate change at the level of the consumer, looking at potential and actual shifts in both behaviour and demand. The chapter begins with an in-depth look at the various ways in which climate change could influence tourist perceptions, motivations and destination choices. Evidence on how tourist behaviour is currently, or could be, influenced by climatic change and related environmental change is then discussed.

Chapter 8 orients the reader toward future issues by providing an overview of knowledge gaps and research needs in tourism and climate change. This is undertaken in two main ways: first by identifying the recognition of tourism in major international climate change reports and its implications for integrating tourism research in broader climate and environmental change research; second by identifying new and emerging research needs specifically within a tourism and climate change context. This chapter also re-emphasizes some of the main themes and issues raised in the book and their importance. It also highlights the need for improved institutional arrangements with respect to climate change and tourism, as well as some of the fundamental ethical issues raised in considering the climate change implications of travel.

CONCLUSION: THE TENSIONS OF TOURISM AND CLIMATE CHANGE

This chapter provides an introduction to some of the issues surrounding tourism and climate change, as well as some of the core concepts of tourism studies and climate change science. One of the main issues in the subject of climate change is the public and media debate around the subject of climate change and its solution. Similarly, although not such a widespread debate, there is also a growing awareness of tourism impacts and the tensions that may exist in seeking to balance economic development with social and environmental goals. In many ways, the subject of tourism and climate change reflects some of the issues faced by other sectors (Parry *et al.* 2007). However, tourism also has specific characteristics and peculiarities that demand its own mitigation and adaptation response. For example, it has been specifically identified by international agencies as an important mechanism for economic development for many less developed countries and small islands that, coincidentally, are also potentially most at risk from the impacts of climate change and climate policy (Gössling *et al.* 2009a). Tourism is of considerable economic and social importance in the majority of developed countries as well, and while tourism overall has continued to grow for many years, there is potentially considerable volatility as a result of climatic and related change that may affect some of the locations that are most dependent on tourism. Tourism also provides one of the main economic justifications for the establishment of national parks and reserves and the conservation of biodiversity and landscapes (Frost and Hall 2009). If tourism is seriously affected by climate change, then the flow-on effects for the maintenance of ecological capital and the natural environment – arguably the core component of sustainability – will be potentially catastrophic, particularly given that biodiversity is already under direct threat from climate and environmental change (Hall 2010c).

The diffuse nature of tourism activities in economic, natural and social systems – and the fact that many people, at least in the developed world, regularly engage in leisure travel – may perhaps mean that tourism is taken for granted, not just by holiday-makers and business travellers, but also by policy-makers and scientists. Tourism also thrives on images of a clean environment, blue skies and waters, and being a smokeless industry. However, as this book stresses, tourism has considerable emissions and they are growing, while the environment on which many destinations and tourist activities depend is being affected by climate change. Tourism cannot be taken for granted by business and government in terms of it being able to fulfil its economic role; by holiday-makers to fulfil its social role; or by those concerned with the environment who want to use tourism to justify the conservation of biodiversity.

As this book illustrates, the relationships between tourism and climate change are complex and occur at multiple scales. Simplistic, top-down managerial 'solutions' of the past that encourage 'management by numbers' will not do. There is no 'one-size-fits-all' solution. Instead, approaches need to be able to integrate the various scales at which the governance of tourism and climate change occurs; the wide range of approaches to mitigation and adaptation that are utilized in different governance, cultural, economic and environmental contexts; as well as the implications of feedback mechanisms that will occur as the tourism system adapts to change.

In following the 25th anniversary of the first publication on climate change and tourism, this book seeks to provide a comprehensive and critical synthesis of knowledge on the direct and indirect effects of climate change on tourism and tourism's contribution to climate change; identify the many ways in which the tourism sector can mitigate and adapt to climate change; and contribute to the process of mainstreaming climate change within tourism policy, planning and management. We also hope that the contents of this book inspire current and future tourism professionals not only to consider the implications of climate change, but to act on them.

FURTHER READING AND WEBSITES

While it is assumed that readers are familiar with the basic workings of the global climate system, those interested in additional resources to improve their understanding of climate science, or more advanced discussions of climate change science, are recommended to consult the IPCC Working Group 1 reports of the Fourth Assessment or their list of frequently asked questions: https://www.ipcc.unibe.ch/publications/wg1-ar4/faq/wg1_faqIndex.html.

The organization Real Climate also provides a list of freely available online resources, conveniently organized into those 'for beginners'; those 'in search of more details'; and those who are well informed but 'seeking serious discussion of common climate change contrarian views': www.realclimate.org/index.php/archives/2007/05/start-here.

International tourism data and policy information are available from:

UN World Tourism Organization: http://unwto.org.

World Travel and Tourism Council (private business body): www.wttc.org.

For those interested in a more detailed yet accessible discussion of climate science and associated politics:

Giddens, A. (2009) *The Politics of Climate Change*. Cambridge: Polity Press.

Hansen, J. (2009) *Storms of my Grandchildren*. New York: Bloomsbury.

For a broader discussion of the interactions of tourism and other forms of global environmental change, the following book is recommended:

Gössling, S. and Hall, C.M. (eds) (2006) *Tourism and Global Environmental Change*. London: Routledge.

Why is climate important for tourism?

The tourism–climate interface

This chapter provides an introduction to the multifaceted relationships between climate and tourism. It outlines the development of this field of study over the past 50 years, and examines the defining characteristics of climate as a resource and constraint for tourism. The specific influences of weather and climate on several major tourism stakeholders – tourists, tourism operators and destinations – and their use of weather and climate information in decision-making are discussed: destination choice, resort design, tourism operations and marketing. The importance of communication of weather and climate information in the tourism sector is also addressed in this chapter.

THE DEVELOPMENT OF CLIMATE AND TOURISM RESEARCH

Climate is an important foundation for human civilization and is a crucial influence on development successes and failures (Diamond 2005; Stehr and von Storch 2010). For that reason, there has been a centuries-long scientific interest in the relationship between weather, climate and society, dating back to Aristotle's *Meteorologica* (*c*. 340 BC).

Modern climate science developed rapidly during the 1960s and 1970s, in what Lamb (2002) called the 'climate revolution'. He indicated that the combination of technological advances (improved radar, satellites, communications and computers) that significantly improved climate modelling, forecasting and archiving, and extensive publicity about several extreme climate events (Sahel drought and famine, Soviet crop failures, very cold North American winters) prompted substantial government investment in climate research programmes. These vast improvements in meteorology and climatology facilitated advances in understanding of the socio-economic impacts of climate and the societal applications of climate information. In the late 1960s, the field of applied climatology became defined as 'the scientific analysis of climate data in light of a useful application for an operational purpose', with operational purpose meaning any specialized socio-economic endeavour, including industrial, agricultural, transportation or technological activities (American Meteorological Society 2011).

In the past two decades, the world's attention has been drawn repeatedly to the death and destruction caused by extreme climate events, as well as an emerging scientific under-standing of anthropogenic climate change and the very significant implications it could have for environmental systems and human societies. This has supported the continued advancement and application of 'climate services', which WMO (2009) defines broadly to include historical data, analyses and assessments based on these data, climate monitoring products, forecasts, predictions, outlooks, advisories, warnings, climate and weather model outputs, model data, and climate change projections and scenarios.

Modern study of the climate–tourism interface emerged during the formative stages of the 'climate revolution', when applied climatologists expanded their research to a wide range of socio-economic sectors (Scott *et al.* 2005a). In 1970, Maunder's book *The Value of the Weather* was one of the first to comprehensively examine the complex relationships between weather and climate and human societies and important economic sectors. At the time, Maunder (1970: 165) stated that:

> It is generally believed that the tourist industry is *highly sensitive* to weather conditions. It is not known in any detail, however, as to what extent any particular tourist region is affected by the weather in that region or by the weather in the tourists' areas of origin. Neither is it generally known as to what extent tourists are influenced by weather forecasts.

Perry's (1972) parallel summary of the available literature on the influence of weather and climate on tourism reached a similar conclusion. This is understandable, considering the recent advances in climate science and the early state of development of tourism monitoring and tourism as a field of studies in the late 1960s.

A surge of climate and tourism research activity occurred in the mid- to late-1970s. Among the first to assess the relationship between climate and tourism were Pégy (1961), Paul (1972), the Canadian Atmospheric Environment Service (Crowe *et al.* 1973), and Besancenot *et al.* (1978), each of whom evaluated the minimum climatic conditions required for a range of tourism activities and examined climate as a determinant of the length of tourism seasons (in France and Canada). Despite this new work, in the early 1980s Coppock (1982) noted that, in terms of leisure and tourism studies, climate remained a curiously neglected theme in western countries, despite its importance to all outdoor pursuits. Smith (1983: 14) echoed this sentiment:

> Considering the potentially large effects of weather on the recreational economy of an area, or of the effects of a reputation (good or bad) of a [tourism] region created by weather, more system-atic studies are needed. Analysis of the perceptions of climate and weather, and of the relative benefits of obtaining better climatic and meteorological information, is especially needed.

Masterton (1980) argued that several factors explained the limited studies on weather, climate and tourism:

- insufficient awareness of the significance of weather and climate for tourism;
- studies on weather and climate are complicated;
- weather and climate data are not available at sufficient spatial and temporal resolu-tion for detailed study.

Interestingly, the availability of tourism and recreation data at appropriate temporal and spatial scales was not identified as a research barrier. This can be attributed, in part, to the fact that several of the studies completed during this period did not use tourism or recreation data to establish relationships with climate and weather.

Publication of research on climate/weather and tourism/recreation almost stopped during the early 1980s. A possible explanation for what Scott *et al.* (2005a) termed a 'period of stagnation' in the 1980s was that climate scientists, who predominantly carried out the early research in this field, were deflected into new, salient and better funded atmospheric science issues, such as acid rain, ozone depletion and the re-emergence of air pollution issues. In addition, the concept of anthropogenic global warming was not yet widely accepted in the early 1980s, and applied climatologists were therefore wary of suggesting implications for economic sectors. Although tourism scholars acknowledged the critical role of climate in tourism seasonality, there appeared to be very limited interest in climate issues. Tourism studies at this time were focused largely on documenting the rapid growth of domestic and international tourism and understanding its economic and social implications. Others suggest this was because climate was considered a more or less stable factor that was not subject to deliberate manipulation by tourism marketers (Abegg *et al.* 1998; Amelung *et al.* 2008).

With the emergence of anthropogenic climate change as a global environmental issue in the late 1980s (see timeline in Figure 1.6) came renewed interest in the socio-economic implications of climate and the future impacts of climate change. In this context, Smith (1993: 403) reviewing the then climate and tourism literature, observed that 'there is a clear need for new insights'. Smith (1993: 398) noted 'there have been comparatively few investigations into the relationships between weather and climate and tourism. One possible reason for this is that meteorologists and leisure specialists rarely communicate with each other.' Perry (1997) also concluded that the interface between climatology and tourism had not been adequately studied, and argued that collaboration between researchers from tourism studies and applied climatology was urgently needed if the potentially far-reaching effects of climate change were to be understood adequately.

The first decade of the twenty-first century represented a turning point for climate and tourism research. Both the volume of publications and the number and disciplinary diversity of researchers involved in this field increased substantially in the 2000s, with the field becoming truly multidisciplinary (Scott *et al.* 2005a). Another very important trend in climate and tourism research has been the increasing collaboration between climate and tourism communities. This cooperation has taken multiple forms, but includes two important international initiatives. The International Society of Biometeorology established its Commission on Climate, Tourism and Recreation during its 14th Congress in 1996 (Ljubljana, Slovenia). This Commission organized three international conferences dedicated to climate and tourism in 2001, 2003 and 2007, with the conference proceedings representing the largest collection of research papers in this field to date. The UNWTO and WMO also began new collaborations to improve the availability and use of weather and climate information in the tourism sector, including the establishment of a new Expert Team on Climate and Tourism in 2005, and the commissioning of a White Paper on

Weather and Climate Information for Tourism in 2009 (Scott and Lemieux 2009). Some 40 years after Maunder (1970) initially identified tourism as a 'highly weather-sensitive' economic sector, this recognition was once again reaffirmed by the climatology–meteorology communities in the WMO-commissioned White Paper, which concluded that accumulating evidence:

> demonstrate[s] the climate sensitivity of the tourism sector … [and] that climate change will be a pivotal issue affecting the medium- and long-term future of tourism development and management … [so that] that the need for climate services will increase throughout the twenty-first century as the magnitude of climate change increases and the ability to rely on previous experience diminishes.
>
> (Scott and Lemieux 2009: 44)

Demands for accurate and increasingly detailed climate information are therefore anticipated to increase substantially in order to allow tourism businesses and destinations to minimize associated risks and capitalize upon new opportunities posed by climate change, in an economically, socially and environmentally sustainable manner.

Critically, accumulating evidence indicates that climate change, particularly high emission scenarios, will be a pivotal issue affecting the medium- and long-term future of tourism development and management (Scott *et al.* 2008a). Consequently, it is recognized that the need for climate services will increase throughout the twenty-first century as the magnitude of climate change increases and the ability to rely on previous experience diminishes.

CLIMATE AS A RESOURCE AND CONSTRAINT FOR TOURISM

As the sample of recent media headlines from around the world in Figure 2.1 reveals, climate represents both a resource and a limiting constraint for tourism activities and development. De Freitas (2003) has defined three facets of climate as a resource for tourism:

- the thermal component relates to the thermal comfort of tourists and considers the integrated effects of air temperature, wind, solar radiation, humidity, metabolic rate, clothing and activity;
- the physical component represents features such as wind and precipitation (rain, snow) that may act as a physical annoyance or limit the possibility for tourist activities;
- the aesthetic component refers to climate features that may influence tourists' appreciation for, or the quality of, a view or landscape, including sunshine, blue skies/cloud cover, visibility and day length.

Tourists respond to the integrated effects of these three facets, which affect them simultaneously in physical, physiological and psychological ways (Gómez-Martín 2006). All three facets are highly important to destination image and tourism marketing as well as the actual tourist experience, while the thermal and physical components are fundamental to tourism infrastructure design and important determinants of operating costs. Gómez-Martín (2006) further articulated several key characteristics of climate resources for tourism, which are expanded on here.

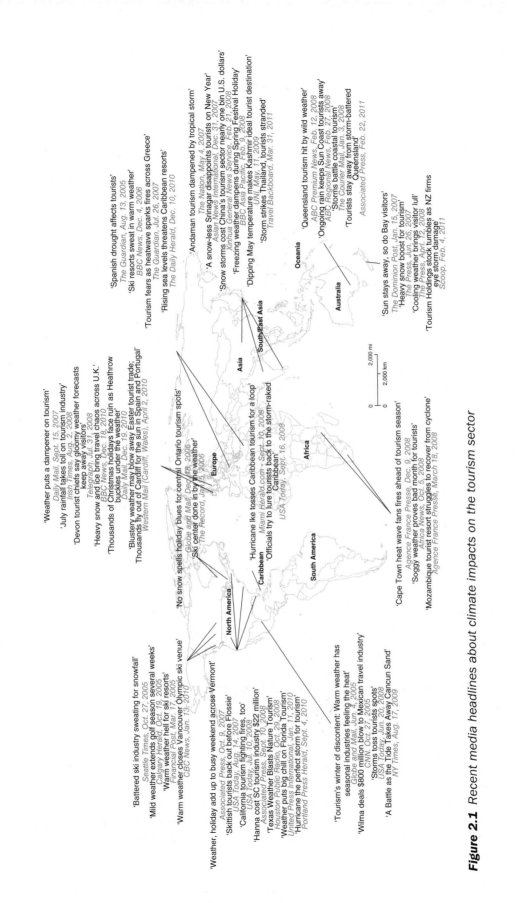

'Weather puts a dampener on tourism'
Daily Mail, Sept. 15, 2007
'July rainfall takes toll on tourism industry'
Irish Times, Aug. 2, 2008
'Devon tourist chiefs say gloomy weather forecasts keep away visitors'
Telegraph, Jul. 31, 2008
'Heavy snow and ice bring travel chaos across U.K.'
BBC News, Dec. 18, 2010
'Thousands of Christmas holidays face ruin as Heathrow buckles under the weather'
Daily Mail, Dec. 19, 2010
'Blustery weather may blow away Easter tourist trade; Thousands fly out of Cardiff for the sun in Spain and Portugal'
Western Mail (Cardiff, Wales), April 2, 2010

'Spanish drought affects tourists'
The Guardian, Aug. 13, 2005
'Ski resorts sweat in warm weather'
BBC News, Dec. 4, 2006
'Tourism fears as heatwave sparks fires across Greece'
The Guardian, Jul. 26, 2007
'Rising sea levels threatens Caribbean resorts'
The Daily Herald, Dec. 10, 2010

'Andaman tourism dampened by tropical storm'
The Nation, May 4, 2007
'A snow-less Srinagar disappoints tourists on New Year'
Asian News International, Dec. 31, 2007
'Snow storms cost China's tourism sector nearly one bin U.S. dollars'
Xinhua General News Service, Feb. 21, 2008
'Freezing weather dampens during Spring Festival Holiday'
BBC Asia Pacific, Feb. 9, 2008
'Dipping May temperature makes Kashmir ideal tourist destination'
UNI, May. 11, 2009
'Storm strikes Thailand, tourists stranded'
Travel Backboard, Mar. 31, 2011

'Queensland tourism hit by wild weather'
ABC Premium News, Feb. 12, 2008
'Ongoing rain keeps Sun Coast tourists away'
ABC Regional News, Feb. 27, 2008
'Storms battle coastal tourism'
The Courier Mail, Jan. 3, 2008
'Tourists stay away from storm-battered Queensland'
Associated Press, Feb. 22, 2011

'Sun stays away, so do Bay visitors'
The Dominion Post, Jan. 15, 2007
'Heavy snow boost for tourism'
The Press, Jun. 26, 2007
'Cooling weather brings visitor lull'
The Press, April 22, 2008
'Tourism Holdings stock tumbles as NZ firms eye storm damage'
Scoop, Feb. 4, 2011

'No snow spells holiday blues for central Ontario tourism spots'
Globe and Mail, Dec. 19, 2006
'Ski center done in by the weather'
The Record, Jan. 15, 2006

'Hurricane Ike tosses Caribbean tourism for a loop'
Miami Herald.com - Sept. 10, 2008
'Officials try to lure tourists back to the storm-raked Caribbean'
USA Today, Sept. 16, 2008

'Cape Town heat wave fans fires ahead of tourism season'
Agence France Presse, Dec. 9, 2008
'Soggy weather proves bad month for tourists'
Africa News, Oct. 5, 2007
'Mozambique tourist resort struggles to recover from cyclone'
Agence France Presse, March 18, 2008

'Battered ski industry sweating for snowfall'
Seattle Times, Oct. 27, 2005
'Mild weather extends golf season several weeks'
Calgary Herald, Oct. 19, 2005
'Warm weather hell for ski resorts'
Financial Post, Mar. 17, 2005
'Warm weather closes Vancouver Olympic ski venue'
CBC News, Jan. 13, 2010

'Weather, holiday add up to busy weekend across Vermont'
Associated Press, Oct. 9, 2007
'Skittish tourists back out before Flossie'
USA Today, Aug. 14, 2007
'California tourism fighting, fires, too'
USA Today, Jul. 10, 2008
'Hanna cost SC tourism industry $22 million'
Associated Press, Sept. 10, 2008
'Texas Weather Blasts Nature Tourism'
Houston Public Radio, Oct. 29, 2008
'Weather puts big chill on Florida Tourism'
United Press International, Jan. 11, 2010
'Hurricane the perfect storm for tourism'
Portland Press Herald, Sept. 4, 2010

'Tourism's winter of discontent: Warm weather has seasonal industries feeling the heat'
Globe and Mail, Jun. 4, 2007
'Wilma deals $800 million blow to Mexican travel industry'
CNN, Oct. 27, 2005
'Storms toss tourists spots'
USA Today, Jun. 20, 2008
'A Battle as the Tide Takes Away Cancun Sand'
NY Times, Aug. 17, 2009

North America

South America

Europe

Caribbean

Africa

Asia

South East Asia

Oceania

Australia

0   2,000 mi
0   2,000 km

Figure 2.1 Recent media headlines about climate impacts on the tourism sector

Climate resources for tourism are culturally and socially defined

Climate resources for tourism thus correspond with Zimmerman's (1951: 814–815) classic definition of a resource:

> Resources are highly dynamic functional concepts; *they are not, they become,* they evolve out of the triune interaction of nature ... and culture, in which nature sets outer limits, but ... culture [is] largely responsible for the portion of physical totality that is made available for human use.

For example, although sun, sand and sea (3S) holidays are one of, if not the, largest world-wide tourism segment today, the pursuit of a suntan as part of heliotheraphy or as a symbol of wealth and leisure time is a phenomenon that emerged in developed countries only in the early decades of the twentieth century (Kevan 1993). Until this shift in fashion in most of the countries that represent the major tourism markets today, a tanned skin was regarded as an indicator of common, outdoor manual work and associated with lower social classes. As Randle (1997: 462) notes, 'In 18th century Europe and the New World, porcelain paleness was the epitome of stylishness.' Without this cultural shift in perceptions of beauty and symbols of wealth, the sunny and warm coastal regions of the Mediterranean and Caribbean could not have become the valuable resources for tourism they are today.

The distribution of climate resources varies in space and time

Climate resources for tourism are not evenly distributed geographically and are subject to inter-annual, seasonal and more rapid temporal variations (daily or even hourly). Indeed, seasonality is arguably one of the hallmarks of the consumption and production of tourism services.

Climate is a free common resource

Climate resources are abundant and do not need markets or regulatory mechanisms for equitable or sustainable allocation among tourism stakeholders or between tourism and other economic sectors.

The value of climate resources fluctuates like other market commodities

Because climate resources for tourism vary in time and space, both demand and geographical supply fluctuate throughout the year, affecting prices for holidays. The value of the same climate resource at the same location can vary substantially throughout the year because of the distribution of supply in other destinations and in tourist source markets. For example, weather conditions can be virtually the same all year in tropical destinations such as Barbados. Conceptually, the climate resource for tourism is uniform year-round; however, tourist arrivals and prices exhibit a very different distribution throughout the year because the climate resource for tourism has much higher perceived value for tourists when the climate resources for tourism are low at their home locations.

Climate is a resource that cannot be transported or stored

As indicated above, climate resources exist at a specific location at a specific time, and the consumer must travel to experience the sought-after climate *in situ*. While freshwater or snow resources from precipitation are possible to capture or transport, once they are on the ground they are considered part of the hydrosphere and cryosphere, respectively, and are no longer atmospheric resources.

Climate is a renewable resource

Climate resources available to a tourist or at a destination in the future are not affected by the amount used by tourists, tourism operators or any other economic sector. Unlike other scarce resources, no conflicts arise from direct use.

Climate is generally considered a non-degradable resource

Minimal or extensive exploitation of climate resources by the tourism sector does not degrade the resource available for the future. However, as discussed in chapter 3, the tourism sector is a contributor to anthropogenic climate change, and thus is altering global and regional climate resources through its broader activities.

As outlined in chapter 1, the tourism sector is characterized by considerable diversity, with major components including transportation (airlines, cruise-ships, rail lines, ground coaches, taxis), accommodation (hotels, apartments, youth hostels), food and hospitality services (restaurants, bars, pubs), travel agents and tour service operators, visitor attractions (cultural or sporting events, casinos, parks, museums) and tourist-focused retail or service providers (insurance, conventions, tourist equipment rentals). Tourism operators within each of these subsectors differ in terms of ownership (government, NGOs, private businesses), size (small and medium-sized enterprises through to multinational conglomerates) and purpose (for-profit or non-profit, as well as conservation, education and community development mandates). Tourism operators have also adapted to provide tourism services in every climatic zone on the planet from deserts and high mountains to the tropical and polar regions. Equally as diverse are the motivations and characteristics of domestic and international travellers from around the world.

With this inherent diversity, the interface between climate and the tourism sector is accordingly multifaceted and complex. Figure 2.2 outlines the temporal scales (weather forecasts, daily weather variations, seasonality, inter-annual variability) at which weather and climate influence each of the major stakeholders within tourism, either directly or indirectly, by affecting other environmental or socio-economic systems that then affect tourism. Importantly, weather and climate are but one macro-scale factor influencing the tourism system, embedded in a matrix and interacting with other macro-scale factors that

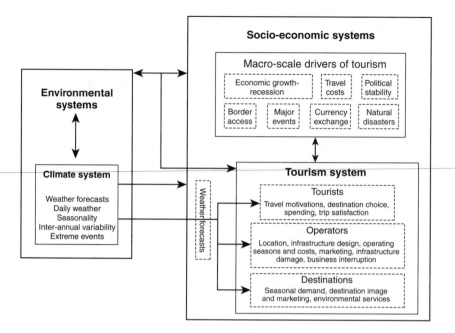

Figure 2.2 *The climate–tourism interface*

influence the travel decisions of tourists as well as operational decisions of the tourism sector.

The remainder of this chapter examines the influence of weather and climate on the specific decision-making processes of major tourism stakeholders: tourists, tourism operators and destinations.

WEATHER, CLIMATE AND TOURISTS

Weather and climate have broad significance for tourists' decision-making and the travel experience. Figure 2.3 conceptualizes the influence of weather and climate information (historical climate, forecasts, real-time weather conditions) on various types of tourist decisions, during the pre-trip, trip and post-trip phases. Climate, both at home and at destinations, is a salient motivator for tourism. Climate is a key factor considered by tourists, either consciously or implicitly during travel planning. For some tourists, climate is considered only in terms of its suitability for intended activities during the planned time of travel. For others, seeking improved climatic conditions is the primary motivation for travel and destination choice. Climate has universal importance in defining a destination's attractiveness. For tourists, destination weather is an intrinsic component of the travel experience and it influences tourist activity choices and expenditures, as well as affecting overall trip satisfaction. The influence of weather and climate on each of these areas of tourist decision-making is reviewed below; however, there are many knowledge gaps with respect to its place in the complex psychological process of travel planning and *in situ*

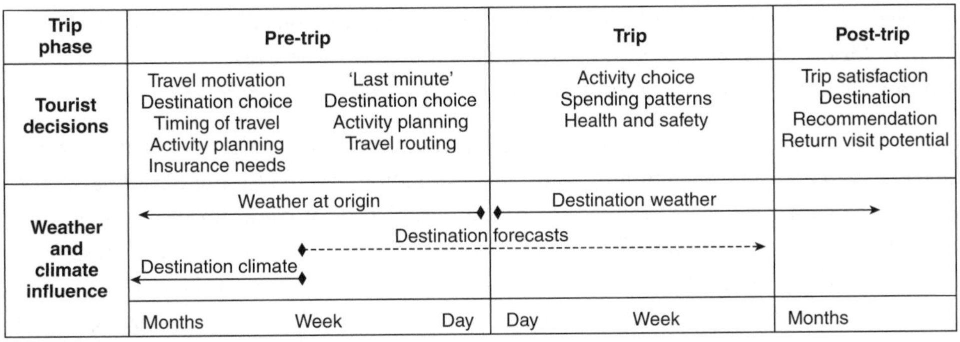

Trip phase	Pre-trip		Trip	Post-trip
Tourist decisions	Travel motivation Destination choice Timing of travel Activity planning Insurance needs	'Last minute' Destination choice Activity planning Travel routing	Activity choice Spending patterns Health and safety	Trip satisfaction Destination Recommendation Return visit potential
Weather and climate influence	Weather at origin ◄────────► ◄─ ─ ─ ─ ─ ─ Destination forecasts ─ ─ ─ ─ ─ ─ ─► ◄─ Destination climate ─►		Destination weather ────────►	
	Months Week Day		Day Week	Months

Figure 2.3 *Influence of weather–climate information on tourist decisions*

tourist decisions, making this an important area for further research (Gössling *et al.* 2011a).

Climate is an important motivator for many tourists, consciously or implicitly, and represents both a 'push and pull' motivational factor (Crompton 1979) for tourists. Seeking the health benefits of improved atmospheric environments (warmer conditions, seaside air, escaping the squalor and smell of cities) is one of the earliest known motivations for tourism in early Roman, Greek, Egyptian and Chinese civilizations (Kevan 1993). As a result of the 'Grand Tour' of Renaissance Europe and mercantile trade, travellers became increasingly aware of the health benefits of climate available in distant lands and at sea (Kevan 1993), and this became further commodified for the wealthy from the 1870s onwards as commercial tourism was developed (Hall 2003). Modern travel surveys conducted in Germany, the United Kingdom and Canada have all found that weather and climate was a salient travel motivation for travellers. Mintel (1991) found that 70–80 per cent of respondents to a UK travel survey cited 'good weather' as the main reason for going abroad. A survey of Canadians (Ontario Ministry of Tourism and Recreation 2002) similarly found that 'escaping winter weather' was the prime travel motivation for 23 per cent of respondents. Kozak (2002) surveyed the motivations of German and British tourists in Mallorca and Turkey, and found that 'enjoying good weather' was the most important motivational factor for their travel.

Jonsson and Devonish (2008) found 'to enjoy good weather' was the fourth highest motivation (out of 14) for a sample of tourists visiting Barbados in December 2006 to March 2007. However, consideration of the survey technique used in this study, which is common among many motivational ranking studies in tourism, raises questions as to whether it adequately conceptualizes the role of weather. In this case, weather was conceptualized as one of four variables to measure the construct of 'relaxation', not a physical parameter or psycho-health variable for the respondents, virtually all of whom were from Canada, the UK and the USA, and would be motivated to escape winter conditions at home. Further, the top three motivations recorded were 'to relax', 'to be emotionally and physically refreshed' and 'to have fun'; all of which could potentially be accomplished by much shorter trips to nearby health and wellness spas, ski resorts, or urban sightseeing or shopping trips. If respondents were asked why these motivations could not be fulfilled by a trip to Montreal, New York or London in the middle of winter, instead of a much longer trip to Barbados, would the salience of 'weather' as a motivator increase?

Other traveller surveys, conducted in a number of countries, have also revealed the importance of climate in the selection of a holiday destination and the timing of holiday travel. There is general consensus in the tourism literature that destination image is a key determinant of destination choice (Pike 2002). Climate is an important aspect of destination image, yet not all studies of destination image include climate as an attribute. For example, of the 25 destination image studies reviewed by Mazanec (1994), only 12 included climate as an attribute. However, many destination image studies that have included climate as an attribute found it to be one of the most important attributes. Hu and Ritchie's (1993) review of destination image studies found that 'natural beauty and climate' were of universal importance in defining destination attractiveness. A survey of German travellers found that 'weather' (or, more accurately, perceived weather or climate) was of major importance in the choice of holiday destinations, third in importance behind only landscape and price (Lohmann and Kaim 1999). Hamilton and Lau (2005) confirmed these findings in their survey of German tourists, which found climate to be the most frequently considered destination attribute in destination decision-making. A large survey of French tourists also found that climate was one of the most important factors in destination choice, behind cost, beauty of landscapes and discovery of new places (Credoc 2009). Gössling *et al.*'s (2006) survey of international tourists in Tanzania found that 53 per cent rated climate as a very important or important factor in destination choice, with 30 per cent stating that climate was not important, perhaps because they were visiting friends or relatives. Climate can also be an important deterrent for some destinations. For example, Hay (1989, in Perry 1997) reported that weather emerged as the most important feature that visitors found unattractive about Scotland.

As indicated, travel patterns are often related to the weather and climate conditions at the point of origin, not just at the destination, variously termed a 'push–pull' (Crompton 1979) or 'double-hurdle' variable (Eugenio-Martin and Campos-Soria 2010). For example, despite the global economic recession in 2008–09 and expectation of reduced travel demand, the very rainy weather throughout much of the early summer in the UK was credited by the Association of British Travel Agents for the increase in foreign holiday bookings (Hill 2009). Eugenio-Martin and Campos-Soria's (2010) study of household travel patterns in EU nations found that the climate of the region of residence is a strong determinant of holiday destination choice and timing of travel. They found that a much higher proportion of residents of Mediterranean countries take holidays in the summer months, while residents of Scandinavian and Northern European countries travel much more often in the winter months. Their analysis also showed that a better climate in the region of residence implied a higher probability of travelling domestically and lower probability of travelling abroad. They argue that climate is a key variable to take into account when investigating the capacity of domestic markets to retain tourists.

In the same way that climate affects the destination choice of travellers, it also highly influences the timing of travel. Seasonal demand is one of the main defining characteristics of global tourism, and is comprised of natural and institutional seasonality (Butler 2001), both of which are influenced by climate. The former is directly related to climate; the school holidays component of the latter is historically related to agricultural labour needs that were also climate-dependent. Seasonal climate fluctuations at tourism destinations and at major outbound markets, particularly at high latitudes, are a key driver of tourism demand at global and regional scales.

Climate variability has been found to influence travel patterns (proportion of domestic and international holidays), activities and tourism expenditures. Agnew and Palutikof's (2006) analysis of the sensitivity of UK tourism to climate variability found that outbound flows of tourists were more responsive to weather of the preceding year, whereas domestic tourism was more responsive to weather conditions within the year of travel. Nadal *et al.* (2008) similarly found temperature, heatwaves, frost and sunshine days were significantly related to the dynamics of outbound tourist flows from the UK. Agnew (1995) also found that, in the UK, tourism spending abroad increased following a cold winter. Giles and Perry (1998) found that the exceptional summer of 1995 in the UK had the opposite effect, resulting in a drop in outbound tourism. Consistent with these results, Smith (1990) found a good statistical relationship between summer visits from the UK to Portugal and summer rainfall over Great Britain the previous summer. In Norway, demand for summertime inclusive tour charters, most of which are to sunshine destinations, is influenced by poor weather conditions in the previous summer (Jorgensen and Solvoll 1996), as travellers avoid further risk of poor domestic weather by travelling abroad to satisfy their need for good summer weather. In Canada, a 1°C warmer than average summer season was found to increase domestic tourism expenditures by 4 per cent (Wilton and Wirjanto 1998). Similar correlations between monthly accommodation demand (bed-nights) and summer temperatures (both current year and the previous summer) have been found in Italy (Bigano *et al.* 2005).

A review of the literature on the effect of weather on leisure activities and physical activity (Tucker and Gilliland 2007) found that weather had a significant impact on these behaviours in cities, and it appears that this influence would also extend to leisure activities while away from home as a tourist. Weather is likely to have a strong influence on the activity patterns of tourists, and thus to affect spending patterns throughout destinations. A number of sector-specific studies have shown significant relationships between weather conditions (daily to weekly timescales) and a range of tourism activities – ski-lift tickets, golf rounds, park attendance, special event attendance (Paul 1972; Meyer and Dewar 1999; Hamilton *et al.* 2003; Ploner and Brandenburg 2003; Jones and Scott 2006; Scott and Jones 2006a, 2007; Moreno *et al.* 2008; Shih *et al.* 2009).

Travellers are interested in the weather forecasts at their intended destination as well as the weather along the way (travel phase). Business travellers are particularly cognizant of how weather causes delays and diversions, and utilize forecasts in routing decisions (Scott and Lemieux 2009).

Weather is an intrinsic component of the travel experience, and for many travellers weather conditions at the destination are central to overall trip satisfaction. Visitor surveys by the Scottish Tourist Board show that, for overseas visitors, weather is identified as the main cause of trip dissatisfaction (Smith 1993). Similar observations have been made in New Zealand, where weather and limited time were the two most frequently mentioned disappointments to travellers (Becken *et al.* 2010). Holiday visitors were more likely to be disappointed with the weather than those visiting friends and relatives, and other categories of tourists. Coghlan and Prideaux (2009) found that adverse weather experience reduced the overall satisfaction of tourists visiting Australia's Great Barrier Reef, and that while the aesthetic components of weather had an important role in creating particularly positive experiences, the physical aspects of weather dominated negative experiences. Spiller *et al.* (1988) likewise found weather conditions significantly affected the satisfaction of sport fishers and boaters in the USA. Weather also has a significant effect on the net benefits of beach tourism (McConnel 1977; Silberman and Klock 1988), where cool and windy conditions, such as those shown at a 3S tourism resort in the Caribbean, have a highly negative impact on holiday satisfaction and perceived value of holiday purchase. Williams *et al.*'s (1997) analysis of the influence of weather conditions on visitor satisfaction at ski resorts concluded that weather was a 'troublesome source of contamination' in studies of holiday satisfaction and destination evaluation. They further noted that, in their review of visitor satisfaction assessments, contextual weather information was typically lacking and had rarely been controlled for.

This negative effect of adverse weather on trip satisfaction probably extends to many other forms of tourism, and probably leads to the greatest dissatisfaction during holidays where improved climate was a specific motivation for the holiday (3S or winter escape), where marginal weather would result in the perception of very poor value for the holiday purchase. A wide gap between promoted weather (via tourism brochures and websites) and the weather experienced may also have a negative impact on tourist satisfaction, but no known study has documented this effect. Importantly, while adverse weather has been shown to have a disproportionately negative impact on tourist satisfaction, evidence suggests that the decision to return to a destination is largely unaffected by past experiences of poor weather (Lohmann and Kaim 1999; Moreno 2010).

The limited studies that have examined the use of weather and climate information by travellers reveal widespread use of such information in holiday planning. Tourism and recreation users were shown to generate the largest demand from automated telephone weather services in Scotland and Britain (Smith 1981). A survey of German outbound tourists found that 73 per cent informed themselves about the climate of their destination, and 42 per cent informed themselves about the climate before booking their travel (Hamilton and Lau 2005). A similar survey of Northern European travellers to the Mediterranean region found that 86 per cent would obtain information on their destination's climate, with 81 per cent doing so before making any travel reservations (Smith 1981). Results for international travellers to New Zealand were consistent with those from Europe, with 94 per cent obtaining climate information prior to arrival (Becken *et al.* 2010). Consideration of current weather conditions or near-term forecasts (next 1–4 days) has been found to be the most important factor in last-minute domestic leisure tourism (Szalai and Ratz 2007). With the trend toward shorter timeframes for travel planning, especially for discounted last-minute bookings made in the week (or day) prior to departure, the value of short- and medium-term forecasts for travel planning is likely to increase

(Scott and Lemieux 2009). In the same way, the influence of media stories about unfavourable weather for tourism and extreme events may also increase. For example, when Northern European travellers were asked if media stories about heatwaves in the Mediterranean would affect their travel plans to a destination in the region, 51 per cent indicated they would alter their travel plans in some way, and a further 15 per cent said they would seek additional information before deciding (Rutty and Scott 2010).

While it is clear that a large proportion of tourists consult climate information of some type on their planned destination, it is not certain whether tourists consider the climate of a destination holistically, or focus on specific attributes of climate (e.g. temperature or precipitation) when making choices regarding destination and timing of travel. Hamilton and Lau's (2005) survey of German tourists found that, of the 73 per cent of tourists who had informed themselves about the climate of their selected destination, 91 per cent sought information on more than one climate variable. A survey of tourists in Tanzania similarly found that tourists' perceptions of climate were shaped by a set of parameters, including temperature, rainfall, humidity and storms, rather than any singular parameter (Gössling et al. 2006). These results are consistent with studies that ask tourists to rank the relative importance of weather/climate parameters, and have found the rank order of temperature, wind, rain and sunshine/cloud cover to vary by tourism environment (Scott et al. 2008b; Rutty and Scott 2010), suggesting that all have importance to tourists in specific circumstances.

Another area of uncertainty regarding tourists' use of weather and climate information is how they interpret weather forecasts and other forms of climatic information. A number of studies have shown that portions of the public do not interpret forecasts by meteorological services correctly (Gigerenzer et al. 2005); however, there have been very few studies specific to tourism. In a study of beach holiday travel, Adams (1973) found that the decision to go on a holiday when faced with an uncertain weather forecast (50:50 probability of poor weather) ('risk taking'), and even the interpretation of forecast itself ('risk manipulation'), was mediated by series of situational variables, with level of commitment to the trip having the most significant affect. Similar types of psychological processes operate at larger scales for outbound trips planned to long-haul destinations, but the entry points of weather and climate information in trip planning, and how perceptions of possible weather conditions are mediated by other situational factors, remain poorly understood.

The production of weather and climate information alone is not sufficient for tourists' decision-making. Information must be delivered to end-users in a form that is relevant to them and that they have the capability to interpret. Scott and Lemieux (2009) outlined the range of communication channels for the delivery of climate information to tourists and the tourism sector (Figure 2.4). Tourists have the greatest diversity of sources, including directly from government or private meteorological services as well as tourism operators and destinations themselves, and communication media, ranging from tourism marketing and guide books, the internet, television, radio, newspapers and hand-held smartphone devices.

How climate information is communicated to tourists is largely unexplored, and the communication channels, especially related to warnings of abrupt and dangerous weather

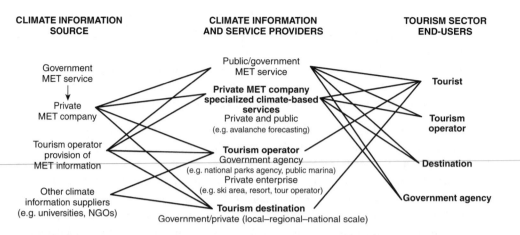

Figure 2.4 *Climate information communication channels in the tourism sector*
Source: Scott and Lemieux (2009)

events, are not widely documented within the sector (Scott and Lemieux 2009). Furthermore, there has been almost no evaluation of what sources of weather and climate information tourists utilize, the quality of these sources (accuracy; are the parameters tourists are interested in provided?), or the effectiveness of different communication pathways and formats. Do the typical idyllic photographs and descriptions of climate in tourism marketing ('blue-sky bias') mislead tourists, as some authors have argued (Smith 1993; Perry 1997; Gómez-Martín 2006); or is Maddison (2001) correct that tourists cannot be fooled because of their access to easily available and detailed climatic information?

There is evidence from repeated media stories (e.g. Barnes 2002) that at least some travellers feel misled by travel operators and destination marketers about the climate of destinations. So perhaps a middle ground exists, where some tourists do have access to plentiful climatic information through meteorological services or the tourism sector, but this general information may not address specific interests (de Freitas 2003; Gómez-Martín 2005; Scott and Lemieux 2009) or may not be interpreted correctly, as studies of public interpretation of climate information and weather forecasts suggest (Gigerenzer *et al.* 2005). Moreover, it is unclear whether there are perceived differences in the reliability of information provided in various media, or whether personal discussions and climate/weather-related travel recommendations have a specifically important role in the understanding of weather and climate in certain destinations.

Tourists can be particularly vulnerable to extreme weather events because they often visit highly dynamic environments, they may have limited familiarity with the places they are visiting and the meteorological hazards that occur there, and remote locations can lack communication channels for public warnings of impending hazards. The tornado that struck the Pine Lake, Alberta (Canada) campground in July 1999, killing nine and injuring over 100, is one such example. Although the Meteorological Service of Canada had issued a tornado warning well in advance of the event, there was a communication breakdown. Virtually all the tourists at the campground were unaware of the warning of imminent severe weather, and there was no on-site emergency warning system available (like the siren systems used in urban areas, for example). Furthermore, because tourists can be unfamiliar with the local language and are less likely to utilize local media information sources (either purposely or unintentionally), tourists are less apt to receive hazard warnings when they are issued, and prompt communication with them of imminent climatic hazards poses a particular challenge (Scott and Lemieux 2009).

Gamble and Leonard's (2005) study is the only known attempt to evaluate tourists' preferences for web-based climate information. Examining the communication of coastal climatology information, the authors found that simple websites with the following characteristics were perceived by tourists to be most effective: (1) a limited amount of information presented; (2) efficient and clear navigation features; (3) limited use of scientific jargon and complex graphics; and (4) limited use of colours and flashy graphics. The need for localized data was also highlighted.

In contrast, the complexity of the 'climate–tourism information scheme' and 'bioclimatological leaflet' proposed by Zaninovic and Matzarakis (2009) appears fundamentally at odds with these findings, and illustrates a continuing disciplinary disconnect between climatologists/meteorologists and tourism scholars and professionals. The leaflet was not based on any survey of what weather information tourists actually sought, or what conditions they perceived as optimal or unacceptable for various types of holiday activities or destinations. The communication tool was not 'market tested' with tourists to determine if they could understand it, or whether individual variables such as 'PET' (physiologically equivalent temperature), cloud cover 'above 5 octas', wind speed above '8 m/second', or sultriness vapour pressure greater than '18 hPa' had any meaning for them at all. Portraying the suitability of climate resources for different types of tourism segments and different types of destination is a very difficult task, and the development of any weather and climate communication tool for tourists must be done in collaboration with the tourism industry and market-tested adequately with tourists.

Finally, the revolution in communication technology, particularly the internet and, more recently, mobile personal data devices (smartphones), as well as the increased development of neogeography (user-generated spatially referenced tourist information), has revolutionized the weather and climate information available to tourists. The research community has yet to examine how this ongoing revolution in climate information and information communication technologies is influencing decision-making by tourists.

Tourists' climate preferences

Conceptual discussion of climate as a resource and a constraint for the tourism sector (Perry 1993; de Freitas 2003; Gómez-Martín 2005) has focused largely on tourists' climate perceptions – what was valued by, or not acceptable to, tourists defined the resource for the entire sector. As discussions in the following two sections of this chapter indicate, this is not necessarily the case, as the climate needs and optimal conditions for tourism operators and destinations can differ from those of tourists. Nonetheless, advancing our understanding of tourists' perceptions of climate, which include physical, physiological and psychological assessments of climate and those conditions that are most preferred or avoided, is very important in order to:

- better understand how weather and climate influence tourism decision-making (including specific thresholds that generate behavioural responses from segments of tourists);
- inform weather and climate information-providers (meteorological services, tourism operators, destinations) specifically about what information tourists want and how the climate aspect of tourism marketing can be made most informative and accurate;
- assist meteorological services (both government and private sector) to develop specialized products tailored to the needs of tourists and the tourism industry (activity- or destination-specific forecasts, financial products such as 'holiday weather guarantees');
- provide input into tourism infrastructure and landscape design to optimize microclimate conditions for tourists;
- assess the implications of future climate change for tourism demand, travel patterns and destination competitiveness.

Figure 2.5 is a schematic of the relationship between climatic range and tourism potential, where potential for visitation to a destination is a function of perceived appeal of climate to tourists (physically, physiologically and psychologically) and its constraints on tourist activities (including health and safety risks). This representation of climate as a resource and constraint for tourists (and, by extension, tourism) was originally developed by Perry (1993) and later modified by de Freitas (2003). It has been further modified here to recognize that climate can act as a constraint on tourism without representing a 'risk' to tourist health or safety. Figure 2.6 indicates that the perceived appeal or value of climate as a resource for tourism (solid line) is low at climate extremes, but peaks at some optimal state. In contrast, climate as a constraint for tourism (dashed line) is greatest at climate

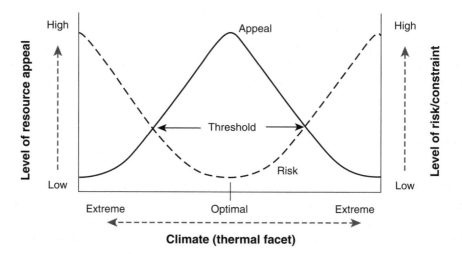

Figure 2.5 *Conceptualization of thermal thresholds for tourism*
Source: adapted from Perry (1993); de Freitas (2003)

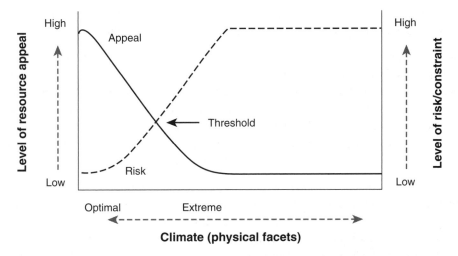

Figure 2.6 *Conceptualization of physical and aesthetic thresholds for tourism*

extremes, where conditions are less suitable for tourism activities and may even pose a health and safety risk for tourists. Perry (1993) argued that there exist optimal weather conditions for specific tourism activities, and critical thresholds outside which participation is no longer possible or satisfaction declines to the point where that specific type of tourism is no longer feasible. The transitions between higher climate appeal and constraint represent climatic thresholds (physical, physiological or psychological) that have implications for tourist behaviour. It is presumed that individual tourists' perceptions of climate as a resource/constraint vary based on their specific characteristics (age, market segment, culture, climatic zone of residence) as well as across specific tourists' activities and destination environments, therefore these transitions are more likely to be zones of transition than precise thresholds.

Although Perry (1993) and de Freitas (2003) developed the schematic of Figure 2.5 as a conceptual representation of climate as a resource and risk factor for tourists, more accurately it is a representation of the thermal facet of climate as a resource for tourists. Other meteorological variables that are important to tourists and comprise the physical facet of climate have a very unidirectional distribution (Figure 2.6), with the resource appeal rapidly declining and level of constraint rapidly increasing even with small amounts of rain or fog, or moderate winds. The distribution of aesthetic parameters such as sunshine–cloud cover is also unidirectional, but with a more gradual decline in the appeal of the resource from the initial optimal condition of clear skies to minimal cloud cover. While the aesthetic facet may diminish tourist satisfaction to an extent, it is limited as a constraining factor unless it is integral to a destination's appeal and does not represent a risk to tourist safety or health.

Attempts have been made to quantify the main components of Figures 2.5 and 2.6: optimal climate conditions for a range of tourism segments and destination environments; unacceptable climate conditions for a range of tourism segments and destination environments; and key thresholds where the tolerance limits of tourists are exceeded. The relative importance of the individual parameters used to define climate have also been explored. The available literature on each of these themes is synthesized below.

Various studies have evaluated climate resources for tourism and identify optimal climate conditions, both for tourism broadly and for specific tourism segments or activities (e.g. sightseeing, beach holidays). The different approaches to examining climate preferences of tourists and defining optimal climates can be grouped into three types: expert-based, revealed preference and stated preference (Scott *et al.* 2008b). An overview is provided of these three distinct approaches, and the range of optimal climate conditions identified by these studies is summarized in Figure 2.7 (temperature).

Expert-based approaches include 'weather typing' (Besancenot *et al.* 1978; Besancenot 1991; Gómez-Martín 2004) and 'climate indices' (Pégy 1961; Mieczkowski 1985; Harlfinger 1991; Becker 2000; Morgan *et al.* 2000), both of which integrate multiple parameters in a holistic rating of the climate resource for tourism. The highest rated conditions for each of the components that make up Besancenot *et al.*'s (1978) 'type 1' climate were: temperature of 25–33°C, no precipitation, nine hours or more of sunshine in a day, less than 25 per cent cloud cover, and wind speeds less than 30 km per hour. The parameters of Mieczkowski's (1985) 'excellent' category were: temperature of 20–27°C, less than 15 mm of rain, ten hours or more of sunshine per day, and wind speeds of under 5 km per hour. From the perspective of identifying optimal climate conditions for tourists,

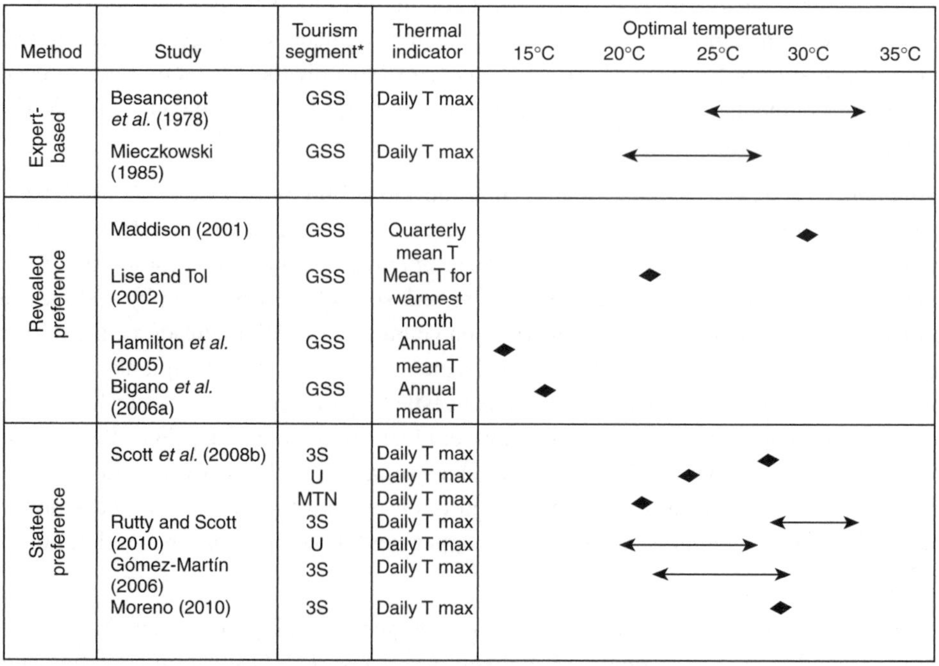

Method	Study	Tourism segment*	Thermal indicator	Optimal temperature (15°C – 20°C – 25°C – 30°C – 35°C)
Expert-based	Besancenot et al. (1978)	GSS	Daily T max	←——→ (≈25–33°C)
Expert-based	Mieczkowski (1985)	GSS	Daily T max	←——→ (≈20–27°C)
Revealed preference	Maddison (2001)	GSS	Quarterly mean T	◆ (≈30°C)
Revealed preference	Lise and Tol (2002)	GSS	Mean T for warmest month	◆ (≈23°C)
Revealed preference	Hamilton et al. (2005)	GSS	Annual mean T	◆ (≈15°C)
Revealed preference	Bigano et al. (2006a)	GSS	Annual mean T	◆ (≈16°C)
Stated preference	Scott et al. (2008b)	3S	Daily T max	◆ (≈28°C)
Stated preference	Scott et al. (2008b)	U	Daily T max	◆ (≈25°C)
Stated preference	Scott et al. (2008b)	MTN	Daily T max	◆ (≈22°C)
Stated preference	Rutty and Scott (2010)	3S	Daily T max	←——→ (≈20–30°C)
Stated preference	Rutty and Scott (2010)	U	Daily T max	←—→ (≈30–33°C)
Stated preference	Gómez-Martín (2006)	3S	Daily T max	←——→ (≈18–30°C)
Stated preference	Moreno (2010)	3S	Daily T max	◆ (≈27°C)

*GSS = general/sightseeing; 3S = sun, sand, sea; U = urban; MTN = mountain

Figure 2.7 *Comparison of optimal temperatures for tourism*

a central limitation of the expert-based approach is that the thresholds that specify optimal conditions are based on subjective expert opinion and have not been validated by tourists (de Freitas 2003; Scott *et al*. 2004; Gómez-Martín 2006).

A second distinct approach to evaluating climate for tourism is revealed preference studies. These studies used statistical relationships between measures of tourism demand (e.g. visits to a tourist attraction, national tourist arrivals, occupancy rates) and climate to infer the climate preferences of tourists and optimal conditions. A strength of this approach is that it is based empirically on indicators of aggregate tourist demand and not on subjective expert opinion. The spatial scale of revealed preference studies varies from local to international.

At the local scale, studies have modelled statistically significant relationships between weather and specific tourist destinations such as parks (Crapo 1970; Meyer and Dewar 1999; Jones and Scott 2006; Nicholls *et al*. 2008); beaches (Van Lier 1973; Moreno *et al*. 2008; Martinez Ibarra 2011); multi-activity resorts and major theme parks (Paul 1972; Emmons *et al*. 1975); and golf resorts (Scott and Jones 2006a, 2007). While illustrative of the influence of weather and climate on specific tourism activities and destination types, the revealed climate preferences in these limited studies cannot be considered broadly representative of comparable tourism destinations or tourism segments without replication in several other locations, and consequently are not included in Figure 2.7.

Another group of studies examine the role of climate in tourism demand and infer optimal climate conditions through analysis of international tourism arrivals data. Using a modified pooled travel-cost model, Maddison (2001) found that demand for a country by tourists from the UK peaked when its quarterly (three-month average) maximum daytime temperature was 30.7°C. This was referred to as the 'optimal temperature' for tourism. Lise and Tol (2002) used a similar model for Dutch tourists, but could not determine an optimal temperature. Using data for a cross-section of tourists from OECD nations, Lise and Tol (2002) estimated that the optimal mean temperature of the warmest month of the year was 21°C. They also concluded that perceived optimal temperatures for the warmest month varied only slightly among tourists from different nations, ranging from 21.8 (French) to 24.2°C (Italians). Using the same approach, Hamilton and Lau (2005) found the optimal mean monthly temperature for German tourists was 24°C. Using an econometric modelling approach, Hamilton *et al*. (2005) used annual arrival and departure data for 207 nations, and defined the optimal annual mean temperature for global tourism as 14°C. Bigano *et al*. (2006a) also used annual international tourism data for 45 nations, and concluded the global optimal temperature for tourism to be an average annual temperature of 16.2°C, arguing that the preferred holiday climate is the same for all tourists regardless of origin. Using similar statistical approaches, Lyons *et al*. (2009) suggest the optimal temperature for Irish travellers is 41.7°C, averaged over the month. This incredibly high temperature and the significant range found in these studies raises questions about this approach to identifying preferred climatic conditions for tourists.

The 'optimal' temperatures that emerge from these international revealed preference studies are summarized in Figure 2.7 for comparison, but there are critical limitations to this approach that must be considered (Gössling and Hall 2006b; Bigano *et al*. 2006b), and from the perspective of identifying optimal conditions, the coarse temporal and spatial resolution of the models is problematic. For example, in the studies of Maddison (2001) and Lise and Tol (2002), the climate of the capital city is taken to represent the climate of the entire nation. Washington DC is therefore representative of the United States, which contains 10 completely distinct climate zones using the Köppen classification scheme. Elsewhere, an average temperature for each nation is utilized, so that California and Florida are considered the same climatically as North Dakota and Michigan in the United States. Can these studies reveal tourists' climate preferences and optimal temperatures, if the indicators of climate used in these analyses (mean annual temperature or average temperature of the hottest month of the year) are meaningless to tourists? What proportion of international tourists travelling to France, the USA or China (three of the top tourism nations by international arrivals) could estimate what the 'mean annual temperature' is, and how many considered this in their choice of destination or when to visit? As Shaw and Loomis (2008) indicate, our understanding of the micro-scale behavioural decisions of tourists must inform, or be consistent with, macro-scale analyses of the influence of climate on global tourism patterns. Currently, the studies that examine these macro-scale patterns utilize constructs of climate that have no meaning to tourists.

Furthermore, de Freitas (2003) and Gómez-Martín (2006) observed a discrepancy between climate suitability and visitation, with peaks in visitation occurring when climate resources were not considered ideal (as defined by visitors to the respective destinations), and lower visitation occurring at times of year when climate resources were considered ideal. This would suggest that, even for highly temperature-sensitive tourism activities such as 3S holidays, there may be other intervening variables that influence tourist demand behaviours, and unless revealed preference approaches are able to account for all of these adequately (which becomes increasingly difficult when the goal is to model the factors that influence all types of tourism), error will be introduced into any calculation of optimal climate.

The final approach used to examine tourists' climate preferences utilizes a direct consultative approach, where tourists are asked to define their perceived ideal weather conditions and thresholds of what they believe to be marginal or unacceptable conditions. The stated preference approach assumes that tourists are able accurately to define climate conditions that guide their holiday decisions as consumers, and that would maximize their level of satisfaction during a specified type of trip.

De Freitas (1990) and Mansfeld *et al.* (2004) surveyed tourists *in situ* (in Australia and Israel, respectively) about their perceptions of current weather conditions in beach environments, in order to compare their satisfaction ratings with simultaneous on-site weather monitoring. Both studies confirmed the importance of multiple weather parameters in determining visitor satisfaction, with de Freitas (1990) demonstrating the overriding effect of even 30 minutes of rain. Mansfeld *et al.* (2004) also argued that domestic Israeli tourists were more sensitive to weather conditions, suggesting climate preferences or tolerances for marginal conditions different from those of international tourists. Gómez-Martín (2006) surveyed tourists in Spain about their climate preferences more generally, and found that the majority of respondents (75 per cent) felt the optimal temperature range for tourism in the region to be 22–28°C. During a summer day, less than one hour of rain was deemed acceptable by 69 per cent of respondents, but three hours was rated unacceptable by 62 per cent. A good sunny day was perceived to be one with the sun shining for 75 per cent of the day. Limitations of *in situ* surveys of visitor satisfaction with current weather conditions include the small range of weather conditions that can be examined without very significant personnel costs, and the potential for response bias because surveys during marginal weather conditions cannot capture the perceptions of those who found conditions unacceptable and are not there to survey. Furthermore, all of these studies focused on beach or 3S tourism, and cannot be generalized to other major tourism environments or destinations.

Scott *et al.* (2008b) conducted the first *ex situ* study of tourists' stated climate preferences, administering surveys to university students in Canada, Sweden and New Zealand in an indoor controlled-climate setting, free of potential bias from existing weather conditions and where respondents could express their perceived satisfaction with a wide range of climate conditions in very different tourism settings. Optimal climate conditions varied across both tourism destination types (Figure 2.7) and by nationality. Optimal temperatures varied from 27°C for beach tourism to 23°C for urban tourism and 21°C for summer mountain tourism. In other respects, the ideal climate for these three different tourism environments was very similar (no precipitation, less than 25 per cent cloud cover, less than 9 km per hour wind speed). Optimal temperatures for beach tourism differed across the three countries, the lowest for respondents from New Zealand (25°C), followed by Canada (27°C) and Sweden (29°C), but were very similar for urban and mountain holidays. Rutty and Scott's (2010) similar survey of the climate preferences for beach and urban tourism among university students in Europe (Austria, Germany, the Netherlands, Sweden and Switzerland) were very consistent, with ideal temperatures for beach tourism between 27 and 32°C and those for urban tourism lower, between 20 and 26°C (Figure 2.7). The preferences for precipitation, wind and cloud cover were identical to Scott *et al.* (2008b). A key limitation of these *ex situ* revealed preference studies is that the surveys were exclusively with university students, and consequently the results can be considered only to represent the young adult traveller market.

Importantly, more recent *ex situ* studies that have overcome this limitation with broader tourist samples have found similar results that confirm hypothesized differences among tourism segments. Wirth (2010) found very similar results between a public sample of German travellers and a young adult sample in the same region. The perceived ideal temperatures for a beach holiday ranged from 25–32°C for those aged 56 and older to 27–32°C for those aged 25 or younger. For urban holidays, ideal temperatures were 20–26°C across all age groups. Preferences for precipitation, wind and cloud cover were very similar to those found by Scott *et al.* (2008b) and Rutty and Scott (2010), and did not differ significantly across age groups. Moreno's (2010) survey at Belgian and Dutch airports found ideal conditions for a 3S holiday were considered to be 28°C, light breeze (1–9 km per hour), 8–10 hours of sunshine and a clear blue sky (0 per cent cloud cover). A survey of national park visitors in Canada (Hewer and Scott 2011) revealed ideal daytime temperatures of 24–30°C and ideal night-time temperatures of 18–22°C. Denstadli *et al.* (2011)

also found inter-cultural differences in summertime weather preferences among tourists in northern Scandinavia.

Fewer studies have attempted to identify climate tolerance thresholds for tourists. Revealed preference studies at a national scale are of little use for this task, as the coarse temporal and spatial scales preclude identification of major seasonal visitation fluctuations in specific market segments or destination types. Local or destination-scale revealed preference studies are able to provide insight into the climatic limits for specific market segments. For example, real-time visitation observations on beaches in Spain by Martinez Ibarra (2011) did not record an upper temperature threshold that led to a decline in beach use. In fact, observations that included the record 2003 European heatwave suggested that higher temperatures stimulated beach visitation, probably in an effort to escape even higher temperatures in urban areas. As with optimal climate conditions, more studies of this nature are needed before generalizable climatic thresholds can be established for specific tourism activities or market segments. These thresholds are mostly likely to be applicable at the regional scale or within specific climate zones.

Stated preference surveys have been used recently to explore the boundaries of tourism climate tolerance. Rutty and Scott (2010) examined thresholds of unacceptable climate conditions for 3S and urban tourism. Figure 2.8 combines their results for tourists' perceptions of optimal and unacceptable temperatures (both too hot and too cold) for 3S tourism. Interestingly, the distribution of optimal temperatures (26–32°C), transition zones and thresholds of tolerance (lower than 22°C being too cold, higher than 36°C being too hot)

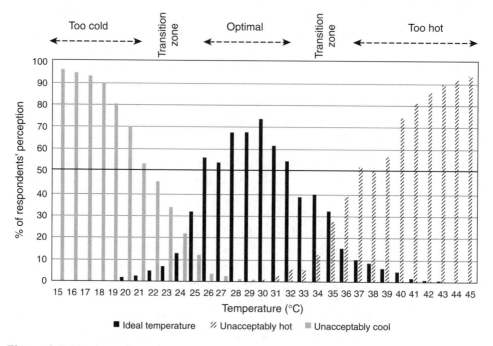

Figure 2.8 *Tourist rating of temperatures for beach holidays*
Source: adapted from Rutty and Scott (2010)

for the majority of respondents closely resembles the conceptual schematic proposed by Perry (1993) and de Freitas (2003) in Figure 2.5. Wirth (2010) found some differences among age cohorts, with 36°C considered by young adults (under 30) as too hot for a beach holiday, and 34°C considered too hot by those 56 and over.

There were no significant differences in perceptions of unacceptably cool conditions for beach holidays, or both too hot and too cold conditions for urban holidays. In a survey commissioned by the Government of France (Credoc 2009), respondents stated that summer tourism would be too hot in France if temperatures exceeded 32°C, with significant differences found by age (18–25-year-olds stated 34°C as unacceptable, ≥60-year-olds stated 30°C) and by activity (beach 34°C, walking 32°C, mountain 30°C). The threshold of too cold for summer tourism in France was 14°C, again with significant differences by age (≥60-year-olds stated 12°C), and by activity (beach 17°C, urban 9°C, mountain 9°C).

Other tourism market segment characteristics are also likely to influence perceptions of thermal or other climatic thresholds. For example, tourists seeking changes in weather conditions (sunshine and warmer temperatures during winter months) are likely to be more tolerant of higher temperatures and less tolerant of lower temperatures. Further investigation is required to quantify key climatic thresholds among diverse tourism segments and across nationalities.

While it is evident that tourists experience and respond to the integrated effects of the parameters that comprise climate, there remains a very incomplete understanding of the relative importance of different climate parameters to tourists and their decision-making. Although most expert-based studies and macro-scale revealed preference studies contend that the thermal facet of climate (or temperature, specifically) is the most important or only climate parameter significant to tourists, contrasting results have emerged in a number of stated preference studies. Table 2.1 provides a summary of the results from stated preference studies that have examined how tourists rank the importance of climate parameters in different types of tourism environment. While temperature has consistently been rated the most important climate variable for urban tourism, in other major tourism environments, sunshine or the absence of rain has been as important or rated even higher. Studies have observed that the occurrence of precipitation or very strong winds can override even the most ideal temperatures or most aesthetic sky conditions (de Freitas *et al.* 2008; Moreno 2010), and this certainly influences tourists' perceptions and behaviours at the local destination scale. Differences in the relative ranking of climate parameters have also been observed between national samples (Morgan *et al.* 2000; Scott *et al.* 2008b) and within national samples (Credoc 2009). Presumably, travellers in wetter locations may value less frequent sunny days more than those where sunshine is plentiful. Similarly, rain occurring in a high-rainfall region is likely to have less impact on visitor numbers than rain of the same intensity and duration occurring in a low-rainfall region.

Further studies are required to confirm these patterns for different types of tourism environment, and to explore differences in tourism market segments and nationalities. If similar finding are confirmed, these empirical results need to be incorporated into climate rating systems (weather typing, climate indices) that aim to assess both current and future

Table 2.1 *Rated importance of climate parameters in major tourism environments**

Type of tourism	Region	Temperature	Sunshine	Absence of rain	Absence of strong winds	Reference
Beach – 3S	UK, Mediterranean	3	2	1	4	Morgan et al. (2000)
	Canada, New Zealand, Sweden	2	1	3	4	Scott (2008)
	Northern Europe	3	1	2	4	Rutty and Scott (2010)
	France	2	1	3	4	Credoc (2009)
	Belgium, Netherlands	2	3	1	4	Moreno (2010)
Urban	Canada, New Zealand, Sweden	1	3	2	4	Scott (2008)
	Northern Europe	1	3	2	4	Rutty and Scott (2010)
	France	1	3	2	4	Credoc (2009)
Mountain (summer season)	Canada, New Zealand, Sweden	1	3	2	4	Scott (2008)
	France	2	3	1	4	Credoc (2009)
Mountain/ ski (winter season)	USA	5	4	1	3	Scott and Vivian (2012)*

*Good visibility was ranked second; falling snow was ranked sixth

climate conditions for tourism. As de Freitas *et al.* (2008) make clear, a single-facet thermal rating system for tourism, regardless of its sophistication, is insufficient to facilitate an assessment of the integrated effects of a range of climatic conditions that are important for tourism.

Similarly, more investigation is required to understand better how tourists trade off climate parameters and predictability. For example, if faced with the following weather conditions for sightseeing in an urban destination such as Rome or New York, which is the most preferred: 20°C and sporadic rain; 15°C and windy and grey; or 10°C and sunny with no wind or rain? Do tourists prefer destinations with more predictable weather conditions, and are they able to identify these locations on the basis of widely available climate information? Typical climate information consists of mean conditions and does not provide insight into variability and probabilities of certain conditions (with possible exception of days with rain), so this is doubtful. Trade-offs between specific climate parameters and other travel determinants is also a useful area of inquiry: for a 3S holiday during the winter months, is the choice a destination with a probability of 24°C daytime high; or a destination with a probability of 29°C daytime high, other climate parameters being equal, but that requires 20 per cent longer travel time and 20 per cent additional cost?

Knowledge remains limited about the process of how tourists integrate weather and climate information into specific decisions. Key knowledge gaps exist with regard to the entry points and decisiveness of weather and climate in tourists' decision-making processes. The proportion of tourists travelling primarily for climate-related motivations or to engage in climate-sensitive activities remains unknown, and is central to understanding the relative climate sensitivity of destinations. Climate information is typically embedded in a matrix with other relevant information, and disentangling the role of climate and its relationship (trade-offs, substitutability) to other situational factors in major decision-making processes in tourism remains an important area for future research.

WEATHER, CLIMATE AND TOURISM OPERATORS

Weather and climate affect tourism operators in several ways and over a wide range of timescales (from hourly to multi-year droughts). Weather conditions affect tourism operators on a short-term basis, from hours to days. For example, a study of the reasons for the failure of special events run by members of the International Festival and Events Association found that weather was ranked first among eight external factors, and second overall, ahead of inadequate marketing, poor strategic planning and poor cost management (Getz 2002).

Tourism operators are also profoundly affected by inter-annual climate variability (see Figure 2.1). Climate defines the length and quality of multi-billion-dollar tourism seasons in different parts of the world. Winter sports tourism is clearly highly sensitive to temperature and snowfall, and Figure 2.9 demonstrates the impact of inter-annual climate variability on the length of ski seasons in three ski regions of the United States, with associated impacts on skier visits and revenues.

Inter-annual variations in precipitation also have important impacts on tourism operators. Drought conditions have led to devastating wildfires in several key tourism regions including Australia, Canada, Greece, Spain and the United States (Hystad and Keller 2006; Sanders et al. 2008). In Greece, after the devastating fires of summer 2000, more than 50 per cent of all bookings from tourists for 2001 were cancelled (IUCN 2007). Drought in the US state of Colorado during the spring and summer of 2002 created dangerous wildfire conditions, and the media coverage of major fires in some parts of the state had a significant impact on summer tourism. Visitor numbers declined by 40 per cent in some areas of Colorado, and reservations in state park campgrounds dropped 30 per cent (Butler 2002). The drought also affected fishing and river-rafting tourism in the state. Anglers were restricted from fishing in many rivers because fish populations were highly stressed by low water levels and high water temperatures. Low water levels also shortened the river-rafting season substantially. Some river-rafting outfitter companies lost 40 per cent of their

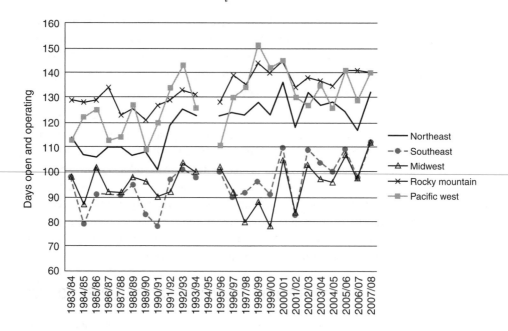

Figure 2.9 *Historic ski-season variability in US ski regions (operating days)*
Source: data from National Ski Area Association – end-of-season survey reports
1983–2008

normal business, and economic losses in the state's river-rafting industry alone exceeded US$50 million (Associated Press 2002a, 2002b). Similarly, in 2008 natural disasters and drought were identified as having a major impact on Texas' US$16 billion a year nature tourism industry, primarily as a result of localized environmental changes affecting fishing, hunting and wild-flower viewing as well as camping opportunities (Johnson 2008).

In Australia, a country noted for its droughts and associated natural hazards in the form of bushfires and dust storms, a survey of tourism operators in 2008 by the Victorian Tourism Industry Council found that the most significant or critical constraint to tourism businesses, after rising oil and petrol prices, was the effects of drought (Sharp 2008). In a report on the effects of bushfires on tourism in Victoria's Alpine National Parks, Sanders *et al.* (2008) reported that, although no operators were affected by direct physical loss, all operators they interviewed experienced significant negative impact on their operations. Earlier research had suggested that the economic impacts of bushfire on tourism could be substantial, with Cioccio and Michael (2007) estimating that the losses by small tourism business in the first month after the 2003 bushfires in north-eastern Victoria were in excess of A$20 million.

Extreme weather and climate-related environmental events also routinely have major impacts on tourism operators, particularly those located in highly dynamic environments such as coastal areas (see Figure 2.1). In the Florida Keys, the ten-day closure and clean-up following Hurricane Georges in 1998 resulted in tourism operator revenue losses of approximately US$32 million (US EPA 1999). The effects on tourism of the 2003 heat-wave in Western Europe also illustrate the complexities of the climate–tourism interface. The 2003 heatwave was exceptional for its temperatures and also its length. Tourism establishments in the Spanish beach destination of Costa Brava reportedly lost an estimated 10 per cent of guest nights during the summer season, with decreased stays at campsites most pronounced, while visitation to inland mountain destinations increased as travellers sought comfortable climatic conditions (Lagadec 2004). Similar shifts in tourist patterns were documented in France, with increased occupancies on northern and north-western shores and in central mountains, and decreases in urban centres and southern regions. Activities and consumption patterns were also modified, as access to some forests with high fire risk was blocked, fishing restricted, and camping and accommodation without adequate space-cooling systems became uncomfortable, while demand for accommodation with pools increased and sales of beverages and ice cream increased substantially (Létard et al. 2004).

In a study of the consequences of algal blooms in the Baltic sea in 2002–07 (events connected to extreme weather situations linked to climate change), results indicate that tourists pay considerable attention to extreme environmental situations, which are shaped by media reports (Nilsson and Gössling 2010). Only a small share of travellers reported to be entirely unaffected by algal blooms, although an equally small proportion reported to have favoured other destinations (3.9 per cent) or to have shortened their holiday because of the algae situation (1.8 per cent). However, the study also indicates that longer-term responses to environmental crises may be far more important than immediate responses, with 16.5 per cent of respondents indicating that they visited other destinations in the years following severe algal blooms. Whether these results are transferable to other tourist populations is unclear, however, as the summer season in the Nordic countries is rather short and associated with high expectations of the 'perfect' holiday.

The diversity of weather and climate impacts means that historical climate information and weather forecasts enter into a number of decision-making contexts for tourism operators, including investment and locational analysis; infrastructure design and construction; daily to annual operational decisions and budgeting; risk assessment; and marketing. Figure 2.10 outlines the timescales of weather and climate information utilized in these major areas of tourism operator decision-making.

Tourism operators function in virtually every climatic region of the planet, from tropical rainforests, to deserts, to highly dynamic coastal and mountain regions. Climate and local-ized micro-climates have important implications for investment in tourism properties, locational analysis, and architectural and landscape design. For example, Gómez-Martín (2005) documents how resorts have prospered by using favourable local climatic conditions (including micro-climates) to their advantage and avoiding areas where conditions could not be readily remediated through design.

Altalo and Hale (2002) found that weather and climate were not usually cited as reasons for development (or not) of new resorts and accommodation, relative to other macro-level factors such as transportation access, source markets, land ownership and coastal access. Resort and hotel operators often 'followed the competition'. They found that climate

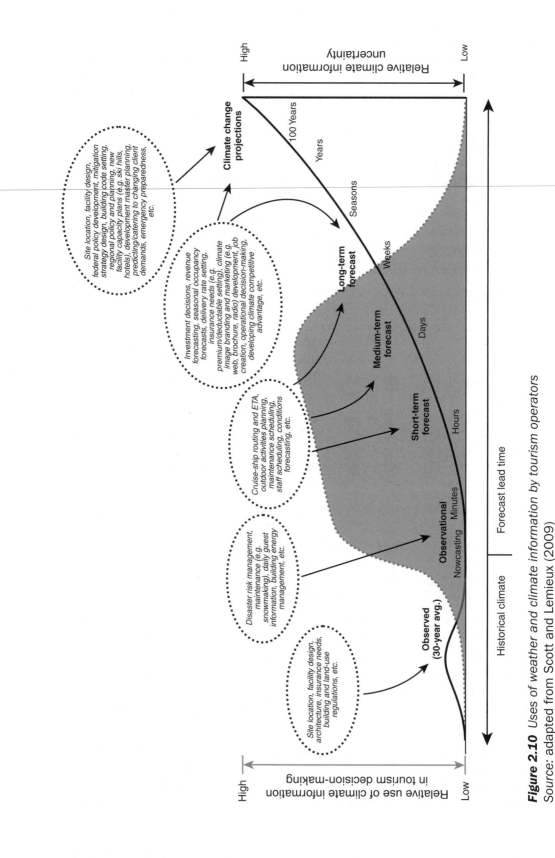

Figure 2.10 Uses of weather and climate information by tourism operators

Source: adapted from Scott and Lemieux (2009)

Table 2.2 *Heating and cooling degree days at European destinations*

City	Country	Heating (degree days)*	Cooling (degree days)*
Oslo	Norway	4714	9
London	UK	2800	58
Paris	France	2702	114
Zürich	Switzerland	3413	77
Athens	Greece	876	1020
Rome	Italy	1253	786
Seville	Spain	931	908

*Calculated using EUROSTAT/ASHRAE method for 1961–90 period

information was utilized more extensively in engineering, construction planning, property design and maintenance, and other post-build decisions (insurance, heating–cooling budgeting, staffing). Numerous authors have outlined how climate parameters such as temperature, humidity, rainfall, snowfall, prevailing winds and sun intensity have a major influence on tourism architectural and landscape design considerations, from construction materials, to colour and orientation of infrastructure, to existence and dimensions of shade and ventilation structures, to the size and type of heating and cooling equipment, to name but a few (Altalo and Hale 2002; Nikolopoulou and Steemers 2003; Gómez-Martín 2005). Table 2.2 outlines the wide range of heating and cooling requirements at different major destinations in Europe, expressed as heating–cooling degree days (a quantitative index of associated energy requirements). The limited availability of historical climate information in many developing nations and remote locations (for example, smaller islands and mountainous areas), at the scale relevant to tourism developers, has been noted as an informational barrier in the past (Scott and Lemieux 2009).

As noted, weather and climate information, particularly real-time observations and short-term forecasts, are used extensively by tourism operators. Recent weather observations are important inputs into a range of decision-support tools, including cruise-ship and aviation routing; automated turf irrigation-management systems used by golf courses and other sports facilities; snow-production systems used by ski areas; fire and avalanche warning systems; and energy-management (heating–cooling) systems used by accommodation providers. Of course, weather and climate information also enter into a wide range of other decisions, including construction and supply scheduling, landscape and infrastructure maintenance, energy cost projections, human resource requirements, visitation/revenue forecasts and business projections.

Weather and climate are also increasingly being incorporated into broader business risk-assessment processes. For example, as tourism operators have begun to assess their weather sensitivities and associated business risks, some have begun to utilize innovative weather derivative and index insurance products to reduce their weather risk. Weather derivatives and weather index insurance are financial instruments that emerged in the late 1990s to protect against weather-related loss of revenue in the energy and agriculture sectors, and differ fundamentally from weather insurance commonly held by tourism operators, typically property damage or business interruption insurance from wind, floods or other extreme events, that pay out only in the event of a documented economic loss and reimburse only the actual amount of loss via established claim procedures. In contrast, weather derivative contracts and weather index insurance pay a fixed amount based on observed weather conditions, regardless of the extent of losses experienced or whether actual damages can be demonstrated. A key advantage of weather derivatives and weather index insurance is that the scale of the products can range from individual businesses to national governments, which therefore make them particularly suited to the small and medium-sized enterprises that dominate the tourism sector.

Participation of the tourism sector in the weather derivatives and index insurance market has remained limited thus far. Nonetheless, as the examples of applications in the tourism sector in Table 2.3 illustrate, there is tremendous potential to develop innovative, highly customized products to reduce weather-related revenue loss and create new destination marketing strategies that deliver a competitive advantage, regardless of actual weather conditions.

Tourism operators also utilize the climate extensively in a variety of marketing approaches, from the imagery contained on websites to real-time and forecast-based marketing strategies. Gómez-Martín (2006) reviewed a number of studies that analysed how climate is used in advertising by tourism operators. These studies document that a high percentage of marketing products feature climate, and conclude that most tourism operators 'managed to combine, with varying degrees of accuracy, idyllic descriptions and numerical values, to compose brochures that both attract tourists and provide a source of information'

Table 2.3 *Selected applications of weather derivatives in the tourism sector*

Tourism operator	Weather challenge	Weather derivative protection
Barbados Tourism Authority	Low temperatures or heavy rain	Barbados Perfect Weather Guarantee gives travellers money back when daytime temperatures do not reach 26°C or when there is more than 5 mm of rain: a recent weather derivative-backed marketing campaign to promote perfect holiday weather on the island
City of Victoria (Canada)	Reputation for seasonal rainy conditions	Offers a 'Sunshine Guarantee' that will refund travellers a set amount if they experience more than 1.25 cm of rain in one day during their holiday

Continued

Table 2.3 *Continued*

Tourism operator	Weather challenge	Weather derivative protection
Bombardier Corp. (Canada)	Limited snowfall	Promised buyers of new snowmobiles that if snowfall in their area was less than 50 per cent of a three-year average, they would receive a set payment
Corney & Barrow (wine bar chain) (UK)	Low temperatures	Contract paid the company for every Thursday and Friday that temperatures did not reach over 24°C – the lower the temperature, the greater the payout
Flagstaff Nordic Center Guaranteed Season Passes (USA)	Lack of snow or cold temperatures required for snowmaking	Purchase of seasonal weather protection for season pass-holders and weekend protection for special events against inadequate precipitation (less than two feet of snow) between 23 November and 23 March
PGA Championship – Greater Hickory Classic (USA)	Rain interruption of special event	Purchase of derivative protection against rain event: payment for rain accumulation over 0.75 inches during the tournament
Taste of Antwerp (Belgium)	Rain interruption of festival with no advance booking of tickets available	Organizers will receive a payment should rainfall during the event exceed 9.7 mm
Priceline.com 'Sunshine Guaranteed' vacation (international)	Rain during holiday tour	Priceline.com offered consumers a 'Sunshine Guarantee' that refunded travellers if it rained (>0.50 inches per day) during half or more of their vacation days at over 100 destinations in the USA, Caribbean, Canada and Europe
Itravel2000.com 'Let it Snow!' weather promotion	Snow-related travel delay	Offered consumers a refund for travel if it snowed five inches or more at Calgary, Halifax, Montreal or Toronto airports on New Year's Day, 1 January 2008

Source: Speedwell Weather Derivatives (2001); WeatherBill (2008, 2009)

(Gómez-Martín 2006: 208). In his earlier review of climate information in tourism marketing, Perry (1993: 411) commented, 'It is hardly surprising that weather expectations by the holiday-maker frequently differ greatly from the weather experienced on the holiday, with obvious frustration and disappointment.' Tourism operators in some regions gained a reputation for 'optimistic' interpretations of local climate conditions. The implementation of live webcams at virtually all ski resorts in North America since 1995 is a direct response to growing distrust of tourism operators' reports of snow conditions, and the desire of skiers for an objective assessment of snow conditions.

Other innovative marketing strategies have emerged recently that utilize forecast information to tailor marketing messages or to target certain markets. For example, a major golf course operator in South Carolina monitors forecasts in nearby regional markets, and when poor conditions are forecast in those markets, targets those locations via prearranged internet marketing arrangements with two-day specials for weekend golf packages in South Carolina's more pleasant weather conditions (Bruce Farren, Director of Grounds and Golf Course Management, Pinehurst Resort and Country Club, 2008, pers. comm.).

Tourism operators have developed a wide range of climate adaptations that allow them to function in virtually every climatic zone on the planet. These climate adaptations have diverse business applications that provide a competitive advantage: improving tourists' comfort and satisfaction (e.g. indoor air conditioning, trees and landscaping to provide shade and wind protection, varied-temperature pools to provide swimming comfort across a range of weather conditions); extending tourism seasons or improving the reliability of tourism products (e.g. irrigation and snowmaking systems); improving operating reliability and lowering operating costs (e.g. systems to capture rainfall for landscaping or other uses, desalination of saltwater, automated and dip irrigation); infrastructure protection (e.g. storm shutters, sea walls, fire-smart landscaping); and marketing (e.g. weather derivative products that guarantee good holiday weather).

Snowmaking is a high-profile example of climate adaptation by tourism operators, and illustrates its effectiveness. Figure 2.11 illustrates the historical development of snowmaking as a climate adaptation in the ski tourism market in the USA since it was first implemented by a single ski area in 1952. Since that time, snowmaking has become an integral climate adaptation by the ski industry worldwide, effectively extending ski tourism seasons and improving the reliability of snow-based tourism products year over year, but particularly during winters characterized by high temperatures and low snowfalls. Scott (2006b) and Steiger (2011a) have found that, because of the widespread implementation of snowmaking, the vulnerability of ski tourism to anomalous warm winter conditions has declined markedly since the 1980s in parts of North America and the European Alps. In eastern Canada, the USA and the Tyrol region, the impact of record warm winters in the 2000s was far less than that of warm winters in the 1980s and early 1990s.

It is important to note that, while these adaptations to past and current climate conditions may also prove useful for coping with climate change in the decades ahead, they should not be confused with purposeful adaptations to future climate change. There is an

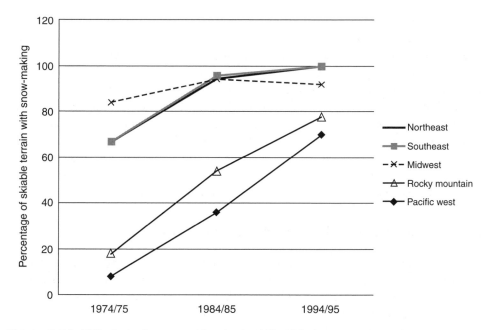

Figure 2.11 *Diffusion of snowmaking in the US ski industry*
Source: data from National Ski Area Association – end-of-season survey reports
1975–1995

important distinction between the two. Climate change adaptations need to be capable of dealing effectively with projected future climate regimes, and there is almost no evidence available that tourism operators have evaluated the design capabilities of climate adaptations against future climate regimes or, in most cases, even historical variability.

The extent to which climate information is used by tourism operators for strategic, operational or marketing purposes has not been evaluated sufficiently to determine its economic value or the degree to which needs of tourism operators are being met (Scott and Lemieux 2009). Studies reveal that weather information is important in decision-making processes, but that few tourism operators use climatological information for longer timescales (a month or longer), and that tailored products would serve as an important resource to improve the utility of climatological information (Altalo and Hale 2002; Gamble and Leonard 2005; Scott and Lemieux 2009).

The tourism sector is virtually absent from the growing literature on the societal benefits and economic value of climate information and forecasts (Katz and Murphy 2000), although Altalo and Hale (2002) contend that the financial benefits of weather and climate information for the sector are likely to be very substantial. One of the few attempts to evaluate the specific financial benefits of improved weather and climate information for the tourism sector was a study by the National Oceanic and Atmospheric Administration of the multi-sector benefits of new observational equipment (NOAA 2002, 2004b; Centrec

Consulting Group 2007). Specific to the tourism sector, the study found that weather information and hurricane forecasts from the new satellite imager and sounder would create US$196 million per year in socio-economic benefits through improved golf safety, irrigation efficiency, grounds maintenance, tournament and personal golf planning, as well as US$31 million per year in economic benefits from damage avoidance in recreational boating, amusement and recreation services (Centrec Consulting Group 2007; NOAA 2009). Based on the success of studies that have applied techniques for economic and social valuation of climate services (e.g. market prices, normative market models, descriptive behavioural response studies and contingent valuation studies) in other economic sectors, uncovering the potential value for the tourism sector and determining how to fully realize that potential remain critical areas for future inquiry.

WEATHER, CLIMATE AND TOURISM DESTINATIONS

Weather and climate are also salient for tourism destinations, influencing their short- or long-term competitiveness and sustainability in diverse ways (see Figure 2.1). Smith (1993) distinguished two types of destination: those that are 'weather-sensitive' and those that are 'climate-dependent'. In weather-sensitive destinations, climate resources do not generate tourism directly, but either facilitate or constrain tourist activities. In climate-dependent destinations, climate is the principal resource on which tourism is predicated (e.g. many tropical islands). However, the influence of climate on destinations extends beyond its influence on tourist demand. All tourism destinations are affected either positively or negatively by natural seasonality in demand: inter-annual climate variability that can bring unseasonable temperatures or substantial fluctuations in precipitation causing drought, flooding, or transportation interruptions, and extreme events that not only affect tourists' comfort and safety (and thereby satisfaction), but also the products that attract tourists and tourism infrastructure. Wall and Badke's (1994) survey of national tourism organizations worldwide found that the majority (81 per cent) felt weather and climate were major determinants of tourism in their nation.

The global distribution of climate resources for tourism varies throughout the year, giving rise to the well documented phenomenon of seasonality (Baum and Lundtorp 2001). Although seasonality is associated primarily with temperate nations (e.g. Western Europe and North America), it is a global phenomenon (Figure 2.12). The seasonality of tourism is driven largely by major market countries of the Northern hemisphere, and 'natural seasonality' is a primary driver of some of the largest regional tourism patterns (northern Europe to the Mediterranean; northern USA and Canada to the Caribbean). Where tourism products cannot be developed to attract sufficient tourists to a destination throughout the year, seasonal shut-downs occur, with attendant implications for local economies and livelihoods.

Scott and McBoyle (2001) theorized that the tourism climate resource of every destination could be classified into one of six annual distributions (Figure 2.13). Their tourism climate typology ranged from 'optimal' year-round tourism climate in destinations such as Barbados or southern California, through to a 'poor' year-round tourism climate in high polar regions. The 'summer peak' and 'winter peak' curves are similar, but distinguished by the season in which more favourable climatic conditions occurs. A summer peak

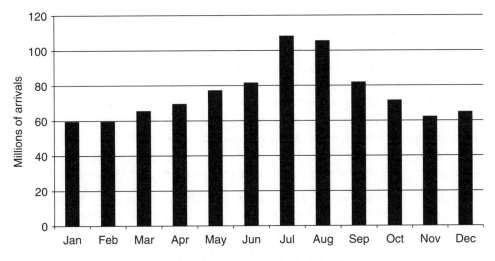

Figure 2.12 *Monthly international tourist arrivals (millions)*
Source: average of UNWTO data for 2008–10

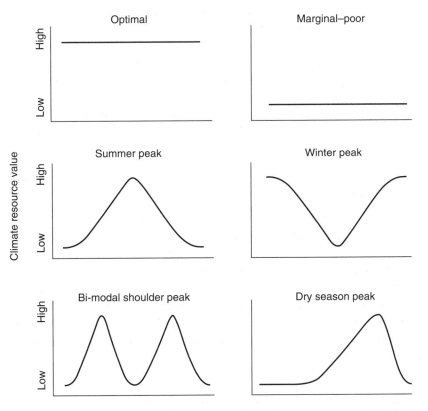

Figure 2.13 *Conceptual typology of annual tourism climate-resource distribution*
Source: adapted from Scott and McBoyle (2001)

is indicative of mid- to high-latitude locations where summer is the most pleasant period of the year for tourism. A winter peak would occur in more equatorial locations, where somewhat cooler and/or lower humidity in winter are more comfortable for tourists compared with hot and/or humid summer conditions. Where spring and autumn months are more suitable for tourist activity, a 'bi-modal shoulder peak' distribution occurs. The tourism climate resource in regions with distinct wet and dry seasons (e.g. monsoon regions of Asia) will be determined to a large extent by precipitation, displaying a peak during the dry season, when the climate is more conducive to tourism activities. The monthly hotel/resort accommodation costs (as a proxy of tourism demand) in many destinations follow a similar annual pattern, suggesting that this typology has some validity in the tourism marketplace.

The impacts of extreme weather events on tourism destinations are well documented in the media almost every year (Figure 2.1), but have been the focus of a limited number of studies. For example, the economic impact of the four hurricanes that struck the state of Florida in 2004 was estimated to be several times larger than the immediate impact, as the storms caused thousands of cancellations. A marketing survey found that 25 per cent of potential visitors were also less likely to visit Florida in the future during the hurricane season (Pack 2004), indicating that major events can have a multi-year impact on visitation and destination reputation. Importantly, these same extreme events in one destination can have a positive impact on other parts of the tourism system, as destinations such as Arizona and California benefited from the transfer of large numbers of visitors and convention business (*USA Today* 2005). That same year, the Government of Mexico estimated that as a result of the late-season Hurricane Wilma and media coverage of damage and stranded tourists, it would lose US$800 million in tourism revenue between October and December. With 26 tropical storms and 14 hurricanes, the 2005 hurricane season was one of the most active and destructive in history, spawning three of the most intense North Atlantic storms on record, including Hurricane Katrina, which caused extensive damage to the tourism infrastructure in New Orleans and coastal Mississippi, where impacts on convention business and gambling are expected for years, perhaps decades, to come (Bhatnagar 2005).

Climate in tourism marketing

One of the main uses of climate information by destinations is for marketing. Like tourism operators, organizations that promote tourism to specific destinations typically provide climate information with two purposes: to market the destination, and to assist travellers to prepare for safe and comfortable travel experiences. Studies examining how climate is used to advertise destinations have consistently found that weather and climate are used in a high percentage of marketing products (de Freitas 2005; Gómez-Martín 2005). According to Besancenot (1991: 208), 'The iconographic analysis of tourist brochures and the careful reading of the accompanying text only confirm the obsessive presence of references, direct or indirect, to the climate.'

The sophistication of climate information within destination marketing communications (brochures, TV, websites) ranges widely and depends on the prominence of climate within the brand image of the destination. At one end of the spectrum are marketing communications that provide absolutely no climate information on the destination, but display a very obvious 'blue-sky bias' in all photography of the destination (Perry 1997; Gómez-Martín 2005). Often, even basic climate information (monthly mean temperature and precipitation) and advice on what to wear and health-related issues (insect repellent, UV index and sunscreen) are not provided. More detailed climate information on sunshine hours, water temperatures, elevational differences in temperature and climate-related hazards remains the exception. At the other end of the spectrum are destinations that provide detailed climate information tailored to the types of tourists who visit and to the activities they typically undertake.

BOX 2.1 CASE STUDY: PROMOTING THE 'WARMEST PLACE IN BRITAIN'

The claim to be the warmest place in a country may provide significant advantage in a competitive tourism market. So much so that destinations may even take formal action to be able to claim such status. In April 2011, tourism officials on the Isles of Scilly announced that they were to lodge a complaint with the Advertising Standards Authority (ASA) over a claim made by the island of Jersey's tourism office in a £1 million TV advertising campaign that the largest of the Channel Islands is 'the warmest place in the British Isles' (Hickman 2011).

The Met Office officially recognizes Scilly as the warmest place in the UK. The small print on Jersey's advert indicates that the claim is based on minimum temperature data supplied by the Jersey Meteorological Department, which uses mean minimums for 1971–2000, the same period used by the Met Office to calculate all its mean temperatures. The mean minimum for Jersey for this period was 8.9°C. *The Guardian* reported that, in comparison, Met Office data for Scilly gave a mean minimum for St Mary's, Scilly's largest island, of 9.4°C over the period (Hickman 2011). However, on other measures, such as hours of sunshine and maximum mean temperatures, Jersey performs marginally better. The challenge to Jersey's promotion is geographical as well as meteorological, as the council of the Isles of Scilly argues that the Bailiwick of Jersey is a British crown dependency but is not part of the UK, and that Jersey is not part of the British Isles archipelago.

Malcolm Bell, head of tourism at VisitCornwall, which works closely with the Council of the Isles of Scilly, commented,

> It seems a bit desperate to base an advert on warmth alone when people who are only interested in that would just go to somewhere such as Dubai instead. Jersey might just about be technically correct in what they say, but it is bordering on unethical to stretch the truth like this when promoting your destination.
>
> (quoted in Hickman 2011)

> This is not the first time that complaints have been made to the ASA about Jersey's tourism promotions. In 1990, the ASA upheld a complaint against Jersey after it failed to substantiate a claim that it had more sunshine than anywhere else in the British Isles. The Jersey weather campaign was so successful that Jersey Tourism won the Best Travel PR Campaign at the 2011 Chartered Institute of Marketing Travel Marketing Awards. In collecting the award, the island's Economic Development Minister, Senator Alan Maclean, said, 'I believe that this campaign … has had a significant impact on wider perceptions of Jersey across the British Isles and that our reputation as the "warmest place in Britain" has been enhanced by its success' (BBC 2011a).

In destinations where seasonal weather is not highly conducive to tourism, and functions more as a constraint, climate is sometimes downplayed in marketing communications (Iceland's 'more solar than polar', for example), or is omitted altogether. Marketing is also used specifically to address unfavourable perceptions about a destination's climate. For example, a tourism brochure for Brittany, France informs tourists that 'Common misconceptions and prejudices have portrayed Brittany as a rainy region when in fact its maritime climate is mild and bracing … [and its] iodine-rich sea air is unique and just to breathe it in is to enjoy its health-giving properties.' In contrast, other destinations have cultivated a destination image and marketing strategy around certain activities or experiences that are largely based on the local climate. The Cayman Islands promotes its 'perpetual summer'; Florida brands itself as 'the Sunshine State'; and a region in south-east Queensland, Australia is promoted as 'the Sunshine Coast' – a destination brand so successful that it became the official name for the area.

Other destinations use what would typically be considered adverse weather conditions to develop distinct tourism products. Historically, tourism has not been an important activity to the local economy of Tarifa, Spain, as the frequent high winds were not suitable for conventional 3S tourism development. However, the high winds of Tarifa are a very valuable resource for water sports such as windsurfing and kitesurfing, and the community now markets itself as the 'windsurfing capital of Europe' or the 'Costa de la Luz y el Viento' ('the Coast of Light and Wind'). Vancouver Island (British Columbia, Canada) promotes a Storm Watching product on its Pacific Ocean coast during the fall and winter months and encourages visitors to take advantage of the special conditions created during La Niña phase of the Pacific Ocean Southern Oscillation.

Although all tourism destinations utilize climate information in marketing and other ways, two frequent laments from the tourism sector are that weather forecasts often do not accurately reflect conditions in tourism areas, and that the media misrepresent these forecasts to the detriment of the tourism industry (de Freitas 2003; Gómez-Martín 2005; Scott and Lemieux 2009; Becken *et al.* 2010). Meteorological stations that provide climate information and weather forecasts for a destination can be many kilometres away from the destination and may not accurately reflect the microclimate where tourism operators are established. For example, the microclimates offered at a beach resort can be very different than the meteorological station located at an airport 10 kilometers inland and a meteorological station in a low-lying valley may have little in common with conditions at a high-elevation ski area. A limited number of studies have confirmed that the climate conditions of the nearest meteorological station do not present an accurate reflection of the microclimate tourists encounter and of beach resorts, parks or ski hills (Höppe and Seidl 1991; Hartz *et al.* 2006; Scott and Lemieux 2009).

'It's a fair assumption to suggest that the weather and the weather forecast – however derived – is a very important part of the (tourist) decision-making process … I think we … want accuracy, but also at the same time, we want it balanced, with a situation where the travelers are not deterred because the off chance of a shower is portrayed as if it was a certainty.'

Chief Executive, Queensland Tourism Industry Corporation, in Nolan (2001)

The perception that climate information, and particularly weather forecasts provided by the media, are 'inaccurate' and deter visitation is a lament heard often from the tourism industry. For example, at the Climate, Weather, and Tourism Workshop in North Carolina (Curtis *et al.* 2009), multiple tourism stakeholders voiced frustration with media reporting of weather and how it unnecessarily affected tourism. Wineries in the region saw visitation decline for an entire season after inaccurate reports that spring frost had wiped out that year's crop. Similar concerns about forecast skill and media coverage of weather forecasts were expressed by other tourism operators and tourism authorities in the region, especially the typical margin of error in early hurricane track forecasts that have been observed to be very damaging to the tourism economies of regions at the low probability edges of the forecast track (Curtis *et al.* 2009). Seasonal forecasts have also recently been shown to have adversely affected travel decisions in the United Kingdom. In 2009, travel agents and tour operators observed that, following the long-range summer forecast of 'unusually warm, dry weather with heatwaves up to 30°C', demand for foreign holidays declined substantially (Hill 2009). However, after a very rainy July and a revised forecast for 'wet weather until September', the Association of British Travel Agents reported an increase in travel bookings of up to 40 per cent and a diminished supply of package holidays to sunshine destinations (Hill 2009).

'Accurate, geographically specific meteorological information is essential for tourism operations. General forecasts, though meteorologically accurate, often have a negative impact on tourism operations because tourist destinations, such as beaches, coastal areas and mountains are often regions with unique and better than average microclimates … [climatic conditions] which would attract tourists may differ substantially from prevailing regional conditions.'

Final Communique, *Secure and Sustainable Living Conference on the Social and Economic Benefits of Weather, Climate and Water Services,* Madrid, 19–22 March 2007

Certain key tourism environments are underserviced by the global meteorological observation system (mountains and small islands) and would benefit from system improvements and improved working partnerships between meteorological services, the tourism sector and the media (Scott and Lemieux 2009). The need for improved accuracy (or skill) of forecasts (short to medium-term and seasonal) has been identified at workshops in the United States, Spain, Jamaica, the Bahamas, Fiji and Greece (Altalo and Hale 2002; Gamble and Leonard 2005; Scott and Lemieux 2009) as a requirement for increased use in operational decision-making in the tourism sector.

IMPROVING CLIMATE SERVICES FOR THE TOURISM SECTOR

Despite the growing global economic importance of the tourism sector and the multiple, complex interactions between climate and tourism, as this chapter has outlined, there have been limited evaluations of the use of climate information or assessments of the needs for climate services within the sector (Altalo and Hale 2002; Scott 2006a). There has been no systematic evaluation of the extent and nature of climate information use by major tourism stakeholders. The role of climate information in specific decision processes within the tourism sector (either demand or supply-side), the economic and non-market value of climate information for tourists and society, and the most effective ways to communicate climate information to diverse tourism end-users are important knowledge gaps that need to be addressed in order to reduce current and future climate risk in the sector.

It is anticipated that the need for more accurate and increasingly detailed weather and climate information will increase in the decades ahead, in order for tourists, tourism businesses and destinations to minimize the associated risks and capitalize on new opportunities posed by climate change in an economically, socially and environmentally sustainable manner. A jointly commissioned WMO and UNWTO White Paper for World Climate Conference-3 in Geneva, Switzerland in 2009 developed 14 recommendations to ensure the future climate service needs of the tourism sector would be realized (Box 2.2).

BOX 2.2 POLICY OVERVIEW: WMO AND UNWTO WHITE PAPER RECOMMENDATIONS FOR ADVANCING CLIMATE SERVICES FOR THE TOURISM SECTOR

1. Investment is necessary to strengthen climate monitoring networks in areas where the tourism sector is vital to local economies, specifically rural areas and many developing countries (particularly small island developing states), in order to improve climate risk management and climate change adaptation in the tourism sector.
2. With the risk in developing countries of permanent loss of historical climate data, which has potentially high value for managing climate risk and informing climate change adaptation, action is urgently needed to establish a coordinated international data rescue initiative.

3. Strengthening of climate monitoring networks is necessary to support the development of and access to innovative financial products (weather derivatives and index insurance) to manage climate risk in the tourism sector.

4. The development of regionally and locally specific climate change scenarios is necessary to facilitate effective climate change adaptation by the tourism industry and tourism-dependent communities. The refinement of near-term climate change predictions (covering the next 25–30 years) that are most relevant to business investment and government policy timeframes is particularly encouraged.

5. Support is necessary for the fundamental multidisciplinary research needed to understand the salience of climate (both in source markets [push factor] and destinations [pull factor]) in different travel decision-making contexts, cross-cultural climate preferences for major destination types, the effect of weather on holiday satisfaction and future travel choices and the climate sensitivity of major tourism activities.

6. Developers of specialized climate products for the tourism sector, whether the private sector, universities or governments, are encouraged to disclose the scientific methodology or market testing results to demonstrate validation in the tourism marketplace.

7. The tourism sector, in collaboration with NMSs [national meteorological services], private meteorological companies and university researchers, are encouraged to develop accepted standards for specialized climate products, to ensure consistent and accurate communication of climate information to international travellers and to facilitate objective destination comparisons and marketing claims in a global tourism marketplace.

8. Collaboration between governments, universities, communities and the private sector (tourism businesses, meteorological service companies, financial services) must be strengthened to drive innovation that connects climate information to the needs of the tourism sector and tourism-dependent communities.

9. The active collaboration of the tourism industry is necessary to support the development of climate services to improve outcomes for the sector, and the industry is strongly encouraged to provide increased access to sectoral data, consult on specific climate information needs and constraints to its use, provide expert review of specialized products and create effective strategies to communicate weather and climate information to tourists.

10. An interdisciplinary initiative should be established to evaluate the economic and non-market societal value of climate information for decision-making by tourists and tourism operators.

11. Greater effort should be made to consult with major tourism end-users about their needs for climate information. This consultation must be done

regionally in order to adequately represent specific information needs and the capabilities of regional providers.

12. An interdisciplinary evaluation of best practices for communication of climate information, particularly specialized products and forecast uncertainty, to tourism end-users is encouraged.

13. A series of multi-objective, capacity-building workshops should be initiated in major tourism regions around the world, in order to foster the direct interactions and partnerships between climate service providers and tourism user groups needed to make significant progress in the application of climate information in the tourism sector.

14. Training the next generation of tourism professionals to utilize climate information to reduce climate risks and adapt to climatic change in the decades ahead is a priority, and it is urged that a Climate Risk Management training module be created for use by tourism and hospitality schools around the world.

(Scott and Lemieux 2009)

THE DEVELOPMENT OF TOURISM AND CLIMATE CHANGE RESEARCH AND PRACTICE

Climate change and tourism as an area of scientific research and sustainable tourism practice marked its twenty-fifth anniversary in 2011 (Figure 2.14). Scholarship on the interactions between tourism and climate change began with the first publications in 1986–87 (Harrison *et al.* 1986; Wall *et al.* 1986; Gable 1987). However, tourism was not mentioned in the IPCC's *First Assessment Report* in 1990. This reflected 'the paucity of attention given by tourism researchers to climate and, similarly, by climatologists to tourism' (Wall 1998: 65).

The IPCC's *Second Assessment Report* in 1995 gave greater attention to tourism, with Wall (1998: 68) concluding that while 'it is encouraging that tourism is receiving greater attention in IPCC reports, it is also apparent that the likely consequences of climate change for tourism and recreation are not well understood'. The contribution of national and international tourism to GHG emissions and the implications of mitigation policies for travel patterns were not raised in the *Second Assessment Report*, but were first discussed the following year by Bach and Gössling (1996).

Climate change and tourism scholarship was still in its very formative stages in the late 1990s; however, the volume of publications grew substantially in this period, increasing threefold between 1990–94 and 1995–99 (Scott *et al.* 2005a). The seminal IPCC Special Report on Climate Change and Aviation (Penner *et al.* 1999) drew attention to the contribution of aviation emissions to global climate change, but made no specific estimate of the emissions generated by millions of international tourists.

Consideration of climate change was negligible in the range of 'trends'-focused publications that appeared at the close of the twentieth century; for example, climate change is not

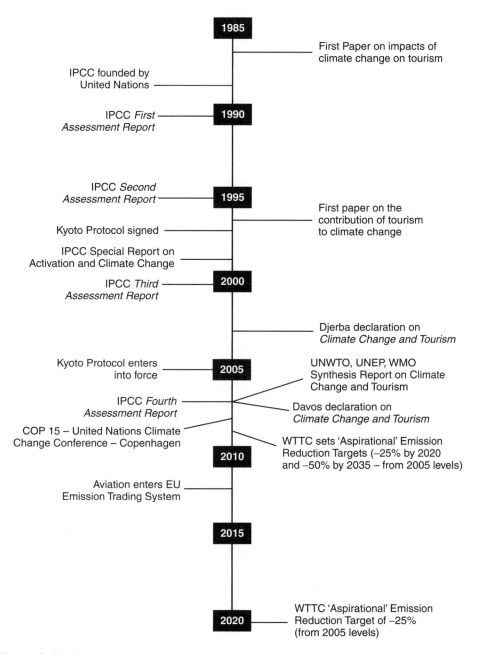

Figure 2.14 *Timeline of important climate change and tourism events*

mentioned by either Gartner and Lime (2000) or Kraus (2000). The position of the tourism community with respect to climate change was summarized by Butler and Jones' (2001: 300) conclusion to the *Tourism and Hospitality in the 21st Century* conference: '[climate change] could have greater effect on tomorrow's world than anything else we've discussed ... The most worrying aspect is that ... to all intents and purposes the tourism and hospitality industries … seem intent on ignoring what could be *the* major problem of the century.'

The climate change and tourism literature continued to grow rapidly, with multidisciplinary contributions doubling the number of publications that examine the two-way interactions of tourism and climate change between 1996–2000 and 2001–05 (Scott *et al.* 2005a). This led to a notable advancement of the place of tourism in the IPCC *Fourth Assessment Report* (AR4) (Amelung *et al.* 2008; Hall 2008a). Tourism was discussed in many regional chapters of the *Impacts, Adaptation and Vulnerability* report (AR4: Parry *et al.* 2007), and, much more briefly, in the *Mitigation of Climate Change* report (Metz *et al.* 2007) that focuses on GHG emissions and possible mitigation strategies (see chapter 8 for further discussion).

In 2007, the UNWTO, in collaboration with UNEP and the WMO, commissioned a state of knowledge report to be prepared for the Second International Conference on Climate Change and Tourism in Davos, Switzerland. This scientific assessment (Scott *et al.* 2008a) identified regional 'hotspots' where the tourism sector was particularly vulnerable, discussed the state of adaptation within the sector, provided a quantitative estimate of the contribution of global tourism to climate change (roughly 5 per cent of CO_2 emissions in 2005), and set out options for decoupling future growth in the tourism sector from GHG emissions. The conclusion of this scientific assessment was that '(climate change) must be considered the greatest challenge to the sustainability of tourism in the twenty-first century' (Scott *et al.* 2008a: 38). This conclusion was retained in the Davos declaration *on Climate Change and Tourism* and the subsequent *Tourism Minister's Summit on Tourism and Climate Change* (London, England).

Engagement in the challenges of climate change by the tourism sector (international and national tourism government agencies, NGOs) and by organizations representative of the tourism industry followed a different timeline from academic interests. The policy momentum generated by the release of the IPCC's *Third Assessment Report* (in 2001) and the Kyoto Protocol entering into force as an international treaty (in 2005) were instrumental in stimulating the first high-level involvement of the tourism sector. Several UN agencies organized the First International Conference on Climate Change and Tourism in Djerba, Tunisia in 2003. This event was a watershed in terms of developing awareness among government administrations, the tourism industry and other tourism stakeholders about the complex inter-linkages between the tourism sector and climate change. The Djerba declaration on *Climate Change and Tourism* (UNWTO 2003) recognized the salience of climate change for the sustainability of the global tourism industry, urged the formulation of appropriate adaptation plans, recognized the two-way relationship between tourism and climate change, and signified the obligation of the tourism industry to be part of the solution by reducing its GHG emissions and subscribing to all relevant inter-governmental and multilateral agreements to mitigate climate change. Recognition that GHG mitigation had become a new global policy reality inspired the travel sector to ensure its voice would be heard in future international climate change policy negotiations.

Building on the momentum created in Djerba, and coinciding with the release of the IPCC AR4, the Davos declaration on *Climate Change and Tourism* (UNWTO *et al.* 2008) urged a collective, expeditious and determined response by the tourism sector in four key directions:

- mitigate GHG emissions from the sector, derived especially from transport and accommodation activities;
- adapt tourism businesses and destinations to changing climate conditions;
- apply existing and new technologies to improve energy efficiency;
- secure financial resources to assist regions and countries in need.

A proposal to place a levy on all international air travellers (International Air Passenger Adaptation Levy, IAPAL) was put forward at COP-13 in Bali by the Group of Least Developed Countries. The IAPAL levy would generate approximately US$8–10 billion per year for climate change adaptation in developing nations. If such a proposal were implemented, then the tourism sector would become one of the world's largest funders of climate change adaptation in developing countries. Not unexpectedly, aviation sector organizations (International Civil Aviation Organization and Air Transport Action Group) strongly opposed proposals that single out international aviation as a source of revenue. The UNWTO Secretary-General Taleb Rifai (2010) likewise expressed unease over any climate change finance proposals targeted at air travel.

The increased response of the tourism community to the challenge of climate change after the release of the IPCC AR4 was very notable and a coordinated response of the tourism industry became visible for the first time. Prior to the COP-15 conference in Copenhagen, Denmark, the World Travel and Tourism Council (WTTC 2009) issued its first position paper on climate change. Most salient among these was the announcement of 'aspirational' emissions reduction targets to cut carbon emissions 50 per cent by 2035 (from 2005 levels).

> While the momentum since 2007 must be considered highly encouraging, there remain varied signals with respect to tourism sector engagement in climate change and a critical question from our perspective is how far has climate change discourse, and more importantly action, really penetrated the tourism sector? One answer to this comes from the business community itself, where KPMG's (2008) assessment of the regulatory, physical, reputational and litigation risks of climate change posed to 18 major economic sectors versus their level of preparedness found tourism to be one of six sectors in the 'danger zone' (along with aviation, transport, health care, the financial sector and oil and gas). Can tourism afford to continue to be one of the least prepared major global economic sectors?
>
> (Rifai 2010)

CONCLUSION

This chapter provides a review of the historical development of climate and tourism research, including the parallel rapid development of applied climatology and tourism studies, and the varied level of scholarly activity over the past four decades. The conceptual discussion of climate as a resource for and a constraint to tourism sets out the defining characteristics of climate and its value for major tourism stakeholders. An expanding body of evidence indicates that weather and climate have broad significance for tourist decision-making and the travel experience. However, important knowledge gaps remain with respect to the psychological processes of travel planning that utilize weather and climate information.

The ongoing revolution in communications technology, particularly the internet and, more recently, mobile personal data devices (smartphones) and social media, demands greater insights into what weather and climate information tourists desire and how it should be delivered to them. We are in the very early stages of the development of tailored weather and climate products for tourism, and better cooperation between academics, tourism professionals and the private meteorological services will be essential. While the tourism industry and meteorological services both identify significant value in weather prediction services for tourism, detailed valuation of these weather services remains an important task. The prominence of weather in tourism marketing is illustrative of its importance to destination image and helps to explain the dangers that inaccurate weather forecasts or, more importantly, poorly communicated forecasts represent to the tourism industry. Overall, this chapter reveals that the relationship between climate and tourism is multi-faceted and complex, and that we have only begun to understand some of the complexities that are central to advancing the capability of the tourism sector to better utilized climate services in an era of climate change. The need for closer collaboration between the climate research community and tourism studies was identified over 20 years ago. While this chapter identifies some signs of progress, examples provided here, and our own collective experience, demonstrate that there remain obvious disjoints between these disciplines. More than ever, there is a need for climate and tourism scholars and professionals to work together!

FURTHER READING AND WEBSITES

A comprehensive, up-to-date bibliography of academic research on weather/climate and tourism/recreation is available from:
> Scott, D. and Matthews, L. (2011) *Climate, Tourism & Recreation: A Bibliography*, 2010 edn. Waterloo: Department of Geography and Environmental Management, University of Waterloo. www.environment.uwaterloo.ca/geography/faculty/danielscott/publications.htm.

For those seeking additional information on the provision and use of weather and climate information in the tourism sector, a comprehensive review was commissioned by the WMO and WTO:
> Scott, D. and Lemieux, C. (2010) 'Weather and climate information for tourism', *Procedia Environmental Sciences*, 1: 146–183.

A broader discussion of the role of climate information services in fostering sustainable development in an era of global climate change is provided in the *WMO Bulletin*:
> Boodhoo, Y. (2005) 'Climate services for sustainable development', *WMO Bulletin*, 54(1): 8–11.

and via the 'Global Framework for Climate Services':
> www.wmo.int/pages/prog/wcp/ccl/documents/ConceptNote_GlobalFramework_ver3.4110309.pdf.

The WMO's website (www.wmo.int) also provides links to national meteorological services.

Further details of tourism industry, government and NGO positions on climate change and the nature of deliberations at the major international conferences on climate change and tourism are available in a report jointly published by the UNWTO, UNEP and WMO (2008):
> *Climate Change and Tourism: Responding to Global Challenges*. www.unwto.org/climate/index. php; www.uneptie.org/shared/publications/pdf/WEBx0142xPA-ClimateChangeandTourismGl obalChallenges.pdf.

For further information on the climate change position of the WTTC, the largest representative organization for the tourism industry, readers should consult:

WTTC (2009) *Leading the Challenge.* London: World Travel and Tourism Council. www.wttc. org/bin/pdf/original_pdf_file/climate_change_final.pdf.

and

WTTC (2010) *Climate Change – A Joint Approach to Addressing the Challenge.* London: World Travel and Tourism Council. www.wttc.org/bin/pdf/original_pdf_file/climate_change_-_a_ joint_appro.pdf.

(Both reports are available via www.wttc.org/eng/Tourism_Initiatives/Environment_Initiative.)

Growth in tourism, mobility and emissions of greenhouse gases

3

Growth in tourism mobility and associated energy use and emissions has been enormous over the past 60 years. The aim of this chapter is to provide an overview of the socio-cultural and economic drivers of growing tourism mobility, as well as the distribution of energy use and emissions by the tourism subsector. Emissions are also discussed from national, individual and per-trip perspectives. The chapter ends with a systems perspective, outlining relationships of relevance for the development of significant emission reduction strategies.

GROWTH IN TOURISM

Tourism has grown rapidly over the past 60 years. From 1950 to 2005, international arrivals increased by 6.5 per cent per year, from an estimated 25 million in 1950 to 806 million in 2005 (UNWTO 2010a), with an even greater increase in domestic tourism, which is about five times the volume of international tourism if calculated in trip numbers (UNWTO *et al.* 2008). In the three years from 2005 to 2008, international tourist arrivals increased by another 100 million, reaching 913 million. By then, the global financial crisis had stopped the strong growth trend, and arrivals dropped to 877 million in 2009. However, growth resumed in 2010, growing to 935 million and thus compensating for the loss experienced in 2008, and prompting the World Tourism Organization to state that: 'In 2010, world tourism recovered more strongly than expected from the shock it suffered in 2008 and 2009 ... The vast majority of destinations worldwide reported positive and often double-digit increases' (UNWTO 2011: 1). Figure 3.1 shows growth over the period 1995–2010, over which international tourist arrivals grew by more than 400 million, despite various global crises including terrorist attacks in a number of regions, SARS and flu pandemics, the increasing cost of oil, and the global financial crisis in 2008/09 (Hall 2010d).

Further growth is expected, and international tourist arrivals are estimated to reach 1.6 billion in 2020 (UNWTO 2001), mostly as a result of strong growth in emerging economies and newly industrialized countries. For example, in 2010 the Asian region recorded a 13 per cent increase in international tourist arrivals of 204 million, Africa a 6 per cent increase to 49 million, and the Middle East a 14 per cent increase to 60 million

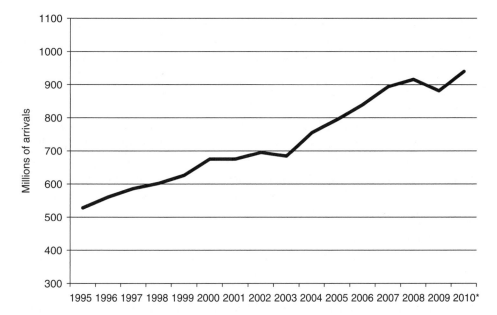

Figure 3.1 *International tourism arrivals, 1995–2010 (*preliminary figure)*
Source: UNWTO (2011)

(UNWTO 2011). Among the top outbound tourism markets in terms of expenditure abroad, emerging economies clearly drive growth. In 2010, China experienced 17 per cent growth, the Russian Federation 26 per cent, Saudi Arabia 28 per cent and Brazil 52 per cent. This compares with the traditional source markets in developed countries such as Australia (9 per cent growth in outbound tourism), Canada (8 per cent), Japan (7 per cent), France (4 per cent), and approximately 2 per cent growth in the USA, Germany and Italy (UNWTO 2011).

Observed strong growth in emerging economies shows that historically, mobility has not been evenly distributed between societies (Hall 2005). People living in industrialized countries are more mobile on an averaged per capita basis than those in developing countries, as mobility is primarily a function of wealth. This has been captured in the concepts of travel time budgets (TTB) and travel money budgets (TMB), originally proposed by Zahavi and Talvitie (1980) and developed by Schafer and Victor (2000). TTB and TMB recognize that travel time remains fairly constant with income growth, at around 1.1 hours per person per day on global average, but as higher incomes allow access to faster transportation, the distances travelled grow. This means that the distance travelled per person per year has continued to increase. For example, the average distance travelled per person per year in Great Britain between 1995–97 and 2005–06 increased by 3 per cent per year to 7200 miles (Office for National Statistics 2011). In Scotland, the National Travel Survey results for the period 2004/05 suggest that an average Scottish resident travelled around 7332 miles per year (or about 20 miles per day) within Great Britain. This is much more than 20 or 30 years earlier: since 1985/86, this average has risen by more than 2500 miles (58 per cent); and there has been an increase of over 3000 miles (75 per cent) since 1975/76. The main cause of this increase does not appear to be people travelling more often (the average number of trips per person per year has risen by only 15 per cent since 1975/76), but people going further when they do travel (the average length of a trip was 53 per cent higher in 2004/05 than in 1975/76) (Scottish Government 2007). Significantly, much of that growth corresponded to an increase in personal car use and a decrease in the use of public transport. Similarly, in Canada the per capita distance travelled by road vehicles increased by 33 per cent between 1980 and 1997. The total distance travelled by road vehicles increased by 117 per cent between 1970 and 1997, reflecting growth in population, growth in the number of vehicles on the road, and greater distances travelled per person (Boyd 2001). This relationship suggests that with any rise in global income levels, overall mobility will grow as well, with consequent implications for fuel demand and increase in emissions.

Growth in mobility has also been accompanied by a trend for the lowering of real transport costs. For instance, the Tata Nano, the world's cheapest mass-produced car, has a price almost an order of magnitude lower than that of other small cars, and made private automobility accessible to vast middle classes in India. In aviation, there has been a considerable decline in airfares because of market liberalization and the rise of low-cost airlines, leading to increased price competition (Gössling and Upham 2009). Marketing concepts have even included 'free seats' or 'all-you-can-fly' passes (Goetz and Vowles 2009). This has encouraged further mobility growth, even in already highly mobile societies. For instance, Goetz and Vowles (2009) present data showing constant growth in passenger numbers in the USA (with the exception of the 2001/02 post-11 September period) – despite the fact that almost one-third of global passenger transport by air, measured in trip numbers, already takes place in the USA (IATA 2008). With the rise of low-cost carriers in other continents such as Asia (Liang and James 2009), strong growth in global air travel and individual mobility is thus increasing rapidly. This is facilitated through rising incomes, and accelerated through the declining real transport costs of mobility.

Growth in aviation is most relevant for the development of mobility, because it allows travellers to cover large distances within short periods. However, the trend in aviation also reflects a more general trend in favour of increasingly energy-intense transport modes. Figure 3.2 shows growth in individual averaged mobility in person kilometres per year for the period 1850–1990, indicating that the average distance walked or bicycled has declined rapidly since the 1930s on a global scale, while averaged per capita car travel has grown by a factor of 10 in the same period. Rail and bus travel still increase because of the globally increasing mobility in many population-rich countries with low average incomes, such as India or China. Air travel is currently seeing the fastest growth rates, even though Airbus (2011) and Boeing (2011) have adjusted projections for global growth in passenger traffic downwards to an annual 4.8 and 4.2 per cent, respectively, up to 2029 (previously about 5 per cent per year in passenger numbers).

Increasingly mobile lifestyles in emerging economies reflect changes in income levels, but also changing cultural, social and political relations (Green *et al.* 1999; Coles *et al.* 2005; Connell and Williams 2005; Hall 2005; Shaw and Thomas 2006; Larsen *et al.* 2007; Frändberg 2008a, 2008b). The globalization of business, for instance, has led to an

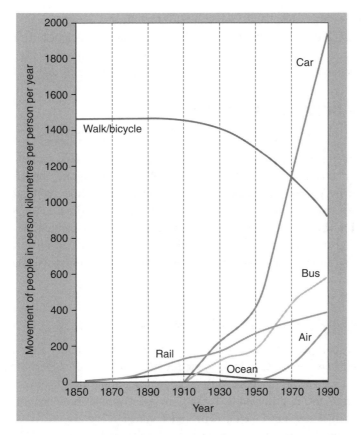

Figure 3.2 *Movement of people (person kilometres per person per year)*
Source: based on Gilbert and Perl (2008)

increase in global mobility, and employees of large companies regularly use aircraft as a transport choice – even though this might not always be the most cost-efficient or fastest option (Lassen 2006). Culturally and generationally, there might be fundamental changes in the way children grow up in contemporary industrialized countries, with socializing processes fostering and establishing highly mobile lifestyles (Frändberg 2008a, 2008b), and where children are already targeted by frequent-flyer programmes (Gössling and Nilsson 2010). Yet other motives can be found in increasingly global social networks and associated mobilities of labour markets, higher education, family life, second homes, migration and diasporic trips (Williams and Hall 2002; Larsen *et al.* 2007). Beyond the more simplistic relationship of income and mobility growth, there are thus more complex interactions fostering growth and establishing mobility patterns. It is important to understand these relations, because there is evidence that much of this mobility becomes firmly enrooted socio-culturally and generationally, in the sense that people are unwilling to give up acquired mobility patterns.

In this context, it is also important to outline that the greatest differences in averaged per capita mobility are found not between countries but within countries, where the continuum between almost 'immobile' and 'hypermobile' travellers is several orders of magnitude (Gössling et al. 2009b; Hall 2010e). This shows how mobile societies can hypothetically become. Moreover, highly mobile lifestyles are often aspired to and, being regularly pursued by the economic, cultural and political elites, represent social status.

EMISSIONS OF GREENHOUSE GASES

Tourism and travel contribute to climate change through emissions of greenhouse gases (GHGs), including in particular CO_2, but as well methane (CH_4), nitrous oxide (N_2O), hydrofluorocarbons (HFCs), perfluorocarbons (PFCs) and sulphur hexafluoride (SF_6). Various short-lived GHGs are important in the context of aviation – although their effect partly disappears within hours and weeks, their combined contribution to global warming is considerable (Lee et al. 2009). Tourism-related emissions include all domestic and international leisure and business travel, and are usually calculated for three subsectors: transport to and from the destination; accommodation; and activities (see UNWTO et al. 2008).

Global perspectives on greenhouse gases

Together, transport, accommodation and activities contributed about 5 per cent to global anthropogenic emissions of CO_2 in the year 2005 (UNWTO et al. 2008). Most CO_2 emissions are associated with transport, with aviation accounting for 40 per cent of tourism's overall carbon footprint, followed by cars (32 per cent) and accommodation (21 per cent) (Table 3.1). Cruise-ships are included in 'other transport' and, with an estimated 19.17 Mt CO_2, account for around 1.5 per cent of global tourism emissions (Eijgelaar et al. 2010).

All calculations in Scott et al. (2008a) represent energy throughput. As the construction of hotels, cars, airports and other infrastructure consumes considerable amounts of energy, a life-cycle perspective accounting for the energy embodied in the tourism system would

Table 3.1 Distribution of emissions from tourism by subsector, 2005

Subsector	CO_2 (Mt)	Percentage
Air transport	515	40
Car transport	420	32
Other transport	45	3
Accommodation	275	21
Activities	48	4
Total	**1,304**	**100**
Total world	26,400	
Tourism contribution		5

Source: Scott et al. (2008a)

lead to significantly higher estimates. Furthermore, tourism also leads to emissions in associated sectors, including tour operators and their offices, or employees commuting to work, as well as food requirements, which appear to lead to higher emissions than food consumed at home (e.g. Gössling *et al.* 2010a).

As outlined in Box 1.2, a more accurate assessment of tourism's contribution to global warming may be made using the concept of radiative forcing (RF). A recent estimate by Scott *et al.* (2010) found the sector to contribute to 5.2–12.5 per cent of all anthropogenic forcing in 2005. These figures are an update on Scott *et al.* (2008a), and can be seen as a better estimate of the impact of tourism on climate than an estimate based on CO_2 alone.

While tourism's contribution to climate change is thus considerable, the challenge lies in the sector's future emissions development. As outlined, emissions from tourism will grow because of several trends, including the growing number of people travelling, increasing frequency of trips, as well as growth in the average length of trips made, and the growing energy intensity of the transport modes used. Based on a business-as-usual scenario for 2035, which considers changes in travel frequency, length of stay, travel distance and technological efficiency gains, Scott *et al.* (2008a) calculate that CO_2 emissions from tourism may grow considerably in the coming 25 years. The scenario shows that emissions will increase by about 135 per cent compared with 2005 (Scott *et al.* 2008a), reaching 3059 Mt CO_2 by 2035. These estimates can be compared with a projection for emission growth by the World Economic Forum (WEF 2009a), which estimates that CO_2 emissions from tourism (excluding aviation) will grow at 2.5 per cent per year until 2035, and emissions from aviation at 2.7 per cent, which suggests emissions of 3164 Mt CO_2 by 2035.

As a result, if travel and tourism remain on a business-as-usual pathway, they will become important sources of GHG emissions in the medium-term future. Even if the per capita per trip contribution of tourists to GHG emissions continues to fall as a result of increased efficiencies from technological and management innovations, along the lines suggested by the UNWTO, WEF, WTTC and IATA, the absolute contribution will continue to grow as a result of increasing tourism mobility (Gössling *et al.* 2010b; Hall 2010e). It is also important to note that estimates of such future contributions of tourism do not include those of emerging forms of tourism such as space tourism, which are potentially extremely significant for climate change (Box 3.1). Therefore, if the world economy embarks on an absolute emission reduction pathway (Scott *et al.* 2010), growing emissions from tourism will be in juxtaposition to declining overall emissions, and an increasing share of the global carbon budget will fall on the sector. This is shown in Figure 3.3. Lines A and B represent emission pathways for the global economy under 3 per cent per year (A) and 6 per cent per year (B) emission reduction scenarios, with emissions peaking in 2015 (A) and 2025 (B), respectively. Both scenarios are based on the objective of avoiding a +2°C warming threshold by 2100 (for details see Scott *et al.* 2010). As indicated, a business-as-usual scenario in tourism, considering current trends in energy efficiency gains, would lead to rapid growth in emissions from the sector (crosses, C). By 2060, the tourism sector would account for emissions exceeding the emissions budget for the entire global economy (intersection of C with line A or B).

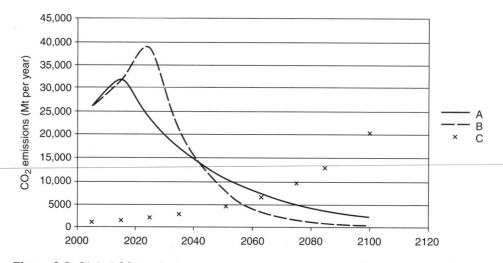

Figure 3.3 Global CO_2 emission pathways versus unrestricted tourism emissions
Source: Scott et al. (2010)

BOX 3.1 FUTURE CLIMATE IMPLICATIONS OF SPACE TOURISM

Space tourism is an emerging area of commercial space flight. The development of suborbital space tourism, led initially by Virgin Galactic, and eventually of orbital tourism, is often portrayed as one of the means by which the commercialization of space will lead to an expansion in launches (Launius and Jenkins 2006). However, there is increasing concern over the environmental implications of space tourism (Ross et al. 2009). Rockets are the only direct source of human-produced compounds above 22 km, and therefore it is important to understand how their exhaust and black carbon emissions affect the atmosphere. Fawkes (2007a, 2007b) estimated that a typical suborbital flight, using technology similar to Bristol Spaceplanes' Ascender, will produce total CO_2 emissions of 6267 kg per flight and therefore 3133 kg per passenger. Virgin Galactic has stated that its spaceport will use renewable energy and may even be a net energy producer, which could make it 'carbon negative', and that its suborbital flights will have emissions equivalent to a London-to-New York business-class flight (Fawkes 2007a, 2007b). However, such claims need much more detailed analysis.

Although the stratospheric emissions from a single suborbital rocket are small compared with an orbital rocket, Ross et al. (2010) believe that total suborbital fleet emissions could become comparable with present-day rocket emissions after a decade of continuous launches. Using a global climate model (the Whole Atmosphere Community Climate Model, WACCM3),

Ross *et al.* (2010) predict that emissions from 1000 suborbital launches using this hybrid engine would create a persistent layer of black carbon particulate in the northern stratosphere that could cause potentially significant changes in the global atmospheric circulation and in distributions of ozone and temperature. Tropical ozone columns are predicted to decline as much as 1 per cent while polar ozone columns increase by up to 6 per cent. Polar surface temperatures rise 1 K regionally and polar summer sea-ice fractions shrink between 5 and 15 per cent. After one decade of continuous launches, Ross *et al.* (2010) forecast that globally averaged RF from the black carbon would exceed the forcing from the emitted CO_2 by a factor of about 140,000, and would be comparable with the RF estimated from current subsonic aviation.

National perspectives on greenhouse gases

Similar insights apply for most nations with significant tourism economies. Assessments are, in chronological order, now available for New Zealand, Norway, Sweden, Australia, Switzerland, the Maldives and the Netherlands, including audit reports (Maldives, Norway, Netherlands, Germany) as well as scientific publications (New Zealand, Sweden, Switzerland). Assessments are generally not comparable due to differences in the system boundaries chosen as well as the methods used, comparing, for instance, tourism in the country or tourism of a country's residents, and including or excluding international air transport. Air transport is again calculated based on a CO_2-only approach, the consideration of other long-lived GHGs, or the use of 'uplift factors' to account for non-CO_2 GHGs. Assessments for these countries conclude that tourism accounts for at least 4 per cent and up to 68 per cent of national emissions, although the relative share also depends on the system boundary chosen, particularly with regard to international aviation. As none of the calculations appears to consider indirect and life-cycle emissions, all estimates nevertheless need to be seen as conservative. This indicates that emissions from tourism in most countries studied are higher both in relative (as a share of national emissions) and absolute terms (measured in t CO_2 per capita of the population) than on world average. Figures also indicate that the lower the national emissions on a per capita basis, the larger the tourism share.

New Zealand

One of the first publications seeking to understand emissions from tourism on a country basis looked at air travel in New Zealand. Becken (2002) concluded that international air travellers visiting New Zealand added 6 per cent (1.9 Mt) to the country's CO_2 emissions compared with the national GHG inventory. Patterson and McDonald (2004) sought to understand the energy intensity of tourism in New Zealand in comparison with other sectors. Combining Tourism Satellite Account (TSA) estimates with environment accounts including direct and indirect emissions from tourism activities in New Zealand in 1997–98, they concluded that tourism was the fifth largest emissions sector when focusing on internal energy use, and the second highest emitter out of 25 sectors when including emissions from overseas travel by inbound tourists (see also Becken and Patterson 2006). Smith and Rodger (2009) calculated that CO_2 emissions attributable to air travel to and from New Zealand from the 2.4 million international tourist arrivals in 2005 are 7.9 Mt CO_2-eq, if calculated using an 'uplift factor' of 1.9 (for discussion of the uplift factor see e.g. Gössling 2009). New Zealand as a country accounts for emissions of 77.2 Gt CO_2-eq, thus they conclude that this corresponds to about 10 per cent of national emissions. Air travel attributable to New Zealand residents amounted to 3.9 Mt CO_2-eq, again based on an uplift factor of 1.9.

Norway

Another approach was developed by Hille *et al.* (2007), focusing on emissions related to tourism and leisure activities by Norwegians (excluding those from visitors in Norway). They included work-related emissions (e.g. conference visits), education (e.g. evening courses or hobbies), free daily time spent (e.g. shopping), as well as organizational work (e.g. in religious contexts) and holidays. GHG inventories focus on CO_2, but include direct and indirect energy use on a life-cycle basis (for a discussion of methods see Aall 2011). Hille *et al.* (2007) conclude that tourism and leisure of Norwegians contribute 4.4 Mt CO_2 to the country's emissions of 33.3 Mt CO_2, that is, 13.3 per cent.

Sweden

In the case of Sweden, Gössling and Hall (2008) calculated 'Kyoto-relevant emissions' on the basis of a bunker fuel approach, that is, including all bunker fuels tanked within the country but excluding travel by Swedes in countries outside Sweden. Although it remains unclear whether emissions from bunker fuels will be allocated on this basis, results show that tourism accounts for 11 per cent of national CO_2 emissions in 2001, a figure that is expected to increase to 16 per cent by 2020 in a business-as-usual scenario considering technological efficiency gains. The comparably high emissions from tourism in Sweden might be explained with the country's otherwise low per capita CO_2 emissions, which are primarily a result of the high share of electricity produced from renewable energy sources.

Australia

In the most recent analysis of the carbon intensity of a national tourism system, Dwyer *et al.* (2010) focus on a TSA approach to calculate emissions on a CO_2-eq basis for the year 2003–04. Dwyer *et al.* (2010) perceive the TSA approach to be more suitable because earlier studies have focused on direct emissions from tourism, not considering indirect impacts associated with supplying inputs to tourism, such as those stemming from travel agencies and tour operator services; taxi, air, rail and water transport and motor vehicle hiring; as well as cafés, restaurants and food outlets, clubs, pubs, bars, food and beverage manufacturing; retail; casinos, libraries and other entertainment services. In estimating emissions from these sectors based on a production approach and an expenditure approach, it is concluded that, depending on the approach chosen, tourism is the fifth- to seventh-ranked Australian industry in terms of emissions, accounting for 21.6 to 29.5 Mt CO_2-eq, or 3.9–5.3 per cent of total industry emissions.

Note that the lower estimate is not including international aviation (corresponding to a 'Kyoto approach', cf. Gössling and Hall 2008), while the higher value includes international aviation of inbound tourists, although only on the basis of CO_2 (not considering the RF contribution of aviation to global warming). It also needs to be noted that emission factors chosen for aviation appear comparably low, and it is unclear whether these consider load factors. Overall, these findings would suggest that on a TSA (carbon intensity) basis, and including international aviation, emissions from tourism in Australia are in line with the global Scott *et al.* (2008a) estimate, while the carbon footprint of Australian tourism is higher, in the order of 54.4 Mt CO_2-eq in 2003–04 (production approach).

Switzerland

Perch-Nielsen *et al.* (2010a) assess the GHG intensity of tourism in Switzerland to determine the sector's GHG intensity, including all long-lived Kyoto greenhouse gases (in addition to CO_2 these are methane, nitrous oxide, HFCs, PFCs and sulphur hexafluoride). Emissions are calculated as CO_2-eq for five sectors (accommodation, foods and beverages; culture, sports and entertainment; travel agencies; other transport; and air transport) based on the Swiss TSA for 1998. Perch-Nielsen *et al.* (2010a) find that tourism is four times more GHG-intense than the Swiss economy on average, accounting for 2.29 million t CO_2-eq, or 5.2 per cent of the overall emissions released by Switzerland. Most of this is attributed to aviation, and the authors emphasize that emissions from tourism would fall to 1.1 per cent of overall Swiss GHG emissions if there was no international aviation.

The Maldives

The Maldives announced in 2009 its aim to become carbon neutral by 2020 (Clark 2009). As a first step to achieving this goal, emissions from tourism, one of the most energy-consuming sectors in the islands, were audited (BeCitizen 2010). Based on IPCC guidelines for carbon audits, the Maldives' emissions were assessed in a combined top-down/bottom-up approach, with the result that these account for 1.3 million t CO_2-eq in 2009. However, the assessment does not include international flights to the islands, which, when included, account for another 1.3 million t CO_2-eq, doubling the islands' emissions. Even though this is not assessed in the report, it can be deduced that tourism accounts for 36 per cent of national emissions excluding international aviation, as well as virtually all of international aviation. Consequently, of the overall emissions associated with the economy of the island, about 68 per cent are a result of tourism.

The Netherlands

The holiday carbon footprint of the Dutch in 2008 was recently calculated by de Bruijn *et al.* (2010). The report shows that there has been a 16.8 per cent increase of the total carbon footprint of Dutch holiday-makers between 2002 and 2008. The 15.6 Mt of CO_2 emissions caused by Dutch holidays thus resulted in a 9 per cent share of total Dutch CO_2 emissions in 2008. Emission growth is caused mainly by the strong increase in the average distance travelled (+33 per cent between 2002 and 2008), resulting from an 82 per cent increase in intercontinental, long-haul holidays during the same period. Domestic tourism, responsible for 48 per cent of all trips, produced 18 per cent of all holiday emissions in 2008 (Table 3.2).

Table 3.2 *The carbon footprint of the Dutch, 2008*

Parameter	Emissions	
CO_2 emissions per average Dutch holidayt	433	kg
CO_2 emissions per average Dutch holiday per day	49.1	kg
Total CO_2 emissions for Dutch holidays	15.6	Mt
Average annual CO_2 emissions per person in the Netherlands	10,369	kg
Average CO_2 emissions per person per day in the Netherlands	28.4	kg
Total Dutch CO_2 emissions*	170.1	Mt

*Excluding LULUCF (forestry and land use)
Source: de Bruijn *et al.* (2010)

Germany

A first attempt to calculate emissions from tourism of German citizens was made by Schmied *et al.* (2009), who suggested that these accounted for 61 Mt CO_2-eq in 2006. Most of this falls on international trips (58 Mt CO_2-eq), with 7 per cent of long-distance journeys accounting for 45 per cent of overall emissions. The calculation does not include 46.3 million short trips (up to three nights), and corresponds to 6 per cent of national emissions of 1002 Mt CO_2-eq in 2005. A potential problem with this calculation is that international emissions from aviation related to the travel of Germans is compared with national emissions not including bunker fuel use. Furthermore, CO_2 equivalents are calculated based on the use of an uplift factor for aviation. A second audit was presented by Gössling *et al.* (2011b), calculating the emissions of Germans including transport and accommodation on the basis of CO_2. This study for the year 2005 concludes that emissions from German tourism are in the order of 37.5 Mt CO_2, including all tourism and short trips for recreational purposes. Although not directly comparable with the national GHG emission inventory, which does not include international aviation, emissions from German tourism would account for an estimated 4.5 per cent of total emissions. The study also develops a business-as-usual scenario, showing that by 2020, German tourism would lead to emissions more than twice as high as the climate policy goal of the federal government (−40 per cent compared with 1990, applying equal emission reductions over all economic sectors). The share of aviation in tourism-related CO_2 emissions is >60 per cent of tourism emissions in 2005.

Individual perspectives on greenhouse gases

Further insights on emissions in tourism can be derived from the analysis of mobility patterns, based on the analysis of frequency of travel and the distance covered. In this context, it is worth remembering that climate change is the result of the consumption activities of currently about 6.9 billion people (July 2011 estimate, CIA 2011). It is thus the combined result of individual choices and lifestyles that determines the overall impact of these consumption activities on climate change.

In 2004, emissions of GHG in CO_2-eq were about 49 Gt, and of this there was about 27.7 Gt CO_2 (IPCC 2007b). Calculated at 6.5 billion people in 2007, average per capita emissions are thus in the order of 7.5 t CO_2-eq, or 4.2 t CO_2 per year. However, there are vast differences in per capita emissions between countries and within countries, most of which are a result of mobility patterns, with an estimate that 'individual emissions associated with food consumption vary by a factor 2–5, by a factor 5–10 for housing, but possibly by a factor of 100–1,000 for mobility' (Gössling et al. 2009b: 147).

Highly mobile travellers (for both business and leisure) are likely to exceed annual emissions of 50 t CO_2 from air travel alone, with studies finding individuals participating in up to 600 individual flights per year (Gössling et al. 2009b; Gössling and Nilsson 2010), corresponding to almost 12 times the global per capita annual average of 4.2 t CO_2. Such high mobility-related emissions are always a result of air travel, with private jets vastly increasing emissions (Table 3.3, see also Box 3.2). Using an A380, the largest privately owned aircraft, would incur emissions 460 times as high to travel over 5000 km as would travelling the same distance in economy class. For highly mobile travellers there may consequently be a 'comfort career', with frequent flyers being increasingly keen on greater comfort, ultimately seeking to fly first class and to then move on to private jets (a trend visible in the USA) (Cohen 2009). Other trends fostering greater energy intensity for highly mobile travellers are the plans of some airlines to provide greater leg-space, wider seats and extra baggage allowances.

Table 3.3 Energy intensity of different air traffic transport choices

Energy use for air travel over 5000 pkm	Emissions (t CO_2)	Factor*
Commercial aircraft (economy)	0.555	4.1
Commercial aircraft (first class)	1.110	8.2
Private jet	2.125	15.7
A380	253.761	1880

*Emissions compared with travelling this distance by train
Source: Gössling (2010), *Carbon Management in Tourism: Mitigating the Impacts on Climate Change*. London: Routledge, p. 96.

In line with the trend towards more energy-intense transport, individual travellers also participate in more frequent and more distant holidays (UNWTO 2008), staying over shorter periods, using more luxurious accommodation and participating in more energy-intense activities (UNWTO *et al.* 2008). These trends are expected to continue into the foreseeable future. According to UNWTO's *Tourism 2020 Vision* (UNWTO 2001), the share of long-haul tourism is projected to increase from 18 per cent in 1995 to 24 per cent in 2020, which, given the overall growth in tourism, implies that the number of long-haul trips will more than triple between 1995 and 2020. Furthermore, average trip distance is also increasing. In the EU, the number of trips is projected to grow by 57 per cent between 2000 and 2020, while the distances travelled are expected to grow by 122 per cent (Peeters *et al.* 2007), also because of trends towards more frequent holidays over shorter periods (Hall 2005; Dubois and Ceron 2006). The individual contribution made by travellers will thus, on a per capita basis, probably increase. This is largely a result of the carbon intensities of the various travel components, multiplied by the number of kilometres travelled or nights stayed, as shown in Box 3.2. For instance, travelling by air rather than by train will often be at least five times as energy intense, if measured on a per passenger kilometre basis. Also, considering that travel by air is at least five times faster, the difference in emissions will become a factor 25 times greater than for a trip identical in terms of travel time.

BOX 3.2 EMISSION INTENSITIES OF TRIP TYPES

Different tourist activities cause different emissions, depending on their energy intensity (the amount of energy needed to carry out a given activity) as well as the carbon intensity of the energy used (the emissions caused because of the use of energy). The following provides an overview of emission intensity ranges in various tourism subsectors.

- Passenger transport, aircraft, in the EU (<500 km): 0.206 kg/pkm
- Passenger transport, aircraft, in the EU (1000–1500 km): 0.130 kg/pkm
- Passenger transport, aircraft, in the EU (>2000 km): 0.111 kg/pkm
- Passenger transport, car, in the EU: 0.133 kg/pkm
- Passenger transport, rail, in the EU: 0.027 kg/pkm
- Passenger transport, coach, in the EU: 0.022 kg/pkm
- Accommodation, eco-lodge, Zanzibar: <0.1 kg CO_2/guestnight
- Accommodation, eco 4* hotel, Germany: 0.1 kg CO_2/guestnight
- Accommodation, campsite, New Zealand: 1.4 kg CO_2/guestnight
- Accommodation, 5* hotel, Oman: 260 kg CO_2/guestnight
- Activities, skiing: 21.4 kg CO_2 per skier day
- Activities, diving: 27–61 kg CO_2 per trip per person
- Food by air: 0.725 kg CO_2 per tkm

(Gössling 2010)

Finally, the distribution of emissions within societies is also important. Of tourism's global CO_2 emissions, 40 per cent are a result of air travel. As only an estimated 2.5 per cent of the world's population participate in international air travel (Gössling and Peeters 2007) and only 5 per cent of the world's population is believed to have ever flown (Worldwatch Institute 2006), a major share of emissions is associated with the travel patterns and life-styles of relatively few people. Taking into account highly mobile travellers, as well as high-energy transport and more luxurious accommodation often used by the highly mobile, at least a quarter of tourism's global CO_2 emissions may be caused by a share of the world's population as small as 1 per cent. Measured in RF, this distributional relation-ship would be even more distinguished. The fact that only very few people are engaged in air travel is also shown in maps (Figure 3.4) showing the distribution of global travel.

Per trip perspectives on greenhouse gases

New insights can also be gained from a per trip perspective, as there are huge differences between various forms of holidays, with, for instance, a fly–cruise from Europe to Antarctica (Lamers and Amelung 2007) entailing emissions of about 6 t CO_2, 1000 times more than a cycling holiday. For comparison, a 14-day holiday from Europe to Thailand might cause emissions of 2.4 t CO_2, and even holidays often perceived as eco-friendly, such as dive holidays, cause emissions in the range 1.2–6.8 t CO_2 (UNWTO *et al.* 2008). These figures show that emissions caused by a single holiday can vastly exceed or account for a significant share of annual per capita emissions of the average world resident (4.2 t CO_2) or the average EU resident (9 t CO_2). In contrast, many holidays cause comparably low emissions, only marginally increasing overall per capita emissions (cf. Figure 3.5).

In summarizing these findings, it can be shown that, globally, CO_2 emissions from tourism may be considered relatively small when compared with other industries such as agricul-ture or construction. However, they become more relevant when looking at RF in a given year, with tourism emissions accounting for an estimated 5.2–12.5 per cent of global RF in 2005 (Scott *et al.* 2010). Nevertheless, as climate change mitigation is addressed in national politics, it might be more pertinent for nations to discuss emissions from tourism on a national basis. As outlined, results will be influenced by the system boundaries chosen for calculation, but estimates presented for Australia, Germany, New Zealand, Norway, the Netherlands, Sweden and Switzerland indicate that in industrialized countries, emis-sions from tourism can be considerably larger than the global average, both on averaged per capita terms and if calculated as a share of national emissions.

IMPLEMENTING LOW-CARBON TOURISM: A SYSTEMS PERSPECTIVE

Any systematic approach to mitigation needs to be based on a review of emission intensi-ties: an assessment of where emissions occur as well as an identification of where further growth occurs, possibly in combination with an evaluation of the underlying reasons for this growth. To focus on the largest emissions subsectors as well as the major growth

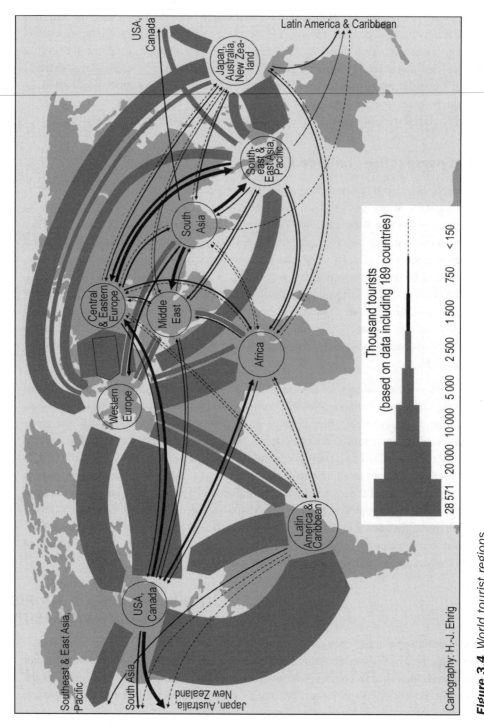

Figure 3.4 *World tourist regions*

Source: UNWTO international arrivals data multiple years

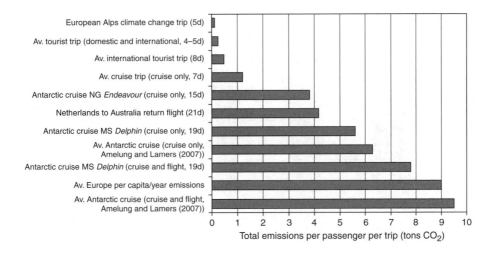

Figure 3.5 *Emissions associated with different holiday forms*
Source: Eijgelaar *et al.* (2010), 'Antarctic cruise tourism: the paradoxes of ambassadorship "last chance tourism" and greenhouse gas emission', *Journal of Sustainable Tourism*, 18(3): 337–354. Reprinted by permission of the publisher Taylor and Francis.

subsectors is important, because even addressing multiple less relevant emission subsectors will not be sufficient to achieve significant emission reductions. Scott *et al.* (2010) calculate, for instance, that even if emissions from accommodation and all transport except aviation fell to zero, overall emissions would still increase, given the strong growth in air travel.

Likewise, out of 26 mitigation scenarios developed by UNWTO *et al.* (2008), just one yields absolute emission reductions. This is a scenario combining high energy-efficiency gains with considerable modal shifts, changes in the choice of destinations, and increases in average length of stay. The results indicate that only strong pressure on the subsectors to become more energy efficient, combined with behavioural change in tourism consumption with respect to where and how they travel, will lead to absolute reductions in emissions (Figure 3.6).

In combination with other issues, such as the distribution of travel distances among travellers and the energy intensity of different transport modes, several relationships of relevance for the development of emission reduction strategies can be identified.

- Aviation is the most important tourism subsector for emissions, because it accounts for 40 per cent of current tourism emissions and up to 85 per cent of the RF caused by them (2005). It is also the sector showing the strongest growth, with growth in emissions of at least 2 per cent per year, even in the most optimistic technology development scenario. This means that, while aviation emissions may be lower on a per capita per trip basis, they are continuing to grow in absolute terms.
- A minor share of trips, including in particular the longest flights, but also cruise-ship journeys, is accounting for the majority of emissions from tourism. For instance, in the EU, 6 per cent of trips cause 47 per cent of CO_2-eq emission (Peeters *et al.* 2004); in France, 2 per cent of the longest flights account for 43 per cent of aviation emissions

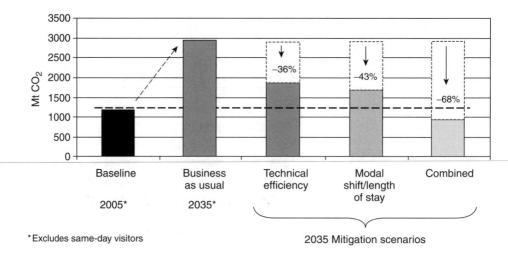

Figure 3.6 *Emission growth and reduction scenarios*
Source: Scott *et al.* (2008a)

(Dubois and Ceron 2009); and in the Netherlands, 4.5 per cent of long-haul trips cause 26.4 per cent of all tourism emissions (de Bruijn *et al.* 2008).

- National studies of tourism emissions show that the tourism sector accounts for a share of at least 4 per cent and up to 68 per cent of total emissions. Figures represent conservative estimates, although these are also dependent on the system boundaries chosen.
- There appears to be a relationship between per capita emissions and the share of tourism in developed economies, in the sense that countries with comparably low per capita emissions tend to have a higher tourism emission share.
- Growth in tourism emissions is primarily a result of increasing wealth and the development of more consumptive mobility lifestyles. With global incomes increasing and consumerism increasing in the newly developed and developing countries, there will be an increase in the number of people travelling, the distances they are travelling, the number of trips made by each individual per year, and the energy intensity of the transport modes chosen. A minority of highly mobile travellers appears responsible for a considerable share of the overall emissions caused.
- The combined emissions from tourism are mainly a result of the travel choices made by a comparably small share of humanity – those using aircraft. Within the group of air travellers, a highly mobile segment of 'hypermobile' travellers, primarily constituting economic, political and cultural elites, can be distinguished. These are likely to also use particularly energy-intense transport modes, and to contribute disproportionally to global emissions from tourism.

CONCLUSION

This chapter provides an overview of the emissions of the tourism industry from a variety of different perspectives. These different national, sectoral and institutional perspectives highlight some of the difficulties in reducing tourism's overall emissions. The fact that tourism emissions are continuing to grow in absolute terms highlights the significance of tourism's contribution to climate change. As a result, there are increasing demands that tourism, like other sectors, should seek to reduce its emissions in real terms. However, several relationships of relevance for the development of emission reduction strategies are identified, which provide a strong framework within which to mitigate tourism's contribution to climate change. These issues, and the various means by which emissions may be reduced, are discussed in more depth in chapter 4.

FURTHER READING AND WEBSITES

Global overviews of tourism's contribution to climate change include:

UNEP, University of Oxford, UNWTO and WMO (2008) *Climate Change Adaptation and Mitigation in the Tourism Sector: Frameworks, Tools and Practice*, United Nations Environment Programme, Oxford University, United Nations World Tourism Organization and World Meteorological Organization. Paris: UNEP.

WEF (2009a) *Towards a Low Carbon Travel & Tourism Sector*. Davos: World Economic Forum.

For overall global contributions to GHG emissions and climate change, see the website of the Intergovernmental Panel on Climate Change (IPCC):

www.ipcc.ch.

4 Carbon management
Climate change mitigation in the tourism sector

Systemic strategies to reduce emissions from tourism need to consider the development of average distances travelled, the frequency of travel per individual, as well as the energy intensity of the transport modes used. Aviation is a key sector to be addressed along with car travel, given that growth in emissions from these subsectors is likely with a rapidly increasing number of people becoming motorized. Accommodation accounts for one-fifth of emissions and is another important subsector because of anticipated rapid growth in global accommodation capacity, as well as the great potential to reduce emissions from buildings. The following sections discuss how policy, carbon management, technology and behavioural change – currently existing, planned for the future or potentially to be developed – can address the mitigation challenges in tourism.

POLICY FRAMEWORKS FOR TOURISM AND CLIMATE CHANGE

Policy frameworks for tourism operate at a number of different scales, and apply to different sectors and elements of the tourism system. Public policy 'is whatever governments choose to do or not to do' (Dye 1992: 2). Following on from this approach, Hall and Jenkins (1995) defined tourism public policy as whatever governments choose to do, or not to do, with respect to tourism. This definition covers government action, inaction, decisions and non-decisions, as it implies a deliberate choice between alternatives. For a policy to be regarded as public policy, it must have been processed, even if only ratified, by public agencies. The latter point is significant as it indicates that the formulation of policy can occur outside government, but then be ratified by government. In the case of climate change policy, this is extremely important because of the role of industry bodies such as the WTTC and IATA in influencing international tourism and climate change policy, and the influence of business associations at all scales of policy-making.

Figure 4.1 outlines the different spatial scales (local to global) and timelines (days to decades) of tourism development issues and related policy responses with respect to global climate change. One of the difficulties in developing climate change mitigation policies is that, while change in the climatic system is something that usually happens over decades and centuries, political decision-making usually operates over the time span

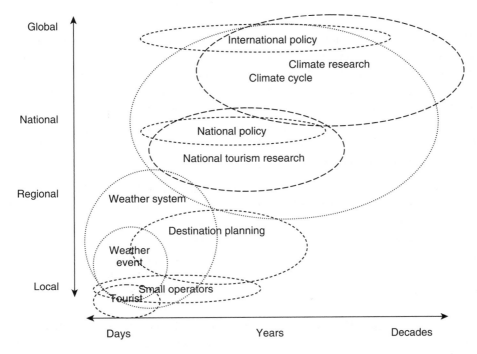

Figure 4.1 *Relativities of scale with respect to tourism and climate change policy*
Source: after Hall and Lew (2009)

between elections. For businesses the timescale is even shorter, with the operations of businesses usually measured by quarterly financial results or, for small tourism business operators, from season to season or from booking to booking. Given this conflation of different scales of change and action, it is little wonder that policy frameworks on climate change have been extremely hard to develop, especially when there have been deliberate attempts by some policy actors to undermine climate change science and associated recommendations for the mitigation of emissions (Giddens 2009).

In the case of tourism, policy-making becomes even more problematic because of the need to differentiate between tourism-specific policy frameworks and policy frameworks that affect tourism (Hall 2008b). As the following discussion suggests, both are significant, although the broader policy frameworks, such as those developed in international climate change agreements, are likely to be the most important in the long run assuming that they cover all tourism sectors and aviation in particular.

The combination of global wicked problems and the desire to develop cross-border public–private collaboration to try and deal with such issues has contributed to the development of the concept of governance. Kooiman (2003: 4) defines governance as 'the totality of theoretical conceptions on governing'. It is therefore 'The pattern or structure that emerges in a socio-political system as a "common" result or outcome of the interacting intervention efforts of all involved actors. This pattern cannot be reduced to one actor or group of actors in particular' (Kooiman 2003: 258), and can be best understood through terms such as steering and guidance. This theme is also picked up by Morales-Moreno (2004: 108–109), who argues that 'we could define governance as the capacity for steering, shaping, and managing, yet leading the impact of transnational flows and relations in a given issue area'. As chapter 3 illustrates with respect to emissions, tourism and climate change is clearly one such policy area that is marked by substantial transnational flows of people, businesses and their associated impacts, as well as new sets of economic, political and environmental relations. To understand some of the issues that emerge in managing the impacts of tourism, it is important that we investigate how they may be managed at various scales and via various means.

Tourism governance occurs at different policy levels, ranging from the international and supranational (a multinational political community to which power has been transferred or delegated by governments of member states) through to the local. At each of these different levels, there will be a range of policy stakeholders seeking to influence climate change policy, decision-making and implementation. Table 4.1 provides examples of some of the policy actors influencing climate change policy and decision-making at various scales. These are identified under the headings of government and inter-governmental organizations, producer (business) organizations, and non-producer organizations (often NGOs and public interest groups). However, to complicate matters further, many of the most significant policy actors and associated frameworks that affect tourism's mitigation of GHG emissions will not be tourism-specific, or may cover only a particular sector. For example, the Kyoto Protocol was not developed as a tourism-specific framework for climate change, and has a differential effect on various sectors. Similarly, many national government climate change frameworks are developed to cover all industries, not just tourism.

Significantly, tourism and climate change policy frameworks are developed not just horizontally (as a result of interaction between policy actors across a scale, e.g. between the member nation states negotiating an international environmental agreement), but also vertically (with policy actors at various scales influencing each other). In the case of such vertical policy influence, the actions of local governments and municipalities, and in federal systems states and provinces, may be especially important for encouraging national governments to be more active on climate change issues. For example, in the USA the state of California passed a law addressing climate emissions that was strategically intended to be in tension with federal law (Carlson 2003). Such a situation led Hodas (2004: 53) to pose the interesting question, 'Is it constitutional to think globally and act locally?' (He concluded that it was!)

In many national legal systems, local and regional governments have considerable regulatory powers that are essential for effective emissions mitigation. For example, Kaswan (2009) observes that as a consequence of continuing urban sprawl, transportation-related emissions in the United States are not expected to decrease despite federal legislation to improve fuel efficiency and increase biofuel use. Ewing *et al.* (2007) suggest that increasing compact development relative to sprawl, which is a function of local planning regulation, would reduce the distance that vehicles travel by 10–14 per cent and reduce the US transportation sector's CO_2 emissions by 7–10 per cent by 2050 (see also Brown *et al.* 2008).

Table 4.1 Examples of different policy actors at different scales of climate change-related decision-making

Scale/scope	Government and intergovernmental organizations		Producer organizations		Non-producer organizations	
	Tourism-related	Non-tourism-specific	Tourism-related	Non-tourism-specific	Tourism-related	Non-tourism-specific
International	United Nations World Tourism Organization (UNWTO)	OECD (Organisation for Economic Co-operation and Development)	World Travel and Tourism Council (WTTC)	World Economic Forum (WEF); Business Foundation for Social Action	Ecumenical Coalition on Third World Tourism; Tourism Concern	IUCN; Fairtrade International; International Federation for Alternative Trade
Regional	Caribbean Tourism Organization; South-Pacific Travel (South Pacific Tourism Organization)	UN Economic and Social Commission for Asia and the Pacific (ESCAP); CARIFORUM; European Union	Pacific Asia Travel Association (PATA); Caribbean Hotel & Tourism Association; Pacific Tourism Leaders Forum	Caribbean Association of Industry and Commerce; European Automobile Manufacturers' Association	European Association for Tourism and Leisure Education (ATLAS); Third World Tourism Ecumenical European Network (TEN)	European Consumer Consultative Group; Pacific Islands Managed and Protected Area Community

Continued

Table 4.1 Continued

Scale/scope	Government and intergovernmental organizations		Producer organizations		Non-producer organizations	
	Tourism-related	Non-tourism-specific	Tourism-related	Non-tourism-specific	Tourism-related	Non-tourism-specific
National	Tourism Australia; Tourism Canada; Irish Tourist Board (Bord Fáilte)	Climate Commission (Australia); Office of the Renewable Energy Regulator (Australia)	Tourism Industry Association of Canada; Australian Tourism Export Council; Ecotourism Australia	Australian Institute of Company Directors; US Council for International Business	Consumer Travel Alliance (USA); Air Transport Users' Council (UK); Tourism Concern (UK); Passenger Focus (UK)	Australian Conservation Foundation; Australian Consumers Association
Regional (provincial and state)	Tourism Alberta; Tourism Western Australia; Scottish Tourist Board; Tourism British Columbia	Conservation Commission (Western Australia); Department for Water (South Australia)	Japan Tourism Association of Queensland; Scottish Confederation of Tourism	Wyoming Business Council; Northwest Environmental Business Council	Forum Advocating Cultural and Eco Tourism (FACET) (Western Australia); Ecotourism Society of Saskatchewan	Consumer Focus Scotland; Wilderness Scotland; Saskatchewan Environmental Society
Local	Tourism Vancouver (Canada)	Kangaroo Island Council (South Australia)	Riverland Tourism Association; Gold Coast Tourism	Sierra Business Council; Los Angeles Business Council	Manas Maozigendri Ecotourism Society (Assam, India)	Save Goa Campaign

Subnational mitigation initiatives may also be especially significant when national and federal governments have failed to provide a framework for climate change mitigation (De Souza 2011) and/or where other subnational actors refuse to participate in emissions trading schemes or similar initiatives. For example, with respect to the Western Climate Initiative (WCI) cap-and-trade system, the non-participating Canadian provinces Alberta and Saskatchewan described it as a 'cash grab' by some of Canada's resource-poor provinces, with Alberta instead seeking to invest in carbon capture and storage (*Ottawa Citizen* 2008).

In North America, ten north-eastern and mid-Atlantic states (Connecticut, Delaware, Maine, Maryland, Massachusetts, New Hampshire, New Jersey, New York, Rhode Island and Vermont) are cooperating through a Regional Greenhouse Gas Initiative. Several mid-western states (Illinois, Indiana, Iowa, Kansas, Michigan, Minnesota, Ohio and South Dakota) and the Canadian Province of Manitoba signed the Midwestern Regional Greenhouse Gas Reduction Accord in 2007, and the WCI includes six western states (California, New Mexico, Montana, Oregon, Utah and Washington) and the Canadian provinces British Columbia, Manitoba, Ontario and Quebec. Arizona was a partner until Governor Jan Brewer rescinded its partnership agreement because of concerns that the implementation of a cap-and-trade programme would damage the economy during the economic downturn (see www.westernclimateinitiative.org).

In addition to the specificity of any policy framework for tourism and climate change, there are also a number of different policy mechanisms that intergovernmental bodies and governments can seek to mitigate climate change. At the international level it is important, for example, to distinguish between 'hard' and 'soft' international law (Shelton 2006). Hard international law refers to international treaties that are legally binding and are enforceable, for example by sanctions for non-compliance. One of the best examples here is the General Agreement on Tariffs and Trade (GATT) that established the World Trade Organization, which has substantial powers to arbitrate between member countries as well as to enforce its decisions. In contrast, most international environmental law, including climate change agreements, is relatively weak and dependent on international pressure and diplomacy, rather than legal actions, to achieve its objectives.

Soft international law refers to non-binding international policy recommendations and declarations. This is not to suggest that soft international law is without value, as it is an important determinant in pushing the international legal and regulatory system for climate and the environment in particular directions. In addition, 'making an agreement non-binding lowers the penalty associated with deviating from the existing legal rules, and thus encourages states with a significant interest in the content of legal rules to unilaterally innovate' (Meyer 2008). Such legal institutions can therefore 'be designed to harness the benefits of differences in state power, and ameliorate the negative effects when states' interests are in too much tension' (Meyer 2008: 941), as in the case of an area such as climate change, where national and regional governments may perceive mitigation as carrying potentially negative economic development consequences.

In addition to distinguishing between the implications of soft and hard international law, there are also different conceptions of governance. Four forms of governance may be identified in western liberal democratic countries: communities, hierarchies, markets and networks (Pierre and Peters 2000). Each of the conceptualizations of governance structures is related to the use of particular sets of policy instruments as well as their implementation (Hall 2009a, 2011a, 2011b). However, while such measures may be used in tandem, one form of governance may be substituted for another depending on government ideology. For example, in April 2011 environmental campaigners were angry that all of Britain's 278 environmental laws, including the Climate Change Act, were included in a list of 'red tape' regulations to be considered by the public in a crowdsourcing exercise launched by the UK government to establish which regulations restrict business. All UK regulation except with respect to tax and national security was included in the exercise. A business department spokesman said:

> It wouldn't look right for [environmental regulations] not to be on there. We are committed to meeting our climate change obligations, but at the same time we did not want to keep certain things off the website because we knew people would want to comment on them ... We've got to look at things from both sides. Yes, there's the environmental side, but businesses have to deal with these regulations on a daily basis and it takes a lot to grow a business.
>
> (Stratton 2011)

Critical to the value of the different modes of governance are the relationships that exist between public and private policy actors, and the steering modes that range from hierarchical top-down steering to non-hierarchical approaches. The main elements of the four models or frameworks of governance are outlined in Table 4.2, which identifies their key characteristics, the policy instruments associated with each concept of governance, and various dimensions with respect to policy-making and implementation. Because of the diversity of governance approaches that exist at different scales, it is important that, as the rest of this chapter discusses, mitigation policy frameworks be identified in international, national, local, industry and individual behavioural contexts.

International climate change policy frameworks

Currently, Annex I Parties included in Annex B to the Kyoto Protocol have to achieve averaged emissions reductions by 5 per cent of 1990 emissions in the period 2008–12. Countries included in Annex B of the Protocol have agreed to reduce their anthropogenic GHG emissions including CO_2, CH_4, N_2O, and the F-gases (hydrofluorocarbons, perfluorocarbons and sulphur hexafluoride), which are calculated as CO_2-equivalent emissions. Further emission cuts have to be achieved to not exceed the 2°C warming target, and negotiations for a post-Kyoto climate policy framework are currently ongoing. Voluntary post-Kyoto emission pledges have already been submitted to the UNFCCC by some countries – for instance, Germany and Sweden both committed to emission cuts by 40 per cent by 2020, and the UK by 80 per cent by 2050, compared with 1990 levels (Bundesregierung 2011; DECC 2011; Regeringskansliet 2011). Emissions from aviation and from cruise-ships and other seaborne transport are not attributed to country targets. Responsibility for emission reductions in these sectors are with the international aviation and shipping organizations, the International Civil Aviation Organization (ICAO) and the International Maritime Organization (IMO).

Table 4.2 *Frameworks of governance and their characteristics*

	Hierarchies	Communities	Networks	Markets
Classificatory type characteristics	• Idealized model of democratic government and public administration • Distinguishes between public and private policy space • Focus on public or common good • Command and control (top-down decision-making) • Hierarchical relations between different levels of the state	• Notion that communities (including business communities) should resolve their common problems with minimum of state involvement • Builds on a consensual image of community and positive involvement of its members in collective concerns • Governance without government • Fostering of public and community spirit	• Facilitate coordination of public and private interests and resource allocation and therefore enhance implementation efficiency • Range from coherent policy communities to single-issue coalitions • Regulate and coordinate policy areas according to preferences of network actors rather than public policy considerations • Mutual dependence between network and state	• Belief in the market as most efficient and just resource-allocative mechanism • Belief in empowerment of citizens via their role as consumers • Employment of monetary criteria to measure efficiency • Policy arena for economic actors where they cooperate to resolve common problems
Governance/ policy themes	Hierarchy, control, compliance	Complexity, local autonomy, devolved power, decentralized problem-solving	Networks, multi-level governance, steering, bargaining, exchange and negotiation	Markets, bargaining, exchange and negotiation

Continued

Table 4.2 Continued

	Hierarchies	Communities	Networks	Markets
Policy standpoint	Top: policy-makers; legislators; central government	Bottom: implementers, 'street-level bureaucrats' and local officials	Where negotiation and bargaining take place	Where bargaining takes place between consumers and producers, and business-to-business
Primary focus	Effectiveness: to what extent are policy goals actually met?	What influences action in an issue area?	Bargained interplay between goals set centrally and actor (often local) innovations and constraints	Efficiency: markets will provide the most efficient outcome
View of non-central (initiating) actors	Passive agents or potential impediments	Potentially policy innovators or trouble-shooters	Tries to account for behaviour of all those who interact in development and implementation of policy	Market participants are best suited to 'solve' policy problems
Criterion of success	When outputs/outcomes are consistent with a priori objectives	Achievement of actors' (often local) goals	Success depends on actors' perspectives	Market efficiency
Implementation gaps/deficits	Occur when outputs/outcomes fall short of a priori objectives	'Deficits' are a sign of policy change, not failure; they are inevitable	All policies are modified as a result of negotiation (there is no benchmark)	Occur when markets are not able to function
Reason for implementation gaps/deficits	Good ideas poorly executed	Bad ideas faithfully executed	'Deficits' are inevitable as abstract policy ideas are made more concrete	Market failure; inappropriate indicator selection
Solution to implementation gaps/deficits	Simplify implementation structure; apply inducements and sanctions	'Deficits' are inevitable	'Deficits' are inevitable	Increase capacity of market

Primary policy instruments	• Law • Regulation • Clear allocation and transfers of power between different levels of the state • Development of clear set of institutional arrangements • Licensing, permits, consents and standards • Removal of property rights • Development guidelines and strategies that reinforce planning law	• Self-regulation • Public meetings/town hall meetings • Public participation • Non-intervention • Voluntary instruments • Information and education • Volunteer associations • Direct democracy (citizen-initiated referenda) • Community opinion polling • Capacity building of social capital	• Self-regulation and coordination • Accreditation schemes • Codes of practice • Industry associations • NGOs	• Corporatization and/or privatization of state bodies • Use of pricing, subsidies and tax incentives to encourage desired behaviours • Use of regulatory and legal instruments to encourage market efficiencies • Voluntary instruments • Non-intervention • Education and training to influence behaviour
Climate change examples	• Mandatory requirements for biofuel use, buildings, carbon offsetting, disclosure, energy efficiency, introduction of technology, waste management and recycling • Educational programmes and awareness-building • Public investment and incentives • Public transport provision • Spatial planning • Taxes	• Community requirements for biofuel use, buildings, carbon offsetting, disclosure, energy efficiency, introduction of technology, waste management and recycling • Educational programmes and awareness-building	• Voluntary requirements for biofuel use, buildings, carbon offsetting, disclosure, energy efficiency, introduction of technology, waste management and recycling • Education of network members	• Emissions trading schemes • Government non-intervention

Source: adapted from Hall (2009a, 2011b)

Aviation

Civil aviation has grown constantly since the 1960s. The sector is now responsible for emissions of about 700 Mt of CO_2 (in 2004), 2.6 per cent of total anthropogenic CO_2 emissions in that year, or 1.3–14.0 per cent of radiative forcing (90 per cent likelihood range) (Lee *et al.* 2009). As outlined by UNWTO *et al.* (2008), about 80 per cent of this share can be attributed to tourism, the remainder being primarily airfreight.

Article 2 of the Kyoto Protocol states that limiting and reducing GHG emissions from international aviation is the responsibility of the ICAO, while emissions from domestic flights are included in national GHG inventories and are part of national emission reduction targets. However, ICAO made little progress on reducing emissions from aviation (Figure 4.2). Even though responsibility for reducing emissions from aviation was assigned to ICAO in 1997 – more than 20 years after interactions between exhaust fumes and atmosphere chemistry had been established (Fabian 1974, 1978), and more than five years after aviation's impact on climate was firmly established – it was not before 2001 that ICAO began to discuss how growth in aviation could be addressed.

Rather than presenting a strategy of how to reduce emissions, ICAO stated that it would oppose fuel taxes and a closed Emissions Trading Scheme (ETS) for aviation, also ruling out GHG emission standards, which could serve as benchmarks for the environmental performance of airlines. In 2004, ICAO imposed a three-year moratorium on charges, discussed as a possible alternative to taxes in 2001, and endorsed inclusion of aviation in existing ETS. In 2007, this decision was recalled, and instead non-binding, aspirational targets were to be formulated by a newly founded Group on International Aviation and Climate Change (GIACC). The Group recommended an 'aspirational' global fuel efficiency target of 2 per cent per annum to 2020, a timeline later extended to 2050. To achieve this, 'new technology, market-based measures, and alternative fuels' are seen as key measures. While none of ICAO's suggested 'commitments' is thus binding, even achieving an ambitious 2 per cent per year efficiency goal would mean that absolute emissions from the sector would increase. Because of expected continued strong growth in traffic volumes of about 4.5 per cent per year, the sector's CO_2 emissions would increase 2.0–3.6-fold over 2000 levels by 2050 (Owens *et al.* 2010).

Similar weaknesses as are outlined for ICAO characterize the mitigation goals of other organizations: highly optimistic assumptions regarding annual efficiency gains, non-binding targets, a missing discussion of growth in absolute emissions, non-consideration of non-CO_2 emissions, and no milestones against which progress in emission reductions based on generic solutions ('fleet renewal', 'biofuels') could be measured. Even though this appears impossible to achieve, the International Air Travel Association (IATA), International Business Aviation Council (IBAC) and Aviation Global Deal (AGD) envisage absolute emission reductions in aviation by at least 50 per cent by 2050 over 2005 (Table 4.3). 'Stabilization' (IATA) refers to no further growth in net emissions, an interim target by 2020. The AGD Group (including Air France–KLM, BAA, British Airways, Cathay Pacific Airways, Finnair, Qatar Airways, Virgin Atlantic and The Climate Group) is more ambitious, foreseeing emission reductions by 50–80 per cent by 2050, and up to a 20 per cent emission reduction by 2020. Non-CO_2 emissions are not mentioned.

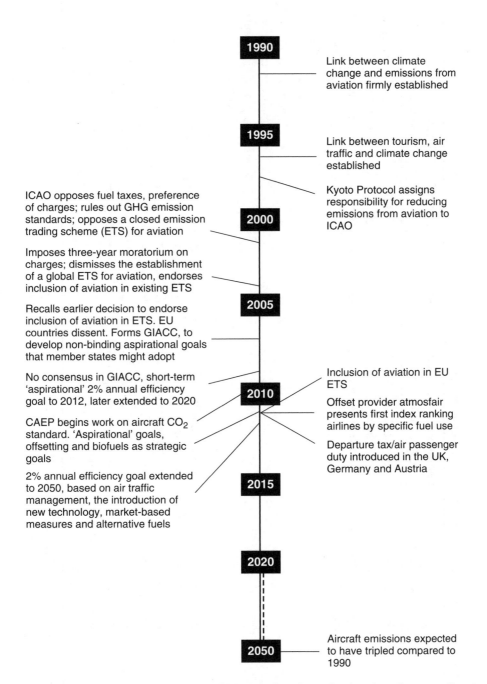

1990
Link between climate change and emissions from aviation firmly established

1995
Link between tourism, air traffic and climate change established

Kyoto Protocol assigns responsibility for reducing emissions from aviation to ICAO

ICAO opposes fuel taxes, preference of charges; rules out GHG emission standards; opposes a closed emission trading scheme (ETS) for aviation

2000

Imposes three-year moratorium on charges; dismisses the establishment of a global ETS for aviation, endorses inclusion of aviation in existing ETS

2005

Recalls earlier decision to endorse inclusion of aviation in ETS. EU countries dissent. Forms GIACC, to develop non-binding aspirational goals that member states might adopt

No consensus in GIACC, short-term 'aspirational' 2% annual efficiency goal to 2012, later extended to 2020

2010

Inclusion of aviation in EU ETS

Offset provider atmosfair presents first index ranking airlines by specific fuel use

CAEP begins work on aircraft CO_2 standard. 'Aspirational' goals, offsetting and biofuels as strategic goals

Departure tax/air passenger duty introduced in the UK, Germany and Austria

2% annual efficiency goal extended to 2050, based on air traffic management, the introduction of new technology, market-based measures and alternative fuels

2015

2020

2050
Aircraft emissions expected to have tripled compared to 1990

Figure 4.2 *Timeline of International Civil Aviation Organization inaction on climate change*

Table 4.3 Emission reduction targets and suggested action in aviation

Target/action	International Civil Aviation Organization (ICAO)	International Air Transport Association (IATA)	Aviation Global Deal (AGD) Group	International Business Aviation Council (IBAC)
Emission reduction goal	Improvements in fuel efficiency of at least 2 per cent per year until 2050	−50 per cent until 2050, stabilization by 2020 (base year 2005)	−50 to −80 per cent by 2050, up to −20 per cent by 2020	• Carbon-neutral growth by 2020 • Improvement in fuel efficiency of average 2 per cent per year until 2020 • Reduction in total CO_2 emissions by 50 per cent by 2050 relative to 2005
GHG considered	CO_2	CO_2	CO_2	CO_2
Suggested measures	• Energy efficiency measures • Air traffic management • Biofuels • Open and unlimited emission trading with other sectors	• Energy efficiency measures • Air traffic management • Biofuels • Open and unlimited emission trading with other sectors	• Energy efficiency measures • Air traffic management • Biofuels • Open and unlimited emission trading with other sectors	• Technology • Infrastructure and operator best practices • Alternative fuels • Market-based measures

Source: Aviation Global Deal Group (2009); IATA (2009); IBAC (2009); ICAO (2009)

IBAC, which represents some of the highest aviation emitters on a per capita basis (IBAC claims that business aviation's total global CO_2 emissions are very small, being approximately 2 per cent of all aviation and 0.04 per cent of global human-made carbon emissions), provided a statement of support for the ICAO Declaration on International Aviation and Climate Change that outlined its own commitments of 'carbon neutral growth' by 2020 (IBAC 2009). In conclusion, plans to address emissions as outlined by ICAO, IATA, IBAC and AGD appear insufficient to meet global climate policy requirements, as they are non-binding, while accepting further growth in absolute emissions from the sector.

International shipping

Unlike many land-based industries, marine emissions have historically remained largely unregulated,

> in part because of the international and multi-jurisdictional nature of ocean-going vessels, and in part because of a lack of empirical data on emissions output (a standardized and recognized emissions inventory) at the individual vessel level, that could provide reliable readings of emissions output.
>
> (Neef 2009: 3–4)

Emissions from shipping are estimated to have been in the order of 1046 Mt CO_2 in 2007, which corresponds to 3.3 per cent of global CO_2 emissions in that year (IMO 2009a). According to the WEF (2009a), global ocean-going cruise emissions for 2005 were estimated at 34 Mt CO_2, less than 5 per cent of global shipping emissions. However, this figure does not accommodate the full range of tourist passenger vessels. For example, according to Lloyd's Register, the world commercial fleet included 46,654 ships in 2006. Of these, 339 (0.7 per cent) were passenger/general cargo ships, 2743 (5.9 per cent) were passenger/roll-on roll-off ships, and 2873 (6.2 per cent) were passenger ships. In addition, there are an estimated 40,000 vessels not included as part of the commercial fleet (Kågeson 2007). Buhaug *et al.* (2009) used a revised estimate of the world fleet for 2007 that was based on ships over 100 Gt. Of the total fleet of 100,243 ships, 6912 were identified as passenger ships, accounting for 4 per cent of gross tonnage.

In addition to CO_2, ocean-going ships are estimated to emit 1.2–1.6 Mt of particulate matter (PM) with aerodynamic diameters of 10 micrometres or less (PM_{10}); 4.7–6.5 Mt of sulphur oxides (SO_x as S); and 5–6.9 Mt of nitrogen oxides (NO_x as N) (Corbett and Koehler 2003; Endresen *et al.* 2003; Eyring *et al.* 2005a; Corbett *et al.* 2007a, 2007b). This means that maritime transportation is also responsible for 15–30 per cent of global NO_x emissions and 5–8 per cent of global SO_x emissions (Corbett *et al.* 2007b). Table 4.4 indicates some of the estimates of NO_x and SO_x along with the estimates of global maritime fleet fuel consumption. The significant variability in estimates also highlights some of the difficulties that exist in accurately assessing GHG emissions in the maritime sector as a result of differences in not only methodology but also base statistical information.

Given that almost 70 per cent of ship emissions occur within 400 km of land, ships contribute significant pollution to coastal communities. Capaldo *et al.* (1999) estimate that

Table 4.4 Estimates of maritime CO_2, NO_x and SO_x emissions

Inventory year	Estimated maritime fleet fuel consumption (million tonnes)	CO_2 (Mt)	NO_x (Mt)	SO_x (Mt)	Source
2001	280	812.63	21.38	12.03	Eyring et al. (2005a,b)
2001	289	912.37	22.57	12.98	Corbett and Köhler (2003)
Average of 1996 and 2000	166	557.32	11.92	6.82	Endresen et al. (2003)
2007	333	1054	25	15	IMO (Buhaug et al. 2009): total shipping*
2007	277	870	20	12	IMO (Buhaug et al. 2009): international shipping*
2012	299	–	24.5	13.7	Corbett et al. (2007b)

*IMO figures are consensus estimates

ship emissions contribute between 5–20 per cent of non-sea-salt sulphate concentrations and 5–30 per cent of SO_2 concentrations in coastal regions. In addition to environmental damage, such localized air pollution may also have considerable negative health effects (Cofala et al. 2007). Corbett et al. (2007b) estimated that shipping-related fine PM emissions and ground-level ozone contribute approximately 60,000 cardiopulmonary and lung cancer deaths annually at a global scale, with most deaths occurring near coastlines in Europe, East Asia and South Asia, and that with the expected growth in shipping activity, annual mortalities could increase by 40 per cent by 2012.

The International Maritime Organization (IMO 2009a, 2009b) anticipates that, in the absence of mitigation policies, emissions from shipping will grow by 1.9–2.7 per cent per year until 2050, leading to overall growth by 150–250 per cent in the period 2007–50. Corbett et al. (2007a) forecast that world shipping activity and energy use will double from 2002 to 2030 (annual growth rate of 4.1 per cent), while the growth rates of four different scenarios in Eyring et al. (2005b) fall in the region of 2.5–4.0 per cent annually. As reported by Eijgelaar et al. (2010), tourism is an important component in this growth: worldwide cruise demand has grown steadily at an average annual rate of 7.4 per cent since 1990 (CLIA 2009), and emission growth from this sector has consequently been faster than from shipping more generally. For the year 2007, IMO estimates the global fuel use of all passenger ferries and cruise-ships at 31.3 Mt, corresponding to 96 Mt CO_2 (IMO 2009a). The WEF (2009a) estimates that emissions for ocean-going cruises will rise by 3.6 per cent per year, reaching 98 Mt CO_2 by 2035. As is the case with aviation, efforts to reduce absolute CO_2 emission levels from international shipping have been unsuccessful (Haites 2009).

The first ever regulation of GHG emissions in shipping was introduced by the IMO in 2011 under amendments to the International Convention for the Prevention of Pollution from Ships (commonly referred to as MARPOL, for MARine POLlution) Annex VI Regulations for the prevention of air pollution from ships, and comes into force in January 2013. This is expected to lead to emission reductions of 45–50 million tonnes a year and fuel savings of US$5 billion by 2020. Under the agreement, all ships over 400 tonnes built after 2013 will be required to improve their efficiency by 10 per cent, rising to 20 per cent between 2020 and 2024, and to 30 per cent for ships delivered after 2024. China, Brazil, Saudi Arabia and South Africa secured a 6.5-year delay for new ships registered in developing countries, which could mean the first guaranteed effective date of the reform will be in 2019. However, if the same standards were applied to the existing fleet of ocean-going ships, it could save approximately US$50 billion a year in fuel and 220 million tonnes of CO_2 by 2020 (Vidal 2011). However, Vidal (2011) also noted that 'The deal is not likely to satisfy the European Commission that the maritime organization is successfully regulating GHG emissions. The EC is therefore expected to proceed with its threat to bring shipping into the Emissions Trading Scheme …'

Kågeson (2007) estimates that a European Maritime Emissions Trading Scheme, initially limited to the ports of the European Union, would result in at least 6200 million tons less CO_2 being emitted between 2012 and 2035 compared with a business-as-usual scenario. However, a great part of this would be reductions in land-based sources paid indirectly by the shipping sector. Kågeson (2007) also notes that any maritime emissions trading scheme must be in line with the United Nations Convention on the Law of the Sea (UNCLOS), which provides a universal legal and regulatory framework for the management of international marine resources and marine-related activities.

UNCLOS Article 212 (Part XII Protection and Preservation of the Marine Environment) regulates the rights and duties of states with respect to pollution from or through the atmosphere.

1. States shall adopt laws and regulations to prevent, reduce and control pollution of the marine environment from or through the atmosphere, applicable to the air space under their sovereignty and to vessels flying their flag or vessels or aircraft of their registry, taking into account internationally agreed rules, standards and recommended practices and procedures and the safety of air navigation.
2. States shall take other measures as may be necessary to prevent, reduce and control such pollution.
3. States, acting especially through competent international organizations or diplomatic conferences, shall endeavour to establish global and regional rules, standards and

recommended practices and procedures to prevent, reduce and control such pollution.

(UNCLOS, 10 December 1982)

Kågeson notes that both the IMO and the UNFCCC could be regarded as competent international organizations under Article 213 of the Convention. Another significant avenue for emissions control is MARPOL, which entered into force on 2 October 1983 (Annex VI with respect to air pollution entered into force in 2005). Regulation 12 of MARPOL Annex VI prohibits deliberate emissions of ozone-depleting substances and also prohibits new installations that use ozone-depleting substances. Regulation 13 limits emissions of NO_x by applying limits on emissions for ship engines built on or after 1 January 2000. This has been supplemented by national actions, as in the introduction of a tax on NO_x emissions from domestic shipping in Norway. Buhaug *et al.* (2009) estimate that the introduction of regulation 13 has resulted in a reduction of about 6 per cent of NO_x emissions from shipping in 2007 compared with a no-regulation scenario.

Regulation 14 of MARPOL caps sulphur emissions globally at 4.5 per cent, and less in SO_x Emission Control Areas (ECAs) (Table 4.4). However, Buhaug *et al.* (2009: 57) note, 'It is widely acknowledged that the global limit of 4.50% of sulphur does not practically reduce global sulphur emissions, since a sulphur content exceeding this level was very rarely found in fuels before this regulation came into force.' In an ECA, the sulphur content of fuel oil used on board ships must not exceed 1.5 per cent by mass and hence may have a considerable impact on sulphur emissions. In Europe, a Baltic Sea ECA has been in force since May 2006 and a North Sea and English Channel ECA since November 2007. Waters off the coast of North America were designated an ECA in 2010, with emissions restrictions becoming enforceable in August 2012. A US Caribbean ECA has been proposed by the US Government covering waters around Puerto Rico and the US Virgin Islands.

> ... the United States estimate that the price impacts of the proposed ECA on a large cruise-ship that travels from the United States East Coast throughout the Caribbean may be US$0.40 per passenger per day; this represents a less than 1 per cent increase in the price of the cruise. The estimated price impacts on a medium-sized cruise-ship that operates a route between the United States and Puerto Rico will be approximately US$0.60 per passenger per day for a 5-day cruise; this represents a less than 1 per cent increase in the price of the cruise. The impacts on a small cruise-ship that spends nearly one-quarter of the time in the proposed ECA is estimated to be approximately US$1.30 per passenger per day for an 8-day cruise; this represents a less than 1 per cent increase in the price of the cruise.
>
> (MEPC 2010: 7)

Table 4.5 outlines international ship engine and fuel standards for sulphur and NO_x until 2020 under MARPOL Annex VI as of early 2011. Further work is required to bring in CO_2 emissions under the standards.

European Union climate policy

The EU is the only supranational region in the world with a legally binding target for emission reductions, imposed on the largest polluters. Current legislation foresees emission

Table 4.5 International ship engine and fuel standards (MARPOL Annex VI)

Standard	Year	Fuel sulphur (ppm)	NO$_x$
Emission control area	>2010	15,000	
	2010	10,000	
	2015	1,000	
	2016		Tier III (after treatment-forcing): 80 per cent NO$_x$ reduction below Tier I
Global	>Jan 2011		Tier I (engine-based controls): 9.8–17 g per kWh, depending on engine speed
	2011		Tier II (engine-based controls): 20 per cent NO$_x$ reduction below Tier I
	>Jan 2012	45,000	
	2012	35,000	
	2020*	5,000	

*Subject to a fuel availability study in 2018, may be extended to 2025
Source: adapted from US EPA (2010)

reductions by 20 per cent by 2020, compared with the base year 1990, but a recent call by ministers in France, Germany and the UK has been to adopt a 30 per cent reduction target (OECD 2010a), with Germany and Sweden having adopted 40 per cent reduction goals by 2020, compared with 1990. Discussions are ongoing about how to control emissions from consumption not currently covered by the EU ETS, which may lead to the introduction of carbon taxes in the EU in the future (EurActiv 2009). Moreover, the EU ETS will set a tighter cap on emissions year-on-year, and in the medium-term future (the next 5–10 years) the consumption of energy-intense products and services, including tourism, can be expected to become perceptibly more expensive.

The EU will also include shipping and aviation in an open ETS from 2012 onwards (European Parliament and Council 2009). Emissions from aviation will be capped at 97 per cent of their average 2004–06 levels by 2012. Airlines will receive 85 per cent of their emission allowances for free in 2012, though this percentage may be reduced from 2013. Likewise, the 97 per cent cap will decline to 95 per cent from 2013 onwards, although this percentage may be reviewed as part of the general review of the Emissions Trading Directive. EU policy will include all flights originating from or ending in the EU with 27 Member States (EU 27), irrespective of the country of origin of airlines and/or aircraft. European Union policy will include all flights originating from or ending in the EU 27, irrespective of the country of origin of airlines and/or aircraft. Airlines are required to auction 15 per cent of all emission permits, while the rest will be distributed based on the principle of grandfathering (based on historical emission reductions). While the EU approach is the only regional policy response worldwide, it is not likely to significantly change tourism flows or to reduce absolute emissions from the aviation sector (Gössling *et al.* 2008; Mayor and Tol 2009; Scott *et al.* 2010).

With regard to tourism more generally, the EU has recently presented its new EU Tourism policy framework, which will concentrate mainly on four strategic objectives: (1) improving the competitiveness of the tourism sector in Europe; (2) promoting the continuous sustainable development of EU tourism; (3) enhancing Europe's image as home to sustainable and high-quality destinations; and (4) maximizing the potential of EU policies and financial instruments for the development of European tourism (CEC 2010). With regard to sustainable tourism development, the EU policy framework mentions 'sustainable management indicators' to promote development respecting 'environmental, social and economic criteria', as well as awareness campaigns for tourists, awards, cooperation with countries outside the EU, and the identification and wider communication of best practice. The document mentions climate change only once, with regard to facilitating 'a better identification of the risks linked to climate change by the European tourism industry in order to avoid unsuccessful investments and to explore opportunities to develop alternative tourist offers'. Mitigation is not mentioned as a sector-specific goal.

National climate policy for tourism

The OECD and UNEP Secretariats conducted a survey on 'Climate Change and Tourism' to understand how well countries are prepared to deal with the climate change challenge for tourism. The survey was sent to all OECD member countries and selected non-members, and 18 members replied to the survey. Results are presented in Table 4.6. OECD and UNEP (2011) note that the development of tourism-related mitigation policy in OECD and selected other countries has hardly begun. As shown in Table 4.6, only one-third of countries have identified tourism-related mitigation strategies, and only six countries have already implemented policies (excluding the EU ETS as a policy framework for aviation). While there are a considerable number of potential strategies – including energy-efficiency gains in buildings, the use of biofuels, carbon offsetting, GHG disclosure, educational programmes and awareness building, energy-efficiency measures, incentives, investments, information technology to facilitate low-carbon choices, low-carbon mobility, mobility management, public transport, development of renewable energies, strategic assessment of emissions to identify mitigation options, sewage treatment, spatial planning, solid waste management, taxation and the enforced use of low-carbon technology – no country has presented a comprehensive strategy to achieve measurable emission reductions in tourism.

Examples of 'serious' climate policy: UK, France and Spain

There are few countries that either foresee or have implemented more ambitious climate policy, in particular with regard to tourism. One exception is the UK, which has, to date, implemented the most far-reaching policy in the world to address emissions from aviation. Likewise France, with its *bonus écologique* (ecological bonus), has introduced a policy that rewards/punishes car ownership based on emission intensities. Spain, although it has not

Table 4.6 *Mitigation action in selected countries**

Country	Mitigation strategies identified†	Policy in place†	Perception as threat/opportunity
Anguilla	E, EE, RE, I	Under development	• Marketing opportunities
Australia	ETS, RE, EE, CO	D, RE, EE, CO under review	• Major political divide between parties • Federal election promoted as a 'referendum' on climate change legislation
Austria	TE, BIO, MM, E, EE, PT, T, I, SP	Under development (EU ETS)	• Increasing costs • Marketing opportunities
Belgium	–	(EU ETS)	–
Canada	T‡, ETS	–	–
Czech Republic	–	(EU ETS)	–
Denmark	–	(EU ETS)	–
Egypt	EE, SW, SEW, TE	–	• Financial constraints • Limited potentials (aviation) • Possibility for fund raising
Estonia	RE, BIO	I, EE (EU ETS)	• Increases long-term competitiveness
France	B, CO, EI, I, INV, PT, RE	EI, I, INV	–
Germany	EE, RE, TE, BIO, T, E, SW	(EU ETS)	–
Greece	B, E, EE, INV, RE, TE	(EU ETS), INV	–
Hungary	–	(EU ETS)	• Transport costs • Investment necessary (accommodation) • New marketing opportunities • Competitive advantage
Ireland	D, SA, E	T, D, BIO, E, I, EE (EU ETS)	• Reduced costs
Italy	–	(EU ETS)	–
Luxembourg	–	(EU ETS)	–

Continued

Table 4.6 Continued

Country	Mitigation strategies identified[†]	Policy in place[†]	Perception as threat/opportunity
Mexico	EE, D, SA, RE, E	Under review	• Additional investment costs • Marketing opportunity
Netherlands	E	(EU ETS)	—
New Zealand	BIO	EE, D, ETS	• Competitiveness of aviation
Norway	ETS, T	—	—
Poland	EE	(EU ETS)	• Transport costs
Portugal	I, INV, EE, TE	(EU ETS)	• New market opportunities • Technological innovation
Romania	—	(EU ETS)	—
Slovak Republic	LCM	(EU ETS)	• Emission reductions pose challenge
Republic of Slovenia	EE, RE, SW, PT, IT	(EU ETS)	• Higher operational costs
South Africa	EE, RE, BIO	T, I, under further development	• Increased costs for travel • Costs for energy efficiency programmes • Reduced arrivals • Technology transfer
Spain	—	(EU ETS)	—
Sweden	TE	(EU ETS)	—
UK	T	(EU ETS)	—

*Directly or indirectly relevant for tourism

[†]B: buildings; BIO: biofuels; CO: carbon offsetting; D: disclosure, i.e. audit of emissions; E: educational programmes and awareness-building; EE: energy efficiency measures; ETS: Emissions Trading Scheme; I: incentives; INV: investments; IT: information technology to facilitate low-carbon choices; LCM: low-carbon mobility such as cycling or walking; MM: mobility management; PT: public transport; RE: development of renewable energies; SA: strategic assessment of emissions to identify mitigation options; SEW: sewage treatment; SP: spatial planning; SW: solid-waste management; T: taxes; TE: introduction of technology; EU ETS: European Union Emission Trading Scheme, relevant for aviation from 2012

[‡]In some states or parts of the country

planned or implemented any tourism-specific climate policy, has recently introduced a new speed limit to reduce emissions from traffic. This is a notable policy, as other countries such as Germany still do not have general speed limits, even though a majority of the population demands this (Umweltbundesamt 2010), and countries such as Denmark have increased their speed limits (Fosgerau 2005).

In November 2008, the UK Parliament enacted the Climate Change Act 2008, which sets a binding target to reduce UK emissions by 80 per cent by 2050, compared with the base year 1990, with an interim target of at least 2 per cent lower CO_2-eq emissions by 2020. The Act focuses on trading schemes for the purpose of limiting GHG emissions or the encouragement of activities that reduce such emissions or remove GHG from the atmosphere. Tourism is not mentioned in the Act, but it is understood that all sectors should be part of emission reductions. As of 1 November 2010, the UK introduced a new air passenger duty (APD) for aviation, which replaced its earlier, two-tiered APD. The new APD distinguishes four geographical bands, representing one-way distances from London to the capital city of the destination country/territory, and based on two rates, one for standard class, the other for other classes of travel (Table 4.7). Although no studies appear to exist as yet regarding the impact of the new APD, it can be assumed that low-cost airlines as well as long-haul travel will be affected by cost increases, particularly if these coincide with increasing oil prices. This is because low-cost airlines may no longer be able to sell tickets based on an understanding that such journeys are bargains and entailing virtually no costs (the APD will add €27 to ticket costs), and long-haul journeys may become too expensive (the APD will add €193 to ticket costs). The duty has been highly criticized by long-haul destinations from the UK, including New Zealand (see Box 4.1).

In comparison, the German Federal Government introduced a departure tax for aviation from 1 January 2011 (Bundesregierung 2010). The tax is expected to yield €1 billion annually, and is meant to function as a substitute for the missing taxation of kerosene in Europe and internationally. The tax includes only commercial passenger travel and is structured at three levels: flights up to 2500 km from Frankfurt/main airport (€8 per passenger

Table 4.7 *UK air passenger duty as of 1 November 2011*

Band (approximate distance in miles from …)	In lowest class of travel (reduced rate, £)		In other than lowest class of travel* (standard rate, £)	
	2009–10	2010–11	2009–10	2010–11
Band A (0–2000)	11	12	22	24
Band B (2001–4000)	45	60	90	120
Band C (4001–6000)	50	75	100	150
Band D (>6000)	55	85	110	170

*If only one class of travel is available and that class provides for seating in excess of 40 then the standard (rather than reduced) rate of APD applies
Source: HM Revenue & Customs (2008)

Table 4.8 *Emission and 'feebate' (fee–rebate) classes in the French bonus–malus system for cars*

Emissions of CO₂ per km (g)	Bonus–malus (€)	Paid by
≤60	5000	Bonus paid by government
61–90	800	
91–110	400	
111–150	0	
151–155	–200	Malus paid by car-owner
156–190	–750	
191–240	–1600	
≥241	–2600	

Source: Government of France (2011)

and flight); flights of 2500–6000 km (€25 per passenger and flight); and flights longer than 6000 km (€45 per passenger and flight). The tax will apply to all flights departing from German airports. However, the tax is implemented primarily with a view to supplementing government finances, its importance for mitigation being considered secondary, and it is unclear whether it will actually affect travel behaviour and lead to significant emission reductions.

France introduced a bonus–malus system in December 2007, rewarding purchases of low-emission cars and punishing purchases of high-emission cars. In 2011, this 'feebate' (fee–rebate) system had four bonus and four malus classes (Table 4.8). Purchases of highly efficient cars are rewarded with up to €5000, while purchases of cars with emissions above 155 g CO₂ per km are fined with payment of €200 up to €2600 (for cars emitting more than 241 g CO₂). In 2012, malus classes will be expanded to six and caps will be tightened. For instance, payments of €200 will apply for cars emitting more than 141 g CO₂ per km. The current system also foresees bonus payments for hybrid cars emitting 110–135 g CO₂ per km (€2000).

Spain introduced a new speed limit on 7 March 2011, reducing the old limit of 120 km per hour to 110 km per hour to the end of June 2011 (Boletín official del Estado 2011). According to BBC (2011b), the speed limit was introduced as a temporary measure to address rising fuel prices because of unrest in several Arab countries, in particular Libya. The government expects two benefits from the new speed limits: first, a declining share of money being spent on oil imports, going along with lower emissions from traffic; second, an increasing share of money spent on Spanish products, benefiting the national economy. While expected fuel-use reductions of 15 per cent were criticized as being unrealistically high (BBC 2011b), it is unclear whether the calculation also included transportation shifts, for instance because high-speed trains become more attractive in comparison with cars, or whether this has impacts on car purchasing behaviour, as it may become less attractive to own high-powered cars.

BOX 4.1 LONG-HAUL DESTINATIONS' REACTIONS TO THE UK AIR PASSENGER DUTY

The introduction of the UK APD has been substantially criticized by a number of long-haul destinations from the UK, which see the tax as potentially affecting their inbound travel from the UK as well as market share. Referring to the UK Government's decision as 'short-sighted', Brian Deeson, President and CEO of the Pacific Asia Travel Association (PATA), stated: 'PATA is an organization committed totally to sustainable development in travel and tourism. This move by the UK government is simply about increasing revenues for the state under the very dubious cover of consolidating its green credentials' (ehotelier. com 2008). Deeson indicated that PATA supported the views expressed by the Australian Tourism Export Council (ATEC), particularly with respect to the threat to tourism markets in emerging markets such as the South Pacific.

ATEC stated that 'this new charge will significantly impact travel to Australia from the United Kingdom, and is a critical trade barrier between the UK and Australia and New Zealand', and also accused the UK of using the climate change debate 'to disguise this protectionist departure tax in "greenwash"'. ATEC also suggested that the APD would 'also impact on developing nations – for example those in the South Pacific – who are reliant on inbound tourism to generate much of their export income' (TavelMole 2008).

In April 2010, PATA organized a meeting of the Pacific Tourism Leaders' Forum (PTLF) that focused on the APD. Matt Hingerty, Managing Director of ATEC and a member of the PTLF, stated: 'It is difficult to understand that a global economic power such as the United Kingdom could act with such insensitivity. This draconian tax serves only to generate revenue with absolutely no benefit to the environment that it purports to protect' (PATA 2010). Indeed, PATA Regional Director – Pacific Chris Flynn claimed that the APD would do more environmental harm than good:

> The long haul tourism industry has helped to preserve natural environments because of the income generated for regional governments ... Threats to tourism revenue will lead to pressure on natural resources as locals look to other sources of income – and that's bound to include the leveraging of these resources.
>
> (PATA 2010)

The APD has also been criticized by Giovanni Bisignani, Director General of IATA:

> I want to know where the money will go. How many trees will the chancellor be planting with GBP 2.5 billion? ... Padding the UK budget at the expense of holiday-makers, business travelers or exporters is not sound environmental policy. Instead of inventing new taxes with convoluted calculation methods, governments must support investment in basic green technology research, assist air navigation

service providers to straighten out routes and allow airlines to operate as fuel efficiently as possible. And when it comes to economic measures, let's focus on a global emissions trading scheme.

(quoted by Browne 2009)

The New Zealand Government also lobbied the UK for several years following the announcement of the introduction of the APD in 2008, and with the election of the Conservative–Liberal Democrat coalition in 2010, which New Zealand hoped would replace the levy (Vass *et al.* 2008). New Zealand Tourism Industry Association Chief Executive Tim Cossar said the tax increase was being pitched as an environment tax, but the money was not being used for environmental purposes, and was putting New Zealand at a competitive disadvantage (TVNZ 2010). Hospitality Association of New Zealand Chief Executive Bruce Robertson said that he considered the passenger tax to 'be a protectionist measure – anti-trade ... they're using the guise of sustainability and conservation as a measure of putting in a trade barrier ... to make it more difficult for Britons to travel long distance and, effectively, will be encouraging them to stay at home' (TVNZ 2010). Similarly, New Zealand Prime Minister John Key described the tax as a revenue-collecting exercise (TVNZ 2010), and had previously stated his concern that the tax would have a contagion effect and would lead to other countries following the UK's lead (Vass *et al.* 2008). A spokeswoman for the Prime Minister said that according to the New Zealand Ministry of Transport, the amount of duty per economy class passenger to New Zealand is around five times what it should be if the tax was simply aimed at offsetting the cost of carbon emissions: 'It's hard to find an environmental justification for this. Travellers to distant locations like New Zealand should not be disproportionately penalised' (Krause 2010). However, a contrary position has been put by the spokesperson for the travel agency House of Travel, a major New Zealand chain, who suggested that any cost increase is never well received by the public, 'but the reality is these are only quite small increases to the overall cost of travelling from Britain to New Zealand' (TVNZ 2010). Both New Zealand and Australia have government-mandated departure taxes, but they are not priced according to distance travelled and instead are flat fees.

LOCAL CLIMATE POLICY

Local and regional governments have an important role in setting policy frameworks to mitigate climate change. Although the majority of these are not tourism-specific, there are a number of subnational initiatives that have a significant impact on tourism and hospitality by virtue of:

- planning and development regulation;
- provision of public transport and sustainable mobility options;

- ownership of tourism infrastructure including, in some cases, airports, as well as a wide range of visitor attractions and facilities such as museums, art galleries, stadia and convention centres.

The need for local climate policy has also been increased by continued urban growth on a global scale and the subsequent ecological footprint of major urban centres. As of 2008, it was estimated that more than half of the world's population lived in urban areas. Cities consume over two-thirds of the world's energy and account for more than 70 per cent of global CO_2 emissions (C40 Cities 2010). Cities also experience significant micro- and meso-climatic effects as a result of the urban heat-island effect, which refers to built-up areas being hotter than nearby rural areas. According to the US Environmental Protection Agency (EPA), the annual mean air temperature of a city with 1 million people or more 'can be 1.8–5.4°F (1–3°C) warmer than its surroundings. In the evening, the difference can be as high as 22°F (12°C). Heat islands can affect communities by increasing summertime peak energy demand, air conditioning costs, air pollution and GHG emissions, heat-related illness and mortality, and water quality' (US EPA 2011). Heat-island effects are also significant in the context of climate change because of the extent to which they may exacerbate the effects of a warming climate, placing even further pressure on urban environmental and energy systems.

There are a number of international and national initiatives to mitigate climate change at the urban and regional planning scales. One of the most high profile of these is the C40 Climate Leadership Group Initiative. In October 2005, London Mayor Ken Livingstone organized the first large cities climate summit, which was attended by representatives of 18 cities. In August 2006, the initiative was strengthened when President Clinton and Ken Livingstone announced a partnership between the Clinton Climate Initiative and the Large Cities Climate Leadership Group (since renamed C40). The C40 is managed in London by a Secretariat and a Steering Committee.

Many of the C40 cities have developed action plans with specific GHG reduction targets (Table 4.9). Mechanisms to achieve these goals include:

- creating building codes and standards that include practical, affordable changes that make buildings cleaner and more energy efficient;
- conducting energy audits and implementing retrofit programmes to improve energy efficiency in municipal and private buildings;
- installing more energy-efficient street and traffic lighting;
- implementing schemes to reduce traffic and developing bus rapid transit and non-motorized transport systems;
- establishing the infrastructure and incentives to promote the use of low-carbon vehicles;
- developing sustainable waste-management solutions, reducing reliance on landfill disposal, and creating waste-to-energy systems.

For example, at the 2007 meeting in New York, President Clinton announced the global Energy Efficiency Building Retrofit Program involving Bangkok, Berlin, Chicago, Houston, Johannesburg, Karachi, London, Melbourne, Mexico City, New York, Rome,

Table 4.9 *C40 cities, greenhouse gas emissions reduction targets*

City	Country	Greenhouse gas reduction targets
Participating cities:		
Buenos Aires	Argentina	32.7 per cent below 2008 baseline levels by 2030
Chicago	USA	25 per cent below 1990 levels by 2020, 80 per cent below 1990 levels by 2050
Hong Kong	Hong Kong SAR	50–60 per cent below 2005 baseline levels by 2020
Houston	USA	60 per cent below 1990 baseline levels by 2025; limiting total CO_2 between now and 2025 to 600 million tonnes
Los Angeles	USA	35 per cent below 1990 baseline levels by 2030
Madrid	Spain	20 per cent below 2004 baseline levels by 2020, 50 per cent below 2004 baseline levels by 2050
Melbourne	Australia	Zero net emissions by 2020
Mexico City	Mexico	Reduction of GHG emissions 12 per cent by 2012
New York	USA	30 per cent below 2007 baseline levels by 2030 (accelerated municipal government target = 30 per cent reduction from 2007 levels by 2017)
Paris	France	75 per cent below 2004 baseline levels by 2050
Philadelphia	USA	At least 10 per cent below 1990 baseline levels by 2010 (11.6 per cent reduction for community 2010 and 12.3 per cent reduction for city government by 2010)
Rome	Italy	6.5 per cent below 1990 baseline levels by 2012
São Paulo	Brazil	30 per cent below 2003 levels by 2012
Seoul	South Korea	40 per cent below 1990 baseline levels by 2030
Sydney	Australia	70 per cent GHG emission reduction target for local government area by 2030 based on 2006 levels
Tokyo	Japan	Reduction of GHG emissions by 25 per cent by 2020

Continued

Table 4.9 Continued

City	Country	Greenhouse gas reduction targets
Toronto	Canada	6 per cent below 1990 baseline levels by 2012, 30 per cent by 2020, 80 per cent by 2050
Affiliate cities:		
Amsterdam	Netherlands	40 per cent below 1990 baseline levels by 2025
Copenhagen	Denmark	20 per cent below 2005 baseline levels by 2010
Portland	USA	40 per cent below 1990 levels by 2030, 80 per cent below 1990 levels by 2050
Rotterdam	Netherlands	50 per cent below 1990 baseline levels by 2025
Salt Lake City	USA	Reducing emissions by 3 per cent annually for 10 years, long-term goal of reducing emissions 70 per cent from present levels by 2040
San Francisco	USA	20 per cent below 1990 baseline levels by 2012
Seattle	USA	7 per cent below 1990 baseline levels by 2012
Stockholm	Sweden	60–80 per cent below 1990 baseline levels by 2050
Yokohama	Japan	30 per cent below 2004 levels by 2025, 60 per cent below 2004 levels by 2050

São Paulo, Seoul, Tokyo and Toronto. The programme aims to improve the energy efficiency of existing municipal buildings by installing new insulation, heating and cooling systems and control devices, and other 'green' retrofits, and is being financed by ABN Amro, Citibank, Deutsche Bank, JPMorgan Chase Bank and UBS. Governments and building owners will repay the loans plus interest with the energy savings generated by the reduced energy costs resulting from the building retrofits (Weiss 2007).

Tourism is explicitly recognized in a number of the C40 city GHG emission reduction action plans. For example, the City of San Francisco notes that

> The health of San Francisco's economy is dependent on the regional economy and depends heavily on its attraction as an international tourism destination. Further, both the regional economy and a good deal of the tourism industry are based in part on regional and local environmental health.
>
> (San Francisco Department of the Environment 2004: 1.15)

Under San Francisco's action plan, the city aims to discourage driving and encourage shifts to alternative modes of transport, leading to an estimated 155,000 t CO_2 reductions. One measure proposed is to collect parking lot taxes from hotels.

> Because the City considers hotel guests "temporary residents," it exempts them from parking lot taxes. By applying these taxes to hotel guests, the City would encourage visitors not to drive while in town, and generate additional revenue that could be used to fund transportation alternatives, such as increased Muni service or a free tourist shuttle.
>
> (San Francisco Department of the Environment 2004: 3.13)

In San Francisco, hotels were also recognized as a sector that required incentives and partnerships to encourage energy efficiency, lighting efficiency and solid waste recycling. In the Amsterdam action plan, hotels were encouraged to undertake Green Key certification in order to improve energy efficiency and reduce emissions (Amsterdam Climate Office 2008).

Other international initiatives include the World Mayors and Local Governments Climate Protection Agreement, which was launched at the UNFCCC in Bali in December 2007 by a coalition of local government organizations, including ICLEI (International Council for Local Environmental Initiatives) – Local Governments for Sustainability, United Cities and Local Governments (UCLG), C40 and the World Mayors Council on Climate Change (WMCCC). The Agreement calls for a number of actions, including the reduction of GHG emissions by 60 per cent from 1990 levels worldwide and by 80 per cent from 1990 levels in industrialized countries by 2050. As of the end of 2009, 112 local governments had signed the agreement (ICLEI 2009).

ICLEI is probably the international local government network with the longest connection to climate change mitigation following the initiation of the Cities for Climate Protection (CCP) campaign in 1993. Local governments join the CCP campaign by passing a resolution pledging to reduce GHG emissions in both their local government operations and their communities. As of 2009, over 1000 local governments worldwide were involved in climate change mitigation via the implementation of ICLEI's 'five milestone' process, which provides a simple, standardized means of calculating GHG emissions, establishing targets to lower emissions, reducing emissions, and monitoring, measuring and reporting performance. Within this frame, following a political commitment statement of the representative of their local governments, participating cities are expected to:

- measure their emissions of GHGs, generated through the actions of their local government administration (government emissions) and through the actions of the community they serve (community emissions);
- commit to an emissions (government or community) reduction target with respect to a base year and a target year;
- plan their actions (e.g. energy efficiency in buildings and transport, introduction of renewable energy, sustainable waste management) at the government and community level to reach this committed reduction target;
- implement their local climate action plan;
- monitor emissions reductions achieved by their mitigation actions (ICLEI 2008).

Despite a lack of policy leadership on climate change mitigation at the national level in the United States during the Presidency of George W. Bush, there was significant local-level activity. The US Conference of Mayors (USCM) has been particularly active in this regard, and launched a Climate Protection Agreement initiative in 2005 to advance the goals of the Kyoto Protocol through leadership and action at the local government level. Under the Agreement, participating cities commit to take following three actions:

- strive to meet or beat the Kyoto Protocol targets in their own communities, through actions ranging from anti-sprawl land-use policies to urban forest restoration projects to public information campaigns;
- urge their state governments, and the federal government, to enact policies and programs to meet or beat the GHG emission reduction target suggested for the United States in the Kyoto Protocol – 7 per cent reduction from 1990 levels by 2012; and
- urge the U.S. Congress to pass the bipartisan GHG reduction legislation, which would establish a national emission trading system.

(Mayors Climate Protection Center 2008)

By the end of 2009, over 1000 mayors had signed the agreement. The initiative has been significant politically and has been an important mechanism to try and gain further federal funds for energy efficiency and climate change initiatives. It has also provided a policy framework for the sharing of best practice and climate change information. Many of the elements of the UCSM's climate protection and 'cool mayors' initiatives have since been taken up by ICLEI USA. The USCM's measures were paralleled by a 'cool counties' initiative launched by 12 counties in conjunction with the Sierra Club at the 2007 National Association of Counties Annual Conference. Participating counties commit to the following.

1. Reduce county operational GHG emissions by creating an inventory of your local emissions and then planning and implementing policies and programs to achieve significant, measurable and sustainable reductions.
2. Work closely with regional and state governments and others to reduce regional GHG emissions to 80 per cent below current levels by 2050. One idea is to engage the nation's metropolitan planning organizations to develop regional GHG emissions inventories and create regional implementation plans that establish short-, mid-, and long-term emissions reduction targets. The goal is to stop the increase in emissions by 2010, and to achieve average reductions of 10 per cent every five years thereafter through to 2050.
3. Urge Congress and the Administration to enact a multi-sector national program of requirements, market-based limits, and incentives for reducing GHG emissions to 80 per cent below current levels by 2050.
4. Identify impacts of climate change on your region, and implement plans to prepare for and build resilience to those impacts.

(King County 2011)

Global tourism industry organizations

The WTTC produced a report in 2009 that sets out a vision for tackling GHG emissions. It includes a commitment, endorsed by more than 40 of the world's largest travel and tourism companies, to cut 2005 carbon emission levels by half by 2035. There is also an interim target of achieving a 30 per cent reduction by 2020 in the presence of an international agreement, or a 25 per cent reduction in the absence of such an agreement

(WTTC 2009, 2010). No specific strategies are mentioned, even though accountability, local community sustainable growth and capacity-building, educating customers and stakeholders, greening supply chains, and innovation, capital investment and infrastructure are mentioned as building blocks. Notably, the WTTC (2010: 7) suggests producing 'an internationally agreed framework of standards to measure progress against GHG emission targets', that is, to monitor emissions development in tourism. Several other strategies suggested by WTTC (2010) are innovative and deserve mention, as their implementation would affect how businesses and governments will be working with the implementation of climate policy.

For businesses:

- taking the lead in setting carbon reduction targets and timelines, monitoring sectoral achievements at home and in destinations;
- collaborating at regional, national and international levels to overcome the difficulties of an industry fragmented by complex supply chains and characterized by SMEs and micro-enterprises – such collaborative efforts will enable clear messages to get through to governments and regulatory authorities;
- investing in energy control systems and staff training to implement monitoring and reporting of CO_2 emissions and energy efficiency.

For governments:

- mainstreaming the tourism sector into national climate change policies and plans to enable the growth of a low-carbon future;
- integrating mitigation and adaptation measures into national climate resilience plans, tourism planning and destination management;
- incentivizing mitigation and adaptation actions, such as retrofits and eco-builds, pilot schemes for testing/embedding new technologies, and research and development through tax schemes and grants;
- establishing clear goals and a framework for implementation for national tourism sector emissions reductions, noting that policies stand most chance of success when worked out in partnership with all key stakeholders;
- collaborating with industry to set measurable targets and appropriate timelines for CO_2 reductions by sector and size of business in accordance with national conditions and broader international obligations;
- offering fiscal incentives (e.g. tax relief, grants, matching funding, benefits in kind) that promote energy-efficiency improvements;
- facilitating and promoting knowledge transfer between research centres and public and private sectors;
- creating a clear framework for sectoral monitoring that is transparent, nationally appropriate, and that quantifies lower emissions;
- developing taxation processes that incentivize and reward good corporate behaviour and the achievement of agreed targets;
- converging emissions reduction actions with reduced fossil-fuel reliance, as well as creative sector responses to new energy sources and the emerging new global energy paradigm.

In June 2008, CEOs from across the world, representing every industrial sector, submitted a set of recommendations to the G8 leaders for inclusion within a post-Kyoto climate framework. In March 2009, in response to the recommendations and to elaborate on how they should be implemented, a World Economic Forum Task Force on Low Carbon Prosperity was launched. In May 2009, the WEF, in collaboration with UNWTO, ICAO, UNEP and travel and tourism business leaders, produced the report *Towards a Low Carbon Travel & Tourism Sector* (WEF 2009a). The WEF suggests as mechanisms to achieve emission reductions:

- a carbon tax on non-renewable fuels;
- economic incentives for low-carbon technologies;
- a cap-and-trade system for developing and developed countries;
- the further development of carbon trading markets.

Support of climate policy and measures to achieve emission reductions, on the basis of identical 'rules for all', appears to be substantial among business leaders. PricewaterhouseCoopers (2010) found, in 700 interviews with company executives in 15 countries, that government leadership is perceived as indispensable in mitigation; that the business community is ready for, and supportive of, government action; that carbon taxes, emission trading and incentives have widespread support in the business community; that existing environmental taxes, regulations and incentives are seen as ineffective, inconsistent and unclear; and that businesses want clear, long-term investment signals.

International organizations: OECD, UNEP and UNWTO

The Organisation for Economic Co-operation and Development has frequently outlined the need to reduce emissions substantially (e.g. OECD 2009). While there are no tourism-specific documents in this regard, OECD (2009) outlines that a policy mix will be needed to reduce emissions economy-wide, including market-based approaches, regulations and standards, research and development, as well as information-based instruments to facilitate consumer choices. Together with the International Energy Agency (IEA), OECD published a review of transport energy use and emissions, concluding that key climate policies should include land-use planning to increase population density and mixed-use development; promotion of teleworking and other information-based substitutes for travel; parking supply (limiting parking space, car-free zones) and pricing (cash-out schemes); car-sharing; road pricing; improved bus transit systems; and encouragement of cycling and walking (IEA and OECD 2009).

The United Nations Environment Programme promotes tourism sustainability through its Tourism and Environment programme, with the aim to facilitate local efforts by tourism stakeholders in integrating climate change into their broader institutional, industry, sectoral, policy and national goals and programs – 'mainstreaming' climate change. A particular focus has been on capacity-building, although UNEP has also contributed to the development of a number of key documents, for instance with regard to biofuel development. Specific strategies to reduce emissions in tourism have, for instance, been outlined in UNEP *et al.* (2008), while issues relating to biofuels have been discussed in UNEP (2009a).

The United Nations World Tourism Organization's Davos declaration, adopted by UNWTO at the global Conference on Climate Change and Tourism in Davos, Switzerland in October 2007, specifies that the tourism sector must respond rapidly to climate change within the evolving UN framework, and progressively reduce its GHG contribution. UNWTO outlines that this will require action by the sector *inter alia* to 'mitigate its GHG emissions, derived especially from transport and accommodation activities'. The Davos declaration calls for a range of actors, including governments and international organizations, to 'collaborate in international strategies in the transport, accommodation and related tourism activities'. Through the 'Davos process', UNWTO is pursuing programmes on both adaptation and mitigation. Mitigation of GHGs from air transport, which is acknowledged as the primary and growing contributor to global emissions of GHGs from tourism, is seen as critical to sustainable tourism development.

With regard to aviation, the most important emissions subsector in tourism, UNWTO asserts that the following principles should be incorporated into ongoing work on mitigation of GHG emissions from air passenger transport, as UNWTO's contribution to the UNFCCC COP-16 meeting in Cancun, Mexico:

- Assessment of mitigation measures in the context of broad-spectrum tourism, including domestic, inbound and outbound flows, rather than for air transport in isolation, considering social and economic costs and benefits in cohesion with the climate change mitigation impact.
- Application of the UNFCCC principle of Common But Differentiated Responsibilities (CBDR) amongst countries.
- Classification of differentiation to alleviate negative impacts on tourism destination markets in developing and particularly least developed and island countries, through differentiated targets, financial transfer mechanisms, and/or reductions in emissions levies or requirements for emissions permits, preferably applied in a framework of traffic flow origin and destination rather than solely according to country.
- Effective performance monitoring, unambiguous and appropriate indicators and targets, transparent and public reporting and auditing processes, at national and global levels.
- Treatment of air passenger transport operations analogously with alternative passenger transport modes where available (for example at short-haul) taking into account such factors as respective taxation and subsidy regimes (including government contributions to infrastructure) and enabling such travel choice criteria as price, comfort, convenience and trip duration to be assessed along with GHG emissions and on a non-discriminatory basis amongst modes.
- Open access for air transport to carbon markets, whether national, regional or global.
- Non-duplication of emissions levies on transport and other tourism-related activities (for example as a result of application by more than one authority or through different regimes such as taxation and emissions trading).
- Earmarking of all revenues from levies and trading of emissions permits to GHG mitigation activities yielding measurable, reportable and verifiable mitigation results, including projects in transport and other tourism-related activities, and financial and other incentives for the earliest possible global introduction of sustainable additional or alternative fuels for air transport.
- Acknowledgement of the pivotal role of the private sector and of the efforts and collective commitments by airports, air navigation service providers, air carriers and manufacturers for

increased fuel efficiency, setting aspirational targets and working towards carbon neutral growth and subsequently substantial absolute reductions in emissions.

- Continued recognition of a key role for ICAO in the fields of airframe and engine technology, air traffic management and operational approaches leading to tighter standards on aircraft emissions and improved operating procedures, and promotion of early certification and acceleration into usage of sustainable additional or alternative fuels for air transport.
- Addressing of reduction targets and economic instruments for aviation emissions in co-operation with all parties representing directly affected sectors, including tourism in particular, and development of any global GHG emissions mitigation framework or globally accepted approach specific to aviation in partnership by all related intergovernmental parties including UNWTO and in close consultation with relevant NGOs and with input from both the public and the private sectors.

(UNWTO 2010c: 2–3)

This review of policy suggestions from the most important tourism-related organizations shows that, while there is agreement on the need to reduce emissions in tourism, neither measurable nor binding targets have been identified. On the contrary, rather than formulating specific policy suggestions, UNWTO in particular defines a range of preconditions that will be difficult to resolve and that can be assumed to delay action for years to come.

The Commission of the European Communities' *Agenda for a Sustainable and Competitive European Tourism* (CEC 2007) is a result of the Commission's adoption of a renewed tourism policy in March 2006, with the main objective to 'improve the competitiveness of the European tourism industry and creating more and better jobs through the sustainable growth of tourism in Europe and globally'. To this end, the Agenda's objectives are to 'deliver economic prosperity, social equity and cohesion and environmental and cultural protection'. Among the challenges to achieve these objectives, 'addressing the environmental impact of transport linked to tourism' is mentioned, along with the statement that 'Policies and actions need to take into account how demand and supply will be affected by environmental challenges – such as climate change and water scarcity.' In its 2010 document, the Commission again emphasizes as two of its four priority actions the need to 'promote the development of sustainable, responsible and high-quality tourism' as well as to 'consolidate the profile of Europe as a collection of sustainable and high-quality destinations'. With regard to mitigation, only a single strategy, 'the use of "clean" energy', is mentioned (CEC 2010). The Commission also states, more generally, that 'the sector has to become more resilient to the impact of climate change' (CEC 2010: 6).

In summary, there appears to be consensus among tourism stakeholders that average global warming should not exceed 2°C by 2100 – as outlined in the Copenhagen Accord and reconfirmed in Mexico in December 2010 – and that the tourism system should decarbonize in line with other economic sectors. This would require the global tourism system to reduce absolute emission levels by about 20 per cent by 2020, compared with 1990. However, plans or ongoing efforts to identify mitigation options for tourism appear to exist in only a very limited number of countries. Moreover, most strategies as currently presented are generic in character ('energy efficiency achievements', 'use of biofuels') and not connected to, or based on, quantified or quantifiable and binding emission reduction goals and the monitoring of progress. On the contrary, tourism is identified as a growth area for emissions by many governments, and policy in many countries is even contradictory to climate policy goals, as exemplified by plans in many European countries to develop long-haul markets further. In contrast, current climate policy does not have a significant effect on tourism (Mayor and Tol 2007; Gössling *et al.* 2008; Pentelow and Scott 2010).

More stringent policy frameworks are thus needed, but appear difficult to discuss and implement. Conditional policy formulation as currently suggested by UNWTO, for instance, needs to be seen as contra-productive in this regard, as it will lead to lasting negotiations. Non-action by ICAO on emissions from aviation further hinders progress in the most important emissions growth sector. With the exception of a few notable initiatives in a few countries, including in particular APDs in the UK and Germany, there are no discernible efforts to curb emissions from tourism in a systematic, coordinated and controlled way.

IMPLEMENTING 'SERIOUS' CLIMATE POLICY IN TOURISM

One of the most fundamental obstacles to restructuring the global tourism system appears to be the lack of knowledge, awareness and interest among key stakeholders regarding energy production and use, as well as climate change and its root causes (Gössling 2010). Any policy advance consequently needs to foster the understanding of energy use as relevant for the future of tourism. One way of generating this knowledge, awareness and interest may be to introduce mandatory carbon auditing. Several global assessments now exist of the contribution made by tourism to climate change (UNWTO *et al.* 2008; WEF 2009a), but few countries have audited their tourism-related energy use and emissions, and energy/carbon audits are far from being a standard in businesses. National tourism emission inventories, as well as policy demanding that companies as well as consumers assess their energy use and related emissions, would both create a better knowledge base for decision-making and raise awareness among stakeholders. Moreover, only on the basis of such databases can progress be made to reduce emissions, involving regular monitoring.

As a starting point for discussions, it may be argued that energy use and emissions have to reflect their environmental costs. Through mandatory carbon audits, businesses will understand their contribution to global climate change and their role in affecting the lives of millions of people, as well as the fact that climate change is affecting the very basis of tourism (e.g. UNWTO *et al.* 2008) and is also undermining some of the rationale for using tourism as a means of economic development in peripheral regions and the less-developed countries. This will help to implement the most important measure in curbing emissions – the taxation of energy use. From a policy perspective, emissions of GHGs represent a market failure. The absence of a price on emissions encourages pollution, prevents innovation, and creates a market situation where there is little incentive to use innovations (OECD 2010b). While governments have a wide range of environmental policy tools at their disposal to address this problem, the fairest and most efficient way of reducing emissions is to increase fuel prices, that is, to introduce a tax on fuel or emissions that is proportional to fuel use and emissions to ensure cost-efficient abatement (Mayor and

Tol 2007, 2008, 2009, 2010a, 2010b; Sterner 2007; Tol 2007; OECD 2009, 2010b; WEF 2009a; PricewaterhouseCoopers 2010). In this context, OECD (2010b: 2) outlines:

> Compared to other environmental instruments, such as regulations concerning emission intensities or technology prescriptions, environmentally related taxation encourages both the lowest cost abatement across polluters and provides incentives for abatement at each unit of pollution. These taxes can also be a highly transparent policy approach, allowing citizens to clearly see if individual sectors or pollution sources are being favoured over others.

It is important for economic instruments significantly to increase the costs of fossil fuels and emissions. Price levels also need to be progressive (increasing at a significant rate per year) and foreseeable (implemented over longer time periods) to allow companies to integrate energy costs in long-term planning and decision-making. Ultimately, taxes represent additional costs to companies that will be passed on to the consumer. Consequently, the consumption of energy-intense holidays can be expected to become more expensive in the medium-term future, particularly if consensus can be reached on worldwide, though nation-specific and binding, GHG-reduction goals and the implementation of a global framework for emissions trading, with a global cap on emissions and specified emission reductions distributed by country.

In line with this, subsidies such as the non-taxation of kerosene or the financial support of fossil fuel consumption also need to be addressed. The IEA (2010) even sees 'eliminating fossil fuel subsidies as the single most effective measure to cut energy demand in countries where they persist'. Energy taxes need to be proportional to emissions, and ideally to be harmonized between countries to avoid 'fuel tourism'. Notably, there appears to be broad support for such policy intervention (WEF 2009a; OECD 2010a; PricewaterhouseCoopers 2010), although actual implementation can be difficult because of the political risks associated with 'increasing taxation'.

While carbon pricing is the most efficient tool to stimulate behavioural change and changes in production, market failures justify additional policy intervention. Energy-intense forms of tourism and transport as well as behavioural change that are difficult to steer through rising energy costs also need to be addressed through other measures, such as speed limits, or bans of jet skis, quads, or other motorized transport at the destination level. Moreover, regulation could include building codes and other minimum standards to reduce emissions. Actual enforcement of existing environmental regulation also needs to be ensured, and may even be the most crucial measure in achieving behavioural change.

Carbon taxes may be feasible for accommodation, car transport and other situations where tourism activities cause environmental problems, but they are more difficult to implement in aviation because of a range of bilateral agreements under the Chicago Convention. For aviation, national fees increasing mobility costs proportional to energy use, for instance in the form of a structured, fuel use-related departure tax, may thus be the most efficient in the current situation, and could be implemented by governments in coordination with concerned ministries. Overall, there is a considerable role for user fees, levies, congestion charges, low-emissions zones and various taxes to steer tourism towards greater sustainability (Font *et al.* 2004; for details on the use of taxes see OECD 2010b).

As taxes are usually less acceptable to businesses, market-based instruments can be combined with incentive structures to support the introduction of low-carbon technologies and to increase the speed of innovation (WTTC 2010). This could include tax breaks, credits or grants to businesses, the redistribution of tax burdens through internalization of environmental costs, or their redistribution between high-level and low-level polluters, as exemplified by bonus–malus systems. Such systems may be designed in particular to involve the millions of small and medium-sized enterprises in the tourism production system.

Finally, biofuels may be one of the few options to make a contribution to emission reductions in the aviation sector, and their rapid introduction would be meaningful once regulation, demand management and efficiency measures have contributed to a reduction in overall energy demand for aviation. However, the production of biofuels faces considerable technical, social, economic and environmental difficulties, including its potential competition with food production and its impact on biodiversity and water use (Fargione *et al.* 2008; Vidal 2010). In 2008, the International Monetary Fund estimated that 20–30 per cent of food price increases in the previous two years were accounted for by biofuels, and that in 2007 they accounted for about half the increase in demand for principal food crops (Borger 2008). It is also unclear what contribution biofuels can make to the growing energy demand in aviation. For example, the UK Nuffield Council on Bioethics (2011) (NCB) concluded that the EU Renewable Energy Directive target that biofuels should account for 10 per cent of transport fuel by 2020, a much criticized mandate originally designed as part of Europe's strategy to combat climate change, was unethical. However, in July 2012, the European Commission approved seven schemes established to ensure that biofuels used in the EU are produced in an environmentally sustainable way. Companies importing or producing biofuels will be required to prove that they meet strict EU sustainability criteria and emit at least 35 per cent less GHGs than fossil fuels such as petrol (BBC 2011c). Similarly, the UK Renewable Transport Fuel Obligation (Amendment) Order (2009) requires that 5 per cent of total transport fuel should originate from renewable sources by 2013. The Council set out six ethical principles that policy-makers should use to evaluate biofuel technologies and guide policy development, as follows:

1. Biofuels development should not be at the expense of people's essential rights (including access to sufficient food and water, health rights, work rights and land entitlements).
2. Biofuels should be environmentally sustainable.
3. Biofuels should contribute to a net reduction of total GHG emissions and not exacerbate global climate change.
4. Biofuels should develop in accordance with trade principles that are fair and recognize the rights of people to just reward (including labour rights and intellectual property rights).
5. Costs and benefits of biofuels should be distributed in an equitable way.
6. If the first five principles are respected and if biofuels can play a crucial role in mitigating dangerous climate change then, depending on certain key considerations (absolute cost; the availability of alternative energy technologies; alternative uses for biofuels feedstocks; the existing degree of uncertainty in their development; their irreversibility; the degree of participation in decision-making; and the overarching notion of proportionate governance), there is a duty to develop such biofuels.

Nuffield Council on Bioethics (2011)

The NCB recommended that these principles be backed by a mandatory and strictly enforced EU certification scheme, which reinforces fair trade principles:

> Given the limitations of current Fairtrade schemes, we propose that trade principles that are fair be developed as part of sustainable biofuels certification requirements by EU and national stakeholders ... This needs to happen in a proportionate and flexible way to acknowledge the differences between countries and production systems, while at the same time strictly maintaining the protection of vulnerable populations.
>
> (Nuffield Council on Bioethics 2011: 100)

The significance of such a stance is illustrated by the fact that, according to the UK Renewable Fuels Agency in 2008, only 19 per cent of the biofuel supplied under the British government's initiative to use fuel from plants to help combat climate change met its green standard. By 2011 the figure had reached 31 per cent (cited in Rowley 2011), although for the remaining 69 per cent of the biofuel, 'suppliers could not say where it came from, or could not prove it was produced in a sustainable way'.

The NCB report's chair Professor Joyce Tait told *BBC News*: 'We clearly need a new overarching ethical standard backed up by certification to improve the way the world produces biofuels' (Harrabin 2011). However, in contrast Robert Palgrave from the Biofuelwatch campaign was extremely critical of the Council's conviction that certification would guarantee productive agricultural land would not be taken over by biofuels:

> There is no scientific credible way of calculating the full climate impacts of agrofuels. Indirect impacts are not just about 'hectare for hectare' displacement; they are also about the interaction between land prices and speculation, about the impacts of roads, ports and other infrastructure on forests, about policy changes which affect land rights, about scarcely-understood interactions between biodiversity, ecosystems and the climate.
>
> (quoted by Harrabin 2011)

Undoubtedly, policy-makers need to address this situation, and to implement frameworks that support and ensure the development of sustainable biofuels. In this regard, important guidelines and tools have already been developed by UNEP, including the Global Bioenergy Partnership, the Roundtable on Sustainable Biofuels and the UN Energy Decision Support Tool for Sustainable Bioenergy.

Carbon management

Many businesses appear to see emission reductions as a necessary response to regulation and policy. However, there are many economic reasons for engaging in carbon management, or what is here understood as the strategic reduction of energy use and emissions based on economic considerations. Rising energy costs are an increasingly important factor in operational costs, accounting, for instance, for 30 per cent of airlines' operational costs in 2008 (up from 15.8 per cent in 2002) (Gössling and Upham 2009). About 15 per cent of a typical Caribbean hotel's operating cost is attributable to energy usage, and management-related reductions in energy use of 20 per cent would correspond to savings of 3 per cent on the overall economic baseline (Pentelow and Scott 2011). This should represent a significant incentive to engage in energy management.

Although the high energy intensity of tourism and the sector's energy dependence has been outlined since the mid-1990s, the issue of 'peak oil' – the point at which the maximum capacity to produce oil is reached – and its consequences for tourism did not emerge until a decade later (Harrison and Winter 2005; Becken 2008). It is currently unclear whether peak oil has already been reached, or will be reached in the near future. Table 4.10 provides a range of projections for the peaking of world oil production. The UK Energy Research Centre concludes in a review of studies (UKERC 2009) that a global peak in oil production is likely before 2030, with a significant risk of a peak before 2020. Like climate change modelling, different models of the forecast of the demand and supply of oil use different parameters.

Nevertheless, as in the case of climate change, although forecasts of when peak demand will occur do differ with respect to the exact date, there is nevertheless broad agreement among stakeholders that it will happen, with the exception of some international oil companies and energy organizations (UKERC 2009). Table 4.11 provides an overview of forecasts by different stakeholders as to when global peaks will occur. The decline in conventional oil production will lead to the development of costlier options, although IEA (2009) emphasizes the need for investments, and oil prices can consequently be expected to rise, particularly with a view on demand growth. At the time of writing in April 2011, oil prices already exceeded US$120 per barrel, again moving towards the maximum of US$145 per barrel experienced in mid-2008, and notably after an intermediate temporary low of US$40 per barrel during the financial crisis in 2008/09.

Significantly higher oil costs could lead to actual changes in travel choices. O'Mahoney *et al.* (2006) reported in an Australian study that fuel prices (and the increase in associated cost of holidays) were negotiated through changes to holiday planning, including changes to duration, location and travel mode, although the actual effect of fuel price increases was greater than that perceived by tourists. Becken and Lennox (2012) argue in a study of New Zealand tourism that rising oil prices have negative income effects, which affect mobility. If oil prices doubled, these could indeed have a significant impact on New Zealand tourism, although price elasticities for transport are also know to vary significantly between markets (Schiff and Becken 2011). Even though it may thus take some time until the cost of oil will lead to significant changes in travel flows on a global scale, the decade from 2002–12 has shown that enormous fluctuations in oil prices are possible, depending on world market demand for oil, hedging strategies and actual oil production.

Another New Zealand study, by Wardell (2010), highlighted the gap between policy-makers and elected officials when it comes to transport policy-making with respect to peak oil. Wardell (2010) conducted a series of case studies involving interviews and surveys with transport policy-makers in three cities of varying size in New Zealand. The results of the case studies established that many technical staff have major concerns about peak oil, but their concerns are not translated into policy because the majority of elected officials, who give the final approval on policy, believe that alternative fuels and new technologies will mitigate any peak oil impacts. The inaction of elected officials is also reinforced by a perception of lack of scientific consensus on peak oil, and a lack of political and financial

Table 4.10 Projections of peak world oil production

Projected date	Source of projection	Background and references
2005	Pickens, T. Boone	Oil and gas investor; Boone Pickens warns of petroleum production peak, *EV World*, 4 May 2005
December 2005	Deffeyes, K.	Retired Princeton professor and retired Shell geologist; www.defense-and-society.org/fcs/crisis_unfolding.htm, 11 February 2006
Close or past (2006)	Herrera, R.	Retired BP geologist; Bailey, A., 'Has oil production peaked?', *Petroleum News*, 4 May 2006
2006–07	Bakhitari, A.M.S.	Iranian Oil Executive; 'World oil production capacity model suggests output peak by 2006–07', *Oil & Gas Journal*, 26 April 2004
2007–09	Simmons, M.R.	Investment banker; Simmons, M.R., ASPO Workshop, 26 May 2003; CFA Society of St Louis, Brentwood, MO, slide 23, 'World should assume we are at peak for oil and gas', 24 May 2006
After 2007	Skrebowski, C.	Petroleum journal editor; 'Oil field mega projects – 2004', *Petroleum Review*, January 2004
At hand	Westervelt, E.T. et al.	US Army Corps of Engineers; *Energy Trends and Implications for US Army Installations*, ERDC/CERL TN-05-1, September 2005
Before 2009	Deffeyes, K.S.	Oil company geologist (ret.); *Hubbert's Peak – The Impending World Oil Shortage*, Princeton University Press, 2003
Very soon	Groppe, H.	Oil/gas expert and businessman; 'Peak oil: myth vs. reality', Denver World Oil Conference, 10 November 2005
Before 2010	Goodstein, D.	Vice Provost, Cal Tech; *Out of Gas – The End of the Age of Oil*, W.W. Norton, 2004
Around 2010	Bentley, R.	University energy analyst; 'The case for peak oil', DOE/EPA Modelling the Oil Transition, 21 April 2006
Around 2010	Campbell, C.J.	Oil company geologist (ret.); 'Industry urged to watch for regular oil production peaks, depletion signals', *Oil & Gas Journal*, 14 July 2003; 'An updated depletion model', Association for the Study of Peak Oil and Gas (ASPO) Newsletter, no. 64, April 2006

Continued

Table 4.10 Continued

Projected date	Source of projection	Background and references
2010 ± 1 year	Skrebowski, C.	Editor of *Petroleum Review*; 'Peak oil – the emerging reality', ASPO-5 Conference, Pisa, Italy, 18 July 2006
After 2010	World Energy Council	NGO; *Drivers of the Energy Scene*, World Energy Council, 2003
A challenge around 2011	Meling, L.M.	Statoil oil company geologist; Statoil ASA, 'Oil supply, is the peak near?', Centre for Global Energy Studies, 29–30 September 2005
Around 2012	Pang, X. et al.	China University of Petroleum; Pang, X., Meng, Q., Zhang, J. and Natori, M., 'The challenges brought by the shortage of oil and gas in China and their countermeasures', ASPO IV International Workshop on Oil and Gas Depletion, 19–20 May 2005
Around 2012	Koppelaar, R.H.E.M.	Dutch oil analyst; *World Oil Production & Peaking Outlook*, Stichting Peakoil-Nederland, 2005
2010–20	Laherrere, J.	Oil company geologist; Seminar, Center of Energy Conversion, Zurich, 7 May 2003
Within a decade	Volvo Trucks	Swedish company; Volvo website, www.volvotrucks.com
Within a decade	De Margerie, C.	Oil company executive; ASPO Newsletter, no. 65, May 2006
Around 2015	al Husseini, S.	Retired Exec. VP of Saudi Aramco; 'End of an era', *Cosmos*, April 2006
2015–20	ECRSAS	*Statements on Oil*, Energy Committee at the Royal Swedish Academy of Sciences, Stockholm, 14 October 2005
2015–20	West, J.R., PFG Energy	Consultants; Energy insecurity, testimony before the Senate Committee on Commerce, Science and Transportation, 21 September 2005
2016	Energy Information Administration (EIA) nominal case	US Department of Energy (DOE) analysis/information; DOE EIA, *Long Term World Oil Supply*, 18 April 2000
Around 2020 or earlier	Maxwell, C.T., Weeden & Co.	Brokerage/financial; 'The gathering storm', *Barron's*, 14 November 2004
Within 15 years	Amarach Consulting	Ireland; *A Baseline Study on Oil Dependence in Ireland*, Amarach Consulting, Ireland, December 2005

Estimate	Source	Reference
Tight balance by 2020	Wood Mackenzie	Energy consulting; *MacroEnergy Long Term Outlook – March 2006*, Wood Mackenzie
Around 2020	Total	French oil company; Bergin, T., 'Total sees 2020 oil output peak, urges less demand', Reuters, 7 June 2006
After 2020	CERA	Energy consultants; Jackson, P et al., 'Triple witching hour for oil arrives early in 2004 – but, as yet, no real witches', *CERA Alert*, 7 April 2004
2025 or later	Shell	Major oil company; Davis, G., 'Meeting future energy needs', *The Bridge*, National Academies Press, Summer 2003
Mid- to late 2020s	UBS	Brokerage/financial; 'Oil output set to peak, but no fuel shortage – UBS', Reuters, 24 August 2006
After 2030	EIA	US DOE analysis; Morehouse, D., private communication, 1 June 2006 (in Hirsch 2007)
After 2030	CERA	Energy consultants; 'Barring unforeseen events there is no reason to believe capacity couldn't meet demand well after 2030, CERA researchers said'; Strahan, A., 'Global petroleum capacity to rise 25 percent by 2015', *Bloomberg*, 8 August 2006
Now–2040	Congressional General Accountability Office	*Uncertainty about Future Oil Supply Makes It Important to Develop a Strategy for Addressing a Peak and Decline in Oil Production*, GAO-07-283, February 2007
No sign of peaking	Exxon Mobil	Major oil company; www.exxonmobil.com/Corporate/Files/Corporate/OpEd_peakoil.pdf (as of March 2007)
No visible peak	Lynch, M.C.	Energy economist; 'Petroleum resources pessimism debunked in Hubbert model and Hubbert modelers' assessment', *Oil and Gas Journal*, 14 July 2003
Impossible to predict	BP CEO	Major oil company CEO; 'As profit rises, BP chief seeks to allay anger', *Washington Post*, 27 July 2006
Does not subscribe to peak oil theory	OPEC	Neil Chatterjee, 'OPEC needs clear demand signals for spare capacity', *Mail & Guardian Online* (South Africa), 11 July 2006

Source: derived from ECRSAS (2005); Hirsch *et al.* (2005); Hirsch (2007); Hall and Lew (2009)

Table 4.11 *Forecasts of date of global peak of oil production by different stakeholders*

Date of global peak	Forecast	Category of stakeholder
2006	LBST	Consultancy
2008	Campbell	Individual (Association for the Study of Peak Oil)
2008–18	University of Uppsala	University
2011–13	Peak Oil Consulting	Consultancy
2013–17 (2019 given unlimited investment)	Miller	Individual
2017	Energyfiles	Consultancy
2020	Total	Oil company
Around 2020	BGR (2006)	National organization
2028 (base case)	Meling (StatoilHydro)	Oil company
Around 2020	Shell ('Scramble' scenario where all oil production declines after 2020)	Oil company
No peak. Conventional oil plateau by 2030	IEA	International organization
Around 2030	Shell ('Blueprints' scenario where all oil production levels off after 2020)	
No peak	OPEC	International organization
No peak	US EIA	National organization
No peak	Exxon Mobil	Oil company
No peak for all liquids	Shell	Oil company

Source: derived from UKERC (2009)

support from central government to plan for peak oil. Wardell (2010) concludes that a change in attitude towards peak oil by central government is a prerequisite for planning decisions for peak oil at local level in New Zealand. However, also of significance is the way inaction on peak oil is mirrored by the similar slow rate of action with respect to climate change adaptation and mitigation.

The impacts of higher costs for oil can be exacerbated when these coincide with more serious climate policy, as outlined in the previous sections. Climate policy is emerging only slowly globally, but is likely to become increasingly relevant nationally. While evidence suggests that current climate policy does not affect tourism, and that even future climate policy as currently discussed will not have significant effects on tourism (Gössling *et al.* 2008; Pentelow and Scott 2010), there are signs that some countries will move towards 'serious' climate policy. For instance, in the UK, government policy already has a considerable effect on the cost of mobility, and it can be assumed that this will be the case even in other countries. High energy intensity will thus represent a key vulnerability issue in the future. A further argument for engaging in carbon management is that tourism is highly wasteful of energy, and there is huge potential to increase the sector's energy efficiency. Finally, consumers appear to be increasingly interested in buying products and services that are environmentally benign, while innovative pro-environmental management is usually recognized in the media and rewarded with awards – both leading to greater public recognition and representing free marketing. Engaging proactively in energy management and emission reductions should thus become an increasingly powerful argument in business operations.

In conclusion, destinations and tour operators, along with providers in the transport chain, such as airlines, as well as place-bound actors such as accommodation establishments, should seek to reduce their dependency on fossil energy. The following sections discuss systemic and structural approaches to reducing energy dependencies and vulnerabilities.

Destinations

Destinations are key units in climate change mitigation, as they can mobilize large stake-holder numbers, which together can achieve significant change, and because they can initiate systemic changes to reduce emissions. This can include measures ranging from the restructuring of markets, to decisions to become car-free, restrictions on individual motor-ized transport, encouragement of public transport, establishment of cycle paths, or addi-tional benefits offered to visitors coming by train, bus or bicycle.

Various destinations have started to work proactively with carbon management in this direction. These include large ski resorts such as Whistler Blackcomb, Canada and Aspen Snowmass, USA; tourist regions such as Davos and south-west England; and networks covering several countries such as the Alpine Pearls (for details see Gössling 2010). For any destination, work with carbon management has to begin with an inventory of energy consumption and associated emissions, followed by an identification of strategies to reduce emissions, and the monitoring of progress towards a specific goal.

To help destinations to reconsider energy use, Becken (2008) suggested ten indicators focusing on transport – the most energy-consuming part of a holiday. These include des-tination-based and origin–destination (O/D) transport, although in the case of the ten most important markets for New Zealand, for which calculations are presented, O/D transport is shown to account for 90.1–97.4 per cent of overall transport-related energy use. Not all of Becken's (2008) oil-use indicators are thus equally relevant (Figure 4.3). As this is the case in most long-haul destinations, including many small island developing states (SIDS), a simplifying approach for destinations may be to calculate average emissions per tourist for different markets, based on transport emissions (Gössling et al. 2008).

A simple indicator is the arrival to emission ratio, based on a comparison of the percentage of arrivals from one market with the emissions caused by this market (Table 4.12). For instance, tourists from the USA account for 67 per cent of arrivals in Anguilla, but cause only 55 per cent of overall emissions. The resultant ratio is 0.82 (55 divided by 67).

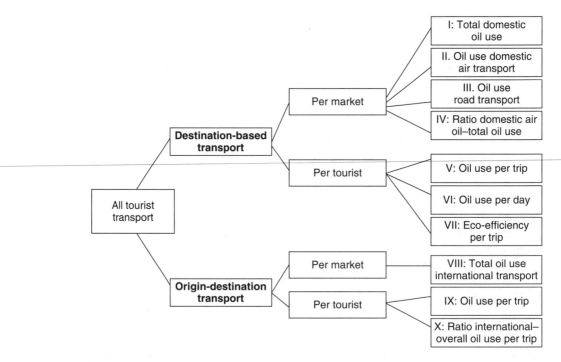

Figure 4.3 *Oil-use indicators*
Source: Becken (2008)

The lower the ratio, the better this market is for the destination in terms of energy intensities, with ratios <1 indicating that the market is causing lower emissions per tourist than the average tourist (and *vice versa*). Arrivals from source markets with a ratio <1 should thus be increased in comparison with the overall composition of the market in order to decrease emissions, while arrivals from markets with a ratio >1 ideally should decline. In the case of Anguilla, the replacement of a tourist with a ratio of >1 in favour of one tourist from the USA (ratio 0.8) would thus, from a GHG emissions point of view, be beneficial. However, as arrivals from the USA already dominate overall arrivals, it may be relevant to discuss whether the destination becomes more vulnerable by increasing its dependence on this market (Gössling *et al.* 2008).

Another approach was recently presented in a report by de Bruijn *et al.* (2010) on the basis of 'emissions per day', showing that holidays by cycle and train, as well as non-organized holidays, have a relatively small carbon footprint, whereas holidays by plane, those spent in hotels, and organized holidays have a relatively high environmental impact. The holiday types with the highest average environmental impact per day are compared with the average footprint of Dutch holidays, 49 kg CO_2 per day, and allow for a relative comparison of their carbon intensity:

- cruises (+265 per cent);
- intercontinental (long-haul) holidays (*c.* +200 per cent);

Table 4.12 *Ratio of arrivals to emissions for selected markets*

Market (emissions ratio)

Anguilla	Bonaire	Comoros	Cuba	Jamaica	Madagascar	St Lucia	Samoa	Seychelles	Sri Lanka
USA (0.8)	USA (0.5)	France (1.4)	Canada (0.4)	USA (0.8)	France (1.2)	USA (0.9)	New Zealand (0.7)	France (1.2)	India (0.3)
UK (2.5)	Netherlands (1.6)	Reunion (0.3)	UK (1.8)	–	Reunion (0.1)	UK (2.0)	American Samoa (0.1)	Italy (1.0)	UK (1.4)
–	–	–	Spain (1.9)	–	Italy (1.0)	Barbados (0.1)	Australia (1.1)	Germany (1.2)	Germany (1.4)
–	–	–	Italy (2.1)	–	–	Canada (1.0)	–	UK (1.2)	USA (2.0)
–	–	–	France (2.0)	–	–	–	–	–	Australia (1.2)
–	–	–	Germany (2.6)	–	–	–	–	–	France (1.4)

Source: Gössling et al. (2008)

- holidays by airplane (+102 per cent);
- holidays in hotels/motels (*c.* +78 per cent);
- organized holidays (+35 per cent);
- outbound holidays (+27 per cent).

In contrast, the holiday types with the lowest environmental impact per day are:

- domestic cycling holidays (–76 per cent);
- outbound holidays by train (–55 per cent);
- all camping holidays with a tent (–50 per cent);
- domestic holidays (–47 per cent);
- all non-organized holidays (–39 per cent);
- all nearby outbound holidays, e.g. in Belgium (–31 per cent).

De Bruijn *et al.* (2010) conclude that these results can provide perspectives for industry stakeholders to develop low-carbon tourism products, and for policy-makers to take a new approach to aviation and outbound tourism development. They also warn that business as usual will interfere with national and EU emission reduction targets.

More comprehensive approaches would seek to combine the focus on emissions with economic indicators, such as the eco-efficiency concept (Gössling *et al.* 2005). On an incoming tourism basis (by nationality), the usefulness of an eco-efficiency approach is illustrated in Figure 4.4 for Amsterdam, where tourist nationalities were found to have substantially varying eco-efficiencies (Gössling *et al.* 2005). Results allow identification of the tourist nationality with the highest spending patterns in relation to emissions and, *vice versa*, those with the highest emissions and the lowest spending pattern. The eco-efficiency approach can also be applied to various tourist types (e.g. day visitors, nationals, overseas tourists), tourism subsectors (hotels, restaurants, retail), or on a product value-chain basis. It thus opens up new opportunities to work strategically with emission reductions, because energy use does not seem directly related to the profitability of tourism products, and there is thus room for optimization, to maintain or increase the profitability of tourism businesses while reducing the environmental impact (Gössling *et al.* 2005; see also Perch-Nielsen *et al.* 2010a).

The eco-efficiency approach can also be scaled down and calculated on a tourism/leisure consumption basis to identify more climatically and economically beneficial consumption. Hille *et al.* (2007) illustrate this for consumption in Norway, showing that there are vast differences in energy use per unit of expenditure. Theatres or restaurants, for instance, entail energy use of just 0.2 MJ per Norwegian crown, while other leisure activities can be more than 12 times as energy intense. In this context, air travel is a major factor in travel costs, but leads to only marginal profits – which will not usually accrue to the destination. Hence increasing average length of stay is a key measure to capture a greater share of travel budgets as profits.

Tourism stakeholders can use indicators such as those presented above to restructure markets and to prepare for low-carbon futures by integrating energy use and emissions in their planning. Such strategies do not necessarily imply declining visitor numbers. On the contrary, recent work done in Barbados (Gössling unpublished) indicates that there is

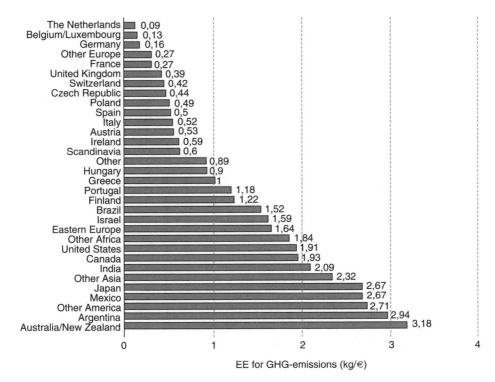

Figure 4.4 *Eco-efficiency by source market for Amsterdam, 2002*
Source: Gössling *et al.* (2005)

considerable scope to increase average length of stay and expenditure within the existing tourism system. This can be combined with further analysis to distinguish revenue–profit ratios, the share of money staying within regional or national economies, as well as the price sensitivity of markets (Schiff and Becken 2011).

Destinations receiving visitors via surface-bound transport can also influence transport mode choices. Werfenweng, a community in the Alps in Austria, for instance, has achieved a considerable modal shift from car to train, based on mobility guarantees and pleasure mobility offers (Gössling 2010). Destinations planning for slow/pleasure mobility with a focus on bicycles, e-mobility and public transport are generally attractive for tourists, not least because average speed levels decline, and noise and air pollution are reduced (Dickinson and Lumsdon 2010). When cities become attractive for walks and bicycle tours, this can also open up the way for new tourism products such as guided tours by bicycle or inline skates, as now offered in many major European cities. Slower travel, associated with a longer average stay, would again lead to greater spending.

Overall, these insights would lead to five key recommendations for destinations:

- flows: develop closer markets;
- travel: encourage low-energy transport;
- length of stay: reward visitors staying longer;

- spending: encourage low-energy spending;
- profits: increase high-profit rather than high-turnover spending within the regional economy.

Ultimately, destinations may even seek to become 'carbon neutral' or 'carbon clean', a goal that has, for instance, been announced by Costa Rica and the Maldives (for a conceptual approach see Gössling 2009). In particular, small destinations seeking to achieve carbon neutrality might profit from such approaches, but these are in reality difficult to achieve. Approaches to carbon neutrality vary widely with regard to the system boundaries chosen, and including or excluding, for instance, emissions associated with the O/D transport of tourists, that is, the major emissions component (cf. Gössling 2010). To date, no destination has convincingly documented carbon neutrality, even based on generous interpretation. This could, however, become a major field of action for destination planners in the near future.

Tour operators and travel agents

Tour operators and travel agents, both those with offices and those that are internet-based, have considerable influence on tourists' choice of destinations, and could become far more important agents in fostering low-carbon holiday choices (e.g. Peeters *et al.* 2009; see also Artal Tur *et al.* 2008). Considerable savings could be realized through the promotion of closer destinations, for instance, with emission savings being proportional to the reduction in flying distance. Convincing customers to focus on closer destinations can be based on social marketing emphasizing shorter flight times or lower transport costs, as well as carbon labelling facilitating more climate-friendly destinations or transport choices. However, travel agents can also advocate longer stays, thus contributing to other objectives in carbon management. Notably, this would also lead to lower costs for accommodation businesses, for instance with regard to cleaning, house-keeping, washing linen, check-in and check-out, or welcome drinks.

As outlined by Peeters *et al.* (2009), global distribution systems (GDS) have become key platforms in facilitating comparisons between competing choices, with three systems – Amadeus, Sabre and Galileo – sharing two-thirds of global air booking. GDS usually compare offers made by airlines or hotels based on standard (first class, economy class); routes (direct, stop-over); travel time (hours from departure to destination); departure time (convenient?); as well as cost – probably the most important parameter in decision-making. A simple measure to facilitate pro-environmental choices is the inclusion in GDS of yet another parameter – emissions of GHG – which could have considerable impact on the choices made by travellers. GDS also have other characteristics that could be changed: for instance, booking complex trips through GDS is difficult, particularly when various transport modes are to be combined. Peeters *et al.* (2009) thus suggest that 'carbon optimization' needs to become a key concept for travel agents – and tour operators – and propose that travel agents:

- favour direct flights wherever possible (for short and medium distances);
- choose airlines with comparatively new fleets and high seat densities;

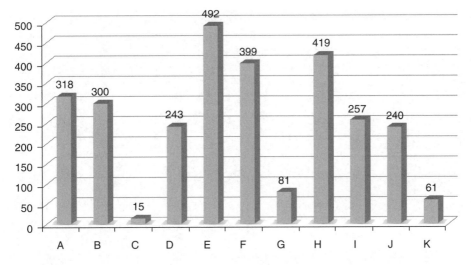

Figure 4.5 *Emissions per customer per day (kg CO$_2$-eq)*
Source: Gössling (2010)

- substitute part of the journey with a trip by train, for instance to reach a hub from which to fly;
- offset remaining emissions.

Tour operators can also use emission-intensity parameters to plan journeys. Figure 4.5 shows, for instance, average carbon intensities for customers of tour operators in Germany, on a per person per day basis. These were calculated from data provided by tour operators within a corporate social responsibility certification developed by a cooperation of German tour operators seeking to offer more sustainable tourism products (Gössling 2010). The range within carbon-intensity figures is considerable, indicating that, while there are tour operators managing to keep emissions as low as 15 kg CO$_2$-eq per day, other tour operators cause emissions approaching half a ton of CO$_2$-eq per day, exhausting annual sustainable carbon budgets within a week. An analysis of package tours on such a basis can be of great help in designing climatically sustainable tourism products.

Airlines

Airlines have considerable opportunities and responsibilities to reduce emissions. For a long time, airlines have advocated hard and soft technology measures such as fleet renewal or 'green inflights' (e.g. IATA 2009) as their primary options to reduce emissions. However, a recent report on the actual efficiency of airlines, based on real data on fuel use and passenger transport, shows that these measures are far less relevant than occupancy rates in achieving low per passenger emissions. The atmosfair Airline Index (atmosfair 2011) rates airlines based on a comparison of their optimal and actual fuel efficiency, considering best available technology, highest possible seat density and a maximum load factor (Figure 4.6). Results indicate that, contrary to common belief, aircraft type

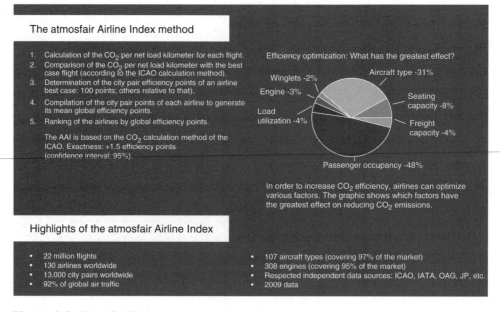

The atmosfair Airline Index method

1. Calculation of the CO_2 per net load kilometer for each flight.
2. Comparison of the CO_2 per net load kilometer with the best case flight (according to the ICAO calculation method).
3. Determination of the city pair efficiency points of an airline best case: 100 points; others relative to that).
4. Compilation of the city pair points of each airline to generate its mean global efficiency points.
5. Ranking of the airlines by global efficiency points.

The AAI is based on the CO_2 calculation method of the ICAO. Exactness: +1.5 efficiency points (confidence interval: 95%).

Efficiency optimization: What has the greatest effect?

Aircraft type -31%
Winglets -2%
Engine -3%
Seating capacity -8%
Load utilization -4%
Freight capacity -4%
Passenger occupancy -48%

In order to increase CO_2 efficiency, airlines can optimize various factors. The graphic shows which factors have the greatest effect on reducing CO_2 emissions.

Highlights of the atmosfair Airline Index

- 22 million flights
- 130 airlines worldwide
- 13.000 city pairs worldwide
- 92% of global air traffic
- 107 aircraft types (covering 97% of the market)
- 308 engines (covering 95% of the market)
- Respected independent data sources: ICAO, IATA, OAG, JP, etc.
- 2009 data

Figure 4.6 *Aircraft efficiency measures*
Source: atmosfair (2011)

(31 per cent) and engine type (3 per cent) are not the primary factors determining fuel use per passenger, compared with passenger occupancy rate (48 per cent). Seating capacity also accounts for 8 per cent of fuel efficiency. These results indicate that airlines should prioritize working with increasing load factors as a measure to reduce fuel use, also because it demands virtually no financial investment. Investments in new aircraft types should, on the other hand, not be delayed, but are more difficult to realize given the low profitability of airlines and major debts of many carriers. Overall, carbon management is consequently more important than investments in new technology.

More generally, airlines need to discuss whether they are transport or air-transport service providers. If a consensus could be reached that airlines are transport service providers, they could start to engage in other transport sectors as well. For instance, high-speed trains have become important links between major European cities. Eurostar has proved to be a strong competitor and a dominant market player in comparison with airlines on the London–Paris route, and in countries such as France and Spain, much of the domestic air traffic between major cities has been replaced in recent years by high-speed train connections. If airlines became engaged in the development of these systems, for instance by using their expertise in electronic boarding and payment systems, they would not have to compete with often more profitable rail systems. The substitution of short-haul air travel by rail is also being pushed by regulatory authorities. For example, in April 2011 the EU Transport Commissioner Siim Kallas announced a series of transport goals, including shifting the majority of flights longer than 300 km to rail, and phasing out the use of petrol cars in city centres by 2050, as part of EU proposals to reduce CO_2 emissions from transport by 60 per cent by 2050 (Milmo 2011).

Car manufacturers

Car ownership is increasing rapidly on a global scale, and car manufacturers consequently have an important role in steering consumer choices through the models they produce. For a considerable time larger cars, in particular SUVs, have been a sales priority, even though car manufacturers now appear to focus on reduced average emissions. Nevertheless, there are still considerable differences between brands and their average fuel economies. For instance, the average model produced by Fiat emitted, according to JATO (2011), some 123 g CO_2 per km, while this value is 172 g CO_2 for Mercedes. Car manufacturers consequently have considerable options to reduce emissions by focusing on the production of cars with lower average emission values (Table 4.13).

Table 4.13 *Average emissions by car manufacturer*

Brand	2009 average CO_2 (g/km)	2010 average CO_2 (g/km)	Difference
Fiat	127.8	123.1	−4.7
Toyota	130.1	128.2	−1.9
Seat	140.9	131.3	−9.6
Peugeot	133.5	131.4	−2.1
Citroën	137.9	131.8	−6.1
Renault	137.6	134.0	−3.6
Ford	140.0	136.9	−3.1
Suzuki	142.4	137.3	−5.1
Hyundai	137.5	137.4	−0.1
Opel/Vauxhall	148.5	139.4	−9.1
Skoda	149.1	139.9	−9.2
Kia	146.8	140.1	−6.7
Volkswagen	150.4	141.0	−9.4
Honda	147.13	147.11	−0.02
Nissan	154.6	147.2	−7.4
Dacia	151.3	147.6	−3.7
BMW	157.2	152.9	−4.4
Audi	160.9	152.9	−8.0
Volvo	171.2	157.4	−13.8
Mercedes	176.4	172.2	−4.2

Source: JATO (2011)

Cruise-ships

As outlined by Dowling (2010), the cruise sector has grown rapidly since the 1970s, from an estimated 600,000 to 13.2 million passengers in 2008, with an additional 5 million passengers anticipated to be added in the period 2008–11. The largest cruise-ship, *Oasis of the Seas* (Royal Caribbean) now accommodates 7000 passengers. There have long

Table 4.14 *Emission calculation methods and results for cruise-ships on a per passenger basis*

Indicator	CO_2 emissions (kg)	kg CO_2 per passenger kilometre	Source
kg CO_2 per available lower berth km	0.330	–	Carnival (2008)
kg CO_2 per available passenger cruise day	139	0.246	RCCL (2010)
kg CO_2 per passenger day	169	0.312	Eijgelaar *et al.* (2010)
kg CO_2 passenger kilometre	0.250–2.200	0.250–2.200	Howitt *et al.* (2010)
kg CO_2 per journey	248–740	0.458–1.366	Atmosfair (2010)

Source: Walnum and Gössling unpublished (2011)

been environmental concerns about cruise-ships discharging untreated waters and other waste at sea, and cruise-ships have received attention because of their high energy intensity on a per passenger basis (Eijgelaar *et al.* 2010). Table 4.14 shows this on the basis of various indicators used by cruise companies and in the scientific literature, indicating that these, for the range of estimates in the literature, may be in the order of 0.246–2.2 kg CO_2 per passenger kilometre (pkm), amounting to 139–169 kg CO_2 per day, or 248–740 kg CO_2 per journey. Adding the flight that may often be needed to get to the port of departure, the combined amount of emissions from flight and cruise may often vastly exceed per capita per year sustainable emissions levels.

As outlined below, technical options to reduce emissions from shipping – the potential of new engine technologies, biofuels or other technical approaches to reduce emissions – may be limited. Even if considerable specific emission reductions are achieved, cruise-ships will remain high emitters on a per kilometre, per day or per journey basis. Given the rapid growth in cruise-ship holidays, this might ultimately mean that the cruise-ship concept has to be reconsidered, with a focus on three aspects:

- the distances travelled, also as a ratio of port to travel times;
- the speed at which ships travel;
- ship design and the use of alternative ship types that can make use of wind propulsion, such as windjammers or yachts.

With regard to distances travelled as well as travel speeds, passengers often move day and night onboard modern cruise-ships – the very purpose of the journey is movement. In the future, cruise-ship trip planners might have to consider shorter routes with more visits to ports to reduce overall travel distances and to overcome the need to move at high speeds, as there is a nonlinear relationship between speed and fuel use (fuel use increases exponentially with speed). Moreover, if alternative ship types were developed, greater use could be made of wind propulsion, an option highly attractive to some tourists (Gössling 2010).

Train systems

Rail and coaches can offer many tangible advantages over aircraft and cars, such as the convenience of being transported instead of driving (cars), which can be relevant in work-related contexts, as time can be used productively; the knowledge of exact arrival times, in principle to the minute (trains); arrival in city centres, rather than the periphery (trains and usually coaches); the relative spaciousness of train compartments; opportunities to stretch legs while travelling; options to take considerable amounts of baggage; opportunities to buy a variety of beverages, snacks or meals at any time (trains); as well as the high speed on major connections, particularly in Europe and Japan (trains). However, world-wide, train travel accounts for only a small share of overall distances travelled (Gilbert and Perl 2008: 66), and innovative service strategies are thus key in attracting travellers to use train systems and coaches. It appears that few train systems systematically seek to improve their services with regard to any of the aspects mentioned above, and virtually no train service provider appears to be focusing on all of these in a comprehensive approach (for further details see Gössling 2010).

With regard to lowering emissions per passenger, even train systems have considerable options. First of all, occupancy rates may often be low, with, for instance, Deutsche Bahn AG reporting 46.3 per cent in long-distance traffic and 25.8 per cent in regional traffic (Arno Seifert, Head Climate Protection, Energy- and Resource-Efficiency, Deutsche Bahn AG). Trains are also often electricity powered and can easily switch to energy derived entirely from renewable energy sources. This is only a solution, however, when it is ensured that power providers do not simply sell a higher share of conventional power to other customers, as most have a share of renewable energy in their mix, and some large utilities are known to have sold this share at a higher premium to environmentally aware customers. Systemically, emissions are reduced only when the capacity of renewable energy production is increased while 'dirty' electricity production is simultaneously phased out.

Accommodation

Accommodation establishments have considerable potential to reduce energy. Ideally, this starts with the construction of new hotels, which theoretically can be built as passive energy structures, or at the very least as low-energy buildings, as it is comparably cheap and cost-effective to consider energy-saving measures when hotels are built. For existing hotels, heating and air conditioning (A/C) are the primary factors in energy consumption, and one of the most important management measures is thus to adjust room temperatures to the lowest (heating) and highest (A/C) levels customers will still perceive as comfortable. This also means adjusting temperatures to optimal levels only shortly before rooms are actually used. With regard to ideal temperatures, the Carbon Trust suggests (2010, based on CIBSE 2006) temperatures of 16–18°C for kitchens and laundries, 19–21°C for corridors and guest bedrooms, 20–22°C in bars and lounges, 22–24°C in restaurants and dining rooms, and 26–27°C in guest bathrooms (in the UK).

It seems unclear, however, why temperatures in restaurants would have to exceed those in bars and lounges, and why guest bathrooms would have to have temperatures as high as 26–27°C, given that heat loss is high in bathrooms because of ventilation, while the time spent in them is comparatively short. Often lower temperatures, for instance in bathrooms, could be accepted by guests without any loss of comfort. This is an insight also applicable to tropical regions, where temperatures in rooms can be kept considerably higher than assumed by many managers. For instance, the Hilton Seychelles experimented with room temperatures and found that 25°C as the (adjustable) standard temperature was accepted without any complaints by guests (Gössling and Schumacher 2010). As a consequence, A/C electricity use could be considerably reduced. Energy use can also be reduced through other measures, including a reduction in hot water use for showers and baths in all climate zones, as well as freshwater treatment more generally, which can be energy intense. For instance, Bermudez-Contreras *et al.* (2008) report energy requirements of 2.6 kWh to desalinate 1 m^3 of seawater. More controversial yet feasible measures include phasing out electrical appliances from hotel rooms, with some upscale hotels having taken fridges (mini-bars) and television sets from hotel rooms, reportedly without any negative reactions from guests (Gössling 2010). The current trend in upscale hotels is otherwise, however, including an increasing number of appliances such as heated bathroom mirrors and heated toilet seats.

With regard to branding and image, initiatives to use renewable energy, participation in certification schemes, or the public display of renewable energy harvested can all be used to create a positive image of accommodation establishments, as the use of renewable energy is generally understood as something positive (Gössling *et al.* 2005; Gössling and Schumacher 2010). Communication of pro-environmental engagement might also allow involvement of tourists, for instance through payments of premiums, and can be an important factor in staff motivation. Hotel Victoria, Germany, for instance, encourages employees to suggest improvements for energy reductions and rewards innovative ideas (Gössling 2010). Accommodation establishments can also make use of their customers' preferences for renewable energy systems or carbon neutrality. For instance, Dalton *et al.* (2008) found that 49 per cent of Australian tourists were willing to pay extra for renewable energy systems, and of these, 92 per cent were willing to pay a premium corresponding to 1–5 per cent above their usual costs. In another study, Gössling and Schumacher (2010) found that 38.5 per cent of a sample of international tourists in the Seychelles expressed positive willingness to pay for carbon neutrality of their accommodation, out of whom 48 per cent stated they would be willing to pay a premium of at least €5 per night. While these values are not representative, they nevertheless indicate that there is considerable potential to involve tourists emotionally and financially in strategies to implement renewable energy schemes.

Advice for energy management in accommodation is available from a range of organizations (EUHOFA *et al.* 2001; International Tourism Partnership 2008). The German Hotel and Restaurant Association (2009) is an example of an organization seeking to involve its members strategically in energy management. Hotels can calculate energy use and emissions on the basis of a wide range of indicators, including various benchmarking tools such as performance reports, regional league tables, hotel profiles, benchmark reports and balance scorecards. Tools such as these can help to illustrate energy consumption, while they are simultaneously strategic management tools allowing for comparison of the performance of hotels within the same chain, as well as individual hotels with the same standard (Bohdanowicz 2009).

Restaurants

By applying food management practices and making more informed choices about the purchasing, preparation and presentation of their food, restaurants can contribute to considerably reduced GHG emissions. A contribution to these processes has also been made by governments, organizations and retailers seeking to support more climatically sustainable food consumption; for example, Tesco (2009), a UK retailer, carbon-labels selected foods. A number of organizations, in some cases on behalf of governments, have also developed food calculators to help inform customers about the options available to them in choosing low-carbon food (e.g. FCFC 2009). Such carbon-labelling approaches have been used with considerable success in restaurants in Sweden.

Measures to reduce GHG emissions in restaurants can pertain to purchases, preparation and presentation (Gössling *et al.* 2011d). With regard to purchases, 'don't-buy' policies include vegetables grown in heated greenhouses, foods involving air transport, problematic species such as lobster, imported beef, and environmentally harmful materials such as aluminium foil (for a discussion of 'food miles' see Defra 2005; Saunders *et al.* 2006; Hall and Mitchell 2008; Weber and Matthews 2008). 'Buy-less' policies would seek generally to reduce the amount of beef, deep-sea and farmed carnivorous fish, rice, seasonal foods out of their season time, and foods with high weight-to-calorie or volume-to-calorie ratios. 'Buy-more' policies would seek to use greater amounts of locally produced foods (transported over short distances, using CO_2-efficient modes), potatoes and grains, pelagic fish, pork and chicken, as well as foodstuffs with a longer shelf life to reduce wastage. It is unclear whether organic food purchases make a significant contribution to climate change mitigation (for discussion see Gössling *et al.* 2011d), but there are other associated benefits, such as organic food production's role in countering trends towards globalization, industrialization and large-scale conversion of ecosystems such as rainforests for soybean or palm oil production, and issues of food safety. Organic farming systems can also help to maintain landscapes and biodiversity (Norton *et al.* 2009), and to ensure that a greater share of the profits of food production remain with the farmer (Hall and Sharples 2008).

Local sourcing of foodstuffs more generally is of particular importance in tourism in developing countries, where imports over large distances are often associated with high emissions. To this end, Hilton Hotels launched the Caribbean Regional Program 'Adopt a Farmer' in cooperation with the World Travel Foundation to strengthen supply chains between farmers and hotels, also advising farmers on quality, desired produce, and funding technology to increase production (Paulina Bohdanowicz, Hilton Sustainability

Manager Europe, pers. comm. 2009). Such strategies appear feasible and are supported by scientific studies: for instance, Torres (2002) reports that international tourists in Yucatan, Mexico appear far more amenable to eating local foods than hotel managers seem to believe; while Bélisle (1984) found that the share of imported foodstuffs varied between 5 and 80 per cent in hotels across Jamaica, suggesting that food import choices are, to a large degree, made by hotel managers rather than by tourists.

Preparation could include purchases of energy from renewable sources, energy-efficient cooking routines, purchases of energy-intensive foods such as bread, use of greater amounts of vegetables in dishes, preparing meals only after orders have been placed, planning purchases to avoid waste, and separating food waste from general waste (see Gössling *et al.* 2011d). With regard to the foodstuffs sourced, companies such as SSP Sweden (2009) also claim to have adopted choice-editing policies, for instance by not purchasing fish species identified by the WWF as being threatened. Likewise, Scandic Hotels has edited out giant shrimps and avoids purchases of bottled water.

Finally, changes in presentation could include reduced portion sizes and plates at buffets, the arrangement of buffets so that less carbon-intensive foods are at the centre, the training of staff to recommend less carbon-intensive meals, and the avoidance of single-use packaging. For instance, Gössling *et al.* (2011d) suggest that buffets may encourage customers to eat greater quantities of food, to eat a higher proportion of climatically relevant foods such as meats, and to choose environmentally harmful foods that they would not consume at home such as prawns. Buffet diners also tend to leave more leftovers. Ideally, buffets could inspire customers to eat only the amount of food they wish to consume, but for this, specific management strategies are necessary. Again, these strategies are supported by evidence that customers' preferences are increasingly turning toward socially and environmentally responsible products (Mitchell 2001; Klein and Dawar 2004), with perceptions of value having considerable importance (Gallastegui and Spain 2002; Gilg *et al.* 2005). Importantly, 'value' can refer to various aspects of food, including perceived quality, taste, health benefits, animal welfare, and responsibilities to future generations. Marketing referring to any of these is more likely to yield emotional responses by customers, creating links between sustainability issues (Hall and Sharples 2008).

Overall it is estimated that, compared with business-as-usual food purchasing, preparation and presentation practices, these measures could save an estimated 50–80 per cent of GHG emissions, indicating that there is considerable scope for food management to make a significant contribution to climate mitigation (Gössling *et al.* 2011d).

Behavioural change

Behaviour is founded in knowledge, awareness and attitudes. Without knowledge of given interrelationships, such as the fact that GHGs cause global warming, there is no reason to engage with the topic to find solutions. Without an awareness of the urgency of mitigation to prevent 'dangerous' levels of climate change, there is no reason to act in the immediate future. And finally, without an attitude that is proactive on climate change, even with knowledge on interrelationships and awareness of the urgency to act, behavioural change is unlikely (Hall 2011c). Consequently, behavioural change is dependent on education to foster knowledge and awareness, which can be provided in schools and universities and through various media, certification schemes or energy/carbon labels. From a psychological viewpoint, attitudes can then be influenced by designing and applying interventions (Steg and Vlek 2009). This will include social marketing strategies, changes in the availability and price of energy-intense products and services, or legal interventions to prohibit certain products and services, as there appears to be a broader understanding that whatever products and services can be purchased are also justified to consume.

At the same time, there is a need to reward 'correct' behaviour, as exemplified in the bonus–malus system for new cars in France (see Table 4.4). Steg and Vlek (2009) outline that rewards are more effective than punishments (but only if they succeed in making pro-environmental behaviour more attractive), and that their effect in time will be dependent on the continuation of rewards. These insights could be used more systematically in tourism, where new low-carbon products could be developed and promoted by focusing not on their carbon properties, but rather on their unique character or the experiences that can be gained through their consumption. Strategies to encourage pro-environmental behaviour will also have to be context-specific, with multiple, combined measures being most successful (Steg and Vlek 2009).

However, in tourism, changes in behaviour may generally be difficult to achieve (see e.g. Becken and Wilson 2007 for an experiment seeking to influence self-drive tourists in New Zealand). There is ample evidence that voluntary behavioural change is not likely to become significant (McKercher et al. 2010; Cohen and Higham 2011; Hall 2011c), also because of a lack of knowledge (Cohen and Higham 2011; Higham and Cohen 2011). As outlined by Peeters et al. (2009: 248): 'Many people have a powerful belief in their personal right and need to travel, coupled at the same time with a contradictory powerful belief that others should be denied that right for the good of the planet.' Such insights are confirmed by studies such as those by Barr et al. (2010) and McKercher et al. (2010), who find that the most environmentally aware tourists might not be more willing to alter travel behaviour – paradoxically, the most climatically aware tourists may also be the most active travellers (e.g. Gössling et al. 2009b; Barr et al. 2010). Climate policy is thus the most important arena in which to achieve change, making use of tourists' stated preferences: for instance, while tourists may not be willing to change their own behaviour, they may be willing to support policy-making behavioural change obligatory for all tourists. Gössling (unpublished) finds, for instance, that a majority of tourists interviewed in a survey in Barbados support the idea of tourist transport based on electric cars.

Insights such as these again point at the importance of raising knowledge and awareness of the underlying reasons for climate change. In this regard, energy, carbon or climate labels are among the easiest-to-implement tools for educating the broader public on energy and emission intensities. The best-known energy label at this moment might be the EU label for white appliances, including light bulbs, refrigerators, freezers, washing machines, tumble dryers, dishwashers, ovens, A/C, water heaters and TV sets. These are rated by energy efficiency class. The grading originally covered a gradient from A (most efficient) to G (least efficient), but has now essentially moved to all appliances being A-graded, currently including A+, A++ and A+++, or alternatively, A minus $x\%$ (cf. EC 2009). The energy label also exists for cars, but displays emissions of CO_2 per pkm, rather than energy use, as well as for flights and hotels (Figure 4.7). The label is easy to understand, even for people with no previous knowledge of energy and emissions, with green bars (starting at A) representing low energy use/emissions and gradating to red bars (starting at F) representing high energy use/emissions.

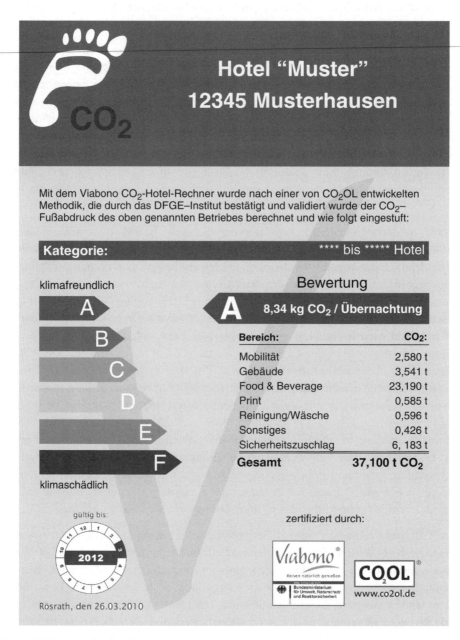

Figure 4.7 *Energy label used in accommodation*
Source: Viabono

The Austrian organization Respect (2010) suggests introducing mandatory carbon labelling of tourism services and products, for three reasons:

- it would allow consumers to develop a deeper understanding about the dimensions of climate impacts associated with their mobility and leisure behaviour;
- it would provide environmentally conscious tourists with transparent product information in order to economise on their personal climate responsibility;
- it would increase climate awareness among tourism businesses as they will be required to establish emission reporting systems and/or standardized benchmarks.

A fourth argument for carbon labelling might be added: carbon labelling allows comparison of the climate performance of tourism businesses and thus stimulates competition. For instance, a recent airline ranking published by German offset provider atmosfair received worldwide media headlines. As the airline ranking will be published annually, customers can now choose the most environmentally friendly airline for their travel (Figure 4.8).

The opportunity to compare performances should steer at least a share of customers towards low-carbon choices, while fostering knowledge and awareness of an abstract topic – climate change, GHGs and their consequences. It appears that this is increasingly realized by governments, with, for instance, France now conducting a national *experimentation* to quantify environmental impacts of products and to communicate this to the consumer (Ministère de L'Écologie, du Développement Durable, des Transports et du Logement 2011). Carbon labels may ultimately help to develop low-carbon cultures, as energy use is clearly embedded in national culture. For instance, Portugal is Europe's most 'fuel-conscious' country with regard to purchases of new cars. Average emissions for new cars are, in 2010, according to JATO (2011) below 130 g CO_2 per km, while in Germany, Sweden and Switzerland, average emissions for new cars exceeded 150 g CO_2 per km, indicating that very different car cultures exist in Europe. Notably, despite a trend towards more efficient cars in Europe, emissions from cars appear to increase as a result of increased mobility. The Swedish Trafikverket (2011) reports, for instance, that 'The record-like efficiency gains achieved for private cars were not enough to compensate for the growth in transport volumes', leading to an absolute growth in emissions.

When knowledge and awareness can be linked to actual change, this may lead to considerable savings. For instance, Hilton Worldwide saved energy and water costs in the order of US$16 million in the period 2005–08, primarily through behavioural change of employees as a result of a training in resource-efficiency (Gössling 2010).

Voluntary emission offsetting

Voluntary carbon offsetting is a management strategy that potentially can be relevant for all tourism actors, in particular airlines and accommodation businesses as well as entire destinations. There is also evidence of considerable support from tourists supporting voluntary offsetting schemes, and significant willingness to pay premiums for mitigating or compensating the climate impact of their travel (e.g. Brouwer *et al.* 2008;

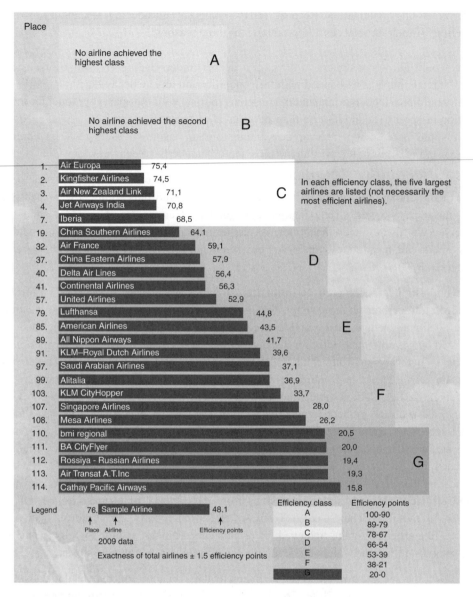

Figure 4.8 *Atmosfair Airline Index ranking airlines by efficiency*
Source: atmosfair (2011)

Gössling *et al.* 2009d; Gössling and Schumacher 2010). While surveys thus indicate substantial willingness to pay for carbon offsets, there appears to be a considerable gap with regard to actual payments. The share of travellers stating that they have paid for offsetting their flight is generally less than 5 per cent, and even lower figures have been reported by airlines and tour operators (cf. Gössling *et al.* 2009c).

Various reasons have been identified for this apparent paradox. For instance, Becken (2007) found that some travellers were unwilling to pay for offsets because other passengers were not – a problem referring to altruistic behaviour. In a survey conducted at Landvetter airport (Gothenburg, Sweden), Gössling *et al.* (2009c) also found that a broad majority of travellers did not feel responsible for emissions, asking that airlines, aircraft manufacturers or governments solve the problem. In this study, travellers also indicated that they were confused about whether flying was making a significant contribution to climate change, or whether their payments would have a significant benefit towards solving the problem, as transport provider Scandinavian Airlines simultaneously communicated that flying was not environmentally harmful. Finally, it has been argued that there might be a 'rebound effect' in traveller behaviour – air travellers might not perceive it as necessary to change travel patterns because they have offset their emissions. Further research is needed to confirm this hypothesis.

In a review of the voluntary carbon market for aviation, Gössling *et al.* (2007a) concluded that there were considerable differences between offset providers with regard to the calculation of emissions caused by identical flights, the reliability of credits sold, and the share of administrative costs compared with project costs, indicating little consistency in the approaches chosen and low degrees of credibility with regard to most actors. The survey also found that most organizations focused partially or entirely on forestry projects, an ambiguous offset strategy (for discussion see Gössling *et al.* 2007a; Broderick 2009). While it is clear that afforestation and halting deforestation need to be part of strategies to address climate change, the role of afforestation/deforestation projects as offsets is more difficult to assess, for instance because of the need to guarantee that carbon stored in biomass can be maintained over decades and centuries (Gössling *et al.* 2007a). It should also be noted that there is no political consensus yet regarding the role of, and mechanisms for, 'reducing emissions from deforestation and degradation' (REDD) in a global emission-reduction framework, even though proposals continue to emerge and implementation of REDD+ activities has been affirmed under COP-16 (see also CBD and GIZ 2011).

Overall, a trustworthy offset proposition would need to consider a wide range of criteria, including a credible calculation of the emissions caused and their climate impact; valid compensation mechanisms; and adequate customer communication. Within these three major categories, further aspects will be relevant. Strasdas *et al.* (2010) list, for instance, 20 criteria (Table 4.15).

Overall, there is considerable potential for carbon offsetting to make a contribution to emission reductions, although it is important to note that these schemes do not prevent emissions. They consequently need to be seen as environmentally risky options, although they can be valuable in introducing technology change and to 'buy time'.

In the future, carbon offsetting could play an important role in the context of developing 'carbon neutral destinations' (for a definition, as well as alternative terms used, see Gössling 2009). A wide range of destinations have announced ambitions in this regard, including Costa Rica, Norway, New Zealand and Sri Lanka (Gössling 2009) as well as the Maldives. Furthermore, there is a wide range of smaller regions and island states now discussing carbon neutrality strategies, including various islands in the Caribbean. However, a closer look at the efforts of these countries and jurisdictions reveals that

Table 4.15 *Criteria for credible offset propositions*

A)	Credible calculation of emissions	1	Accuracy of calculations
		2	Consideration of all greenhouse gases
		3	Multiplier of at least 2.0 to account for non-CO_2 radiative forcing
B)	Valid compensation mechanisms	1	Proof of additionality
		2	Calculation of baseline/reference scenario
		3	Guarantee/permanence of compensation
		4	Time between release of greenhouse gas emissions and compensation
		5	Consideration of carbon leakage
		6	Registration/cancellation of certificates
		7	Transparent and independent verification and certification
		8	Sustainable development (social aspects, biodiversity)
		9	Exclusion of projects for environmental/ social reasons
C)	Customer relations and communication	1	Readable introduction to climate change and offsetting
		2	Information about low-carbon behaviour
		3	Transparency of emission calculations
		4	Transparency of compensation
		5	Transparency of company working processes
		6	Transparency of prices/share of money used for projects
		7	User-friendliness (carbon calculator, payment)
		8	Formal aspects (terms and conditions, data protection)

Source: Strasdas *et al.* (2010)

the system boundaries chosen often exclude air travel to/from the destination (the major factor causing emissions), while headlines in newspapers suggest greater action on emissions than is actually implemented or planned (Gössling 2009). The Maldives, for instance, appear to focus on emissions arising within the country in a recent carbon audit, although there appears to be awareness that inclusion of emissions from international aviation on a bunker-fuel basis would double the amount of emissions caused by the islands (BeCitizen 2010). Nevertheless, the islands' focus is on reducing emissions of hotels, while they pursue an arrival growth strategy focused on medium-income markets. Consequently, the only option for the Maldives to become carbon neutral will be to buy offsets, while emissions from the islands' tourism system are poised to grow.

In Norway, national policy foresees considerable reduction of GHG emissions in the near future, but simultaneously there are ongoing efforts to expand the tourism system, in particular by attracting long-haul visitors and highly energy-intense cruise-ship tourism. There is also considerable reluctance to engage in measures to curb emissions from air travel, and it consequently appears unlikely that Norway will achieve a de-carbonization of its tourism system. New Zealand never had plans to become entirely carbon neutral, as reported in the media, rather than to reduce emissions in some governmental sectors. Scotland, also receiving considerable media attention for its climate neutral strategy (cf. Gössling 2009), appears to have abandoned plans in this regard. Costa Rica and Sri Lanka, also hailed for their announcements to become carbon neutral, seem to rely on reforestation and afforestation schemes to save emissions, although not engaging in any systemic changes that would reduce energy use. Overall, the current situation of carbon neutral destinations is thus one of convenient system boundaries, usually excluding international aviation; impasse with regard to credible and significant strategies to reduce emissions; conflicting tourism development goals focusing on further growth in arrivals; and offsetting strategies lacking credibility.

This does not mean that carbon neutrality could not become an important new paradigm, and it is closely related to the development of 'steady-state' approaches to sustainable tourism. Credible approaches would, however, command wide-ranging changes in tourism systems to reduce energy use, ultimately forcing thousands of stakeholders to work proactively with carbon management. While this would present an advantage in terms of meeting emerging customers' expectations of 'green' tourism operations, and reduce vulnerabilities linked to rising oil prices and climate policy, carbon neutrality nevertheless presents a major obstacle in terms of stakeholder involvement and innovation, ultimately requiring new business models. However, if developed, and perhaps even resulting in 'plus carbon' strategies (requiring tour operators, businesses and tourists not only to compensate, but to overcompensate, their impact on climate change), tourism potentially could become a major force in restructuring societies. This, however, will demand considerable commitment by businesses, politicians and tourists to support the concept, including behavioural change and additional payments to finance low-carbon infrastructure.

Technology

Technology innovation is usually presented as the solution to climate change mitigation, even though, as the order of themes in this book (policy, management, behavioural change, technology) is intended to suggest, technology may be part of the solution, but is not the most relevant strategy to pursue. Nevertheless, in public opinion and among stakeholders, technology change appears to dominate understanding of suitable strategies, which may be seen as partly the result of discourses strategically launched by tourism actors with an interest in business-as-usual operations (cf. Gössling and Peeters 2007) as well as the inconvenience associated with policies, and in particular fees and taxation, as well as behavioural change (Stoll-Kleemann *et al.* 2001). The following sections describe the potential of technological innovation to contribute to emission reductions, focusing on the most relevant emissions subsectors.

Aviation

In terms of emissions, aviation is the most important subsector of the tourism system, in terms of both its current and future contributions to emissions. Part of the aviation sector, including IATA, ICAO and most of the major airlines, have outlined a range of technology-based strategies to achieve emission reductions, including new airframes and engines (to be realized through fleet renewal), improved air traffic management (ATM), biofuels, and 'economic measures', ultimately referring to energy and climate policies. These are seen eventually to lead to absolute reductions in GHG emissions about 30 years from now. To this end, IATA suggests that absolute emissions will increase only slightly over the next 20 years because of energy efficiency gains through fleet renewal and ATM system improvements, followed by absolute emission reductions in the medium-term future (by 2035–40), and reaching a stated −50 per cent target in 2050 with respect to current aviation emissions. Importantly, the purchase of emissions credits from other economic sectors is not identified as a component of this strategy (Scott *et al.* 2010). IATA (2009) simultaneously suggests that efficiency gains will be in the order of 1.5 per cent per year up to 2020, with most overall gains achieved through fleet renewal (Figure 4.9), although evidence provided by atmosfair (2011) suggests that aircraft type (31 per cent), seat capacity (8 per cent), engine (3 per cent) and winglets (2 per cent) together hold a potential of 44 per cent of efficiency gains achievable for the current world fleet, while optimizing the load factor accounts for 48 per cent. A potential problem with fleet renewal is also that, given long development times for new aircraft of one to two decades, it is questionable whether coming aircraft types will be able to achieve similar leaps in fuel efficiency, while existing aircraft have lifetimes of about 25 years, leading to slow rates of fleet renewal (Peeters and Middel 2007).

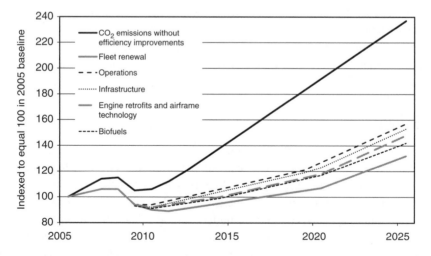

Figure 4.9 *Airline emission growth (CO$_2$) and technology innovation*
Source: IATA (2011)

Given the overall growth in air travel, technology-related measures will in any case yield efficiency gains considerably below growth in transport volumes, and absolute emissions from air travel will continue to increase. Up to 2020, there would thus be considerable growth in emissions from aviation, which is confirmed in other studies, with projected growth rates ranging between 0.8 and 6.2 per cent per year (various periods within 1995–2025; for a review see Mayor and Tol 2010a).

Biofuels are a second option to reduce emissions from aviation that has received much attention in recent years, and they are mentioned by virtually all airlines as a key strategy for mitigation (e.g. Aviation Global Deal Group 2009). Clearly, the production of biofuels in commercial-scale volumes after 2030 seems to be the only option to achieve *absolute* emission reductions in the sector, as they do not demand the development of new aircraft and fuel infrastructure, but can be used in existing commercial aircraft (Upham *et al.* 2009a, 2009b; Vera-Morales and Schäfer 2009).

Several biofuels have been tested in commercial aircraft so far. Virgin Atlantic used a 20 per cent biofuel to 80 per cent kerosene blend in 2008 to fly a Boeing 747 from London to Amsterdam (Virgin Atlantic 2008). In another test, Air New Zealand flew one engine of a Boeing 747-400 with a biofuel blend derived from jatropha (Air New Zealand 2008). Continental Airlines used a 50 per cent kerosene, 50 per cent biofuel blend (from algae, 2.5 per cent and jatropha, 47.5 per cent) to fly a 737-800 in 2009 (Continental Airlines 2009; Vera-Morales and Schäfer 2009). Even though the details of these experimental flights have not been published by the airlines, it might be concluded that biofuels are potentially feasible, at least as a blend with kerosene, as for instance foreseen by Aviation Global Deal Group (2009). Nevertheless, a number of problems remain to be solved. Vera-Morales and Schäfer (2009) show that natural gas-based synthetic oil products would not lead to significant emission reductions, while synthetic oils from cellulosic biomass could reduce life-cycle emissions by up to 85 per cent, but are not feasible at significant scale due to vast land requirements. Microalgae-based fuels demand less, but still considerable, land areas, and need to be supplied with concentrated CO_2, thus potentially increasing life-cycle GHG emissions over those from petroleum-derived jet fuel.

Other problems are incurred by the lower volumetric energy-density (MJ per litre) of current biofuels than Jet A fuel. Vera-Morales and Schäfer (2009: 6) state, for instance, that storing fuel energy of 34.7 MJ requires 1 litre of fossil fuel-based jet fuel, but 1.5 litres of ethanol, 2.2 litres of methanol, and significantly larger volumes of hydrogen. The implication is that aircraft using biofuels would have a reduced range and, for some long-haul routes, even blends of biofuels might consequently be impractical. Furthermore, it is unclear which biofuels might be the most feasible. Jatropha was considered the most promising biofuel as recently as 2007, but it appears already to have been abandoned as a suitable energy crop, while new concerns have arisen about its water requirements (Gerbens-Leenes *et al.* 2009). This is accompanied by failed investments and socio-economic expectations – for instance, Lahiri (2009) reports how jatropha was seen as a major opportunity in India, with millions of US dollars invested by the government to develop plantations. Lahiri (2009) suggests, however, that few saplings have survived, and that it is questionable whether there will ever be any biofuel production. Biofuels also compete with food production (Harvey and Pilgrim 2011; Murphy *et al.* 2011), and sustainability criteria and indicators need to be established to ensure the sustainable development of this sector (for a review of certification developments see Scarlat and Dallemand 2011).

Other biofuels are currently being developed, with algae being the new focus of investors. Yet again, it appears that expectations on algae exceed their potential, with many technical and biological problems remaining unsolved. For instance, Vera-Morales and Schäfer (2009) report that growing 1 kg of algae requires 2.2 kg of CO_2 (as fertilizer), of which only a small fraction can be absorbed from the atmosphere. Consequently, CO_2 would ideally have to be provided from fossil-fuel burning, leading to a recycling of emissions otherwise released to the atmosphere, but incurring a dependency on fossil-fuel combustion. For further discussion regarding the capacity of biofuels to replace conventional jet fuel, and on wider sustainability issues, see also Nygren *et al.* (2009) and Upham *et al.* (2009a, 2009b).

In 2005, global consumption of jet fuel was 232 Mt (Lee *et al.* 2009). Replacing conventional fuels with advanced biofuels would, irrespective of the source of the biofuel, lead to considerable area requirements (see Murphy *et al.* 2011). For instance, Scott *et al.* (2010) calculated that in order to use jatropha, an area of more than 1 million km^2 would be required, roughly the size of Germany, France, the Netherlands and Belgium combined. Based on projected growth in air travel, this area would then grow by a factor of two over the next 15 years. Finding such an immense area for energy crop cultivation, even on the marginal lands on which some jatropha species can be grown, is likely to be difficult politically (for global scenarios on area use, see Murphy *et al.* 2011). Microalgae use smaller, but still considerable areas. For instance, Vera-Morales and Schäfer (2009) conclude that a land area the size of 11 per cent of peninsular Spain or the state of Colorado would be needed to match current fuel demand in aviation, notably in a maximum productivity scenario for biomass. Other problems pertain to the costs of production, as well as the radiative forcing caused by non-CO_2 GHG emissions, and appear as yet not to have been discussed in the literature. Overall, the role of biofuels in solving the emissions problem in aviation – or indeed other transport modes – is thus uncertain and, as acknowledged by ATAG (2011: 2), progress in replacing conventional fuel with biofuels might be slow: 'We are striving to practically replace 6 per cent of our fuel in 2020 with biofuel.'

Cars

The car is the most widely used mode of surface-bound transport for tourism (UNWTO *et al.* 2008). Most cars used by tourists are privately owned, but rental cars can be of considerable importance at destinations. Virtually all cars have in common that they are powered by combustion engines running on fossil fuels. These engines, as well as the car design, have become more efficient over time, but most gains have been lost to more powerful engines and higher comfort standards, including A/C or heated seats. Recently, there might be a significant trend towards declining specific emissions per kilometre, at least in Europe, possibly because of increasing fuel prices and greater environmental consciousness, as well as other mechanisms penalizing or rewarding car choice.

As shown by the Council of the European Union (2010), average emissions from new passenger cars in the EU have declined from 0.1722 kg CO_2 per km in 2000 to 0.1535 kg CO_2 per km in 2008, with petrol vehicles becoming 11 per cent more efficient and diesel vehicles 6 per cent in this period. In 2007–08, emissions dropped by 3.3 per cent, the largest drop in specific emissions since monitoring began (Council of the European Union 2010). Paradoxically, however, more fuel-efficient cars can increase mobility in line with the Jevons paradox (cf. OECD 1996), because of greater efficiency and thus lower costs for travel, travel distances and absolute energy use increase. Gilbert and Perl (2008) also show that owners of new cars are destined to travel significantly more than owners of older cars, based on six surveys from 1969 to 2001 in the USA.

The Council of the European Union (2010) also remarks that the economic crisis has not led to substantial downsizing of the car fleet and that average engine power has stayed the same. There has been growing interest in alternative fuel vehicles, however, which almost doubled in number from 2007 to 2008. Yet they account for only 1.3 per cent of new passenger car registrations (Council of the European Union 2010). Another aspect of importance in this context is the difference in car cultures between countries. These are reflected in national emission averages from newly registered cars, with, for instance, recent EU statistics showing that cars in Portugal have, on average, emissions of 0.138 kg CO_2/km, while the rates are 0.165 kg CO_2/km in Germany and 0.181 kg CO_2/km in Latvia (Council of the European Union 2010), representing a difference of up to 30 per cent in new car registrations within countries in the EU. This indicates that, in terms of technological innovation, the most substantial reductions in fuel use could be achieved by the use of more efficient small cars: a combination of lower weights, less power and reduced speeds. This, however, would have to be accompanied by rising fuel prices in order to maintain stable fuel price budgets to prevent people from travelling more – technology innovation has to go along with policy intervention. Notably, Gilbert and Perl (2008) show that the most important principle to reduce mobility is the prevention of car ownership, an argument supporting equal advantage for non-ownership (EANO) principles.

Future alternative car engine technologies might include electricity-powered vehicles, hybrid cars, and cars using biofuels or hydrogen. Electric cars are more energy efficient than cars with internal combustion engines, and cause no tailpipe emissions. Their overall performance is, however, dependent on how power is generated, with only renewable energies guaranteeing that car use is low-carbon and risk-free, that is, not leading to emission growth or carrying the possibility of nuclear accidents. Electric cars cannot as yet replace fuel-based cars, however, because battery capacity is limited and loading times are considerable. An exception to these limitations in the use of electric vehicles more generally might be small tourist destinations. For instance, Werfenweng, Austria has shown that it is feasible and, in fact, a considerable attraction for visitors to use electric vehicles for tourist movements within the destination – notably with the side effect that visitors arrive predominantly by train.

Hybrid vehicles use both electricity and fuel, recharging their batteries when running the combustion engine. Hybrid cars can use up to 50 per cent less fuel than cars with combustion engines, but the life-cycle emissions of batteries are as yet not fully understood. Biofuels for cars have existed for a long time, and in particular E85, ethanol blended with petrol, has been used in Brazil. However, considerable problems are connected with the use of ethanol and other biofuels, for instance regarding land use competing with food production. The Earth Policy Institute (2010) concludes that the USA's focus on biofuel production from grain has had major implications for global food prices, as more and more grain is retained in the USA for domestic consumption as biofuel:

> The 107 million tons of grain that went to U.S. ethanol distilleries in 2009 was enough to feed 330 million people for one year at average world consumption levels. More than a quarter of the total U.S. grain crop was turned into ethanol to fuel cars last year.
>
> (Earth Policy Institute 2010)

A new report by the Environmental and Health Administration of Sweden (EHAS 2010) also concludes that sustainable production of biofuels for cars is a key issue, but outlines that biofuels could lead to reductions of GHG emissions by somewhere between 4 and 79 per cent: their potential is highly dependent on the type of biofuel used and associated life-cycle emissions (for the USA, see also DOE/NETL 2009). Another technology to be developed for cars is fuel cells using hydrogen. Difficulties remain, however, with regard to the production of hydrogen, which requires electricity, as well as this fuel's storage and distribution. Overall, there is considerable potential to reduce specific emissions from cars, even though the development of travel distances will, as outlined above, remain a key question in this context: if people continue to travel more, this might outweigh efficiency gains.

Cruise-ships

IMO (2009b: 3) states that there is 'significant' potential for reduction of GHGs through technical and operational measures, with a specific emission reduction potential in the order of 25–75 per cent, depending on the ship type and operating pattern. IMO (2009b) details four strategies to reduce emissions from shipping: improved energy efficiencies; the use of renewable energy sources (sun and wind); the use of fuels with lower life-cycle emissions, including biofuels and natural gas; and the use of 'emission-reduction technologies' including chemical conversion and capture and storage. As most of these options to reduce emissions refer to technology, but sometimes refer to operational and thus management issues, both technology and management changes are discussed in this section.

With regard to improving energy efficiencies, IMO (2009b) distinguishes ship design and operational measures. To this end, an Energy Efficiency Design Index (EEDI) for new ships, an Energy Efficiency Operational Indicator (EEOI) for all ships, and guidance principles on best practice for the entire shipping industry are currently being developed. The EEDI determines energy use of new ships, considering intended cargo and speed, and can be used to model different ship choices with the ultimate goal of reducing fuel use requirements. This can include designs of improved hull shape, air lubrication systems to reduce hull resistance in the water, improved engines, propellers and the use of diesel–electric systems as well as wind-assisted propulsion power, solar energy systems to provide lighting, or waste heat recovery systems and shaft generators to improve efficiency. In contrast, the EEOI works as a benchmarking system by measuring how efficiently a ship operates in terms of fuel use per unit of cargo moved over a given distance. IMO (2009b: 4) suggests that the index be used by ports to differentiate their fees, and by 'charterers or cargo owners in connection with energy efficiency branding or in negotiating sub-contracts'.

The use of renewable energy and biofuels is a second pillar of IMO's strategy to reduce emissions, including the use of liquefied natural gas (LNG), fuel cells, fuels with low life-cycle CO_2 emissions, biofuels, as well as sails and kites. Solar power and wind energy are seen as potentially useful to help meet ancillary requirements such as lighting on board ships, but are not seen as promising as primary propulsion options. IMO (2009b: 6) outlines that:

> Current solar-cell technology is sufficient to meet only a fraction of the auxiliary power requirements of a tanker, even if the entire deck area were to be covered with photovoltaic cells. Wind-assisted power, on the other hand, has a promising potential for fuel-saving in the medium and long term but, as present-day trial experiences of these technologies on board large vessels is limited, it is difficult to assess their full potential and further trials and development should be encouraged. ... it seems inevitable, however, that fossil fuels will probably continue to be the predominant source of power for the majority of the shipping industry for the foreseeable future.

Options for reducing emissions of CO_2 through capture are not seen as promising, but there is some potential to reduce other gases, including NO_x, SO_x, CH_4, non-methane volatile organic compounds (NMVOC) and particulate matter (PM). IMO (2009b: 6) acknowledges that projections on efficiency gains are difficult, but suggests that by 2020 a combination of 'regulatory, design and operational measures' might lead to relative fuel savings of around 17–32 per cent per tonne per mile of cargo transported. In other words, absolute emissions from shipping would continue to increase as efficiency gains are outpaced by growth: 'it would be almost impossible to guarantee any absolute reduction by shipping as a whole, due to the projected growth in demand for shipping worldwide arising from the growing world population and global economy' (IMO 2009b: 7). This certainly is true for cruise-ships as well, given above-average growth rates in this shipping subsector.

Other transport

Electric trains, trams, metros and trolley buses that are directly connected to the grid are very energy-efficient transport modes causing low levels of emissions, particularly when purchases of renewable electricity are made. Swedish Railways (Svenska Järnvägen), for instance, like Swiss Railways, sources electricity exclusively from renewable power. According to the company, emissions from one person travelling by train over a distance of 1000 km based on a life-cycle analysis will amount to as little as 0.0021 kg CO_2, which can be compared with emissions of 0.133 kg CO_2 from cars or short-distance emissions of

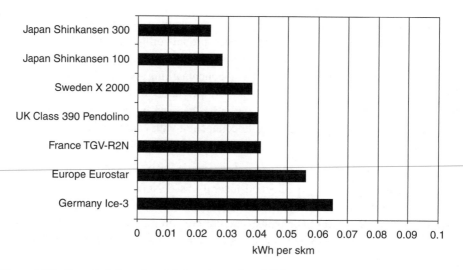

Figure 4.10 *Energy intensity of inter-city trains (kWh per skm)*
Source: Kemp (2009)

0.154 kg CO_2 from aircraft (per pkm) (Svenska Järnvägen 2010). This indicates that train systems could be operated virtually carbon-free.

There are, however, considerable differences between train systems, as shown in Figure 4.10 for inter-city trains, mostly due to differences in capacity. Notably, Shinkansen trains achieve the lowest values even though they run at the highest speeds (270–300 km per hour, compared with around 200 km per hour for trains in Europe). Values of 0.03–0.06 kWh per seat kilometre (skm) correspond to emissions of 0.015–0.030 kg CO_2 per skm at an assumed energy-mix emission of 0.5 kg CO_2 per kWh, and are even lower when renewable energy is used.

The actual operational energy consumption for trains depends primarily on speed, landscape relief and the number of accelerations. With faster trains covering greater distances, energy use for train systems is likely to increase, but a number of technological developments might outweigh growth in demand, including the use of hybrid locomotives, regenerative braking and kinetic energy storage systems (UIC 2007). In order to meet increasing demand in a low-carbon world, railways could also use carriers with greater capacity. For instance, UIC (2007) reports that the French double-decker TGV uses almost the same amount of energy as its conventional trains, but can carry 40 per cent more passengers.

Urban public transport systems include light-rail transit and metro or suburban rail, and increasingly large-capacity buses. Bus rapid transit (BRT) systems have been developed in Curitiba, Brazil and are now in place in many South American cities (e.g. Machado-Filho 2009). BRT offers the opportunity to provide high-quality, state-of-the-art mass transit at a fraction of the cost of other options. BRT utilizes buses on segregated busways and can incorporate services such as pre-board fare collection, user-friendly transit stations, simplified transfers and routings, and highly developed customer services. Other destinations in Asia, Australia, Europe and North America also use BRT.

Accommodation

The accommodation sector represents, globally, approximately 21 per cent of emissions from tourism. Initiatives in this sector are important, as many hotels have considerable options to reduce energy use, which are usually economical. More generally, buildings have been identified as having the greatest potential to reduce emissions (IPCC 2007b), and substantial efficiency gains have been made in recent years in hotels (Bohdanowicz and Martinac 2007; Butler 2008; Bohdanowicz 2009). Reducing energy can lead to considerable cost savings. For instance, Bohdanowicz (pers. comm. 2009) estimates that 15–20 per cent of energy demand can be provided by solar thermal in most tropical and subtropical locations. The Carbon Trust (2010) suggests that 20 per cent energy efficiency reductions can be achieved by a range of simple measures, with another 12 per cent of savings possible based on measures with payback times of less than two years. In the USA, Energy Star (2010: 1) reports that: 'On average, America's 47,000 hotels spend $2,196 per available room each year on energy. This represents about 6 percent of all operating costs.' While hotels thus have much to earn from carbon management, they might also be the sector with the highest support by governments, organizations and customers to restructure and retrofit towards low-carbon operations. For instance, the EU co-financed the Hotel Energy Solutions scheme (www.hotelenergysolutions.net), coordinated with the UNWTO, UNEP, International Hotel & Restaurant Association, European Renewable Energy Council and the French Environment and Energy Management Agency. The scheme helps small and medium-sized hotels in the EU to increase energy efficiency and use renewable technologies. Notably, there is now also a rapidly growing number of companies specializing in technology-based energy management solutions.

The most important measure in existing accommodation establishments is eventually to reduce energy use. Where most energy is used for heating or cooling, adjusting room temperatures is a key measure. This can be facilitated through technology, including digital measurement and control units, which for instance heat/cool rooms only shortly before they are used, or shut down A/C when balcony doors are opened. Building design, including positioning, materials and insulation, can provide an important precondition for maintaining temperatures in the desired range and considerably reduce overall energy use (e.g. Chan and Lam 2003). Likewise, it is crucial to have A/C units and heating in the right locations to avoid inefficient use, and to service units regularly (cleaning filters and coils for A/C) (Becken and Hay 2007; UNWTO *et al.* 2008). Gössling (2010) shows that wherever A/C is used, there are innovative solutions to reduce electricity use – by 40–98 per cent. Hilton Hotels in Prague, Malta and Tel Aviv have also experimented with using riverwater and seawater for cooling (Paulina Bohdanowicz, Hilton Europe Sustainability Manager, pers. comm. 2009). Other technical measures can include energy-efficient lighting, electricity-saving devices such as master switches and key cards, intelligent space temperature controls, heat recovery systems (for instance in kitchen ventilation), and building management systems more generally.

Other technology-based measures include the adjustment of boilers to maintain water temperatures for guest showers at just under 60°C, while limiting overall water use by installing low-flow showerheads. Further measures, such as heat exchangers to recover heat from wastewater pipes, are technically feasible (UNWTO et al. 2008). Energy use for hotel pools can be reduced by installing solar water heaters and heat pumps, while reducing heat loss by using pool covers. Often, such measures have very short payback times. New light technology, including energy-efficient light bulbs with long lifetimes, as well as LED-based lighting, can also considerably reduce energy use, as can room card systems shutting down energy use in rooms when guests are leaving. More recent innovative approaches include water-bottling systems to reduce transport, as shown by Scandic Hotels, which can be economically profitable and have considerable environmental benefits. Food management more generally also offers wide-ranging opportunities to reduce emissions.

Investments in energy efficiency might be complemented with a focus on renewable energy, which can be purchased from power providers, often at no or only small price differences (e.g. Gössling et al. 2005), or through investments in renewable energy technology more generally. When renewable energy is purchased directly from power providers, it is essential to ensure that additional renewable energy capacity corresponding to the amount of energy purchased is actually installed by the power provider. To illustrate this: when a company decides to buy 1 million kWh of renewable power, the power provider should invest in additional capacity to generate this amount. A critical approach to 'green' power purchases is necessary, because some power providers have started to sell the renewable electricity that is in their portfolio at a premium to environmentally aware customers – with the consequence that other customers purchase higher shares of coal, nuclear or other power, thus having no effect on sustainability.

Another option for accommodation establishments to reduce their emissions is to generate their own renewable energy. An increasing number of studies have concluded that renewable energy systems for small to medium-sized accommodation establishments are feasible. For instance, Bakos and Soursos (2003) have shown that photovoltaic cell (PV) installations for small-scale tourist operations in Greece are economically viable with up to 10 years payback time, and considerably lower payback times if government subsidies are provided. Likewise, Dalton et al. (2009), in their analysis of three case studies with stand-alone renewable energy systems in Australia, conclude that PV-based and wind-energy conversion systems are all economically viable, but wind-energy conversion systems had shorter payback times (3–4 years) and were thus economically preferable to PV systems with payback times of 6–7 years. Other studies arrive at similar conclusions for larger-scale wind–hydro energy systems, with Dalton et al. (2008) finding that larger-scale wind-energy systems (>1000 kW) are more economical than multiple small-scale systems at 0.1–100 kW. In a review of grid-connected renewable energy systems, Dalton et al. (2009) report payback times ranging from 5–8 years for PV systems and 4–30 years for wind-energy conversion systems. Overall, these studies indicate that renewable energy systems can be economical, and that it can be worthwhile for accommodation establishments to carry out feasibility studies to assess their potential, particularly in warm and sunny climates or in areas with moderate to high average wind speeds.

In countries such as Australia, a significant share of accommodation establishments appear to already use renewable energy systems: a survey by Dalton *et al.* (2007) found that more than 9 per cent of polled accommodation establishments in Queensland reported use of renewable energy systems. With regard to future solutions, Dalton *et al.* (2009) conclude that hydrogen fuel cells and storage systems are technically feasible but not as yet economically viable, a result confirmed by Bechrakis *et al.* (2006), who estimate that the costs of wind–hydrogen systems are in the order of US$1.05–1.11 per kWh, considerably higher than current electricity prices. Renewable energy sources can also be relevant for other uses. For instance, Bermudez-Contreras *et al.* (2008) discuss renewable energy-powered desalination systems for water-scarce areas. They conclude that such investments can be profitable in tropical destinations, where amortization horizons can often be in the order of a few years. Overall, studies indicate that in many countries there are government-backed programmes, rebates and tax credits, which help to pay part of the costs for setting up renewable energy systems. These, and many other technology-based efficiency measures, are documented by various organizations working with energy audits, benchmarking and retrofitting in accommodation.

CONCLUSION

This chapter outlines some of the mechanisms by which mitigation of the emissions that contribute to climate change are being tackled. Four main areas were outlined. First, there are policies at various scales of governance, ranging from the international through to the regional. Indeed, many local, municipal and provincial governments are often taking the lead on mitigation to the extent that their activities may encourage national action. Second, carbon management is an approach being promoted in various sectors, both as the result of government and regulatory encouragement, and through voluntary actions. Third, changes to consumer behaviour are also regarded as a means of reducing emissions from travel in terms of either engaging in forms of low-emission travel and tourism, or by voluntary offsetting, although the voluntary engagement in such programmes is not encouraging. Finally, technology is being heavily promoted as a means to cut GHG emissions. However, as this chapter indicates, there are serious issues as to whether such efficiency gains are enough to counter overall growth in travel and tourism activity.

FURTHER READING AND WEBSITES

A book dedicated to identifying strategies for mitigation is:
   Gössling, S. (2010) *Carbon Management in Tourism: Mitigating the Impacts on Climate Change.* London: Routledge.
For information, carbon calculators and offsetting, see:
   atmosfair, highest ranked carbon offset provider: www.atmosfair.de.
   UK Department for Environment, Food and Rural Affairs, Carbon Calculator: http://carboncalculator.direct.gov.uk/index.html.

US Environmental Protection Agency, Emissions Calculator: www.epa.gov/climatechange/emisssions/ind_calculator.html.

United Nations Environment Programme (UNEP), Climate Neutral Network: www.unep.org/climateneutral/Topics/TourismandHospitality/tabid/151/Default.aspx.

UNEP biofuel guidelines and further information about biofuel issues are available from: www.unep.fr/energy/bioenergy.

Climate change impacts on
5 destinations

'Weather can ruin a holiday, but climate change can ruin a destination.'
Daniel Scott, Second International Conference on Tourism
and Climate Change, Davos, Switzerland (2007)

As discussed in chapter 2, the earliest studies on the implications of climate change for tourism that took place in the mid-1980s focused on the potential impacts of a changing climate on tourism destinations and specific tourism operations common to those destinations. This chapter provides a synthesis of the large and rapidly growing literature on the impacts of climate change on tourism destinations that has developed over the past 25 years. The chapter begins by defining key concepts used in understanding place-specific impacts of climate change and provides a conceptual framework of the different pathways by which climate change can affect tourism destinations. This framework provides the structure for the discussion of climate change impacts throughout the rest of the chapter. Because so few studies of climate change impacts on tourism have incorporated adaptive responses, most of the impacts discussed in this chapter must be considered 'potential climate change impacts' rather than 'residual climate change impacts' (see definitions in Box 5.1). The exceptions where adaptations have been accounted for are noted. The chapter concludes with a discussion of the recent phenomena variously termed by the media and now some academics as 'disappearing destinations', 'tourism of doom' or 'last-chance tourism', and considers whether such terms accurately represent the impacts of climate change on tourism destinations.

CLIMATE CHANGE IMPACT PATHWAYS

Chapter 2 demonstrated the many ways in which tourism operators and destinations are sensitive (see Box 5.1) to climate variability and extremes. Climate also has very important indirect impacts on the tourism sector through its effects on a wide range of environmental resources that are critical attractions for tourism, such as snow conditions, wildlife productivity and biodiversity, water levels and quality, and environmental conditions that can deter tourists, including infectious disease, wildfires, insect or waterborne pests (e.g. jellyfish, algal blooms). Climate variability and extreme events also influence

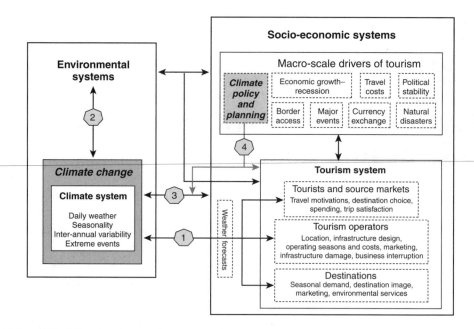

Figure 5.1 *Climate change impact pathways*

a number of other economic sectors, including transportation systems, agriculture and insurance, that closely interact with and influence the tourism system.

The impacts of climate change on tourism are anticipated to be widespread, with no destination unaffected. There are four broad categories of potential climate change impact (see Box 5.1) that will affect the competitiveness and sustainability of tourism destinations. These four impact pathways and their interactions are illustrated in Figure 5.1 (numbered 1 to 4), and are described below.

1 Direct impacts from changing climate regimes

These include changes in mean climate conditions (temperature, precipitation, humidity and winds) and extremes (both frequency and severity of storm events or droughts) that directly affect all parts of the tourism system. Climate co-determines the suitability of locations for a wide range of tourist activities, defines the length and quality of multi-billion-dollar tourism seasons, and for some destinations is the principal resource that attracts tourists. Climate also influences various facets of tourism operations: insurance, food and water supply, heating–cooling costs. Changes in the length and quality of climate-dependent tourism seasons (winter sports holidays) and in operating costs could have considerable implications for competitive relationships between destinations and the profitability of tourism enterprises.

Extreme events, such as hurricanes, floods or drought, can cause heavy damage to tourism infrastructure, and economic losses through business interruption and damage to a destination's reputation. The IPCC (Solomon *et al.* 2007) has concluded that changes in frequency and severity of certain types of weather extremes are probable under climate change, including higher maximum temperatures and more hot days over nearly all land areas, greater tropical storm intensity and peak winds, more intense precipitation events over many land areas, and longer and more severe droughts in many mid-latitude continental interiors (see chapter 1). Such changes will affect the tourism industry through increased infrastructure damage, additional emergency-preparedness requirements, higher operating expenses (insurance, evacuations), and business interruptions.

2 Indirect environmental change and cultural heritage impacts

Climate change also affects many aspects of the natural environment that are very important in determining the attractiveness of a destination and provide essential environmental services. Consequently, as climate change causes lasting alterations to the natural environment of a destination, the tourism product and environmental services can be diminished, with implications for tourism activities, destination image, and the capacity of tourism firms to do business sustainably. Climate-induced environmental changes affecting tourism destinations will include water availability, terrestrial and marine biodiversity loss (e.g. coral reefs), altered wildlife productivity and distribution (e.g. sport fish, bird migrations), altered landscape aesthetic (e.g. loss of glaciers), altered agricultural production (e.g. wine tourism), coastal erosion and inundation, and the increasing incidence of vector-borne diseases. In the same way that climate change will have an impact on some environmental resources that are vital to some tourism destinations, climate change will also affect some cultural heritage assets that are also essential for tourism in some destinations.

3 Indirect impacts associated with societal change

It is well established that climate change will affect many nations and economic sectors and, for some, poses a potential risk to future economic growth and political stability. As the late 2000s economic recession illustrated, tourism is highly sensitive to economic conditions. Global international arrivals declined from 922 million in 2008 to 880 million in 2009 (UNWTO 2010b). Any reduction of GDP resulting from climate change in tourism-generating areas would reduce the discretionary income available to tourism consumers and have negative repercussions for anticipated strong future growth in tourism worldwide.

Climate change is also considered a national and international security risk that will intensify steadily, particularly under greater warming scenarios. Climate change-associated security risks have been identified in a number of regions where tourism is highly important to local and national economies (Raleigh and Urdal 2007). As demonstrated by their historical response to terrorism, armed conflict or even political unrest, tourists, particularly international tourists, are highly averse to political instability and social unrest (Cohen and Neal 2010; Hall 2010d). A security-related decline in tourism would exacerbate deteriorating economic conditions in destinations afflicted by climate change-induced unrest, with the potential to further undermine development objectives in some of the least developed countries.

4 Impacts induced by climate change mitigation and adaptation in other sectors

The international, national and sectoral responses to the challenges of climate change will also have consequences for tourism. As discussed in chapter 4, mitigation policies are likely to have important impacts on the transport sector and consequently on tourist flows, as changes in cost structures cause tourists to reconsider transportation modes and the distances they travel for tourism experiences. Mitigation policies may also contribute to greater environmental awareness of emissions related to travel, which may also foster changes in travel patterns. There has been substantial recent media coverage on this topic, specifically as it relates to air travel. Long-haul destinations could be particularly affected, and officials in Southeast Asia, Oceania and the Caribbean have expressed concern that mitigation policies could have an adverse impact on their national tourism economy. On the other hand, opportunities would arise for low carbon-emission transport modes such as coach and rail, representing an opportunity for destinations nearer to main markets. As the mitigation policy regime is currently being negotiated at the international level and being implemented in some national jurisdictions, if substantial emission reduction targets are adopted, and stringent regulations emerge that have an impact on transport prices over the next 10–15 years, then the impacts related to mitigation policy would be among the first significant effects experienced by the tourism sector.

As chapter 1 indicates, climate change adaptation has taken on new significance during the 2000s, as the scientific community and governments recognize that the planet is committed to further climate change as a result of past GHG emissions, and also realize the political and economic difficulties of making progress on the large emission reductions required to stabilize GHG concentrations and avoid 'dangerous' climate change. All nations and economic sectors are in the process of assessing their adaptation options and adaptive capacity. The state of climate change adaptation in the tourism sector is discussed in chapter 6. Decisions by governments and other economic sectors will have potentially important implications for tourism. For example, in water-stressed areas, governments will have to decide how to distribute water resources among competing users, and tourism may not be a priority user in some locations. The insurance industry will determine whether properties in high-risk areas such as coasts or estuaries can be insured, or will set high premiums, with major implications for tourism development and operating costs.

All of the major components of the global tourism system (tourists, source markets, transport systems, destinations) will be affected by the four distinct impact pathways, and consequently the combined effects of climate change are anticipated to be far-reaching. It must be emphasized that the regional and local-scale manifestations of climate change will generate both negative and positive impacts, although the literature concentrates largely on the former, and the aggregate impact (see definition in Box 5.1) of climate change on tourism will vary substantially among tourism subsectors and geographical regions. There will be relative 'winners and losers' at the enterprise, destination and nation scales.

It is also important to recall from chapter 1 that climate change is a part of broader contemporary environmental change (deforestation, biodiversity loss) with which it interacts – sometimes synergistically. Climate change is also but one of the macro-scale social, economic, technological and political factors that influence the future of the tourism system. As the IPCC identified, 'as a result of the complex nature of the interactions that exist between tourism, the climate system, the environment and society, it is difficult to isolate the direct observed impacts of climate change upon tourism activity' (Rosenzweig *et al.* 2007: 111). The potential implications of the long-term interactions with these other macro-scale factors remains largely unknown (Scott 2006a; Hall and Lew 2009).

BOX 5.1 IPCC CLIMATE IMPACT ASSESSMENT TERMINOLOGY

Sensitivity: the degree to which a system is affected, either adversely or beneficially, by climate variability or climate change. The effect may be direct (e.g. a change in crop yield in response to a change in the mean, range or variability of temperature) or indirect (e.g. damages caused by an increase in the frequency of coastal flooding due to sea-level rise).

Climate impacts: the effects of climate change on natural and human systems. Depending on the consideration of adaptation, one can distinguish between potential impacts and residual impacts:

- Potential impacts: all impacts that may occur given a projected change in climate, *without considering adaptation*.
- Residual impacts: the impacts of climate change that would occur *after adaptation*.

Aggregate impacts: total impacts summed up across sectors and/or regions. The aggregation of impacts requires knowledge of (or assumptions about) the relative importance of impacts in different sectors and regions. Measures of aggregate impacts include, for example, the total number of people affected, change in net primary productivity, number of systems undergoing change, and total economic costs.

Market impacts: impacts that can be quantified in monetary terms and directly affect GDP, for example, changes in the price of agricultural inputs and/or goods.

Non-market impacts: impacts that affect *ecosystems* or human welfare, but that are not easily expressed in monetary terms, for example, an increased risk of premature death or increases in the number of people at risk of hunger.

Climate impact assessment: the practice of identifying and evaluating the detrimental and beneficial consequences of climate change on natural and human systems. Others use the term climate risk assessment to refer to climate impact assessment. Climate risk management is a generic term referring to an approach to climate-sensitive decision-making, that attempts to integrate disaster management and development strategies to maximizing positive and minimizing negative outcomes of climate change. It is not a term utilized by the IPCC, but is used by some international organizations such as the World Bank.

Vulnerability: the degree to which a system is susceptible to, or unable to cope with, adverse effects of climate change, including climate variability and extremes. Vulnerability is a function of the character, magnitude and rate of climate variation to which a system is *exposed*, its *sensitivity*, and its *adaptive capacity*.

Integrated assessment: a method of analysis that combines results and models from the physical, biological, economic and social sciences, and the interactions between these components, in a consistent framework, to evaluate the status and the consequences of environmental change and the policy responses to it.

(IPCC 2007c)

DIRECT IMPACTS FROM CHANGING CLIMATE REGIMES

Changes in climate will directly affect tourism in four main ways, by altering the geographical and temporal distribution of climate resources as a push–pull factor for tourism; the length and quality of climate-dependent tourism seasons; operating costs; and damage caused by extreme events.

The redistribution of global climate resources for tourism

The redistribution of climatic resources will be one of the most direct impacts of climate change on tourism destinations. A number of the scholars who first considered the implications of climate change for tourism speculated that seasonal shifts in climate resources for tourism would benefit higher latitudes, such as northern Europe and Canada, and be detrimental to destinations such as the Mediterranean and the southern USA's 'sunbelt' (Smith 1990; Wall 1992; Perry 1997). Examining the potential changes in the length and quality of tourism operating seasons and shifts in natural seasonality was identified as a research priority by Butler (2001: 17), who regarded the small number of publications on the ramifications of climatic change on tourism seasonality as 'a disturbing fact'.

The same year, the first assessment appeared of how climate change could alter the temporal and geographical distribution of climate resources for tourism. Scott and McBoyle (2001) utilized a modified version of Mieczkowski's (1985) 'tourism climatic index' (TCI), which combines seven climate variables into an easily interpretable scale (essentially 0–100, where 50–59 is 'acceptable', 80–89 'excellent' and 90–100 'ideal') to examine the potential seasonal changes in climate resources across major tourism destinations in Canada. They found that, under projected climate conditions, the currently limited warm-weather tourism season in Canada would both lengthen and improve, thus increasing the tourism 'pull factor' and enhancing the competitive position of Canadian destinations in the international tourism marketplace. Furthermore, with a decreased tourism 'push factor' reducing the need for winter getaway holidays as winters become shorter and less severe, there was the potential for Canada's international tourism trade deficit to diminish under climate change.

This approach was then expanded to destinations across Canada, the USA and parts of the Caribbean, to assess the broader implications for competitive relationships between North American destinations, and the study concluded there was the 'potential for a substantive redistribution of climate resources for tourism in the later decades of the 21st century', with salient implications for competitive relationships among destinations by altering the push–pull factors influencing decisions about tourist destinations, particularly in the shoulder and winter seasons (Scott *et al.* 2004: 116). The number of cities in the USA with 'excellent' or 'ideal' ratings in the winter months was projected to increase, so that southern Florida and Arizona could face increasing competition for winter sun holiday travellers and the seasonal 'snowbird' market (originating from Canada and the northern states of the USA). In contrast, lower winter ratings in Mexico suggest it could become less competitive as a winter sun–sand–sea (3S) destination.

Since then, several studies have similarly used some version of the TCI to examine the future redistribution of climate resources for tourism under climate change at the national or global scale (others have used the TCI to examine how climate resources have evolved over the past 50 years, but are not discussed here). The findings are summarized in Table 5.1 from an annual perspective (net seasonal gains and losses); readers should consult the original publications for more details on specific seasonal patterns. These studies reveal generally consistent temporal and geographical patterns, and Amelung *et al.* (2007) are particularly illustrative of broad regional patterns, as they provide a global analysis. They conclude that there is a pronounced movement in tourism comfort to higher latitudes as time progresses (for the 2020s, 2050s and 2080s).

> By the 2080s, the most ideal conditions for tourism activity in the northern hemisphere will have shifted to the countries of northern Europe and Canada. ... [and the] only region that appears to maintain its "ideal" level of tourism comfort throughout the entire period is that centered on the country of Tajikistan in central Asia.
>
> (Amelung *et al.* 2007: 289)

Most of the studies in Table 5.1 explored the implications of a range of climate change scenarios, and the impacts on the redistribution of climate resources for tourism were much greater under high emission scenarios. In several cases, the impacts of the high emission scenarios in the 2050s exceeded the impacts of the low emission scenarios in the 2080s.

Table 5.1 Projected temporal and geographical shifts in climate resources for tourism under climate change

Time periods compared*	Destinations with improving climate resources	Destinations with deteriorating climate resources	Study
1970s	2050s, 2080s	Canada	Some urban destinations in summer season	Scott and McBoyle (2001)
1970s	2050s, 2080s	Canada, northern USA, mid-USA in spring–autumn	Southern USA, Caribbean in summer season	Scott et al. (2004)
1970s	2080s	Northern Europe, southern Scandinavia, UK	Mediterranean, North Africa, Spain, Portugal, Black Sea	Amelung and Viner (2006)
1970s	2050s, 2080s	Northern Europe, southern Scandinavia, south-central Russia, northern and eastern China, northern USA, Canada, southern Australia, South Africa	Mediterranean, Black Sea, Middle East, northern Africa, southern and central USA, South Korea, southern Japan, southern Indonesia, northern Australia	Amelung et al. (2007)
1970s	2020s, 2050s, 2080s	North-west Europe, Low Countries, southern Scandinavia, UK	Mediterranean, Spain, Portugal, Black Sea	Nicholls and Amelung (2008)
1970s	2060s	Atlantic coast of Spain and France, Baltic Sea	Mediterranean coast of Spain	Moreno and Amelung (2009) – beach tourism only
1970s	2070s	Northern Europe, southern Scandinavia, UK	Mediterranean, Spain, Portugal, Black Sea	Hein et al. (2009)
1970s	2020s, 2050s	South-west England		Whittlesea and Amelung (2010)
1970s	2080s	Northern and central Europe	Southern Europe, Mediterranean Basin	Perch-Nielsen et al. (2010b)
1970s	Not specified	North America, Western Europe, Eastern Europe	Caribbean, Central America, South America	Moore et al. (2010)

*Each typically refers to a 30-year climatological period centred on the decade indicated: 1970s, 1961–90; 2020s, 2010–39; 2050s, 2040–69; 2080s, 2070–99

While the empirical findings in Table 5.1 are somewhat intuitive and consistent with the nature of changes anticipated by the earliest commentaries on climate change and tourism, they are not without criticism. Mieczkowski's (1985) TCI has been subject to a range of critiques. Primary among these is that the rating and weighting schemes are subjective and not empirically tested against the preferences of tourists (Scott and McBoyle 2001; de Freitas 2003; Gómez-Martín 2005; de Freitas *et al.* 2008; Scott *et al.* 2008b). Other noted limitations have included insufficient temporal and spatial scale (although many of the studies in Table 5.1 have overcome this limitation), a reliance on climate means without consideration of variability or probability of key weather conditions, a lack of consideration for the overriding effect of physical parameters (rain and wind) under certain conditions (de Freitas *et al.* 2008), and the generalizability of the rating scheme to represent the varied climate preferences of tourism market segments and the climatic needs at diverse destinations (e.g. coastal, urban and mountain destinations) (Scott *et al.* 2008b). As our understanding of the climate preferences of tourists and specific market segments improves (e.g. Morgan *et al.* 2000; Scott *et al.* 2008b; Moreno 2010; Rutty and Scott 2010), this knowledge can be used to modify (or validate) existing tourism climate indices, and to create specialized indices or rating systems that determine the probability that preferred or acceptable weather conditions will occur. These improved and tourism-validated climate rating systems will provide a more accurate assessment of the future redistribution of climate resources for tourism and the implications for destinations.

More recently, Denstadli *et al.* (2011) have questioned the findings of Amelung and Viner (2006) and Amelung *et al.* (2007) that rate much of Norway and northern Scandinavia as climatically 'unfavourable' for tourism in the summer season. They contend this rating is at odds with the results of tourist surveys conducted on weather expectations and satisfaction, while such results are also at odds with the actual visitation patterns for the region (Hall *et al.* 2008). A reason for this apparent contrast could be that Amelung and Viner (2006) and Amelung *et al.* (2007) use gridded climate data, which combine the weather conditions of available stations in a single value for an area 0.5° latitude by 0.5° longitude (roughly 2030 km² at 49°N latitude). This process involves combining multiple climate stations that, in varied and complex terrain such as the Norwegian Fjords or the Alps of Austria and Switzerland (which are also rated as 'unfavourable' in these three studies), may be situated in a range of elevations. As a result, the very pleasant valley conditions, where most tourism activities take place, may not be well represented in gridded climate data, which are even more abstracted from the weather conditions of tourism areas than climate stations can be (see discussion of the adequacies of the current climate station network for tourism in chapter 2). One of the ironies of such an assessment is that many of the alpine valleys are extremely popular in summer because they provide a more pleasant climate than some of the hotter lowland regions and major cities in Europe. Further analysis is required to better assess the adequacy of using gridded climate data to represent tourism destinations in such regions. Only two of six Mediterranean destinations were classified differently when using station and gridded data, with one being rated higher (Larnaca, Cyprus) and one lower (Antalya, Turkey) using station data. The challenges of adequately representing coastal microclimates with gridded data are elaborated on in the next section.

The Mediterranean Basin

The Mediterranean Basin has been one of the most studied tourism regions from the perspective of changing climate resources, and the attention paid to the related analyses by the media and government climate change assessment reports is useful to further articulate the limitations of the TCI approach. That the Mediterranean has been a focus for tourism-related climate change impact assessments is understandable, because it is considered among the world's leading tourism destinations in terms of international arrivals, attracting almost 20 per cent of total international tourism arrivals (UNWTO 2010a), and one of the principal reasons behind the popularity of the region is the demand for predictable sunny and warm coastal destinations.

With climate as one of the principal attractions of the Mediterranean region, and the projection for hotter and drier summer conditions in the decades ahead, with 'likely' (>66 per cent) increases in the risk of heatwaves (Alcamo *et al.* 2007), much speculation has occurred with regard to the impact of climate change on the climate resources for tourism in the region (Smith 1990; Perry 1997). Perry (2006: 368) contends that 'many Mediterranean beach resorts may simply be too hot to be comfortable in the peak season ... [and that] single-product beach destinations are likely to be the most vulnerable'. As illustrated in Table 5.1, a number of studies using the TCI approach have similarly concluded that climate change will push Mediterranean temperatures above the threshold for tourist comfort during the summer peak season, and that the best climatic conditions for tourism will occur in the spring and fall shoulder seasons.

The assertion that Mediterranean summer tourism will be adversely affected by temperature increases has also appeared in prominent assessment reports by the IPCC (Alcamo *et al.* 2007; Wilbanks *et al.* 2007) as well as by a number of other international agencies (Landau *et al.* 2008; Scott *et al.* 2008a), governments (Whittlesea and Amelung 2010) and industry (Deutsche Bank Research 2008; KPMG 2008). Pronouncements in the media and by travel writers have often been alarmist, declaring that 'the likelihood [is] that Mediterranean summers may be too hot for tourists after 2020' (*The Guardian* 2006, based on Amelung and Viner 2006); and that by 2030 'the traditional British package holiday to a Mediterranean beach resort may be consigned to the "scrap-heap of history"' (BBC 2006, based on the *Holiday 2030* report produced by Halifax Travel Insurance 2006).

How accurate are such statements? As Scott *et al.* (2008a) point out, while the concern has been that the Mediterranean may become 'too hot' for tourism during the peak summer tourism season, at no point has 'too hot for tourists' ever been defined on the basis of the literature or quantified on the basis of consultations with either tourists or tourism operators in the region. Other important questions have also not been asked: how do perceptions of what is too hot differ among major tourism segments in the region (travellers from northern Europe seeking a sunshine beach holiday versus a seniors tour visiting the cultural attractions in a major urban centre such as Rome or Venice); to what extent do coastal microclimates and urban heat islands influence the thermal conditions at tourism destinations; do the projections change when more sophisticated tools to measure human thermal conditions (e.g. physiological equivalent temperature) are used in place of the TCI?

Moreno and Amelung (2009) modified the TCI to the specific preferences and requirements of beach tourists on the basis of surveys done by Morgan *et al.* (2000), and reassessed the redistribution of climatic resources for coastal tourism across Europe. With the TCI tailored to beach tourism and assessing only the peak tourism season of July and August, they found that, under moderate emission scenarios, the Mediterranean remained a climatically 'very good to excellent' destination for beach tourism through the 2060s. Under the warmest scenarios, parts of the Mediterranean (Spanish coast) were projected to suffer decreased climate suitability for beach tourism, while other regions on the Mediterranean retained high ratings. Notable climatic improvements were projected along the Atlantic coasts of Spain and France and the Baltic Sea. In contrast to previous claims, this study concluded that, climatically, the Mediterranean is likely to remain Europe's prime region for summertime beach tourism for the next 50 years, and that improvements in other non-Mediterranean coasts would not be enough to attain the climate suitability of the Mediterranean.

In their reassessment of whether the Mediterranean would be too hot for tourism, Rutty and Scott (2010) utilized surveys of university students in six European countries (thus representative of only the young adult traveller market) to determine tourists' perceptions of 'ideal' conditions as well as thresholds of 'too hot/cold' for the two leading destination types (3S holidays and urban–cultural sightseeing holidays). The thresholds identified by tourists as too hot for beach (>36°C) and urban (>30°C) holidays were then used to assess if ten high-profile Mediterranean destinations would become too hot under climate change scenarios. Table 5.2 indicates when the average July–August temperatures at each of the ten destinations exceeded the too hot thresholds established by the tourist survey. While average summertime temperatures at some, but not all, of the selected destinations did enter the temperature range that was considered to be too hot in the warmest scenarios for the mid- to late twenty-first century, none that was not already classified as too hot surpassed this threshold in the early part of the century (2020s and 2030s) as proclaimed in the media. Thus previous statements that the Mediterranean would become too hot for tourism or beach tourism appear both temporally and geographically inaccurate. Indeed, urban destinations, not coastal destinations, have greater potential for temperature-related impacts.

Further to this point, if we consider the location of climate stations (typically at airports or other urban areas) in relation to the microclimate conditions at coastal and urban destinations, they are probably reasonably representative of temperatures in urban destinations, although they may slightly underrepresent temperatures that are enhanced by urban heat islands when located away from urban core areas. The extent to which these inland, urban climate stations represent the temperatures of coastal destinations and resorts in small towns or on coasts far from major urban areas is uncertain (a concern raised in chapter 2). Modifying Rutty and Scott's (2010) analysis to account for the microclimate offered in coastal areas [the thermal moderating influence of the sea and the development of coastal breezes in the afternoon – ranging from 4 to 8°C in the early to late afternoon, according to studies in coastal tourism areas of Italy (Höppe and Seidl 1991) and Greece (Papanastasiou *et al.* 2010)] illustrates that the average daytime high temperatures in the summer months would not be considered too hot at three of the five coastal destinations,

Table 5.2 Could Mediterranean destinations become 'too hot' for summer tourism?

Urban destinations	Summer* temperatures become 'too hot'[†]	Beach destinations	Summer* temperatures become 'too hot'[†]	Adjusted for coastal microclimate[†]
Istanbul, Turkey	1961–90	Antalya, Turkey	1961–90	1961–90
Athens, Greece	1961–90	Larnaca, Cyprus	1961–90	1961–90
Venice, Italy	1961–90	Milos, Greece	2046–65	Threshold not exceeded
Marseilles, France	2080–99	Nice, France	2080–99	Threshold not exceeded
Barcelona, Spain	2046–65	Costa Brava, Spain	Threshold not exceeded	Threshold not exceeded

*Average daytime high temperatures for July and August
[†]Analysis includes combined influence of temperature and humidity using Environment Canada's (2009) 'Humidex' formula
[†]The upper range of estimates in the literature was utilized to account for the physiological cooling effect of coastal breezes (6°C cooling effect), not just the temperature difference between coastal and inland areas

even under the warmest climate change scenario. As these exploratory results clearly suggest, more detailed analysis is needed to understand the future evolution of climate resources for destinations in this region and elsewhere.

Changes in tourism season length

The second major theme that scholars, and to some extent the tourism industry itself (either individual tourism operators or industry associations), have examined with respect to the direct impacts of a changing climate on tourism has been the destination or sectoral implications for the season length of key activities. More specifically, this has included golf tourism, skiing and winter sports tourism.

Assessment of changes in the length of the warm-weather tourism season has been limited and concentrated in countries where these tourism activities are currently constrained by climate, and thus where there are potential opportunities for tourism to exploit. For example, in Canada the length of the golf season was projected to increase across the country, but the largest gains were expected in the Great Lakes region (27–86 days longer in the 2050s) and along the coast of the Atlantic Ocean (28–56 days longer in the 2050s) (Scott and Jones 2007). Under the warmer climate change scenarios, the extended golf season was very notable to the golf industry and regional tourism marketing organizations that have recently created golf tourism products in both regions, but particularly to Prince Edward Island, which has marketed itself as a golf destination.

The extension of golf seasons does not come without operational challenges. As the climate warms, there will be greater demand for irrigation in many regions (Rodríguez-Díaz et al. 2010). For example, in the Great Lakes region, soil moisture deficits in the summer are projected to increase 24 per cent in the 2020s and 40 per cent in the 2050s (Scott and Jones 2006b). Water resource-management challenges are discussed in greater detail later in this chapter. Aspects of pest and disease management could also be affected by projected changes in the climate, posing challenges to the maintenance of playing conditions and the perception of what a healthy golf course resembles. In some regions, insect pests that currently have only one life-cycle per year could begin to have two life-cycles under warmer conditions. Turf grass diseases and pests currently limited to latitudes that are more southerly could also expand northward and require management interventions. It remains uncertain how changes in irrigation and turf grass disease/insect management would affect the ability of golf course managers in different regions to take advantage of opportunities for a longer and more intensive golf season under climate change.

In Finland, the extension of the summer season as a result of climate change has been identified by tourism entrepreneurs as providing new business opportunities, particularly in the lake districts in the south of the country. According to Marttila et al. (2005), the major impacts on Finnish summer tourism were mainly positive: a longer, sunnier and warmer season, and warmer waters. The noted negative effects were mainly related to increasing algal growth in coastal and shallow lake waters. Although concerns were expressed by entrepreneurs as to whether an increase in summer precipitation would affect tourists' willingness to participate in nature-based tourism activities, such as canoeing or boating, entrepreneurs found the lengthening of summer season to be a positive. However, they also noted that a critical issue for being able to take advantage of the extended summer season would be the institutional holiday periods that were regarded as the decisive factor for the timing of the peak season (Saarinen and Tervo 2006).

In high-latitude and, to a lesser extent, high-altitude destinations, climate change is also lengthening the summer season in time and extending it in space. Climate change has increased the period over which cruise and expedition ships operate in the summer. This has occurred because changes in conditions have made some locations, such as the Antarctic Peninsula, Greenland and Baffin Island, more accessible, while temperature increases have also increased the comfort levels of tourists (Hall and Saarinen 2010a, 2010b). However, in high-altitude and alpine destinations, more concern has usually been paid to the reduced length of the winter season rather than the extension of the summer season.

Skiing and winter tourism

In contrast, the risks posed by climate change to the length of the ski season and winter sports in general have received considerable attention. The ski industry was the first and the most studied aspect of climate change impacts on tourism, with over 30 known studies in 13 countries. The early focus on skiing and winter sports tourism is understandable, even strategic, as this subsector is the most directly and most immediately affected by climatic change (Scott 2011). This geographically and methodologically diverse literature has consistently projected that, from an operations perspective, the ski tourism industry is at risk from climate change through decreased reliability of natural snow cover ('natural snow reliability'), shortened and more variable ski seasons, increased snow-making requirements and decreasing snow-making opportunities ('technical snow reliability'), and a contraction in the number of ski areas. The extent and timing of impacts are highly dependent on the magnitude of climate change and whether the study accounted for snow-making (see discussion of limitations below). As the contrasting statements below illustrate, perspectives within the ski industry on the relative risk posed by climate change remain varied (ski industry perspectives on climate change are discussed further in chapter 6).

'Climate change is the most pressing issue facing the ski industry today.'
Patrick O'Donnell, Chief Executive Officer of Aspen Skiing Company

'We are confronted with the jostling between conflicting opinions. A few years ago they said that we would have less snow, we had a lot of snow and they told us this was an exceptional case. We had five exceptional cases in a row. They also said it would become warmer and then we had a freezing cold winter. This doesn't increase our trust in studies. We can't evaluate the long-term effects of climate change with the results we have been given so far.'
CEO of a cable car company in Tyrol (from Trawöger 2010)

'… whether it's actual climate change or whether it's a glitch in the weather cycle or not, who knows because I don't think we've been keeping records long enough.'
CEO of Australian ski resort (from Bicknell and McManus 2006)

Climate change assessments by the ski industry have utilized two main methodologies to estimate future changes in snow conditions and operational indicators: ski-season length and snow-reliable ski areas. The regional results for both approaches are summarized in Table 5.3.

Modelling-based approaches are far more common. A number of these studies suffer from important limitations that limit their validity and relevance for decision-makers. The use of inappropriate impact indicators for ski operations is a first limitation. Some studies simply model natural snow conditions near ski areas, with no specific indicators for ski operations – for example, what happens to the length of the ski season, opening–closing dates (Harrison *et al.* 1999, 2001; Uhlmann *et al.* 2009; Endler and Matzarakis 2011)? Other studies have utilized indicators that are not relevant to ski-area operations, such as snow cover (which is a meteorological variable defined as 2.5 cm of snow, when ski operators require 30–100 cm of snow to open a ski run, depending on terrain) (McBoyle *et al.* 1986; Lamothe and Periard Consultants 1988); or snow-water equivalent on the first day of April (which is a widely used hydrological indicator for summer water supply in

Table 5.3 *Regional climate change impacts in the ski industry*

Region/study	Number of ski areas	Projected impacts/ timeframe[*]	Limitations[§]	Source
North America				
Michigan	1	−12 to −65[†]		Scott et al. (2006)
Mid-west region (USA)	44	−7		Scott et al. (2012)
New England	14	−8 to −38[†]		Scott et al. (2008c)
North-east region (USA)	45	−12		Dawson et al. (2009)
Ontario	3	−8 to −46[†]		Scott et al. (2003)
Quebec	3	−4 to −39[†]		Scott et al. (2003)
Quebec	6	−10		Scott et al. (2011)
South-east region (USA)	25	−17		Scott et al. (2012)
Alberta	3	Low elevation −1 to −43[†]; high elevation 0 to −6[†]		Scott and Jones (2006b)
Rocky Mountain region (USA)	57	−4		Scott et al (2012)
Pacific region (USA)	48	+5[†]		Scott et al. (2012)
Washington (USA)	3	Low elevation −42 per cent ski days (2040s); high elevation −10 per cent ski days (2040s)	2	Casola et al. (2005)
California	34	−24 to −52	2	Hayhoe et al. (2004)
Europe				
Alps region	609	Snow reliable[¶]; ski areas decline from 609 to between 404 and 202	1, 2	Abegg et al. (2007)

Continued

Table 5.3 *Continued*

Region/study	Number of ski areas	Projected impacts/ timeframe[*]	Limitations[§]	Source
Austria	3	−5 to −29[†]		Steiger (2010)
Austria/Italy	111	Snow reliable[¶]; ski areas decline from 111 in 2030s to between 105 and 85 in 2050s		Steiger (2011a)
Austria		−7 to −12		Steiger (2011b)
Australia				
South-east Australia	9	−15 to −99[†]		Hennessy et al. (2008)

[*]In percentage ski days, 2050s analogue, unless otherwise stated
[†]Also provides projections on snowmaking
[‡]The 2050s temperature analogue winter also had increased snowfall
[¶]Using '100-day' criteria
[§]Limitations: 1 = no site-specific physical snow model used; 2 = snowmaking not considered

mountainous regions, but one that provides no insight into snow accumulation in the vital start of the ski season or the number of days with necessary operational snow depth) (Zimmerman *et al.* 2006). Other studies have provided insight into factors relevant to ski-area management, such as potential timing of wet avalanches as an indicator of safety hazard to skiers, or density of upper layer of snowpack as indicator of snow quality (Lazar and Williams 2008), but do not provide information on the factors pertinent to ski-area operations and profitability, such as season length. Yet others estimate ski-season length solely on the basis of temperature and rainfall, disregarding any measurement of snow depth and thus whether there is a snow product to ski on (Heo and Lee 2008). These studies are not included in Table 5.3.

Other studies have not used physical snow models to estimate future snow conditions, but rather use statistical relationships between snow depth (or key operational snow-depth thresholds for skiing, such as 30 cm) and other climatological parameters such as monthly average temperature and precipitation (Galloway 1988) and days with snow fall (Moen and Fredman 2007). However, such studies do not allow snow-making to be incorporated in the analysis because there is no actual measurement of snow depth, and the amount of snow required to supplement natural snow to make ski areas operational remains unknown. These studies are not included in Table 5.3.

A third limitation, common to most climate change assessments of the ski industry, has been the omission of snow-making in regions where snow-making has been an integral part of ski operations for many years (or decades in some cases). As illustrated in Figure 2.11, snow-making has been almost universal (covering nearly 100 per cent of skiable terrain) to ski areas throughout eastern Canada and the USA for 20 years, and today is widespread throughout the major ski regions of North America, the European Alps, Japan and Australia, and increasingly in China, New Zealand and Scandinavia (Scott 2003, 2006b; Abegg *et al.* 2007; Hennessy *et al.* 2008). Consequently, analyses that do not account for snow-making do not reflect the current operating realities of many ski operations, let alone their adaptive capacity 25 years from now, when studies have found most ski-area managers plan to have enhanced snow-making capabilities (Scott and Jones 2005, 2006a; Wolfsegger *et al.* 2008; Hoffman *et al.* 2009). Studies that have fully incorporated snow-making in physical snow models (Scott *et al.* 2003, 2007a, 2008c, 2011; Hennessy *et al.* 2008; Steiger 2010, 2011b) have all found that the impact of climate change on ski areas was substantially lower than reported in previous studies that did not include snow-making. This core climate adaptation and an evaluation of its sustainability (in terms of snow-production days, costs, water and energy usage) should be incorporated into all future studies.

A common sentiment heard from ski area managers in Canada, USA and parts of Europe, illustrated in the quote below, is that in terms of assessing the potential impact of climate change, while there is interest in projections of changes in natural snow fall, what they really want to know is the future of temperatures suitable for efficient snow-making. Studies that focus solely on the impact of climate change on natural snow, and do not account for snow-making or provide information on temperatures needed for snow-making, particularly early in the season when snow-making is concentrated, are of little use to the tourism industry.

'In fact we don't need snow, we make snow. Too much natural snow is bad for our business because it means higher costs for the preparation. The skiers only complain about natural snow pistes, they want smooth slopes, which we can only provide with the help of artificial snow. It sounds absurd but the best scenario for us is less natural snow, cold temperatures for the snow production and lots of sun.'

CEO of a cable car company in Tyrol (from Trawöger 2010)

The second approach used to examine the impact of climate change on the ski industry has been climate change analogues (see Box 5.2). Despite the number of record warm winters in the past decade, the analogue approach remains under-utilized (see Scott 2006b), with three studies completed in North America (Scott 2006b; Dawson *et al.* 2009; Scott *et al.* 2011), and the only study in Europe limited to the Tyrol region (Steiger 2011b). Future analogue studies should be a priority as they provide opportunities to gain insight into the types and effectiveness of adaptive responses (from ski-area operators, various market segments of ski tourists, and communities and marketing organizations); to identify key impact thresholds; to understand the relative vulnerability of different ski destinations and types of ski operator; and to provide useful comparisons to validate modelling studies.

BOX 5.2 CLIMATE CHANGE ANALOGUES AS SOCIAL EXPERIMENTS

Climate change analogues for understanding the vulnerability of social systems are of two types.

> Temporal analogues use past and present experiences and responses to climatic variability, change and extremes to provide insights for vulnerability to climate change; spatial analogues involve conducting research in one region and identifying parallels to how another region might be affected by climate change.
>
> (Ford *et al.* 2010: 374)

Temporal analogues, such as a record warm winter or major heatwave, have been referred to as 'natural social experiments', where real events provide opportunities to examine associated impacts and the effectiveness of the full range of adaptive responses within social systems, including insights into adaptation processes, barriers and failures.

A limitation of the analogue methodology is the inability to reflect future social conditions, including technological advances, changing behavioural responses (e.g. tourists, tourism operators, investors, regulators), changing demographics and policy conditions. In addition, analogues are typically available only for short- to medium-term timeframes, as few analogue situations have occurred that are representative of long-range modelled climate futures under high GHG emission scenarios.

The analogue approach requires that data on impact indicators be available for both climate change analogue periods and climatically normal periods (baselines). The careful selection of analogue and baseline climate periods can help to control for other external influencing factors and isolate the influence of climate variability as much as possible.

The climate change analogue approach has been under-utilized in tourism studies (Giles and Perry 1998; Scott 2006b; Dawson *et al.* 2009; Scott *et al.* 2011; Steiger 2011b), although it has tremendous potential to offer new insights into future vulnerability, because it focuses on the observed responses of the entire tourism marketplace to real climatic conditions and captures the integrated effects of simultaneous supply- and demand-side adaptations.

Table 5.3 provides a summary of the most recent impact assessment results for each major ski region. As indicated, the first climate change impact work was done on the ski industry of eastern Canada in the mid-1980s. These studies projected massive reductions in the length of the ski season (season losses of 40–100 per cent in Ontario, 50–70 per cent in Quebec) under a doubled-CO_2 scenario (*c.* 2050s) (McBoyle *et al.* 1986; Lamothe and Périard Consultants 1988). However, these studies did not account for snow-making, which was beginning to become prevalent and was growing rapidly at the time. When a second generation of studies were completed 15 years later, which were able to account for snow-making, the predicted impact of climate change was substantially reduced, with season losses of 8–46 per cent in Ontario (Scott *et al.* 2003), and 4–34 per cent in Quebec (Scott *et al.* 2007a). Most of the ski-season losses in Quebec occurred in the early season and late season. In the eastern USA (New England), with advanced snow-making capabilities, only one of 14 locations was projected to lose more than 25 per cent of its ski season under low emission scenarios by mid-century (Scott *et al.* 2008c). However, high emission scenarios had a much greater impact, with eight locations projected to lose 25 per cent or more of their ski season by mid-century, and half losing 45 per cent or more of their ski season by the end of the twenty-first century.

In order to limit ski-season losses to the levels described above, mid-century snow-making requirements were projected to increase by 62–151 per cent in Ontario (Scott *et al.* 2003); 18–150 per cent in Quebec (Scott *et al.* 2007a); and 3–86 per cent in New England (Scott *et al.* 2008c). At the end of the century, several locations required more than double the machine-made snow of today, while in other locations warm temperatures made snow-making unfeasible during parts of winter.

The record warm winter of 2001–02 is representative of projected future average winter climate conditions in eastern Canada and the USA under a medium- to high-GHG emission scenario for the 2040–69 period. Comparison of ski-area performance indicators during the analogue and climatically normal (for 1961–90 period) winters revealed substantial differences. The ski-season length in the climate change analogue was shortened in all regional ski tourism markets [10 per cent in Quebec (Scott *et al.* 2011); 11 per cent in New England (Dawson *et al.* 2009)] and snow-making requirements increased. When examining the differential vulnerability of ski businesses, smaller resorts consistently experienced the greatest loss in season length during analogue years followed by medium, large and then extra-large ski areas (Dawson *et al.* 2009).

The 2001–02 climate change analogue also provided valuable insight into the performance of models used to assess future impacts on ski-season length. Scott *et al.* (2008c) modelled ski-season lengths in the New England region for the 1961–90 baseline and the 2050s time period under six climate change scenarios. The regionally averaged 1961–90 baseline modelled ski season was 132 days, which compares well with the 133 days reported for the two climatically average years (for 1961–90) reported by Dawson *et al.* (2009). Ski seasons were projected by Scott *et al.* (2008c) to decline to 116 days in the mid-range GHG scenario for the 2050s, which is a greater reduction than analogue winter (128 days). The differences between modelled season length and analogue winter can be explained by the inability of physically based snow-modelling techniques to account for business decisions (opening under even very marginal conditions because of staffing inflexibility and to provide some level of skiing, perhaps only one or two open runs, to guests staying in resort accommodations), which can lead to the over-estimation of season-length losses.

Although the Rocky and Sierra Nevada Mountains of western North America are home to some of the continent's most widely known ski tourism destinations, the implications for major ski areas in the region have not yet been comprehensively examined. A 2050s analogue winter (medium to low emission scenario) in the Rocky Mountain region saw the ski season shorten by only 4 per cent (Scott *et al.* 2011). With snow-making, ski seasons at Banff (Alberta, Canada) decrease by 1–43 per cent at low elevations and 0–6 per cent at high elevations under 2050s scenarios (Scott and Jones 2005). A 2050s temperature analogue winter with increased precipitation (medium emission scenario) in the Pacific Northwest region produced a 5 per cent longer ski season due to increased snow fall at high elevations (Scott *et al.* 2011). Considering only changes in natural snow conditions, the ski season in the Sierra Nevada Mountains of California was projected to decrease 24–52 per cent by mid-century (Hayhoe *et al.* 2004). Without snow-making, similar season-length reductions (42 per cent at low elevations and 10 per cent at high elevations) were projected for ski areas in the State of Washington (Casola *et al.* 2005).

A considerable number of studies have been completed on the impact of climate change on skiing in the European Alps (König and Abegg 1997; Breiling and Charamaza 1999; Elsasser and Messerli 2001; Elsasser and Bürki 2002; Abegg *et al.* 2007; Uhlmann *et al.* 2009), with the most recent and those that incorporate snow-making summarized in Table 5.3. The broadest regional analysis, conducted for the OECD, determined that the number of ski areas that were considered naturally snow-reliable declined from 609 (91 per cent) to 404 (61 per cent) under a +2°C warming scenario and further declined to 202 (30 per cent) under a +4°C warming scenario (Abegg *et al.* 2007). By comparison, climate change scenarios for the European Alps project winter warming of 0.75 to 2.6°C by mid-century and 3.5 to 4.4°C by the end of the twenty-first century, with more pronounced warming in winter months (Kotlarski *et al.* 2010). The impacts varied among the five nations examined, with Germany projected to experience the most significant impacts and Switzerland the least (60 and 10 per cent reduction in naturally snow-reliable ski areas, respectively, under a +1°C scenario). Importantly, this heavily cited report did not account for snow-making, even though it noted that over 50 per cent of skiable terrain in Austria, and somewhat lesser proportions in other nations (40 per cent in Italy, 18 per cent in Switzerland, 15 per cent in the French Alps, 11 per cent in Germany), already utilizes snow-making.

More recent studies in the Tyrol region found the potential impact of snow-making on ski-season length is considerable when incorporated (Steiger 2010, 2011a, 2011b). This finding is consistent with second-generation modelling studies in North America. Ski areas near Innsbruck, Austria were found to lose between 5 and 29 per cent of the ski season by the 2050s, but remain snow-reliable (maintain at least a 100-day ski season) through mid-century with significantly increased snow-making (up to 300 per cent in some cases) (Steiger 2010). After mid-century, the technical limits of snow-making were reached in some portions of these ski areas, and a 100-day season could be maintained only on 50 per cent of skiable terrain at all three ski areas until the 2050s under a high emission scenario, and until the 2070s under a lower emission scenario. A broader analysis of 11 ski areas across the Tyrol region of Austria and Italy found that the number of snow-reliable ski areas (using the 100-day rule) decline from 111 in the 2030s to between 105 and 85 in the 2050s (Steiger 2011a). Examining the impact of the record warm winter of 2006–07 in the Tyrol region of Austria, Steiger (2011b) found that ski-operations days at 60 ski areas declined on average 7 per cent (12 per cent at low elevations, 5 per cent at high elevations).

As in North America, the observed reduction in the length of the ski season during climate change analogue winters was not as large as projected by ski-operations models. The available studies in other regions of Europe do not use ski-operation indicators (Harrison *et al.* 1999, 2001 in Scotland), or do not use physically based snow models (Moen and Fredman 2007 in Sweden), and because they are not comparable with other literature, have not been included in Table 5.3.

Australia is the final region where climate change assessments of ski tourism operations have been conducted. Here, the average ski season in the 2050s was projected to decrease 30–99 per cent at lower elevation sites and 15–95 per cent at higher elevation sites among the nine ski areas examined by Hennessy *et al.* (2008). As in North America and Europe, snow-making requirements were projected to increase substantially throughout the Australian ski industry (Bicknell and McManus 2006). Based on Hennessy *et al.*'s (2008) finding that natural snow cover will become inadequate at 65 per cent of sites in the Australian ski resorts by 2020, Pickering and Buckley (2010) suggest that additional snow-making for the six main resorts would require over 700 additional snow guns by 2020, at a capital investment of A\$100 million, and 2500–3300 million litres of water per month. As Pickering and Buckley (2010: 430) note, 'This is not practically feasible, especially as less water will be available.'

Because the literature on the potential impacts of climate change on the ski industry is better developed than for most other areas of the tourism sector, some summary points are worth making. First, the preceding summary of projected impacts on ski-area operations (Table 5.3) focuses on average changes to ski seasons, not future inter-annual variability in ski seasons. Inter-annual variability will be more pronounced under climate change, creating increasingly challenging business conditions. It may not matter to ski-area operators if every ski season by mid-century is a couple of weeks shorter, as much of the season loss will occur at the beginning and end of the season, when skier visits are relatively low. Conversely, three or more consecutive extremely warm winters could cause substantive economic losses that smaller tourism operators may not be able to absorb. Only ski tourism operators can interpret the business implications of change in the average length and variability of ski seasons and, understandably, they assess these business risks confidentially. The impacts of reduced average ski seasons and greater variability in year-to-year ski conditions on skier perceptions and demand will be explored in chapter 7.

Second, studies in all of the ski tourism regions found the impact of climatic change would be more pronounced at lower-elevation ski areas and among smaller ski areas. This has potentially important implications for the development of skiing demand, as lower elevations are often used for beginner ski runs. Whether fewer such slopes would discourage beginner skiers, possibly diminishing the industry's client base over time, remains uncertain. There are divergent views on how best to manage this risk. Some regard market-based contraction of the sector as healthy, while others contend it is essential to retain these nursery ski areas through subsidies, for regional economic reasons and the future of the ski industry (McBoyle and Wall 1992; Bürki *et al.* 2003; Scott *et al.* 2002, 2008c). In North America and Europe, climate change analogues revealed that the impacts of anomalously warm weather on season length were concentrated in the smaller ski areas. These tourism operators are also likely to have less financial capacity to absorb a series of poor years and are considered more vulnerable to climate change (Scott *et al.* 2008c).

Third, it is clear from every ski tourism region where a large number of ski areas have been evaluated that, with the possible exception of the Australian and Scottish ski industries, it is not the entire ski tourism market that is at risk from climate change, but rather individual ski businesses and communities that rely on ski tourism. The probable consequence of climate change will be a contraction in the number of ski operators in most regional markets. Determining where the ski industry is most likely to contract and remain sustainable will be of particular interest to the ski industry, investors, real estate developers, insurers, governments and communities in the next two decades. Although climate change would contribute to the demise of many ski businesses, it could advantage some of the ski operations that remain. If skier demand remained relatively stable, as it has in some climate change analogue years (see chapter 7), remaining ski areas would be in a position to gain market share through diminished marketplace competition. The socio-economic implications of a climate change-induced contraction of the ski tourism marketplace for communities have yet to be examined, but it is clear that, from a tourism planning perspective, communities that are at risk of losing ski resorts, and those where ski resorts are likely to persist, will both need to adapt to climate change, although for very different reasons. The former will need to adjust to reduced winter tourism spending, lost employment and potentially declining real estate prices, and to consider strategies for economic diversification and rebranding as a tourism destination (including new tourism product development). The latter will need to plan for increased tourism development pressures associated with the concentration of ski tourism in fewer areas, including water use for snow-making, real estate development, slope expansion, traffic congestion and tourist crowding. Adaptation in alpine communities will require significant economic investment and careful environmental planning. Adaptation options for the ski industry are discussed further in chapter 6.

A final point with respect to the climate change risks posed to the ski industry is that, although the risks are real, and are increasingly better understood, there is a wide gap between the risk perceptions of the ski industry and what is portrayed in the media. The media in all major ski tourism regions have repeatedly pronounced the impending doom of the ski industry. Many of these headlines have been based on flawed studies. For example, a media release by Colorado College in the USA (which profiled the findings of Zimmerman *et al.* 2006) proclaimed that the 'Ski industry in Rockies may be shut down by 2050'. However, the analysis on which this and other subsequent media headlines ('Rocky future for Colorado ski resorts'; 'Global warming poses major threat to ski industry by 2050'; 'Is Colorado's ski industry doomed due to global warming?') were based did not examine the impact of climate change on *any* performance indicator relevant to the ski industry (operational ski days/season length, capacity to operate during economically key holiday periods, snow-making requirements and costs, or skier visits). The only impact indicator modelled was the natural snowpack (snow-water equivalent) on the first day of April, which provides absolutely no insight into the start or length of the ski season.

Furthermore, the snow-water equivalent model outputs for the first of April in the 2085 scenario (Zimmerman *et al.* 2006) revealed that, while there is substantial loss of snow-pack compared with the 1970s, snow depths at 11 of the 14 study areas exceed 50 cm, suggesting that even without snow-making, ski resorts at all but three locations would still be operational at the beginning of April. In other cases, publications have specifically identified major ski destinations as being particularly vulnerable to climate change without any scientific basis. For example, a report by Halifax Travel Insurance (2006) went so far as to state, 'Whistler [Resort] in Canada ... could be wiped out', with related statements appearing in multiple media outlets, even though no impact assessment of ski operations or even natural snow conditions had ever been conducted at this location.

There are real dangers with this type of misinformation. The Halifax Travel Insurance (2006) report has been cited by articles in a travel magazine and on an international real estate investment website that discussed 'where to buy ski properties to avoid global warming'. Readers are undoubtedly unaware that the statements of this report on some ski areas are scientifically baseless. Investors have become concerned by media stories on the risks of climate change for ski-area operators. Some European banks have imposed financing restrictions on low-lying ski areas, based partially on the findings of early studies that did not account for snow-making. The point here is that studies of the impacts of climate change on ski tourism or other types of tourism and specific destinations are not just academic navel-gazing, but are influencing investment and other types of decision, as well as destinations' reputations. Scholars and practitioners have a strong onus and professional ethic to provide sound science so that misinformation that could be damaging to the ski industry and destination communities is not propagated to investors and skiers.

> 'We try to ... state that snow-deficient winters have always been and will occur again and again and show that we have taken precautions in the form of snow production. The worst part about all this was the banking industry that made use of altitudinal limits of snow reliability in their risk assessments. That was really dangerous!'
>
> CEO of a cable car company in Tyrol (from Trawöger 2010)

Climate change impact assessments of winter tourism have focused almost exclusively on alpine skiing. The vulnerability of other economically important components of winter tourism, such as snowmobiling, has been largely overlooked.

Snowmobiling is a US$25 billion annual business in North America (International Snowmobile Manufacturers Association 2006), and is also significant in the Nordic countries (Tervo 2008). Studies of the snowmobiling season in North America (Scott *et al.* 2002; McBoyle *et al.* 2007) have found it to be much more vulnerable to climate change than is the ski industry, because it is completely reliant on natural snow fall. Unlike the alpine ski industry, the linear nature and long distances of snowmobile trails make widespread implementation of snow-making systems technically and economically impractical. A study of the impact of diminished natural snowpack on the length of snowmobiling seasons in 13 study sites in Canada's non-mountainous regions found much greater reductions than for alpine skiing. In the provinces of Ontario and Quebec, which encompass the densest network of snowmobile trails and the largest number of registered snowmobiles in the country, average snowmobile seasons in the 2020s were projected to be reduced between 11 and 44 per cent under a low-emission climate change scenario, and between 39 and 68 per cent under a high-emission climate change scenario. Under the high-emission scenario for the 2050s, a reliable snowmobiling season would essentially be eliminated from Canada's non-mountainous regions, and this would be the last generation of snowmobile tourists. Very similar conclusions were reached in a comparable study of snowmobile seasons in the New England region of the USA (Scott *et al.* 2008c), where, under high-emission scenarios, reliable snowmobile seasons (longer than 50 days) were restricted to some high elevations and northernmost parts of the region.

Changes in operating costs

A third dimension of the direct impacts of climatic change on tourism is the implications for operating costs of tourism operators, such as annual heating and cooling, snow-making, irrigation, and water supply and treatment costs. Very little has been done to explore these potential changes in business costs in the tourism sector, although the results of studies in the energy (World Bank 2011b) and water sectors (Gude *et al.* 2010) can provide insight into what tourism operators and destinations should anticipate in the decades ahead.

A change in heating and cooling degree-days (HDD, CDD) – climatological indicators designed to reflect the demand for energy needed to heat or cool a home or business – is one visible example. Although the energy expenses of the accommodation sector vary by location and by type of accommodation, it has been estimated that energy costs expressed in terms of gross hotel revenues typically range from 3–6 per cent, but can be as high as 10 per cent for some historical and luxury hotels (Pateman 2001). A large portion of overall energy consumption in the accommodation sector is related to space heating–cooling (Bohdanowicz 2002; Gössling 2010) and therefore changes in heating–cooling degree-days have considerable implications for energy costs in some regions.

In temperate destinations, the implications for energy costs will be mixed, with reduced heating costs and increased cooling costs that may compensate for each other, or in cooler areas may even offer potential energy savings (e.g. see Scott and Jones' 2005 analysis of Banff, Canada). However, in warmer subtropical and tropical destinations, the cost implications are likely to be unidirectional as a result of increased cooling costs. For example, in a study of the impact of projected seasonal temperature changes on electricity demand in four major cities of Australia, Howden and Crimp (2007) estimated that lower winter HDD and increased summer CDD would largely compensate electricity demand in the cities of Sydney and Melbourne, where electricity demand increased only 1–2 per cent. However, in the subtropical cities of Brisbane and Adelaide, increased CDD led to increased electricity use of 10 and 28 per cent, respectively. Importantly, peak demand for electricity also increased, implying needs for additional generating capacity (even if populations and other demand factors remain the same). The impacts at these Australian destinations provide insight into what should be anticipated at most subtropical and tropical island destinations, where electricity is typically produced by diesel generators – both via the common electricity grid and often on-site as main power source or back-up power (Gössling 2010). The implications for the costs of electricity for island tourism operators are potentially substantial when the increases in cooling costs are combined with projected increased oil costs (see chapter 3), let alone if costs to upgrade generation capacity at the destinations are also factored in.

Weather extremes

A fourth dimension of the direct impacts of climatic change on tourism relates to the multiple implications of changes in extreme weather events. The IPCC (Solomon *et al.* 2007) has concluded that changes in a number of weather extremes are very probable, including higher maximum temperatures and more hot days (heatwaves) over nearly all land areas (very likely); greater tropical storm intensity and peak winds (likely); more intense precipitation events over many land areas (very likely); and longer and more severe droughts in many mid-latitude continental interiors (likely). More recent studies have indicated that anthropogenic climate change is already having effects on weather extremes, specifically the intensification of heavy rainfall events across the Northern hemisphere in the past 50 years (Min *et al.* 2011) and a related increased probability of severe flooding like that in England and Wales in 2000 (Pall *et al.* 2011), as well as increases in the probability of major heatwaves, such as that in Europe in 2003 (Stott *et al.* 2004). Changes in extreme events are expected to be among the most immediate and significant consequences of climate change (Solomon *et al.* 2007).

As the headlines in Figure 2.1 reveal, tourism destinations around the world are affected by different types of weather extremes every year. Little has been done to assess how projected changes in the frequency and intensity of extreme events could affect tourism destinations, tourism operators or tourists' safety.

Insurer Munich Re's natural catastrophe database, the most comprehensive of its kind in the world, indicates a marked increase in the number of weather-related events. According to Ward and Ranger (2010), globally there has been a more than threefold increase in loss-related floods from 1980 to 2010, and more than double the number of wind storm natural catastrophes, with particularly heavy losses as a result of Atlantic hurricanes. As Munich Re notes, the rise in natural catastrophe losses is primarily due to socio-economic factors, with rising populations and more people moving into exposed areas, while at the same time higher property values also increase the scale of insurance loss. Nevertheless, 'it would seem that the only plausible explanation for the rise in weather-related catastrophes is climate change. The view that weather extremes are more frequent and intense due to global warming coincides with the current state of scientific knowledge as set out in the Fourth IPCC Assessment Report' (Ward and Ranger 2010).

One of the tourism regions most affected by extreme weather events is the Caribbean. Table 5.4 outlines some of the implications for tourism of hurricanes in the Caribbean region from 2004 to 2010.

Table 5.4 *Implications of hurricanes for tourism in the Caribbean region, 2004–10*

Year	Hurricane	Implications for tourism
2004	Ivan	Grenada's main economic sectors and sources of employment – agriculture and tourism – suffered severely, with 60–90 per cent of agricultural products destroyed and 90 per cent of hotel rooms damaged or destroyed
2005	Wilma	Damage in the hotel sector of the Cancun area was estimated to reach $1.5 billion. Major beach erosion has required millions of dollars to replenish beach assets in the destination
2005	Katrina	Caused hundreds of millions of dollars in damage to tourism infrastructure in Louisiana and Mississippi. In 2010, New Orleans attracted 8.3 million visitors, up from 3.7 million in 2006, the first full year after Katrina
2007	Dean	Thousands of tourists were evacuated from countries in the Lesser and Greater Antilles, Mexico and Belize. Costa Maya's cruise port was closed to cruise-ships for almost a year
2008	Ike	Hundreds of hotel rooms on Turks and Caicos Islands were significantly damaged or destroyed. Carnival Cruise Line's US$60 million cruise-ship terminal on Grand Turk was also damaged
2010	Tomas	Brought St Lucia tourism to a standstill as a result of major damage to coastal and inland roads that connect the island's airports and tourism resorts

Because the impacts of extreme events are highly important to this region now, and are anticipated to become even more so in the decades ahead, it is not surprising that attempts have been made to estimate the potential economic impacts of changing weather extremes on tourism in this region.

Focusing on the island of Barbados, Moore *et al.* (2010) contend that the potential economic losses due to changes in the frequency of category 3–5 hurricanes in the Caribbean region (assuming historical probabilities of striking Barbados, existing number hotels/rooms, current revenues per visit) could increase 36 per cent (US$483 million) to over 500 per cent (US$1969 million) in the latter 30 years of the twenty-first century. The basis for damage estimates and change in major hurricane frequency is not outlined.

In a report from the World Bank Latin America and the Caribbean Region Sustainable Development Department, Toba (2009) argues that, with the assumption of a 27 per cent increase in annual hurricanes in the region, additional annual tourist expenditure losses (resulting from business interruption) will exceed US$446 in 2080 (in 2007 dollars) within

Caribbean Community (CARICOM) countries. Information on damages to tourism infrastructure was not included in this estimate. Extending Tobas' (2009) analysis to the Caribbean region, the UN Economic Commission for Latin America and the Caribbean estimated that the total costs of extreme events occurring in the region, aggregated to 2100, would be US$18.5–19.3 billion (in 2007 dollars) (UNECLAC 2010).

However, the damage estimates and assumptions regarding hurricane intensity and occurrence used in these studies are highly uncertain. There is little but anecdotal evidence about the impact of individual hurricanes on the tourism sector in this region. Thus, extrapolating tourism sector damages to the entire region on the basis of impacts from two single hurricanes that occurred in 1995 in non-major regional tourism destinations (Hurricane Luis – a category 3 that caused extensive damage in the Leeward Islands of Guadeloupe, Antigua & Barbuda and St Martin, and Hurricane Marilyn – a lesser storm that affected many of the same countries) (Haites *et al.* 2002) would seem questionable. Given the frequency with which hurricanes have a major impact on tourism in this region, the lack of information on their impacts is rather astonishing, and would be a highly useful area of research. The insurance implications of changing extremes in the region are also salient for tourism.

Furthermore, the relationship between climate change and hurricanes in the Caribbean is not settled, as hurricane activity is governed not only by air and ocean temperatures, but also by other factors, including ocean currents and the speed and direction of wind in different layers of atmosphere. A review by Knutson *et al.* (2010) concluded that studies consistently project the global average intensity of tropical cyclones will shift to stronger storms and the frequency of tropical cyclones will decrease by 2100, but that change will not become clearly detectable until the second half of the century. Substantial increases in the frequency of the most intense storms (category 4 and 5 storms) and important regional differences were also noted.

INDIRECT ENVIRONMENTAL CHANGE AND CULTURAL-HERITAGE IMPACTS

The attractiveness of destinations is strongly influenced by both the natural (or relatively unmodified) and human-shaped landscapes. The natural environment is particularly important for nature-based tourism, but also has a major role in mass tourism, such as 3S holidays, and provides a range of essential environmental services that support tourism. The effect climate change has on other environmental systems is projected to be widespread and in some cases will be transformative on the regional and even global scales (see chapter 1). The consequences of large-scale changes in environmental systems for tourism are inescapable, but have not been thoroughly examined.

This section examines some of these changes in natural assets for tourism, including changes and degradation of destination landscapes aesthetics; sea-level rise (SLR) and coastal tourism beach assets and infrastructure; changes in water availability; biodiversity losses and diminished wildlife availability; a more accessible Arctic; and altered agricultural production. This is followed by a short discussion of the implications of these changes in natural systems and processes for cultural-heritage assets that have historical, archaeological, architectural or indigenous value and provide the foundation of tourism in many other destinations.

Changing destination landscapes

Climate change is transforming, and will further transform, terrestrial and marine environments on every continent. The IPCC (Parry *et al.* 2007) indicated with high confidence that the most vulnerable ecosystems include coral reefs, the sea-ice biome, high-latitude ecosystems such as boreal forests, mountain ecosystems and Mediterranean-climate ecosystems. Changes in some of these ecosystems will radically alter the aesthetics of landscapes for tourism in some destinations, both positively and negatively, depending on the nature of environmental change. Three important tourism landscapes where substantive impacts are anticipated are discussed: alpine areas, coral reefs and forests.

Alpine areas

Warming climate conditions over the past 40 years have led to the retreat of alpine glaciers worldwide, and climate change projections would drastically reshape alpine landscapes in the decades ahead. The World Glacier Monitoring Service (WGMS 2008) has recorded 19 consecutive years of regionally average glacial retreat (negative mass balance). Since 1980, glacier retreat has become increasingly rapid, so much so that it threatens the existence of many of the glaciers of the world (Pelto 2010). Mid-latitude mountain ranges such as the Himalayas, Alps, Rocky Mountains, Cascade Range and southern Andes, as well as isolated tropical summits such as Mount Kilimanjaro (Thompson *et al.* 2009), are showing some of the largest proportionate loss. The observed rate of glacial retreat in these areas is expected to increase in the decades ahead, further altering the physical landscapes of many well known mountain destinations. For example, it is projected that by the year 2030, the vast majority of glaciers, if not all, in Glacier National Park (USA) will be gone, so that this destination will have lost its namesake (US Geological Survey 2003).

Perhaps less obvious changes in alpine landscapes will be the tremendous changes in mountain ecosystems and natural hazards. Many alpine species have limited capacity to move to higher altitudes in response to warming temperatures, changed snow cover regimes and displacement by lowland species. This is especially true of isolated populations on 'mountain islands', where, with nowhere to go, the danger of localized extinctions is considerable. The loss of colourful mountain meadows and upslope migration of the tree line have been observed in mountain ranges of North America (Saunders *et al.* 2007) and in the European Alps, and some plants previously found only on mountaintops have disappeared (Walther *et al.* 2005). Climate change will also alter the frequency and magnitude of natural hazards in mountain regions via the processes of glacier retreat and the melting of permafrost, which increases slope instability (e.g. rockfalls, glacial lake outburst floods, large-scale debris flow events).

Changes in natural hazards in mountain destinations (increased landslides, flash floods, debris flows, glacial lake outbursts) also pose an increased risk to tourists and tourism infrastructure. For example, the unprecedented melting of the Belvedere Glacier during the summer of 2002 created a new glacial lake that Italian government engineers feared would destroy the alpine resort near the town of Macugnaga (UNESCO 2002).

Nyaupane and Chhetri's (2009) case study of protected mountain areas in Nepal demonstrates that the physical impacts of climate change in mountain environments interact with Jodha's (1991) 'mountain specificities' (inaccessibility, fragility, marginality, diversity and niche) to exacerbate the vulnerability of mountain destinations in developing nations. The quality of the alpine environment is essential for successful tourism in mountain regions, and a number of authors have warned of the potential negative effects of landscape change for mountain tourism (Wall 1992; Elsasser and Bürki 2002; Scott 2003). Recent tourist surveys in high-profile mountain park destinations tend to support those concerns (see chapter 7).

Forests

The IPCC (Parry *et al.* 2007) reported with very high confidence that forests worldwide were already being affected by recent climate changes (the past 30–50 years), including earlier spring and summer phenology, longer growing seasons in mid- and higher latitudes, range expansions at higher elevations and latitudes, population declines at lower elevational or latitudinal limits to species ranges, and vulnerability of species with restricted ranges, leading to local extinctions. More recent reports have also indicated that the frequency and extent of wildfires have increased in Canada, Russia and the USA as a result of recent climate changes (Goetz *et al.* 2007). The increased frequency of such fires can change the usual fire regime, thereby favouring some species over others, and potentially enabling the spread of fire-adapted exotic species. For example, in parts of Europe, especially the Iberian Peninsula, increased wildfires are potentially helping to spread acacia and eucalyptus species, thereby radically changing the woodland landscapes (Lorenzo *et al.* 2010), while also assisting in the expansion of Mediterranean heath species.

Large areas of the world's remaining forests are expected to be greatly affected by future climate change, including regional-scale transitions in forest location and composition, for example, replacement of tropical forest by savannahs in eastern Amazon, upslope shifts of tropical cloud forests and alpine forests, increased risk of disease and pest outbreaks, and large increases in wildfire incidence and severity (Parry *et al.* 2007). Such changes further exacerbate abiotic impacts on forest health (FAO 2010). While all of these types of ongoing and future forest change have implications for tourism aesthetics and safety, little specific information is available to understand what the impact could be at the destination scale and the effects on tourists' perceptions of the landscape (Gössling and Hickler 2006; Hall 2011a).

In European forests, heavy storms can create significant economic, ecological and social problems, and together with fire are likely to be the most important, large-scale disturbance to stands of both natural and managed forests. According to Gardiner *et al.* (2010), storms are responsible for more than 50 per cent of all primary abiotic and biotic damage by volume to European forests from catastrophic events. Historically, catastrophic storms have tended to occur every five to ten years in Europe (FAO 2010), with more than 130 separate windstorms identified as causing noticeable damage to European forests from 1950 to 2010. However, due to the effects of climate change, changes in wind patterns and/ or oceanic currents and general increased variability in meteorological events, the period between destructive storms is expected to change in the coming years and decades (FAO 2010).

There is some evidence that storm intensity is already increasing and that storm tracks are penetrating further into mainland Europe and along a wider swathe, therefore also increasing the risk to forests in Eastern Europe (Gardiner *et al.* 2010). With climate change, higher temperatures will lead to longer periods of unfrozen soils during European winters, leading to a potential increase in forest damage, particularly in the Nordic countries. Storms will also tend to be accompanied by heavier rainfall leading to more saturated soils and increased risk of wind damage. Gardiner *et al.* (2010) also conclude that, if the current expansion of growing stock continues together with predicted changes to the climate, damage levels are then expected to at least double, and possibly quadruple, by the end of the century. Such a situation will also affect the capacity of forests to store carbon, with current estimates suggesting that storm damage to European forests will result in an annual reduction of 2 per cent in the carbon sequestration by forests, with this figure potentially exceeding 5 per cent by the end of the century if the current build-up of growing stock continues (Gardiner *et al.* 2010).

In addition to the direct impacts of climate change on trees and forest ecosystems, such changes can have devastating effects and can increase forests' susceptibility to other disturbances (FAO 2010). For example, a major storm in January 2005, and again in 2007, caused severe windthrow in southern Sweden, especially in middle-aged and old spruce stands. This resulted in increased populations of insects, notably the European spruce bark beetle, *Ips typographus*. Severe storms were also experienced in several other countries in Europe, including Germany and Slovakia, where the storm of 2004/05 affected forests in national parks, resulting in a severe bark beetle outbreak. Such interactions clearly make the prediction of future impacts of climate change on forest disturbances, and their subsequent effects on recreation and tourism, more difficult (Müller and Job 2009).

Increased fuel loads, longer fire seasons, and the occurrence of more extreme weather conditions as a consequence of a changing climate are expected to result in increased forest fire activity. As the headlines in Figure 2.1 illustrate, wildfires have had an impact on tourism destinations in the USA, Greece, Spain, Portugal, Australia, Russia and Canada in recent years. In parts of Greece, after the devastating fires of summer 2000, more than 50 per cent of all bookings from tourists for 2001 were cancelled (IUCN 2007). Dangerous wildfire conditions in parts of the western USA in the summer of 2002, and the media coverage of major fires – including a misstatement by a senior government official who stated that 'it felt like the whole state was on fire' – also had a significant impact on tourism, including reservation cancellations and reduced visitation to parks in the region (even those unaffected by the fires) (Butler 2002). Sanders *et al.*'s (2008) analysis of the impact of the 2006–07 wildfires on tourism in and near the national parks of south-eastern Australia revealed similar findings with respect to the importance of media reporting of the fires and statements by public officials, with generalization of impacts from specific locations to the entire tourism region. Wildfires may deter tourists out of concerns for safety, health effects of air pollution, loss of recreation opportunities (access to certain areas, ban on open fires, damaged infrastructure) and the loss of attractions (landscape and wildlife).

In each destination, governments and the tourism industry have invested additional resources in tourism marketing strategies specifically to counteract negative publicity generated by the fire events. Interestingly, while the immediate impact of wildfires on tourism is highly negative (reduced visitation and damage to tourism assets and infrastructure), in some high-profile nature-based tourism destinations (e.g. Yellowstone National Park in the USA), severe wildfires have been observed to increase tourism in the years following fires, as visitors come to see the devastation and nature's rebirth, and tourism operators develop new products such as 'fire recovery' tours. However, long-term rebound in any forest area affected by wildfire is likely to be dependent on the rate of fire event recurrence, media reporting, promotion and marketing, and the quality of interpretation. In addition, from a climate change impact perspective, the effects of these recent wildfires provide a useful analogue of potential future relationships between environmental change and tourism, and deserve further analysis.

The impact of wildfires on tourism may be particularly acute for parks: fires may be perceived as degrading the social value of parks, where the public perception is often of a healthy environment that is protected in perpetuity (Scott 2003). In British Columbia, Canada, the largest recorded outbreak of mountain pine beetle has affected an estimated 130,000 km^2 (close to the total area of England) between 1993 and 2006 (CBC 2008). The infestation is partly related to the lack of cold winters over this period, which normally curb populations of the beetle. The infestation has resulted in widespread mortality of lodgepole pine, one of the area's most abundant tree species, with the dead trees rendering the forest in entire valleys a rust-red colour. Many forest values are put at risk, including the value of the forest landscape for tourism. The Council of Tourism Associations of British Columbia (COTA 2009) has begun to develop a tourism action plan to respond to the mountain pine beetle damage, which may provide lessons for other destinations facing similar challenges, and highlights the need to identify strategic opportunities for investment in response to the negative impacts of climate change.

In other forests, the impacts of climate change may include a gradual shift in species composition. While in most cases these changes will be innocuous for tourism, in some destinations impacts will occur. In parts of the north-eastern USA (mainly the New England region) and parts of south-eastern Canada, there is a US$400 million tourism industry that draws visitors from around the region and North America to see the varied mosaic of fall foliage (leaf colours) (Rathke 2008). Vegetation modelling has projected that the maple–beech–birch forest type that currently dominates the region would be replaced by the oak–hickory forest type under climate change conditions (Frumhoff *et al.* 2007), replacing species that provide the colour essential to spectacular fall landscapes with a greater abundance of less colourful tree species. How tourists would respond to such changes in the palette of fall forest landscapes remains uncertain, but no such tourism industry exists in less colourful forest landscapes to the south and west.

Marine Environments

'Ocean acidification directly follows the accelerating trend in world CO_2 emissions, and the magnitude of ocean acidification can be ascertained with a high level of certainty based on the predictable marine carbonate chemistry reactions and cycles within the ocean. It is predicted that by 2050 ocean acidity could increase by 150%. This significant increase is 100 times faster than any change in acidity experienced in the marine environment over the last 20 million years, giving little time for evolutionary adaptation within biological systems. ... Ocean acidification is irreversible on timescales of at least tens of thousands of years, and substantial damage to ocean ecosystems can only be avoided by urgent and rapid reductions in global emissions of CO_2.'

SCBD (2009a: 9)

In the same way as terrestrial environments, marine environments are also subject to large-scale degradation from climate change that would alter their appearance. Coral reefs, sometimes referred to as the 'rainforests of the marine world' for their biological diversity and productivity (Hoegh-Guldberg *et al.* 2007), are one of the most important marine ecosystems for tourism and are also considered one of the most vulnerable to climate change (Hughes *et al.* 2010). Reefs are vulnerable to several climate change-related impacts: ocean acidification, coral bleaching, and for those coral reefs located close to shore, greater land runoff as a result of increased storm events.

Ocean acidification has been described as global warming's 'equally evil twin' by the head of the US National Oceanic and Atmospheric Administration, Jane Lubchenco (cited in McKie 2011). Data-based estimates indicate that, globally, the oceans have accumulated about 29 per cent of the total CO_2 emissions from burning fossil fuels, land-use change and cement production, among other activities, within the past 250 years (SCBD 2009a). Between 1751 and 1994, surface ocean pH is estimated to have decreased from approximately 8.25 to 8.14 (Jacobson 2005). The observed annual uptake of anthropogenic CO_2 by the oceans for 1990–99 of 2.2 (± 0.4) Pg C per year led to an estimate that the ocean sink accounted for 24 per cent of total anthropogenic emissions from 2000–06 (SCBD 2009a). This increased acidity affects levels of calcium carbonate, which forms the shells and skeletons of many sea creatures, and also disrupts reproductive and physiological activity.

Coral bleaching – the whitening of corals due to stress-induced expulsion or death of their symbiotic protozoa, or to loss of pigmentation within the protozoa – can occur for multiple reasons, but temperature change is the primary cause, and changes in water chemistry (acidification) can be a strong contributor. Mass coral bleaching is a visually spectacular phenomenon, transforming large reef areas from a mosaic of colour (if healthy) to a stark white. The temperature-induced 1998 mass-bleaching event devastated coral systems worldwide (Wilkinson 1998) and is often considered a precursor to what is expected under climate change in the twenty-first century. With very high confidence, the IPCC (Schneider *et al.* 2007) concluded that a warming of 2°C above 1990 levels would result in mass mortality of coral reefs globally. Hoegh-Guldberg *et al.*'s (2007: 1742) review of the future of coral reefs under warmer ocean temperatures and ocean acidification found that, even under lower-range IPCC scenarios, 'serious if not devastating ramifications for coral reefs [would occur]'. Like ski areas, the vulnerability of reefs to the impacts of climate change will vary spatially and temporally, with shallow reefs, reefs with species closest to their thermal maximum threshold, and those closest to sources of pollution or other human impacts the most vulnerable and where impacts will be visible to tourists the soonest.

Even moderate further warming will consequently affect the attractiveness of coral reefs to tourists in some destinations. Furthermore, coral reefs provide other important services to the tourism sector, including an important resource for sport fishery and supply of fish to tourists in many destinations, and coastal protection against storms. For destinations where reefs are the key attraction for tourists, such as the Great Barrier Reef (GBR) of Australia, which attracts 1.6–2 million visitors per year, the long-term damage arising from bleaching incidents is likely to have important implications for the quality of tourist experiences (snorkelling and scuba-diving) and eventually for the sustainability of tourism operators. Tourism to the GBR contributes A\$5.8 billion to the Australian economy per annum and sustains 55,000 jobs (Access Economics 2007), therefore any significant loss to the tourism experience as a result of climate change may have major economic impacts. Tourist perceptions of coral reef bleaching are explored in chapter 7. The closure of several marine parks and dive sites due to a severe bleaching event in Southeast Asia and the Indian Ocean in 2010, in an attempt to try and reduce further damage (Sarnsamak 2011), may provide an analogue for what reef tourism operators and managers will increasingly face in the future.

For all the aforementioned types of notable environmental change, there are opportunities to apply concepts from tourism studies (e.g. limits of acceptable change) to better understand the implications of these forthcoming changes on tourists and tourism destinations. How tourists are expected to respond to the above types of landscape change is explored further in chapter 7.

Sea-level rise and coastal tourism

Coastal tourism, which includes 'the full range of tourism, leisure, and recreationally oriented activities that take place in the coastal zone and the offshore coastal waters' (Hall 2001: 602), has been identified as the largest tourism market segment globally (Hall 2001; Honey and Krantz 2007; Phillips and House 2009). Despite the massive and growing investment in coastal tourism properties, and recognition that SLR is one of the most salient impacts of climate change, there has been remarkably little analysis of the implications for the tourism sector.

> 'The last time the world was three degrees warmer than today [for a prolonged period of time] – which is what we expect later this century – sea levels were 25 m higher. So that is what we can look forward [in the long term] … None of the current climate and ice models predict this. But I prefer the evidence from the Earth's history and my own eyes. I think sea-level rise is going to be the big issue soon, more even than warming itself.'
>
> James Hansen (2006)

As indicated in chapter 1, while the exact magnitude of global SLR and regional variability remains uncertain, SLR is considered one of the most certain consequences of human-induced climate change (Solomon *et al.* 2007). The impacts of SLR on coastal areas include erosion, inundation, impeded drainage and increased risk of riverine flooding, salinity intrusion into freshwater supplies, coastal habitat loss through the process of 'coastal squeeze' (Box 5.3), and higher water tables which can adversely affect the stability of foundations of coastal infrastructure. The IPCC (Nicholls *et al.* 2007) concluded that anthropogenic climate change contributed to accelerated SLR in the latter half of the twentieth century and to widespread beach erosion, with 70 per cent of the world's beaches receding (Bird 1985) attributed to ongoing SLR. SLR is a unidirectional hazard that, once set in motion, will continue for centuries, if not millennia, even under moderate scenarios of global warming.

SLR is generally considered to be one of the most prominent impacts of climate change. Consequently, a considerable literature of subnational, national, regional and global studies has developed over the past 20 years that examines its potential impacts on ecosystems (e.g. wetland loss) and society, including the number of vulnerable people (environmental refugees), implications for economic activities (infrastructure, agricultural land and productivity losses), and even the physical existence of some sovereign nations (Nicholls *et al.* 2007, 2011). Although the subnational and national details vary from study to study, this extensive literature has generally identified small islands and deltaic areas as being the most vulnerable to SLR.

A common critique of engineering and geomatics-based SLR studies is that they represent potential impacts and have not adequately accounted for adaptation. History teaches us that societies will not sit idly by and watch high-value land, infrastructure and cultural assets be swallowed by the sea. Nicholls and Tol (2006) contend that cost–benefit analyses suggest that widespread coastal protection will be an economically prudent response to avoid land loss and infrastructure damage from SLR. Nicholls *et al.* (2011) conclude that, while the potential impacts of a 0.5 m or a 2 m SLR over the twenty-first century would be severe, these can be partially avoided through widespread upgrade of coastal protection. Importantly, the authors also note that, because of the costly nature of coastal protection, the likelihood of protection being implemented successfully varies substantially between regions, and is estimated to be lowest in small islands, Africa and parts of Asia. The application and success of coastal adaptations are large uncertainties that require more assessment and consideration (Nicholls *et al.* 2011).

None of the regional- and global-scale studies of the impacts of SLR specifically examine potential damages in the tourism sector or compare the relative vulnerability of coastal tourism with other economic sectors. This is because, despite the high value of tourism properties and economic activity in the coastal zone, unlike population or agricultural land classifications and other impact indicators, no geospatial data sets of coastal tourism assets (e.g. accommodation or transport infrastructure, beaches) exist at the regional or global scale. As a result, tourism has not featured prominently in discussions of potential impacts of SLR. For example, although tourism was noted by the 2009 Australian House of Representatives Standing Committee on Climate Change, Water, Environment and the Arts inquiry into climate change and the coastal zone as a key industry sector requiring attention, and despite 80 per cent of the Australian population living in the coastal zone, the Committee recognized that there was a substantial paucity of information on climate change impacts, including SLR, on coastal infrastructure. Accordingly, they recommended that there was a need for a comprehensive national assessment of coastal infrastructure vulnerability to SLR as well as a national coastal zone database to improve access to information.

Although interest in the impacts of SLR on coastal tourism extends over 20 years back to the work of Gable (1990), a number of scholars have remarked that the very limited analysis of the impacts of SLR on coastal tourism represents one of the core knowledge gaps in this field (Gössling and Hall 2006b; Scott *et al.* 2008a; Weaver 2011). A number of studies have identified the types of potential consequences of SLR for coastal tourism (e.g. loss of high-value beaches, destruction of tourism infrastructure, biodiversity, increased need for engineered shore protection, changed coastal aesthetics) and the linkages to coastal zone management and planning (Gable 1990; Phillips and Jones 2006; Scott *et al.* 2008a; Moreno and Becken 2009; Jones and Phillips 2011). Jennings (2004) identified SLR as one of four major issues changing the relationship between coastal tourism and shoreline management. Nonetheless, Buckley (2008: 72–73) notes that, 'Most coastal tourism destinations ... seem to have remained remarkably blasé about rising sea levels, even though these are one of the best-documented aspects of global change.'

The impacts of SLR on tourism infrastructure and coastal resources (mainly beaches) have been examined in a number of country- and state-level studies. A study of flood frequency in the culturally important city of Venice, Italy found a fivefold increase in flooding events in Piazza San Marco between 1900 and 1990 (from seven to 40 times per year) and estimated that with a further 30 cm SLR, flooding events would increase to 360 days of the year (Francia and Juhasz 1993). Similarly, large areas of the historical city of Alexandria, Egypt were found to be at risk of a 0.5 m SLR, including a series of high-value cultural sites and beach areas that are fundamental to tourism in this principal Mediterranean summer resort (El-Raey *et al.* 1999). Similar threats to high-value beach areas in Port Said were also projected.

One of the best recognized locations in which tourism will be affected by climate change is the low-lying islands of the Maldives. Over 80 per cent of its land area is less than one metre above mean sea level. Tourism is the mainstay of the Maldivian economy, with 70 per cent of visitors coming primarily for beach holidays. In 2001, the Government of the Maldives reported that 50 per cent of all inhabited islands and 45 per cent of its tourist resorts face varying degrees of beach erosion, and it was expected that even a one metre SLR would cause the loss of the entire land area of the Maldives (Hall 2011b).

A study of coastal properties at risk of SLR on Prince Edward Island, Canada determined that flooding associated with sea levels projected by the IPCC for later decades of the twenty-first century and a repeat of recent major storms would put at risk over 300 designated heritage properties (principal tourism attractions and, in some cases, accommodation) in the City of Charlottetown (Environment Canada 2002). With projected increase in erosion due to SLR, 10 per cent of coastal tourism properties would be lost within 20 years, and nearly 50 per cent in the latter decades of this century.

Analysis of coastal inundation and erosion impacts on tourism beaches and hotels on the island of Martinique (Schleupner 2008) concluded that the attractiveness of the island destination was likely to decline as a result of SLR. A large majority of tourist beaches (83 per cent) were anticipated to be affected by 'coastal squeeze' (see Box 5.3) and up to 62 per cent of coastal infrastructure (tourism and non-tourism) would be at risk to damage from SLR-induced erosion.

BOX 5.3 THE 'COASTAL SQUEEZE' OF TOURISM

Coastal habitats such as wetlands and beaches will naturally adapt to rising sea levels by migrating inland. Coastal squeeze is an environmental situation where the coastal margin is squeezed between some fixed natural or human-made landward boundary (e.g. a rocky cliff, sea wall or building/road) and the rising sea level, thereby reducing its former area. Most studies have analysed coastal squeeze with respect to coastal habitats, where both the area and functioning of habitat can be altered, but the concept can also be used to examine the impacts for other coastal lands, such as tourist beaches. As Figure 5.2 illustrates, the construction of a sea wall to protect resort structures prevents the beach from shifting inland as the sea level rises (other physical barriers can include very near-shore resort infrastructure). This process squeezes the beach area until it is eventually lost. In this situation, structural protection may allow the resort to stay in its current location, but the value of the property, its ability to attract guests, and the price guests are willing to pay will all be degraded with the loss of beach and the reduced aesthetics of coastal protection works.

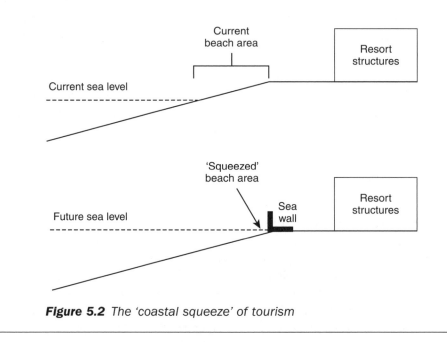

Figure 5.2 The 'coastal squeeze' of tourism

Tourism is a major part of the economy in the State of Florida. Using maps produced by the US Geological Survey of the areas of Florida's coastline that would be inundated by a 68 cm SLR, Stanton and Ackerman (2007) estimated impacts on half of existing beaches, 74 airports, 1362 hotels/motels, and over 19,000 historical structures. The resultant losses in tourism properties and visitation were estimated to be in the tens of billions of dollars by 2050. The US EPA (1999) reported that a 60 cm SLR would erode beaches in parts of south Florida 30–60 m unless beach nourishment efforts were expanded. The cumulative cost of sand replenishment to protect Florida's coast from a 50 cm SLR was estimated at US$1.7–8.8 billion (US EPA 1999).

The broadest multinational study of SLR impacts on tourism examined potential inundation and erosion impacts for major coastal tourism resorts and resort-front beach areas in 19 nations of the Caribbean. Using a georeferenced database of over 906 major tourism resort properties, Scott *et al.* (2012) estimated that 266 would be vulnerable to partial or full inundation by a one metre SLR. A greater proportion of resort properties were at risk in Belize (73 per cent), St Kitts and Nevis (64 per cent), Anguilla (63 per cent), Turks and Caicos (63 per cent), British Virgin Islands (57 per cent), Haiti (46 per cent) and the Bahamas (36 per cent) (Table 5.5). A far higher number of major coastal resort properties (440–546) would be vulnerable to coastal erosion associated with a one metre SLR, applying a conservative interpretation of the Bruun Rule (inland retreat of approximately 50–100 times the vertical increase in sea level; Bruun 1962) to coastlines with unconsolidated beach materials (Table 5.5). A much greater proportion of resort-front beaches would be lost to inundation and accelerated erosion, as beaches would essentially have disappeared prior to damage to tourism resort infrastructure. Such impacts will transform coastal tourism in the region, with implications for property values, insurance costs, destination competitiveness and marketing, and pose a significant challenge to the economy of many of these nations, where tourism represents a mainstay of the economy.

As indicated, a common critique of SLR studies is that they estimate only potential impacts without determining the extent to which damage could be offset through adaptation, including coastal protection schemes. All of the aforementioned studies of SLR and tourism suffer from this key limitation. While Nicholls and Tol (2006) conclude that widespread coastal protection will be an economically sensible response to avoid land loss and infrastructure damage due to SLR, coastal tourism resorts pose some special challenges for structural protection works.

Typical structural coastal protection is not well suited to the key objectives of coastal resorts: providing unobstructed views of the sea, maintaining unhindered access to the beach and sea, and the visual perception of a pristine beach environment. Furthermore, while structural protection can easily be designed to protect resort buildings, coastal squeeze (Box 5.3) will mean that the resort will lose its beach unless it is also willing to invest heavily in beach nourishment. Contemporary beach erosion illustrates the negative economic impact of beach loss on resort attractiveness, room rates and property values (Houston 2002; Cowell *et al.* 2006) and serves as a useful analogue for the impacts of SLR without beach nourishment. So while structural protection will make sense for certain tourism sector assets, such as airports and cruise-ship terminals, and for cities that function

Table 5.5 Potential impacts of sea-level rise on coastal tourism properties in the Caribbean

Country	Number of resort properties	Resorts partially inundated by 1 m SLR	Resorts damaged by 50 m erosion	Resorts damaged by 100 m erosion
Anguilla	60	38	35	42
Antigua and Barbuda	99	10	34	44
Bahamas	133	48	77	93
Barbados	75	6	42	50
Belize	44	32	42	44
British Virgin Islands	14	8	5	6
Cayman Islands	63	11	15	25
Dominica	17	0	5	6
Grenada	45	5	14	19
Guyana	10	0	0	0
Haiti	28	13	14	17
Jamaica	105	8	34	52
Montserrat	1	0	0	0
St Kitts and Nevis	22	14	15	18
St Lucia	30	2	5	9
St Vincent and the Grenadines	21	2	8	16
Suriname	19	1	2	2
Trinidad and Tobago	24	8	15	16
Turks and Caicos Islands	96	60	78	87
Totals	906	266	440	546

Source: Scott *et al.* (2012)

as important tourism destinations, the suggestion that structural protection is the obvious and economical choice to reduce the impacts of SLR is not so straightforward for coastal resorts that must maintain sufficient beach area and aesthetics to attract tourism clientele. This is particularly so as some measures of coastal zone adaptation to climate change and SLR, such as a 'managed retreat' in low-lying areas that 'returns' land to the sea, may end up creating new tourist attractions such as saltwater marshes that attract nature-based tourism (Barkham 2008).

Future water security challenges

'The world faces an unprecedented crisis in water resources management, with profound implications for global food security, protection of human health, and maintenance of aquatic ecosystems.'

US AID (2010)

Global water use is estimated to have grown at more than twice the rate of population increase over the past century (United Nations 2011) and is presently doubling every 21 years (US AID 2010; United Nations 2011). Water stress (i.e. not enough water available for all uses: agricultural, industrial, domestic water supply, ecosystem maintenance) already affects a large and growing share of humanity. Some 1.4–2.1 billion people live in water-stressed basins in northern Africa, the Mediterranean region, the Middle East, the Near East, southern Asia, northern China, Australia, the USA, Mexico, north-eastern Brazil and the west coast of South America (Vörösmarty *et al.* 2000).

Global water use is increasing due to population and economic growth, changes in lifestyles, technologies and international trade, and the expansion of water supply systems. Between 2010 and 2030, water withdrawals are projected to increase by 50 per cent in developing countries, and 18 per cent in developed countries (United Nations 2011). Consequently, by 2025 it is estimated that some 1.8 billion people will be living in countries or regions with absolute water scarcity, and two-thirds of the world population could be under stress conditions (United Nations 2011). Water stress will be more prevalent among poorer countries where water resources are limited and population growth is rapid; of the 48 countries expected to experience chronic water shortages by 2025, 40 are either in the Middle East and North Africa or in sub-Saharan Africa. Many have predicted that water would become the source of future conflicts, but more recent analyses question this supposition based on the limited history of water-based conflict and record of international cooperation (Barnaby 2009). Water security is also increasingly becoming a strategic business concern for companies operating in some parts of the world (Sarni 2011). The Carbon Disclosure Project (2010a, 2010b) reports that more than half of the firms responding to a survey expected problems with water in the next five years, including disruption from drought or flooding, declining water quality, increases in water prices, and fines and litigation relating to pollution incidents. As the projections of precipitation and soil moisture set out in chapter 1 reveal, climate change is anticipated to exacerbate water stress challenges, in some places severely, in the decades ahead. Up to 3.2 billion people could face water stress by 2100 under a 4°C global climate change scenario (Parry *et al.* 2009a, 2009b).

Tourism is both dependent on freshwater resources and an important factor in freshwater use (Gössling *et al.* 2010c). Tourists need and consume water when washing or using the toilet, when participating in activities such as ski or golf tourism (snow-making and irrigation), and when using spas, wellness areas or swimming pools. Freshwater is also needed to maintain the gardens and landscaping of hotels and attractions, and is embodied in tourism infrastructure development, food and fuel production. Recreational activities such as swimming, sailing, kayaking, canoeing, diving and fishing all take place at lakes and rivers, which also form important elements of the landscapes visited by tourists (Gössling 2006; Hall and Härkönen 2006). Many forms of tourism are also indirectly dependent on water, including winter sports tourism, agritourism and wildlife tourism.

The availability of freshwater is currently an important issue for tourism operations in some locations, and water availability challenges for tourism are expected to intensify in many areas as demand for water from other users (industry, agriculture, cities) increases. As the review by Gössling *et al.* (2010c) of global water consumption within the tourism sector reveals, great differences exist in terms of renewable water resources, desalination capacity, use of treated wastewater, and overall water use among the most important tourism countries. The study concludes that direct tourism-related water use is less than 1 per cent of global consumption and typically less than 5 per cent of domestic water use, but that this can be substantially higher in some countries, particularly small islands with high seasonal tourist arrival numbers and limited water resources (Table 5.6). Consequently, national-scale discussions of water security should not overlook tourism as a sector, particularly as water demand from tourism is expected to increase because of increased tourist numbers, higher hotel standards and the increased water-intensity of tourism activities (Gössling *et al.* 2010c). Table 5.6 indicates that nine of the top 25 tourism countries (by international arrivals) have high or extreme water security risks. In the future, tourism businesses in water-scarce regions will face considerably greater problems with regard to water availability and quality due to increasing competition for water use.

Changes in the availability or quality of water resources resulting from climate change will also have detrimental impacts on tourism in many regions; however, the implications of climate change-altered water resource availability (both geographically and seasonally) for tourism operations and future development have not yet been thoroughly assessed (Gössling 2006; Scott *et al.* 2008a; Gössling *et al.* 2010c). Of the top 25 tourism nations in Table 5.6, ten are projected to have lower runoff by the 2050s and 14 are projected to have reduced soil moisture in the 2080s (according to a 15-model ensemble mean for the *SRES* A1B scenario; see chapter 1). Throughout many of the regions where reduced runoff and soil moisture are projected, drought frequency and intensity are also projected (not shown in Table 5.6), which will further test water supply systems in extreme years. For example, in large parts of the European Mediterranean, the current one in 100-year drought is projected to occur once a decade in the latter part of the twenty-first century (Solomon *et al.* 2007). While further analysis is needed to examine whether the seasonal timing of increased or reduced precipitation and runoff coincides with peak tourism arrivals, in the majority of countries where tourism is a more significant sectoral water user, climate change is projected to exacerbate current water demand and scarcity problems (all of these nations currently have medium to high water security risks).

Many of the water resource challenges brought about by climate change in the tourism sector will be most acute at the subnational and destination levels. Subnational-scale analysis revealed additional climate change-induced water challenges emerge for some prominent and emerging tourism destinations, including in the south-western USA, southern Australia, central-coastal Brazil, the Middle East, and central and southern China (Gössling *et al.* 2011a). The range of water-related challenges is illustrated in the selected examples below.

In the western USA, low water levels on lakes, streams and reservoirs are restricting tourism and recreation in western regions of the country, and serve as an important analogue

Table 5.6 *Current and future water stress in major tourism regions*

Country*	Tourism freshwater consumption (per cent total domestic)[†]	Water security risk index[‡]	Runoff available (per cent change) 2050s[¶]	Soil moisture (per cent change) 2080s[¶]
France	10.0	Medium	−5 to −20	−5 to −20
Spain	12.0	Medium	−10 to −30	−5 to −20
United States	2.3	Medium	20 to −20	10 to −15
China	3.4	High	−5 to 50	5 to −10
Italy	5.7	Medium	−2 to −30	−5 to −15
UK	5.9	Low	10 to −2	0 to −10
Germany	4.6	Medium	2 to −5	0 to −5
Ukraine	0.2	High	−10 to −20	5 to −5
Turkey	3.9	Medium	−20 to −30	−5 to −15
Mexico	1.6	Medium	2 to −20	−10 to −20
Malaysia	2.1	Low	10 to 20	5 to −5
Austria	2.5	Low	−10 to −20	0 to −5
Russia	0.6	Low	50 to −2	5 to −15
Canada	0.3	Low	50 to −5	10 to −15
Hong Kong, China	nd	High	10 to 20	0 to −5
Greece	8.8	Medium	−10 to −30	−10 to −20
Poland	1.1	Medium	2 to −5	0 to 5
Thailand	5.9	High	10 to 20	5 to −5
Macau, China	nd	High	10 to 20	0 to −5
Portugal	4.8	Medium	−10 to −30	−5 to −20
Saudi Arabia	2.7	High	50 to −10	5 to −5
Netherlands	2.9	High	2 to −2	0 to 5
Egypt	1.9	Extreme	50 to −2	5 to −5
Croatia	nd	Low	2 to −20	0 to −10
South Africa	4.0	High	2 to −20	0 to −20
Lebanon	nd	High	2 to −30	−10 to −20
Malta	14.2	Medium	−20 to −30	−10 to −15
Cyprus	19.4	Medium	−20 to −30	−10 to −15
Bahrain	3.72	High	40 to 50	0 to 10

*Top 25 nations by international arrivals in 2007, followed by four illustrative nations where tourism represents a high proportion of GDP. Some of the most vulnerable countries from a water perspective are island nations of the Caribbean and Pacific, but none is ranked by water security or water poverty indices

[†]Gössling *et al.* (2011d)

[‡]Maplecroft (2010)

[¶]Solomon *et al.* (2007)

of the potential impacts of climate change on tourism. Drought conditions in Colorado during the spring and summer of 2002 required that sport fishing be restricted from many rivers because the fish populations were highly stressed by low water levels and higher water temperatures, and the river-rafting season was substantially shortened due to low water levels and exposed rock hazards. Tourism losses were substantial (Associated Press 2002a, 2002b). Research on the thermal habitat for many prominent sport fish species in the Rocky Mountain region of the United States under climate change projected substantial losses of habitat area (Keleher and Rahel 1996). Furthermore, climate change in tandem with land use can also impact the runoff of nutrients into lake systems, leading to issues of eutrophication and littoral algal production. In severe cases, eutrophication can have a substantial impact on tourist and recreational activities, not only because of impact of recreational fishing, but because of issues associated with the smell of algae washed up on to the shoreline as well as the potential for toxic algal blooms that can make the water unsafe for swimming (Hall and Härkönen 2006).

The prolonged drought in western regions of the United States also negatively affected reservoirs, which are a major tourism resource in the region. Lake Mead is the largest functional reservoir in the western USA and is visited by over 7 million people annually (National Park Service 2009). Water levels in the reservoir have dropped nearly 30 metres since 1999 due to reduced flows in the Colorado River (US Department of the Interior 2010). Boaters have been exposed to new navigation hazards, while a number of launch ramps have been closed because they no longer extend to the water line.

In addition to being tourism destinations themselves, these massive reservoirs also supply water to cities and agricultural lands throughout the region. A report by the US Department of the Interior (2003) indicated that water supplies in parts of the region were inadequate to meet the estimated demands of cities, agriculture, recreation and the environment in 2025 (without consideration of the impacts of climate change for water availability). A number of important tourism destinations are located in areas where the potential for future water-use conflict was rated as 'high' or 'substantial'. Las Vegas, one of the largest per capita users of water in the world, in large part because of its tourism industry, is in a high-risk zone. Difficult decisions regarding the regulation of existing water-intensive tourism operations (golf courses, resorts with massive fountains and landscaping) and limiting of further tourism development are forthcoming in the next two decades (Ward 2003). Similar challenges will exist for Phoenix, an important winter getaway and golf holiday destination. Importantly, because of the recency of tourism development and in particular the growth of the golf industry, older water-use rights held by cities, agriculture and government (for fisheries and ecosystems) will take precedence when water supplies are limited (Hirt *et al.* 2008).

By the 2070s, it is expected that approximately 35 per cent of Europe will be suffering from water stress, with summer flows being reduced by up to 80 per cent in some parts of southern and central Europe (Alcamo *et al.* 2007). For example, Lake La Nava in Castile-León, Spain and the Nestos Lakes in Hrysoupolis, Greece, both important recreational resources, are expected to be subject to a temperature increase of up to 5°C and a 35 per cent decline in annual precipitation, resulting in substantial effects on biodiversity, unless there are drastic reductions in the amount of GHG emissions (Hulme *et al.* 2003). In these locations, like many others in southern and central Europe, the effects of climate change are exacerbated by the lack of effective regulations governing the extraction of water for industrial uses and the maintenance of landscape and natural vegetation (Hall and Lew 2009).

These same challenges – maintaining water rights as demand from other users increases, and making strategic choices about the type of future tourism development possible given limited water resources – will occur in many destinations with limited current water resources, particularly in areas where there has been strong growth of new golf courses or other water-intense tourism (e.g. Arizona, Texas, Spain, China, Middle East, South Africa) (Rodríguez-Díaz *et al.* 2007; Gössling *et al.* 2010c).

In addition to important changes in water levels, thermal changes in lakes and reservoirs can also influence water quality, affecting the attractiveness of many water resources for tourism. Under warmer climate conditions during the summer, the oxygen-carrying capacity of water bodies is diminished, which can contribute to enhanced algal growth and other water pollution. Bacterial contamination can degrade the aesthetics of beaches and pose a health risk to swimmers. In Europe, the water quality at many popular recreation lakes has been negatively affected by the influx of nutrients, which has contributed to eutrophication and littoral algae production (Scheidleder *et al.* 1999). For example, of 171 freshwater bathing areas in Spain, only 42 per cent met stricter recommended water-quality standards in 2003. In Italy, 58 per cent of 775 freshwater bathing areas met recommended quality standards, and 71 per cent met mandatory standards. Under climate change, there is a higher likelihood that beach closures or restricted use could become more common in many popular tourism areas. Algal growth might also affect increasingly popular recreation sports such as diving. Particularly clear lakes are usually well known to diving tour operators and divers. Although no research has been carried out in the context of lakes, reduced visibility brought about by micro-algae has been documented as an important parameter negatively affecting diving experiences in the tropics (cf. Gössling *et al.* 2007b for a case study in Mauritius).

Island destinations will be particularly vulnerable to changes in water resources as a result of climatic changes, but also as a result of SLR, which will increase saltwater intrusion in coastal freshwater aquifers. As indicated, small island developing states (SIDS) are some of the most data-poor nations in terms of water resources, and few could be included in Table 5.3 even though their tourism industries are likely to be among the most affected by changing water availability.

Emmanuel and Spence (2009) review the climate change implications for water resource management on the Caribbean island of Barbados, where the economy has become increasingly dependent on tourism over the past 20 years. The government estimates that 98 per cent of renewable freshwater resources are already being used, and that industrial and commercial uses (including tourism) had increased from 20 per cent in 1996 to 44 per cent by 2007. Tourism-specific demand (hotels, cruise-ships, golf courses), which represented approximately one-sixth of total demand in 1996, is projected to represent one-third by 2016. Although government officials interviewed for the study felt that water scarcity had not yet affected tourism, the capacity of the island to sustain this redistribution of water resources is questioned, particularly continuation of the golf tourism strategy and the social equity of providing limited potable water to cruise-ships. With mean annual precipitation in the area projected to decline up to 20 per cent under some climate change scenarios by mid-century, there is further impetus to reconsider the long-term tourism development strategy for the island in light of water resources. Many inland destinations in the region and around the world will need to examine how to balance tourism ambitions and available and affordable water resources. Desalination is an adaptation that is already in use in many island destinations, but it is costly and could become even more so under strict GHG emission policies (see chapter 4), with potentially important implications for the costs of water to tourism operators (Gössling *et al.* 2010c).

Even in relatively water-rich mountain regions, water availability for tourism has been questioned (Abegg *et al.* 2007). As discussed above, snow-making is an integral climate adaptation utilized by the ski industry to make up for the lack of natural snow. Snow-making is very water-intensive and has drawn criticism from a number of environmental organizations (Scott and McBoyle 2007). Although the number of ski areas and area of skiable terrain have increased slowly over the past 25 years (with the exception of some emerging markets such as China), the proportion of skiable terrain covered by snow-making has grown substantially, and this trend is expected to continue as ski areas seek to reduce their vulnerability to recent climate variability (Scott and McBoyle 2007; Wolfsegger *et al.* 2008; Steiger 2011b). The sustainability of this adaptation strategy is uncertain if the ski industry is unable to access sufficient water supplies. In one of the few studies specific to water availability and tourism, Vanham *et al.* (2009) examine whether increased snow-making would pose a threat to other water users in the Austrian Province of Tyrol. They conclude that no threat exists at the provincial level, but that on a local basis periodic early winter water stress could be triggered by snow-making if not properly regulated. Similar local analyses will be needed at many destinations where large-scale snow-making occurs.

BIODIVERSITY

The loss of biodiversity is one of the most significant aspects of global environmental change, given the extent to which it underpins the global economy and human welfare (Martens *et al.* 2003). According to the SCBD (2009c), *c.* 10 per cent of assessed species are at an increasingly higher risk of extinction for every 1°C rise in global mean temperature. Given the observed temperature rise, this now could place 6–8 per cent of the species studied at an increasingly high risk of extinction, although it has been estimated that each degree of warming could yield an upward nonlinear increase in bird extinctions of about 100–500 species (SCBD 2009d). The current speed of species extinction through human intervention is estimated to be *c.* 100–1000 times faster than the natural speed of extinction (Martens *et al.* 2003).

Biodiversity is of critical direct and indirect importance for tourism (Christ *et al.* 2003; Hall 2006a, 2010a, 2010b, 2010d), although there is 'an apparent lack of awareness of the links—positive and negative—between tourism development and biodiversity conservation' (Christ *et al.* 2003: 41). National parks are often the focus for tourism and biodiversity relationships. Although 'the sustainability of protected areas will be severely threatened by human-caused climate change' (Millennium Ecosystem Assessment 2005: 12), the significance of biodiversity is much wider than in protected areas. Biological diversity (biodiversity) refers to the total sum of biotic variation, ranging from the genetic level, through the species level, and on to the ecosystem level. The concept indicates diversity within and between species, as well as the diversity of ecosystems. The extent or quantity of diversity can be expressed in terms of the size of a population, the abundance of different species, the size of an ecosystem (area) and the number of ecosystems in a given area. The integrity or quality of biodiversity can be expressed in terms of the extent of diversity at the genetic level and resilience at the species and ecosystem level (Martens *et al.* 2003), with assessments also being undertaken at various scales of governance as well as specific locations and environments (Hall 2006a).

Evidence for the impact of climate change on biodiversity comes from three main sources (Campbell *et al.* 2009). First, from direct observation of changes in components of biodiversity in nature that can be clearly related to changes in climatic variables, such as changes in species distribution or phenological changes in bird arrivals, grape harvest dates, first leaf dates and flowering times (Schwartz *et al.* 2006). Second, from experimental studies using manipulations to elucidate responses to climate change, such as changes in air, soil and water temperatures and/or examining the effect of addition of CO_2 on plant communities (Hovenden *et al.* 2008). Third, and most widely, from modelling (correlative, mechanistic and analogue) studies where understanding of the requirements and constraints on the distributions of species and ecosystems is combined with modelled changes in climatic variables to project the impacts of climate change and predict future distributions and population change (Campbell *et al.* 2009). Of the three modelling approaches, correlative modelling is the most common (Botkin *et al.* 2007).

The biodiversity and climate change literature has only marginally acknowledged the implications for tourism, and these mainly in the context of ecosystem services rather than direct impacts on specific destinations or attractions (see Campbell *et al.* 2009). Nevertheless, experimental and modelling studies have been significant for identifying possible implications for tourism of climate change impacts on biodiversity such as ocean acidification (see below). Direct observational change has been useful for identifying the potential implications of climate change for specific species or ecosystems that are important for tourism (for example, note the above discussion on changes to forest landscapes and the distribution of sports fish).

Climate change has a number of direct and indirect effects on biodiversity:

- increases in average temperature;
- temperature changes linked to changes in precipitation regimes that may increase, decrease and/or change their seasonal distribution as well as increase the occurrence and severity of extreme weather events;
- SLR as well as the melting of sea and land ice;

- changes in the atmospheric, marine and terrestrial concentrations of CO_2;
- climate change effects on biodiversity also interact with other pressures such as land-use change, changes in fire regimes, ecosystem fragmentation, pollution and the introduction of invasive species.

Such changes in climatic regimes will therefore affect the survival of ecosystems and species. Overall, those species that have the smallest ranges are therefore the most specialized climatically and will be the ones most at risk. However, studies of the relationship between species distribution and future climate change scenarios tend to make a number of critical assumptions (Gaston 2003):

- correlations between climate and species occurrence reflect causal relationships;
- any influence of other factors on observed relationships between climate and the occurrence of a species, such as competitors, diseases, predators, parasites and resources will remain constant;
- temporally generalized climatic conditions (e.g. seasonal means, annual means, medians) are more important influences on the distribution of species than rates of climatic change and extreme events;
- spatially generalized climatic conditions derived or interpolated from the nearest climate stations sufficiently characterize the conditions that individuals of a species actually experience;
- climate change will be relatively simple in that its influence on species distributions can be summarized in terms of the projected changes in one or a few variables;
- there is no physiological capacity to withstand environmental conditions which are not components of those existing conditions in areas in which a species is presently distributed;
- range shifts, expansions or contractions are not accompanied by physiological changes other than local non-genetic acclimatization;
- dispersal limit is unimportant in determining the present distribution of species and in their ability to respond to changes in climate.

As Gaston (2003: 185) points out, the reality is that a number of these assumptions will be, and already have been, 'severely violated'. Of critical importance to understanding the effects of both climate change and tourism on species range is that 'human activities impose a marked influence on the distribution of species, and how these alter with changes in climate is alone likely to be extremely complicated, and dependent on social pressures and technological developments' (Gaston 2003: 183). Therefore any understanding of the interactions between biodiversity, climate change and tourism needs to include not only their interrelationships, but also their interplay with other anthropogenic, biological and physical environmental factors.

As noted above, biodiversity is usually expressed at genetic, species and ecosystem scales (Hall 2006a). This section on tourism and biodiversity focuses on the ecosystem scale, but also notes the impact of climate change on species of importance to tourism. Before briefly outlining the changes to specific ecosystems, an overview is provided of the broad changes that are occurring at an ecosystem level.

Ecosystem change

In general, significant changes include shifts in the distribution of some ecosystems, primarily as a result of increasing temperatures and changed precipitation regimes. In a meta-analysis of existing data, Chen *et al.* (2011) estimated that the distributions of species had shifted to higher elevations at a median rate of 11.0 m per decade, and to higher latitudes at a median rate of 16.9 km per decade. These rates are approximately two and three times faster than previously reported (e.g. Beckage *et al.* 2008). Moreover, the distances moved by species are greatest in studies showing the highest levels of warming. However, critical to such shifts is the capacity of constituent species to migrate, thus there is substantial variation between species in their rates of change. Tropical areas are expected to experience substantial changes in ecosystem distribution, with savannah ecosystems moving into equatorial regions now occupied by forests as a result of drier conditions (Salazar *et al.* 2007). Some ecosystems that have either an extremely limited distribution, as at the top of mountain ranges, or are potentially subject to extremely rapid change, such as some coastal ecosystems, are likely to have extremely limited or no capacity to move. This may be significant for niche nature-based tourism products based in vulnerable environments. In the case of Australia, for example, this would include such significant tourism areas as the tropical and temperate rainforests as well the relatively low-lying alpine areas in southeast Australia and the Stirling and Porongup Ranges in Western Australia. A summary of the climate-related changes to specific ecosystem types is provided in Table 5.7

In the tropics, species less tolerant of the new environmental conditions are being replaced by those with greater tolerance for warmer and drier conditions and increased fire occurrence (Campbell *et al.* 2009). Overall, generalist species are more likely to survive than those that are more specialized, but the overall rate of change is dependent on the pool of available species and their migration rate (Colwell *et al.* 2008). Significantly, climate change, along with tourism, also appears to facilitate the spread and establishment of invasive species, which can have major effects on ecosystem composition. For example, in a study of the socio-economic parameters influencing plant invasions in Europe and North Africa, Vilà and Pujadas (2001a) found that the density of naturalized species was positively correlated to the number of tourists who visit a country ($r = 0.49$), with Vilà and Pujadas (2001b: 399) commenting, 'Mediterranean basin countries such as France, Italy and Spain that receive many tourists are the ones with the highest density of alien species. Contrary to our expectations the correlation between the density of alien plants and the percentage cover of protected land was positive.' Indeed, changes to community dynamics and successional dynamics under climate change are likely to provide numerous opportunities for invasive species.

In some ecosystems, changes in species composition will also lead to changes in the physical and trophic (position of an organism in the food web) structure of ecosystems. In some systems, this will mean that woody plants will invade temperate grasslands. In other systems, trees may disappear as a result of drought. In either case, such changes will have dramatic effects on the landscape as well as on the animal species in those areas. Such species are often extremely significant for safari and wildlife tours and their loss may therefore greatly affect the tourism product offering as well as the visitor experience.

Table 5.7 Climate change and ecosystem types

Ecosystem type	IPCC AR4 projections	Climate and ecosystem change
Deserts and arid ecosystems	Deserts are likely to experience more episodic climate events, and inter-annual variability may increase in future, although there is substantial disagreement between projections and across different regions	Continental deserts are likely to experience more severe and persistent droughts, but productivity impacts may be partially offset by the effects of increased atmospheric CO_2 concentrations during wetter periods. Reduced desert biomass is likely to increase the fragility of soils and erosion vulnerability. Many desert species are vulnerable to temperature increases. Alteration of the rainfall regime will put at risk species that depend on rainfall events to initiate breeding
Grasslands and savannahs	Grasslands are sensitive to variability and changes in climate, which are likely to have strong effects on the balance between different life forms and functional types in these systems. The mixture of functional types (C_3 and C_4 photosynthetic systems) and their differential responses to climate variables and CO_2 fertilization mean that non-linear and rapid changes in ecosystem structure and carbon stocks are likely but difficult to predict with any certainty	Rising temperatures are likely to increase the importance of C_4 grasses, but CO_2 fertilization may promote C_3 species and the expansion of trees into grasslands. The major climatic effect on the composition and function of grassland and savannah systems is likely to be in precipitation and associated changes in fire and disturbance regimes. Modelling shows major rainfall reductions as a result of large-scale changes in savannah vegetation cover, suggesting positive feedbacks between human disturbance and climate change. The role of temperate grasslands in carbon storage is highly dependent on rainfall. The proportion of threatened mammal species may increase by 10–40 per cent, and migration routes are threatened. Large reductions in species' range size have been projected
Mediterranean systems	Mediterranean-type ecosystems are vulnerable to desertification and the expansion of adjacent arid and semi-arid systems expected under minor warming and drying scenarios	Likely to suffer some of the strongest impacts of global climate change, compounded by the effects of other pressures, including land use, fire and fragmentation. The effects of increased CO_2 concentration are inconsistent and are moderated by the growth limitation of increased drought. Desertification and arid ecosystem expansion are likely to induce substantial range shifts at rates greater than migration capability for many endemic species. Loss of biodiversity is likely overall, including substantial changes to species richness as well as the extinction of some species

Forests and woodlands	Modelling approaches predict that major changes in forest cover are likely to occur at temperature rises over 3°C. They mostly predict significant loss of forest towards 2100, particularly in boreal, mountain and tropical regions, but some climate-limited forests are expected to expand, especially where water is not limited	Recent moderate climate changes have been linked to improved forest productivity, but these gains are likely to be offset by the effects of increasing drought, fire and insect outbreaks. Estimates of the ability of tree species to migrate are uncertain, but high latitude and altitude shifts appear likely. Losses of species diversity have been projected, particularly in tropical forest diversity hotspots. Mountain forests appear particularly vulnerable. Extinctions of amphibian species in montane forests have already been attributed to climate change, and in most cases extinction risks are projected to increase
Tundra and polar	Tundra and polar ecosystems are likely to be the most vulnerable to climate change and may be turned from net carbon sinks to net carbon sources, with significant feedbacks to climate through both carbon emission and changes to albedo	Tundra climates will shift rapidly polewards and vegetation change is likely to follow, but with significant lags on tundra movement into polar desert and taiga encroachment on tundra due to slow growth and dispersal rates. Changes in temperature alter species dominance and therefore species composition. Food availability may increase for some vertebrates in summer, but decrease in winter. Endemic species, such as the polar bear and arctic breeding birds, are likely to experience large population declines and elevated extinction risks
Mountains	Mountain regions have already experienced above-average warming; its impacts, including water shortages and reduced extent of glaciers, are likely to be exacerbated by other pressures causing ecosystem degradation, such as land-use change, over-grazing and pollution	There is a disproportionately high risk of extinction for endemic mountain biota, partly because of their restricted geographical ranges and possibilities for migration, which can result in genetic isolation and stochastic extinctions. A reshuffling of species along altitudinal gradients is to be expected from their differential capacities to respond to change. Warming is expected to produce drying due to higher evapotranspiration in many mountain systems, and this will reduce the feasibility of upward movement of treelines. Tropical montane cloud forests and their biota are particularly vulnerable to drying trends, and species of mountain ecosystems are potentially subject to sharp declines
Inland waters	Inland aquatic ecosystems are highly vulnerable to climate change, especially in Africa. Higher temperatures will cause water quality to deteriorate and will have negative impacts on microorganisms and benthic invertebrates	Plankton communities and their associated food webs are likely to change. Distributions of fish and other aquatic organisms are likely to shift towards higher latitudes, and some extinctions are likely. Changes in hydrology and abiotic processes induced by precipitation change, as well as other anthropogenic pressures, will greatly impact aquatic ecosystems. Boreal peatlands will suffer major changes in species composition. Many lakes will dry out. Increases in precipitation variability may cause biodiversity loss in some wetlands.

Continued

Table 5.7 *Continued*

Ecosystem type	IPCC AR4 projections	Climate and ecosystem change
		Seasonal migration patterns of wetland species will be disrupted. The impacts of increased CO_2 will differ among wetland types, but may increase net primary production in some systems and stimulate methane production in others. On the whole, ecosystem goods and services from aquatic systems are expected to deteriorate
Marine and coastal	The most vulnerable marine ecosystems include warm-water coral reefs, cold-water corals, the Southern Ocean and sea-ice ecosystems	Ocean uptake of CO_2 reduces the pH of surface waters and their concentrations of carbonate ions and aragonite, which are vital to the formation of the shells and skeletons of many marine organisms, including corals. Other impacts include warming, increasing thermal stratification and reduced upwelling, sea-level rise, increase in wave height and storm surges, and loss of sea ice. The productive sea ice biome is projected to contract substantially by 2050. Changes in planktonic, benthic and pelagic community compositions have been observed and associated with climate change. Marine mammals, birds, cetaceans and pinnipeds are vulnerable to climate-related changes in prey populations. Melting ice sheets will reduce salinity, disrupt food webs and cause poleward shifts in community distributions. Both coral reefs and warm-water corals will suffer serious adverse effects from ocean acidification

C_3 plants: photosynthesis, carbon fixation and Calvin cycle all occur in a single chloroplast. In hot, dry conditions, their photosynthetic efficiency suffers because of photorespiration. About 85 per cent of plants are C_3 species, including most trees, important economic species such as cotton, and food species such as cereal grains, peanuts, beets, spinach and soya beans.

C_4 plants: almost never saturate with light, and under hot, dry conditions much outperform C_3 plants. They use a two-stage process where CO_2 is fixed in thin-walled mesophyll cells to form a 4-carbon intermediate, typically malate (malic acid). Only about 0.4 per cent of the 260,000 known species of plants are C_4 species, but this includes the important food crops maize, sorghum, sugarcane and millet, as well as many tropical grasses and sedges.

Source: derived from Campbell *et al.* (2009): 13–22; see also Fischlin *et al.* (2007)

Similarly, a shift from a coral reef to a sea-grass ecosystem as a result of ocean acidification would also have dramatic implications for tourism (Hoegh-Guldberg *et al.* 2007).

Changes in ecosystem composition and structure in combination with climate change will also lead to changes in ecosystem function. Changes in net primary production (the rate at which all the plants in an ecosystem produce net useful chemical energy) are significant both for ecosystem resilience and for the capacity of ecosystems to store carbon. Such changes also highlight the effects of climate change on the capacity of ecosystems to deliver services to people, including regulation of water flow, timber and fisheries, as well as more specific economic services such as tourism (see Box 5.4).

BOX 5.4 ECOSYSTEM SERVICES

Although the notion that the environment provides services to humanity has existed in the conservation and resource management literature since the 1950s, and arguably back to the 1860s and the work of George Perkins Marsh (1864) (Hall 1998), the concept became popularized by the Millennium Ecosystem Assessment (2005). The Millennium Ecosystem Assessment recognizes four basic categories of ecosystem service that nature provides to humanity:

- provisioning – e.g. providing timber, wheat, fish;
- regulating – e.g. disposing of pollutants, regulating rainfall, storing carbon;
- cultural – e.g. sacred sites, tourism, recreation, enjoyment of countryside;
- supporting – e.g. maintaining nutrient cycles, soils, crop pollination and plant growth.

The value of such services may be substantial. Ecosystem services are regarded as especially important for poverty alleviation, with tourism being a major contributor (SCBD 2009b). For example, the SCBD (2009b: 7) highlights some of the contributions of tourism and biodiversity towards human well-being, as follows:

- Namibia's protected areas contribute 6 per cent of GDP in tourism, with a significant potential for growth. Income from Namibia's conservancies (and conservancy-related activities): US$4.1 million. Percentage of total export from foreign tourist spending: estimated 24 per cent.
- Years of Mexico's (2004) carbon dioxide emissions offset by its protected areas: more than 5. Value of this service: US$12.2 billion.
- Nearly one-sixth of the world's population depend on protected areas for a significant percentage of their livelihoods.
- Cost of global network of marine protected areas conserving 20–30 per cent of the world's seas: up to US$19 billion annually, creating around 1 million jobs.

- Wetlands of Okavango Delta generate US$32 million per year to local households of Botswana, mainly through tourism. Total economic output: US$145 million (2.6 per cent of Botswana GNP).
- Number of times more likely a person living in a poor country is to be hit by a climate change-related disaster than someone from a rich country: 79.

Desert and arid ecosystems

The different levels of drought susceptibility among animal and plant species is likely to lead to major changes in both composition and structure of desert ecosystems. Flooding caused by occasional catastrophic rainfall may affect the mortality rates of some species (Thibault and Brown 2008), as well as affecting tourism infrastructure. Desert and arid ecosystems are not major focal points of tourism activity. However, ecotourism operations in the Saudi Peninsula and the Kalahari Desert in Namibia will potentially be affected by species loss. In the Succulent Karoo biome of South Africa, 2800 plant species face potential extinction as bioclimatically suitable habitat is reduced by 80 per cent with a global warming of 1.5–2.7°C above pre-industrial levels (Fischlin et al. 2007). In addition, increasing aridity and wind speed due to climate change may increase dust emissions from deserts (Campbell et al. 2009), leading to a lowering of amenity values for desert tourists as well as effects on other ecosystem services.

Grasslands and savannahs

The changed composition of grasslands in response to climate change is likely to have significant impacts on tourism in a number of locations. In the savannahs of the Mara–Serengeti in East Africa, which provide much of the resource base for Kenyan and Tanzanian safari tourism, rising temperatures and declining rainfall throughout the 1990s and early 2000s, combined with prolonged and strong El Niño Southern Oscillation (ENSO) episodes, caused progressive habitat desiccation and reduction in vegetation production in the ecosystem (Ogutu et al. 2008). This exacerbated the debilitating effects of adverse weather on local plant and animal communities, resulting in high mortalities of ungulates (Ogutu et al. 2008). As Campbell et al. (2009: 16) note, 'Declines in populations and diversity of savannah mammals may have significant implications for potential revenues from nature-based tourism.'

Climate change, via increased temperatures and/or shifts in the timing and intensity of monsoonal rainfall, will probably increase future fire risk in savannah grasslands (Parry et al. 2007). In Kakadu National Park in the Northern Territory of Australia, which is extensively featured in Tourism Australia's international promotions and Northern Territory's international and domestic promotions, atmospheric CO_2 appears to be favouring the growth of woody C_3 species in the savannah grasslands that make up the majority of the park. However, the potential for a changed fire regime as a result of climate change is posing a major challenge to park management as there is some evidence that high fire frequency is associated with the decline of some species (Brook 2009).

Mediterranean-type ecosystems

Mediterranean-type ecosystems are regarded as demonstrating high biodiversity and being attractive for ecotourism (Fischlin *et al.* 2007). These ecosystems face projected threats from desertification due to expansion of adjacent semi-arid and arid systems under relatively minor warming and drying scenarios. Areas that would be affected by such a process include Western Australia, South Africa, California and the Mediterranean.

The bioclimatic zone of the Cape Fynbos biome in South Africa, best known to tourists through Table Mountain and the Western Cape, could lose 65 per cent of its area under warming of 1.8°C relative to 1961–90 (2.3°C, pre-industrial), with a long-term species extinction of 23 per cent (Thomas *et al.* 2004). The conservation and tourism significance of the area is reinforced by Table Mountain in Cape Town supporting 2200 species, more than the entire United Kingdom and about 20 per cent of all plant species in Africa. The high degree of endemism also means that the Table Mountain Range has one of the highest concentrations of threatened species per area than anywhere in the world. Wildflower tourism, which is significant in both south-west Western Australia and South Africa, is also likely to suffer as species become increasingly affected by increased drought, CO_2 levels and wildfires (Fischlin *et al.* 2007).

Mediterranean-climate south-west Western Australia has experienced a 10–20 per cent decline in autumn and early winter rainfall since the 1970s together with temperature increases, and this has been attributed, at least in part, to global climate change (Yates 2009). The region's biodiversity is a major component of international and domestic tourism. 'There are more than 7400 named plant taxa and an estimated 6500 species of vascular flora in the south-west, of which >50% occur nowhere else' (Lindenmayer and Burbidge 2009: 69). Limited research suggests that further climate change will contract the ranges of many species in south-western Australia, but there are many uncertainties, as some species have persisted *in situ* through climate change in the Pleistocene and their present distributions may not reflect their climate tolerances (Yates *et al.* 2007). However, the high degree of endemism in a biodiversity hotspot, combined with more frequent fires, lower rainfall, extensive habitat fragmentation, limited dispersal ability, and the presence of invasive weeds and introduced diseases, would appear 'likely to reinforce current declines in biodiversity and lead to tipping points much sooner than hitherto realised' (Yates 2009: 126).

Forests and woodlands

Forests are ecosystems with a dense tree canopy (woodlands have a largely open canopy), covering a total of 41.6 million km^2 (about 30 per cent of all land), with 42 per cent in the tropics, 25 per cent in temperate zones, and 33 per cent in boreal zones (Fischlin *et al.* 2007). Forests are among the most productive terrestrial ecosystems, which makes them attractive for both climate change mitigation and forestry as well as tourism (Gössling and Hickler 2006).

Forest ecosystems often require long timeframes to observe changes in distribution. However, evidence is already starting to mount that such changes are already happening (Campbell *et al.* 2009). Temperate and boreal forests are expected to expand polewards and upwards at the expense of tundra and alpine communities. Uphill migration of treelines in Scandinavia has been observed on the order of 150–200 m (Kullman 2007). Similarly, the upward advance of alpine treeline has been observed through encroachment of woody vegetation into alpine meadows in Yunan, China (Baker and Moseley 2007). Substantial uncertainty remains over the extent of climate-related loss of Amazon forest, with projections ranging from 18 per cent (Salazar *et al.* 2007) to 70 per cent (Cook and Vizy 2008).

The rainforests of the North Queensland Wet Tropics are a World Heritage Site that is under substantial threat from climate change (Steffen *et al.* 2009). The region is dominated by mountain ranges varying from sea level to 1600 m, with altitude being the strongest environmental gradient affecting species composition and patterns of biodiversity (Williams 2009). Tropical mountain systems, such as those in northern Queensland, are extremely vulnerable to climate change due to their fragmentary nature; small area; high rates of endemism and species specialization (often as the result of acting as a long-term refugia (Hilbert *et al.* 2007); and the compression of climatic zones over the elevation gradient (Williams *et al.* 2003). Such montane forests contain the highest numbers of endemic species in the Wet Tropics and are the most threatened by climate change. Bioclimatic modelling of the distributions of regionally endemic rainforests vertebrates in the Wet Tropics by Williams *et al.* (2003) predicted that, with a >2°C temperature increase, most species will undergo dramatic declines in distribution, with some completely losing their current climatic environment. Cloud forests and other highland rainforests types are predicted to become greatly reduced in area and more fragmented across the Wet Tropics, even under a conservative scenario of 1°C temperature increase and a small reduction in rainfall (Hilbert *et al.* 2007).

Tundra and polar ecosystems

Biodiversity at both ecosystem and species levels is critical for polar tourism (Hall and Saarinen 2010a; Forbes 2011). As noted in chapter 1, the Arctic is undergoing rapid climatic change that is affecting both terrestrial and marine polar ecosystems. More than any other species, the polar bear (*Ursus maritimus*) has become an evocative symbol of global climate change, evidenced by images in the popular media of polar bears 'struggling' to survive in a warming Arctic climate. Although some of these images may be misleading, it is clear that, across the Arctic, many polar bear populations are under threat due to significant decreases in the extent, thickness and increased variability of sea-ice (Dawson *et al.* 2010). The polar bear tourism industry that has developed in Churchill, Canada is threatened by declining sea-ice conditions on Hudson Bay. Projections are that over the next 30 years, sea-ice conditions may deteriorate to the point that the polar bear population may collapse in this region (Dawson *et al.* 2010). However, the polar bear is not the only polar species threatened by climate change. Other terrestrial iconic species such as musk ox, caribou and reindeer may also experience a decline in numbers as a result of climate change (Tyler 2010), while seal and penguin populations are also affected by the changing abundance and dynamics of sea-ice.

There is some suggestion that the Artic tundra is reducing in size as taiga forest advances northwards as a result of warming and the loss of permafrost (Wolf *et al.* 2008). Some tundra ecosystems recognized as priority conservation habitat by the EU – such as the palsa mires of Lapland, wetland complexes characterized by permanently frozen peat hummocks and a rich diversity of bird species – will almost certainly disappear as the permafrost melts. The Arctic coastal zone provides habitat to an estimated 500 million seabirds. Arctic river deltas are biological hotspots on the circumpolar Arctic coast: they have high biodiversity and are extremely productive in relation to adjacent landscapes (Forbes 2011). Yet Arctic coastal and estuarine habitats are extremely vulnerable to SLR. Many polar species are particularly vulnerable to climate change because they are highly specialized and have adapted to their environment in ways that are likely to make them poor competitors with potential immigrants from environmentally more benign regions as conditions change (Anisimov *et al.* 2007).

> The adaptive capacity of current Arctic ecosystems is small because their extent is likely to be reduced substantially by compression between the general northwards expansion of forest, the current coastline and longer term flooding of northern coastal wetlands as the sea-level rises, and also as habitat is lost to land use.
>
> (Anisimov *et al.* 2007: 659)

Furthermore, stress brought about by increased human contact via tourism may only add to a series of anthropogenic impacts including not only climate change, but also increasing contaminant loads, increased ultraviolet-B radiation, and further habitat loss and fragmentation. Polar ecosystems are also particularly vulnerable to environmental change as their species richness prior to the current period of anthropogenic-induced change is low, with correspondingly low levels of redundancy making it relatively easier for new species to outcompete existing species in the same ecological niche. The general vulnerability of Arctic ecosystems to warming, and the lack of adaptive capacity of Arctic species and ecosystems, are therefore likely to lead, where possible, to relocation rather than rapid adaptation to new climates (Hall 2010a; Hall *et al.* 2010).

The limited species diversity of the Arctic coastal zone means that the ecosystem is extremely vulnerable to rapid change. Flooding is already occurring in some areas. Changes in species composition due to SLR is being experienced most in sandy and silty supratidal meadows, mud flats and marshes that are periodically inundated at high tides and in storm events. Although many salt marshes in temperate regions can keep pace with slow SLR through inorganic sedimentation and organic production, 'there are many observations of flooded tundra along Arctic coasts, where vertical accretion is clearly not keeping pace' (Forbes 2011: 44).

In the Antarctic, there is also substantial evidence indicating that major regional changes are occurring in terrestrial and marine ecosystems in areas that have experienced warming. These include increasing abundance of shallow-water sponges and their predators, and declining abundances of krill, Adelie and Emperor penguins, and Weddell seals. On the Antarctic continent, the abundance and distribution of the only two species of native flowering plant, the Antarctic pearlwort (*Colobanthus quitensis*) and the Antarctic hair grass (*Deschampsia antarctica*), are growing as the amount of ice-free habitats increase (Fowbert and Smith 1994). Similarly, climate change is also affecting Antarctic algae, lichens and mosses, with further changes expected as temperature, water and nutrient availability increase. Importantly, the relative simplicity of Antarctic terrestrial ecosystems makes them extremely vulnerable to the introduction of exotic species, particularly as climate change creates more opportunities for successful introductions (Hall 2010a).

Mountains

As has been noted several times in this chapter, mountains are already experiencing climate change impacts. Glacial retreat is a global phenomenon: treelines have been moving higher in altitude and alpine species' distributions in general have moved upwards over the past 100 years (Walther *et al.* 2005), apparently at an increasing rate (Campbell *et al.* 2009). For every 1°C increase in temperature, it is estimated that the snowline of the European Alps rises by about 150 m (EEA 2010). 'Pioneer' species are moving upwards and plants that are cold-adapted are now being driven out of their natural ranges. European plant species might have shifted hundreds of kilometres northwards by the late twenty-first century and 60 per cent of European mountain plant species may face extinction. Observational data on species declines in mountain ecosystems, principally in alpine systems (Pauli *et al.* 2007), when combined with observations of distributional changes, suggest that the composition of mountain ecosystems is changing appreciably in response to climate change (Campbell *et al.* 2009).

Inland waters

Inland waters are extremely significant for tourism and recreation (Hall and Härkönen 2006). However, they are also highly vulnerable to climate change as a result of warming and changed precipitation regimes. In the Arctic and sub-Arctic, reduced ice-cover duration on lakes, increased and more rapid stratification, earlier and increased primary production, and decreased oxygenation at depth will probably result in a reduction in the quality and quantity of habitat for significant fishing species such as lake trout. Decreased water flow in summer is also likely to decrease habitat availability and possibly deny or shift access for migrating fish (Anisimov *et al.* 2007). In Yellowstone National Park, ponds and even lakes that have been considered permanent since first reported as early as the 1850s have disappeared or become temporary bodies of water (Di Silvestro 2009). McMenamin *et al.* (2008) linked dramatic declines in amphibian diversity populations in Yellowstone to the fourfold increase in permanently dry kettle ponds (glacier-formed depressions that fill in spring with groundwater and snow melt) between 1992 and 2008. The remaining wet ponds were also supporting smaller numbers of amphibians.

Observational evidence supports the notion that the composition of fish communities is changing as water temperatures increase. In France, species richness, proportions of warm water species and total abundance have all increased in fish communities since the 1980s (Daufresne and Boet 2007), although some native fish species, including salmon, appear negatively affected by climate change, especially in headwaters and at higher altitudes (Buisson *et al.* 2008). In North America, it is predicted that the range of cool-water fish (e.g. walleye, perch) and particularly of warm-water fish species (e.g. bass) will expand northward as a result of increased water temperatures, and this will alter the composition of recreational fish catches (Jones *et al.* 2005). Other studies have shown that warmer water temperatures would eliminate the sport trout fishery from most North Carolina streams (Ahn *et al.* 2000) and reduce the habitat for several popular cool-water sport fish in eastern Canada, including walleye, northern pike and white fish (Minns and Moore 1992).

Marine and coastal

Marine and coastal ecosystems are extremely vulnerable to climate change. Moreover, many of these impacts are synergistic and interact with each other via various positive- and negative-feedback mechanisms. For example, extreme events can destroy ecosystems, such as mangrove forests or coral reefs, that provide protection against SLR and further high-magnitude events, which will in turn have consequences both for the coastal ecosystem itself and for the buffering of associated ecosystems. A review of threats to mangrove by Alongi (2008) suggested that climate change may lead to a global loss of 10–15 per cent of mangrove forest, and SLR is an important component of that threat because sediment accretion is not keeping pace with it. This is extremely significant for some tropical and subtropical coastal systems where mangroves are an important coastal protection as well as delivering other ecological services. However, climate change is also having significant impacts on the coastal systems of temperate locations. On the Baltic coast, an area that is extremely significant for tourism as well as second homes, a combination of strong storms, high sea levels induced by storm surge, ice-free seas and unfrozen sediments is associated with marked coastal change and increased instability of wetlands (Kont *et al.* 2007).

Some of the impacts of SLR on tourism have already been noted above. However, SLR also clearly has an impact on biodiversity in coastal areas, and coastal wetlands in particular, that may also be significant attractions. For example, the wetlands of Kakadu National Park in Australia are a major attraction that supports a rich and spectacular biota, including flocks of magpie geese *Anseranas semipalmata*, which congregate in millions to feed on the floodplains. These wetlands were formed around 6000 years ago after sea-level stabilization, following a post-glacial rise of 120 m (Woodroffe *et al.* 1987). However, they are threatened by further SLR associated with anthropogenic global warming (Brook and Whitehead 2006). Rising sea levels, in combination with tropical storm surges, increase the regularity and severity with which saline water flows into the low-lying freshwater wetlands. The situation is also complicated by the presence of invasive species (Brook *et al.* 2008).

A complex network of low-lying natural drainage channels, enlarged or cross-connected by movement of feral animals such as Asian water buffalo *Bubalus bubalis* and pigs *Sus scrofa*, means that even a few tens of centimetres of additional SLR may be sufficient to degrade or eliminate a large fraction of the floodplain communities.

(Brook 2009: 128)

Coral reefs are a crucial resource for tourism and other sectors. In many destinations, reefs are a key visitor attraction and therefore a major economic asset. In the western Indian Ocean region, a 30 per cent loss of corals resulted in reduced tourism in Mombasa and Zanzibar, and caused financial losses of about US$12–18 million (Payet and Obura 2004). Hoegh-Guldberg and Hoegh-Guldberg (2008), in a supplementary paper to the Garnaut Report, estimated that 62 per cent of visitor nights in regions along the GBR represented 'reef-interested tourism', with such tourism accounting for 90 per cent of visitor nights in Tropical North Queensland. However, the GBR, like many other reef systems, is increasingly subject to coral bleaching and has experienced eight mass bleaching events between 1979 and 2009 (1980, 1982, 1987, 1992, 1994, 1998, 2002 and 2006). The most widespread and intense events occurred in 1998 and 2002, with about 42 and 54 per cent of reefs affected, respectively (Berkelmans *et al.* 2004). The economic impacts of coral bleaching on tourism in the GBR were calculated by Oxford Economics for the Great Barrier Reef Foundation, which concluded:

> At a preferred discount rate of 2.65%, streamed over 100 years, holding present day values constant, it is estimated that the present value (PV) of the GBR as a whole (excluding indigenous values) is [A$]51.4 billion, with a value of $17.9 billion estimated for the Cairns area.
>
> From this, an estimate of the cost of bleaching for the Cairns area and the GBR can be derived. If a total and permanent bleaching of the GBR were to occur today, then (holding present day values constant over 100 years, at a discount rate of 2.65%) the costs (in PV terms) are estimated at $37.7 billion with an estimate of $16.3 billion for the Cairns area.
>
> Put another way, the bleaching cost for the whole of the GBR is roughly equivalent to a constant $1.08 billion per annum over the course of a century.

(Oxford Economics 2009: 2)

The Great Barrier Reef Foundation has also provided estimates of the impact of different climate change scenarios on the future condition of the GBR (Table 5.8).

As noted above, ocean acidification is having a major impact on coastal and ocean marine systems and is irreversible on timescales of at least tens of thousands of years (SCBD 2009a). Ocean acidification reduces the availability of carbonate in seawater, with carbonate ion concentrations now lower than at any other time during the past 800,000 years. Given current carbon emission rates, it is predicted that the surface waters of the Arctic Ocean will become under-saturated with respect to essential carbonate minerals by the year 2032, and the Southern Ocean by 2050, with disruptions to large components of the marine food web (SCBD 2009a). By 2100, it is predicted that 70 per cent of cold-water corals will be exposed to corrosive waters (Turley *et al.* 2007).

Tropical waters are also experiencing rapid decline in carbonate ions, reducing rates of net warm water coral reef accretion and leaving biologically diverse reefs outpaced by erosion and SLR (SCBD 2009a). Tropical coral reefs provide in excess of US$30 billion annually in global goods and services, such as shoreline protection, tourism and food security (SCBD 2009a). Examples of the ecosystem services that tropical coral reefs provide

Table 5.8 *Climate change scenarios and the condition of the Great Barrier Reef*

Year (scenario)	Atmospheric CO_2 (ppm) concentration	Temperature changes (°C) of up to:	Reef impacts
Pre-industrial (baseline)	260		• Coral reef condition in its natural state
2009 (maintaining current CO_2 concentrations)	380	+0.74	• Coral reefs will continue to change but will remain coral-dominated • Decreased coral growth and skeletal density
2100 (IPCC scenario of no action taken)	800	+6.4	• Non-existent reef growth • Drowned reefs as sea levels rise • Half or more of coral-associated fauna becoming rare or extinct • Macroalgae/phytoplankton will dominate
2100 (Australian Government target of 5 per cent reduction in emissions from 2000 levels)	550	+3	• Decreases coral calcification and growth by up to 40 per cent • Rapid contraction of coral reefs, reef ecosystems reduced to crumbling frameworks with few calcareous corals • Loss of coral-associated fauna
2100 (Australian Government target of 15 per cent reduction in emissions from 2000 levels)	510	+2.6	• Density and diversity of corals likely to decline • Reef erosion will surpass reef-building processes • Reduced habitat complexity and loss of biodiversity • Reefs at greater risk from other factors (overfishing, declining water quality)
2100 (Garnaut recommendation of 25 per cent reduction in emissions from 2000 levels)	450	+2	• Thermally tolerant, rapidly growing corals dominate • Reef erosion equal to reef-building processes • Increasing frequency and severity of mass coral bleaching, disease and mortality

Source: Garnaut (2008); Great Barrier Reef Foundation (2009)

include building materials and reef-based tourism (direct-use values); habitat and nursery functions for commercial and recreational fisheries and coastal protection (indirect-use values); and the welfare associated with the existence of diverse natural ecosystems (preservation values) (SCBD 2009a). Because of the significance of coral reefs and fishing for tourism, many countries that have reef-based tourism industries are vulnerable to the threats of ocean acidification (Hoegh-Guldberg *et al.* 2007). In Hawaii alone, reef-related tourism and fishing generate US$360 million per year (SCBD 2009a).

Reefs in acidified waters are predicted to decline in the following sequence (SCBD 2009a: 53):

- loss of coralline algae causing decreased reef consolidation;
- loss of carbonate production by corals resulting in loss of habitat;
- loss of biodiversity with extinctions.

The failure of coral communities to compete with algae communities that will not be similarly affected by ocean acidification will result in an ecological phase shift to a stable new ecosystem state that, although not without ecological value, will not support the same species as a coral reef or provide the same degree of protection for coastal areas with respect to high-magnitude storm events. According to Brander *et al.* (2009), the annual economic damage of ocean acidification-induced coral reef loss is estimated to reach US$870 billion by 2100, under the A1 *SRES* global emissions scenario. Such effects will only be amplified by the interaction of the effects of ocean acidification with other dimensions of climate change, including storms, SLR, increasing water temperatures and bio-erosion, as well as anthropogenic change in the form of pollution and over-fishing (Hall 2010c).

A more accessible Arctic

As noted above and in chapter 1, the melting of the polar sea-ice and increasing temperatures is making the Arctic increasingly attractive for tourism. The extent of recent warming is such that it has been recognized as the warmest period in the Arctic for the past 2000 years, with four of the five warmest decades in that period occurring since 1950 (Kaufman *et al.* 2009). The extent of Arctic sea-ice has reduced substantially since the 1950s and there is no indication that long-term trends are reversing (Schiermeier 2009). As a group of scientists associated with the International Polar Year 2009 concluded, 'The entire Arctic system is evolving to a new super interglacial stage seasonally ice free, and this will have profound consequences for all the elements of the Arctic cryosphere, marine and terrestrial ecosystems and human activities' (cited in Hall and Saarinen 2010b).

Climatic conditions are extremely important for tourism because of the extent to which they influence the relative accessibility and attractiveness of a given location (see chapters 1 and 2). Climate change influences the seasonality of a tourism location or attraction because of the extent to which access is economically and geographically feasible in a polar environment, as well as determining the local environmental conditions that may prove appealing to visitors. For example, climate change is regarded as having enabled the lengthening of the northern polar cruise season as well as providing access to hitherto inaccessible locations (Hall *et al.* 2010; Hall and Saarinen 2010a, 2010b). Indeed, there is potentially something of a paradox given that, while tourism is a significant contributor to climate change, it is also a beneficiary because greater access is now possible for tourists to some polar areas as a result of reduced sea-ice and warmer weather.

The numbers of tourists travelling in the Arctic region is substantial, of the order of over 5 million visitors per year (Hall and Saarinen 2010a). Such figures run counter to the perspective of Frigg Jorgensen, general secretary of the Arctic Expedition Cruise Operators, who commented, 'compared to national parks in Alaska where many thousands visit, for example, the number of Arctic tourists [is] minimal' (quoted in Round 2008: 47). Similarly, Round (2008: 46) states, 'do we need just a little more perspective? Only a few thousand travelers visit the Arctic every year compared to the hundreds of thousands of people that cross the manicured grass of New York's Central Park everyday.' Apart from the geographical challenge of not including Alaska as part of the Arctic, there still remains the issue that the number of tourists is continuing to grow and is concentrated in a small number of accessible areas in space and time. The number of fly-in tourists per year to Greenland now exceeds its permanent population, with the number of cruise guests already being over half. A similar situation with respect to number of visitors per year in relation to permanent population also exists in Iceland, Svalbard and northern Norway, Sweden and Finland above the Arctic Circle (Hall and Saarinen 2010a). Although the tourist numbers for the Arctic would appear to be low if they were calculated on a tourist per square kilometre basis (approximately one tourist per 3 km^2), the reality is that sites of permanent settlement and tourist accommodation, attraction and transport hubs are usually co-located, and therefore increases in visitor arrivals can place significant pressure on permanent infrastructure as well as the environment (Hall and Saarinen 2010b).

Both the *Arctic Climate Impacts Assessment* (ACIA 2005) and Anisimov *et al.* (2007) note the potential economic benefits of reduced sea-ice for the lengthening of the ship navigation season and increased marine access, including the opening up of sea routes along the Northwest Passage and the Northern Sea Route. Northwest Passage cruises are already promoted by high-profile travel companies such as Hapag-Lloyd cruises, Quark Expeditions and Peregrine Adventures (Hall *et al.* 2010). Instanes *et al.* (2005) suggest that by 2050, the Northern Sea Route will have 125 days per year with less than 75 per cent sea-ice cover, which represents favourable conditions for navigation by ice-strengthened ships. However, while this is regarded as a potentially positive benefit of climate change for some northern communities, the effects of such changes will be substantial for northern cryogenic landscape, landforms and biodiversity (Forbes 2011).

The biodiversity of the polar regions plays a major role in their attractiveness to tourists, whether it be the attraction of individual charismatic megafauna such as polar bears, or polar ecosystems and landscapes in general (Stewart *et al.* 2005; Hall and Saarinen 2010a). However, tourism can also potentially affect biodiversity as a result of either direct effects, such as disturbance, trampling or the construction of infrastructure, or via acting as a vector for biological invasion (Hall 2010a). Cruise-ships and yachts, as well as site visitation, have been recognized as particularly significant avenues of species introduction given that they are one of the major means of transport in getting tourists to many polar sites (Hall 2010a; Hall *et al.* 2010; Hall and Wilson 2010). Four trends in polar tourism may be of high significance for the introduction and spread of alien organisms to and within the region (Hall *et al.* 2010):

- tourists are disproportionately attracted to sites of relative high/medium biological and landscape diversity;
- the intensity of visitor use is increasing in both absolute terms and, in general, over time and space;
- sites of high popularity are not consistent over time, meaning that the potential for human impact is not contained to a number of specific sites but varies as a result of tourist trends and changing fashions;
- the range of tourist activities is expanding – in addition to being able to land on beaches and observe immediately accessible wildlife, options now include extensive walks, kayaking trips and, in some Arctic locations, fishing and hunting.

Given the growing number of visitors to the Arctic, tourism is regarded as a key component of the economy. Climate change, rather than having a negative impact on the regional economy, is often regarded as being a major beneficiary along with maritime transport, as access to many northern areas is improved (ACIA 2005). Similarly, Antarctica and the sub-Antarctic are also receiving increasing numbers of tourists, which, although not on the scale of the Arctic, has significant economic benefits for the sub-Antarctic communities and gateway communities in Australia, New Zealand and South America (Hall and Wilson 2010). And, given the much smaller amount of visitor access to ice-free areas, tourism is arguably of proportionally even greater significance in terms of direct environmental impact in the Antarctic than the Arctic (Hall 2010a).

Tourism's role in polar economic development when well planned and managed therefore goes well beyond that of tourism alone, as it provides a major 'enabling' role via transport, accommodation and other infrastructure that may also contribute to local quality of life and economic development. Nevertheless, the environmental costs of transport do need to be considered in any assessment of the relationship between tourism and climate change in the Arctic. Indeed, Hall *et al.* (2010) identified that a large majority of the cruise companies that operate in the Arctic do not have mandatory carbon offsetting in their product offerings, with many not even promoting voluntary schemes. As of 2009, only 14 of the 31 cruise companies operating in Atlantic polar and sub-polar waters operated with a code of conduct with respect to minimizing their direct visitor impacts (those that were members of the Association of Arctic Expedition Cruise Operators), while 15 had a publicly available environmental policy (Hall *et al.* 2010). Although this is undoubtedly significant in a site-specific context with respect to ameliorating the immediate direct environmental impacts of tourism, existing codes of conduct do not deal adequately with the broader issues of tourism-related environmental change, including the introduction of biologically invasive species (Hall 2010a; Hall *et al.* 2010). Given that tourism is such a significant economic activity, and even a 'desired industry in some communities' (Stewart *et al.* 2005: 383), it is clearly vital that a deeper understanding of the complexity of polar tourism be achieved in terms that are useful for policy-makers, especially when tourism and the Arctic environment is also integral to climate change adaptation and mitigation.

AGRICULTURE AND TOURISM

Ensuring that food production is not threatened is an explicit criterion of UNFCCC Article 2. In early IPCC reports (1996, 2001b, 2001c), it was generally concluded that, in the near to medium term, world food production is not threatened. However, there is increasing concern over the impacts of climate change on agriculture and associated issues of food security (Burton and Lim 2005). Low-latitude areas are most at risk of having decreased crop yields, while mid- and high-latitude areas may, although not in all locations, see increases in crop yields for temperature increases of up to 1–3°C (Schneider *et al.* 2007). Taken together, IPCC (Schneider *et al.* 2007: 790) concluded that there is low to medium confidence that:

- global agricultural production could increase up to approximately 3°C of warming;
- for temperature increases beyond 1–3°C, yields of many crops in temperate regions are projected to decline;
- beyond 3°C warming, global production would decline because of climate change;
- the decline would continue as global mean temperature increases.

Significantly, Schneider *et al.* (2007) note that most studies of the effects of climate change on global agriculture have not yet incorporated a number of critical factors, including changes in extreme events or the spread of pests and diseases, nor have they considered the development of specific practices or technologies to aid adaptation. Also of significance is that of the four main elements of food security – availability, stability, utilization and access – only the first is routinely addressed in climate change simulation studies (Schmidhuber and Tubiello 2007).

Agriculture and tourism are related in several ways. First, agricultural landscapes act as attractive environments for rural tourism. Second, agriculture clearly is important for the supply of food and drink to the tourism and hospitality industries. Third, agriculture can be a tourist attraction in its own right.

Agricultural and rural landscapes are continually changing in response to economic, social and environmental factors. Climate change is therefore another layer of change in an already complex system. Climate change will have a direct impact on agricultural productivity as a result of such factors as soil moisture, precipitation regimes, and soil and air temperature. But it will also affect the selection of different crops to take advantage of the new environmental and economic opportunities provided by climate change, as well as associated water demands. Such effects will have enormous impacts on the agricultural landscape, which will probably see major transformations as the environments in which the cultural landscapes of agriculture are subject to the same stresses that will lead to the changes in composition and distribution of natural ecosystems discussed above. Unfortunately, little research has been undertaken on the implications of rural landscape change for tourism, but, given the role of the agricultural landscape in the promotion of tourism to regions such as the Napa Valley, Provence and Tuscany, it is clear that the relative attractiveness of rural locations may shift as a result of climate change (Pincus 2003).

The supply of food to tourism and hospitality occurs over a wide range of supply chains. Increasing concerns over food miles may result in some food products being regarded as either uneconomic or unethical by some producers and consumers. A shift towards local food products and systems is clearly one response to issues of sustainability in food supply, which also affects tourism. The effects of climate change on agriculture may potentially lead to some destinations losing access to specific products with which they have been historically associated. This may occur as a result of either species loss or decline, or species migration. One of the clear lessons for destinations is that, while they are rooted in space, many of their food sources are likely to shift over space as a result of a changing climate. A second aspect of food supply, at least in areas in which there is poor food security, is the possibility for increased competition between tourists and locals for scarce food sources. In some developing countries, the relative wealth of visitors may give them improved access to some foods that local people can no longer afford.

Finally, some agricultural sectors are visitor attractions in their own right. To a limited extent, this includes properties such as farm and ranch stays as well as orchards, vegetable growers and farm shops/stalls. However, particular food producers, such as cheese and wine producers, are probably the most recognized with respect to the significance of direct sales to visitors. Hall *et al.* (2003) reported that surveys of food and wine tourism operators in Australia (Victoria and Western Australia), Canada (Okanagan Valley) and New Zealand all indicated that climate change and biosecurity had been identified as a major threat to long-term business viability.

In countries such as Australia and New Zealand, *c.* 80–90 per cent of wineries in the various wine regions are open to visitors (Mitchell and Hall 2006). Such cellar-door sales are integral to the business plans of many smaller wineries, while also important for rural and peri-urban destinations (Hall and Mitchell 2008). Climate change is recognized as already starting to affect grape production and the selection of varieties (Webb *et al.* 2008). A shift towards higher latitudes is occurring with grape production as with other temperate climate species. Regions that are currently producing high-quality grapes at the margins of their optimal climatic zone may be thrust into a climate that is no longer suited to the grapes now grown. Areas of France, Australia and California that are renowned for high-quality wines are projected to see grape-growing conditions impaired by mid- to late-century (Pincus 2003; Jones *et al.* 2005; White *et al.* 2006). Conversely other more poleward wine-growing regions (southern England, southern New Zealand, southern British Columbia, Canada) are projected to be able to produce higher-quality vintages and may benefit from a shift in wine tourism over time.

CULTURAL HERITAGE ASSETS

Cultural heritage consists of built heritage (historical and architectural), archaeological heritage and socio-cultural heritage, including parks and gardens, all of which can be affected directly or indirectly by climate change (McIntyre-Tamwoy 2008). Built heritage may be damaged as a result of climate change-related flooding, SLR and associated coastal flooding and erosion, and increased exposure to extreme events with potentially greater wind speeds. These can be particularly threatening to heritage structures, such as buildings, that are already in a fragile state. Thermal stress can affect building materials (Bonazza *et al.* 2009b), while increased chemical dissolution of carbonate stones, via the karst effect, will increase with future CO_2 concentrations, and will come to dominate over sulphur deposition and acid rain effects on monuments and buildings (Bonazza *et al.* 2009a). Ancient buildings were designed for a specific climate, and are less isolated from the ground than newer buildings that utilize geomembranes and other technologies to limit exposure to water and other soil processes. Changes in water levels or salt mobilization may increase deterioration of foundations of ancient structures. Timber structures may also be exposed to new and increased insect infestations or microorganisms. In addition, changes in wetting and drying cycles can induce crystallization and dissolution of salts, and affect other heritage assets such as wall paintings, frescos and rock art (UNESCO 2007).

Archaeological evidence is preserved in the ground because of specific localized hydro-logical, chemical and biological processes, each of which are subject to change under climate change (Mitchell 2008). Changes may decrease the level of survival of materials. The stratigraphic integrity (organization of levels and types) of soils where buried archae-ological evidence exists could also be rapidly lost in some areas due to changes in pre-cipitation levels, permafrost melting and floods.

Socio-cultural heritage is an ever more popular attraction for tourists, in destinations such as Australia, Africa, Asia and parts of North and South America. However, cultural activities and indigenous and folk traditions are among some of the most fragile aspects of a society. Many have already disappeared through processes of globalization, mechaniza-tion, urbanization, emigration and other factors, and the cultural consequences of physical climate change impacts on landscapes and buildings may reinforce current trends towards the abandonment and break-up of communities leading to the loss of rituals and cultural memory.

In 2005, UNESCO initiated an assessment of the implications of climate change for world heritage. As part of this assessment, it conducted a survey of all State Parties to the World Heritage Convention to assess the extent and nature of the potential impacts of climate change on World Heritage properties. Of the 110 responses, 72 per cent indicated that climate change had an impact on their natural and cultural heritage. The most frequent climate change risks identified were extreme events (hurricanes, storms), SLR, erosion (from flooding or wind), flooding, and changes in rainfall (increases or droughts). A total of 125 World Heritage properties were identified as specifically threatened by climate change (79 Natural or Mixed Natural–Cultural Heritage properties and 46 Cultural Heritage properties). Risks posed to Natural World Heritage properties, such as the GBR and Glacier National Park, are discussed above in the sections on landscape change and biodiversity. Key Cultural Heritage sites identified to be at risk include the Tower of London, the city of Venice, the Great Mosques of Timbuktu and the prehistoric megalithic temples of Hagar Qim in Malta.

UNESCO (2007) concluded that climate change poses a threat to the core values that the World Heritage Convention seeks to protect, and also noted several legal implications for the World Heritage Convention itself.

- Should a site be inscribed on the World Heritage List while knowing that its potential outstanding universal value (OUV) may disappear due to climate change impacts?
- Should a site be inscribed on the List of World Heritage in Danger or deleted from the World Heritage List due to the impacts that are beyond the control of the concerned State Party? If Glacier National Park was inscribed because of its glaciated landscape and the glaciers melt – should World Heritage status be revoked? International NGOs have pressured UNESCO to list five World Heritage properties to be listed as 'in danger' because of the threat posed by climate change [Australia's GBR, the Belize Barrier Reef, Peru's Huascaran National Parks, Nepal's Sagarmatha (Mount Everest) National Parks, and Waterton-Glacier International Peace Park on the Canadian–US border].
- Could a particular State Party blame another State Party (or group of Parties) for the loss of World Heritage property because of their responsibilities for climate change?
- Should the Convention consider adopting 'evolving assessments of OUV' to recognize the fact that for some natural properties it will be impossible to maintain the original OUV for which the site was inscribed on the World Heritage List, even if adaptation is attempted?

Several countries indicated they were undertaking research and monitoring to understand better the nature of the threat posed by climate change; however, knowledge of the threat of climate change to World Heritage properties remains limited and a priority for State Parties (Shearing 2007; UNESCO 2007; Terrill 2008). Australia is the only country known to have completed a preliminary assessment of the implications of climate change for its entire set of World Heritage properties. In one of the few empirical studies of impacts in World Heritage Sites, Beniston (2008) undertook a climatological analysis of the ability to maintain the vineyard-dominated cultural landscape of the Lake Geneva region. The study concluded that, while climatic changes would eventually preclude the cultivation of the same grape varieties grown today, if vineyards are able to adapt to other grape varieties successfully, the visual aspect of vineyards that dominated the landscape since the thirteenth century could be maintained. While, from a wine tourism perspective, the destinations will have changed – perhaps not for the better, depending on how successfully new varieties of grapes can be adapted to the regional climate and soil conditions – the landscape should remain essentially unchanged from the perspective of sightseeing tourists.

INDIRECT IMPACTS ASSOCIATED WITH SOCIETAL CHANGE

> 'Business cannot succeed in societies that fail. There is no future for successful business if the societies that surround it are not working.'
>
> WBCSD (undated)

Tourism depends on economic prosperity and socio-political stability. Recent economic recessions and political upheavals caused by war, terrorism or revolution illustrate vividly the implications for tourism (Hall 2010d). Climate change is anticipated to pose a risk to future economic growth and to the political stability of some nations and thereby to threaten tourism development in these regions.

Any reductions of global GDP due to climate change would reduce the discretionary wealth available for tourism and have negative implications for anticipated future growth in tourism. Understanding the highly complex economic implications of climate change, and our possible responses to it, has challenged fundamental notions of risk, uncertainty, discounting and equity (Weitzman 2007). A broad literature has emerged (1) to examine the costs of mitigation to avoid dangerous climate change or achieve international mitigation commitments (e.g. the Kyoto Protocol) and (2) to evaluate the damages related to non-response or limited mitigation response. The combined answers to these questions determine whether no action, gradual action, or considered immediate action is economically prudent.

As Heal (2009) points out, the range of answers to the first question can be substantial at national scales, but are not particularly controversial at the global scale. Recent reviews of the economic effects of achieving emission reduction pledges made in the Copenhagen Accord estimate costs to be less than 0.5 per cent of estimated global GDP through 2020 (Peterson *et al.* 2011). The Stern Review (Stern 2007) estimated the costs of keeping CO_2 equivalent concentrations at less than 500–550 ppm at less than 3 per cent annual global GDP. The IPCC (Metz *et al.* 2007) estimated the cost of keeping CO_2 equivalent concentrations below 450 ppm as less than 3 per cent of world GDP through 2030 and less than 6 per cent in the 2050s.

Answers to the second question on the potential damages of climate change to the global economy, which is more germane to this discussion of the availability of discretionary wealth to be used for tourism, are much more contentious.

Most of the integrated assessment models used to estimate climate change damages suggest that the costs would be of the order of 1–2 per cent of national income (Heal 2009). The Stern Review (Stern 2007) arrives at a very different conclusion, suggesting that the costs of climate change could be equivalent to an average reduction in global per capita consumption of at least 5 per cent under a business-as-usual emission scenario. The estimated damages would be much higher (up to 20 per cent later in the twenty-first or early twenty-second century) if non-market impacts, the possibility of greater climate sensitivity and regional distributional issues were accounted for. The Stern Review therefore frames strong mitigation action as a pro-growth strategy, with the benefits of strong, early action considerably outweighing the costs of inaction.

> 'Our actions over the coming few decades could create risks of major disruption to economic and social activity, later in this century and in the next, on a scale similar to those associated with the great wars and economic depression of the first half of the 20th century.'
>
> Stern (2007: xv)

The assumptions and omissions of the varying analyses are the reasons for the very different outcomes and controversy created by the Stern Review. Non-market impacts such as biodiversity and ecosystem services, even though absolutely essential for human well-being (Millennium Ecosystem Assessment 2005), are generally not accounted for in these economic analyses, and the sophistication of how market impacts (and which sectors or impact indicators) are incorporated, differs substantially. In some models, the damages associated with widely discussed climate impacts, such as SLR and extreme weather events, are surprisingly negligible (Ackerman and Munitz 2001). The non-mainstream discount rates of the Stern Review, as part of inter-generational equity considerations, were also highly controversial. Weitzman (2009) eloquently outlines how conventional economic analyses have yet to adequately consider the damages of low-probability, extreme impacts associated with high temperature-change scenarios. The economic impact of the upper bounds of climate change have not yet been explored. As Weitzman (2009: 24) notes, 'The economist's case for a carbon tax [i.e. immediate action to reduce emissions] is traditionally made without explicit reference to extreme [scenarios]. This argument is presumably strengthened when extreme events are considered.' For these reasons, we are inclined to agree with Heal, who believes:

> Stern is much nearer the mark: it is impossible to read the IPCC reports and believe that the consequences of climate change along the business as usual path are only 1 per cent or 2 per cent of national income. 1 per cent is almost within the margin of accounting error, and the IPCC certainly gives the impression that climate change will have a far-reaching impact on many human activities, which is not consistent with so small a value.
>
> (Heal 2009: 292–293)

Importantly, tourism has not been a formal component of any of these global economic assessments of climate change damages. The Stern Review gave only minor consideration to tourism and is represented by a selective, non-comprehensive choice of evidence. The most substantive discussion of tourism was in relation to some of the positive effects that climate change will bring for a few developed countries for moderate amounts of warming, although it is also noted that they will become very damaging at the higher temperatures expected in the second half of this century. Stern then went on to note the potentially harmful effects of climate change to alpine and winter tourism and to tourism resources such as the GBR (Hall 2008a).

There are no explicit model outputs that provide insight into the future implications of climate change-related economic change for the global tourism economy. Similarly, no detailed interpretation of the outcomes of the Stern Review or similar economic assessment has been completed specifically for the tourism sector, in part because of the large uncertainties involved and the mismatch between tourism forecasting timeframes (typically 10–20 years as a maximum) and the timeframes of these economic projections (50–100 years).

Arguably, the nearest model outputs that seek to relate climate change-related economic change to changes in tourism is the Ciscar *et al.* (2011) model that focuses on the economic and physical consequences of climate change in Europe. The model integrates climate data with high space–time resolution; detailed modelling tools specific for each impact category considered; and a multisectoral, multiregional economic model. Tourism is one of five impact categories assessed in the model, along with agriculture, river basins, coastal systems and human health. Ciscar *et al.* (2011) modelled the economic effects of future climate change projected for the 2080s on the basis of the current economy as of 2010 – the equivalent of having the 2080s climate in today's economy. They argue that such an approach 'has the advantage that hypotheses on the future evolution of the economy over the next eight decades are not needed, thereby minimizing the number of assumptions. Moreover, the interpretation of the results becomes simpler' (Ciscar *et al.* 2011: 2). One of the problems with such an approach in terms of trying to assess the inter-relationships between tourism and the economic effects of climate change is that the model assessed only the major international tourism flows within Europe. This means that the implications of climate change on non-intra-European travel flows (visitors from other regions of the world, just over 25 per cent of all international visitor arrivals in Europe in 2009) has been left out of the analysis. In all climate scenarios modelled by Ciscar *et al.* (2011), there would be additional tourism expenditures, with a relatively small EU-wide positive impact of €4–18 billion, depending on the scenario and climate model used. Chapter 7 includes further discussion of some of the results of the model.

> 'If climate protection policy fails (mitigation) [...] it is likely that from the mid 21st century local and regional conflicts will proliferate and the international system will be destabilized, threatening global economic development and overstretching global governance structures.'
> German Advisory Council on Global Change (2007)

Contemporary evidence following acts of terrorism, the outbreak of war, or social unrest and political revolutions illustrates very clearly that international tourists are highly averse to political and social instability (Hall 2010d). Scholars and security experts alike have examined climate change as a security risk (Diamond 2005; German Advisory Council on Global Change 2007; Barnett and Adger 2007; Raleigh and Urdal 2007; Zhang *et al.* 2007). The IPCC (Yohe *et al.* 2007) concluded with very high confidence that climate change would impede the ability of many nations to achieve sustainable development by mid-century and become a security risk that would steadily intensify, particularly under greater warming scenarios. Climate change is considered a national and international security risk because it will reduce access to and the quality of certain natural resources (e.g. food and water security) in some areas, cause major natural disasters, lead to large-scale migrations of people (both within nations and across national borders), and destabilize fragile governments.

Although regional climate change-associated security sensitivities vary across the range of available assessments, risks have been identified in a number of regions where tourism is highly important to national economies. Examining the top 25 tourism nations (by inter-national arrivals in 2007), climate change was estimated to pose a 'political instability' risk in eight countries (Ukraine, Turkey, Mexico, Russia, Thailand, Saudi Arabia, Egypt and South Africa) according to the regional ratings by Alert International (Vivekananda 2007), and a risk for large-scale environmental migrations in four countries (China, Mexico, Egypt and South Africa) (German Advisory Council on Global Change 2007). While the climate change security and migration risks were not as widely available for the smaller countries where tourism represents the highest proportion of the national economy or where tourism is looked at as a major future development pathway, regional security hotspots (rated as 'political instability' or 'potential for armed conflict') covered nations in the Middle East, northern and sub-Sahel Africa, the Indian subcontinent, northern and central South America, Southeast Asia and the Caribbean (Vivekananda 2007) and climate change-induced migrations were consistently projected for all SIDS, major deltaic areas, parts of China and India, and a number of central and northern African nations.

The significance for future tourism development in these regions is not underestimated by the tourism industry: 'Given that destination stability is a foundation for successful tourism, WTTC understands the impact that climate change has on food and water security and its influence on a nation's political stability' (WTTC 2010: 3). Nonetheless, to date there has been no attempt to quantify how regional tourism development would be affected by climate change-related security challenges, nor how a significant decline or cessation of tourism would compound deterioration of economic performance and difficult socio-economic conditions in the countries looking to tourism as a future economic engine for their economy.

IMPACTS INDUCED BY CLIMATE CHANGE MITIGATION AND ADAPTATION IN OTHER SECTORS

The responses of other major economic sectors to climate change or the policy frameworks established by government have the potential for major impacts on tourism destinations. The implications of the aviation and insurance sectors are particularly salient and are the focus of this section.

Impacts of aviation policy on destinations

Climate change mitigation in tourism was the subject of chapter 4. The extensive discussion of the stated position of the tourism sector on reducing its GHG emissions; the many technical and management behavioural opportunities to reduce emissions for all major tourism stakeholders; and the international, national and subnational mitigation policy frameworks within which tourism must operate are the foundation for examining the potential impact of mitigation policies on tourism destinations. Discussions of the impact of mitigation policy on tourism have focused almost exclusively on the aviation sector (and bunker fuels more broadly). While international aviation was not included in the Kyoto Protocol, as outlined in chapter 4, there has been considerable dialogue on policy frameworks to facilitate or require future emission reductions. With carbon-justified levies on international flights (e.g. UK air passenger duty) and international aviation entering the EU Emission Trading System in 2012, the reality of the impact of mitigation policies on air travel costs has become unsettling for tourism destinations that are dependent on long-haul tourism. Several have expressed concerns about the impact on the costs of air travel, tourist mobility and arrivals to their countries (e.g. Caribbean, Australia, New Zealand, Indonesia). For example, the Caribbean Hotels Association and Caribbean Tourism Organization (CHA–CTO 2007: 3) state that 'every effort must be made to ensure that future consumer movements and government action in the EU to address climate change … do not deter potential European travellers from taking vacations in the Caribbean'. The UNWTO (2010c) has also cautioned against mitigation policy for air transport that does not consider the negative impacts on tourism in developing countries.

A number of studies have examined the potential impact of a range of aviation sector-targeted mitigation policies on the future of international tourism. At a global scale, two analyses examine the impact of very different climate policies. One analysis examined the impact of a global carbon tax of US$1000 per t C on aviation fuel, and found international air travel (passenger kilometres) declined by about 0.8 per cent (Mayor and Tol 2007). Long-haul flights, because of high total emissions; and short-haul flights, because of the relatively larger emissions related to takeoff and landing, were most affected, implying that destinations that rely heavily on short-haul flights or on intercontinental flights would see a modest decline in tourism numbers. It is argued that regionally, the Americas, Africa and western Europe would suffer the greatest losses, while countries near to China and India would gain. Also at the global scale, the Group of Least Developed Countries proposed an International Air Passenger Adaptation Levy in 2008, as a mechanism to establish a predictable source of funding for adaptation in developing countries (estimated to be US$8–10 billion annually). The differentiated levy system (US$6–64 per flight) was estimated to result in a 0.05 per cent decline in air travel demand and, because it would be universal, was not deemed to have a differential impact for any air carrier or destination (IAPAL 2008).

Other analyses have focused on the implications of aviation mitigation policy for destinations in developing nations. Gössling *et al.* (2008) examined the impact of the EU Emissions Trading Scheme (ETS) and future oil prices on flight costs and tourist arrivals to ten SIDS around the world. They found that the EU-ETS would marginally affect demand for air travel to these developing countries, but with only a slight delay in growth in arrival numbers through to 2020 (arrivals would be 0–6 per cent lower relative to a business-as-usual scenario) (Gössling *et al.* 2008). The consequence of a much more stringent global aviation mitigation policy regime on these same countries (with CO_2 emission costs of €230 per t) was potentially much greater for some individual nations, where arrivals were projected to decline by 4–72 per cent compared with the business-as-usual growth scenario. A similar analysis of the implications of the EU-ETS and an identical implementation in the USA and Canada for Caribbean nations similarly found that reductions in tourist arrivals from the major markets would be negligible versus business-as-usual growth projections (–1 to –4 per cent region-wide through 2020) (Pentelow and Scott 2010). Only under a very stringent mitigation policy scenario, which could portend a post-2020 policy regime, was a significant decrease in tourist arrivals projected (–24 per cent versus business as usual) (Pentelow and Scott 2011).

Consequently, there is no evidence that mitigation policies for international aviation, as currently proposed, would have any substantial impact on tourism arrivals through 2020, with growth in arrivals declining only marginally versus existing growth projections (generally less than 5 per cent, although some notably higher reductions were observed in individual countries as a result of differences in major international markets). Nonetheless, it is clear that, in order to achieve the large emission reductions needed to avoid 'dangerous' climate change, absolute emission reductions will eventually be required of the aviation sector. Based on available evidence (see chapter 4), we do not envision a scenario where peripheral tourism destinations would continue to be serviced by air transport into the 2040s to the same extent as they are now, without a high degree of government support or much higher costs.

Post-2020 mitigation policy for international aviation is therefore very likely to alter the competitive marketplace and to have considerable consequences for tourism-dependent destinations, including many SIDS. Bunker fuel policy deliberations will need to determine whether compensation for these countries is appropriate. Future policy research is needed to assess more comprehensively the implications of proposed regulatory frameworks for aviation, and to determine the implications for air travel demand, routing, modal shifts and destination competitiveness, and what policies might be the most effective in achieving emission reductions without undermining tourist development (particularly in developing countries), if this is indeed possible. Further policy modelling is also required in tourism regions such as the Caribbean that have a rapidly growing cruise tourism segment, which will also become subject to regulations of international marine transport GHG emissions in the years ahead (Chamber of Shipping *et al.* 2009). Most cruise tourists fly to major ports of departure (e.g. Miami) before embarking on their cruise, and with GHG emissions from both flight and cruise segments, the price sensitivity of this tourism segment to increased emissions regulations remains uncertain and a priority for future work. Unfortunately, much of the tourism sector and international development agencies that promote tourism have not adequately considered the manifest long-term implications of aviation and shipping mitigation policy for transforming international travel patterns and destination competitiveness.

Changing insurance practices and tourism

A second economic sector where adaptation to climate change is likely to have major implications for tourism is insurance and the broader financial services sector. While businesses face many types of risk from climate change (e.g. current and future international and national mitigation regulations, changing consumer behaviour, changing supplier requirements, litigation), the implications of physical risks for property value as well as operations is an area that few businesses have considered in business planning (Sussman and Freed 2008).

Munich Re and other large insurance companies have observed that worldwide economic losses due to natural disasters have been doubling every ten years over the past three decades (IPCC 2001c). More specifically, annual economic losses worldwide from weather-related events show an even stronger upward trend, more than tripling between 1980 and 2009 (or an increase in losses of about US$2.7 billion per year) (Ward and Ranger 2010). Facing the prospect of increased number and size of damage claims as a consequence of increased frequency and magnitude of extreme events under climate change in many regions, the insurance industry has expressed concern about the implications of climate change for the provision of insurance coverage for over 20 years (Mills 2005; Ward and Ranger 2010). Although all major insurers are assessing the business implications of climate change (Allianz Group and World Wildlife Fund 2006), few have presented the potential implications for insurance premiums or the insurability of certain high-risk regions. One exception is a report by the Association of British Insurers (2004) that estimated even small increases in the intensity of hurricanes in the Gulf of Mexico–Caribbean, typhoons in Japan, and windstorms in Europe could increase annual damage costs in these major insurance markets by two-thirds by the end of the twenty-first century. To reflect the changed risk, regional insurance premiums would need to increase an estimated 20–80 per cent in the Gulf of Mexico–Caribbean, 20–80 per cent in Japan, and 0–15 per cent in Western Europe. Since then, the Association of British Insurers has concluded that the consequences of changes in weather patterns have been underestimated, and that these estimates are probably conservative and insurance premiums would double in many high-risk areas (Davis 2009). The cost implications for tourism infrastructure in high-risk areas (e.g. hurricane-prone coastlines, floodplains) is obvious and could make insurance unaffordable for many smaller tourism operators. In some regions, insurance coverage may no longer even be available, exacerbating the impacts of climatic extreme events, as some tourism operators may not be able to rebuild (leaving dilapidated properties) and restricting new investment in high-risk regions.

In the same way that insurance risk is increasingly being re-evaluated to consider the implications of climate change, climate risk and the implications of mitigation are beginning to be considered in credit assessments, investment strategies and equity valuation. A survey of financial institutions found that just over one-quarter of the major lenders surveyed 'systematically always' integrate direct effects of climate change in their credit risk assessments (UNEP 2011a). However, 80 per cent expected such risks to become increasingly relevant to lending in the future. The majority of asset managers who participated in the study indicated they now integrate climate change into their portfolio management (37 per cent 'always', 47 per cent 'in exceptional cases'). A separate survey of financial service providers by the Climate Service Center (2009) found that 80 per cent indicated they were poorly informed and would like to be better informed about the implications of climate change for the tourism sector. Nevertheless, there is little evidence that the direct implications of climate change for portfolios of tourism properties and risk assessment of tourism developments have yet been incorporated into the investment and management strategies of tourism corporations (with ski tourism being one area where exceptions exist) or tourism master plans of destinations and countries. However, there are substantial demands for greater certainty from companies, and airlines in particular, as to the regulatory environment that businesses will have to face as a result of GHG mitigation strategies.

CONCLUSION

This chapter provides a conceptual framework (Figure 5.1) of the multiple pathways by which climate change can affect tourism destinations, and a comprehensive review of the large and rapidly growing literature. Based on this synthesis, a number of important observations can be made about the state of climate change impact assessment for tourism destinations. Importantly, the destination impacts literature has improved markedly over the past ten years. There is much to be positive about with respect to the diversity of research approaches and geography of study areas now represented in this literature. That being said, there are a number of areas where this literature needs to be advanced, and these are set out below.

First, all the major components of the global tourism system (tourists, source markets, transport systems, destinations) will be affected by the four distinct impact pathways. The impacts will be both negative and positive, although the literature concentrates largely on climate change risks, and the magnitude and timing of impacts will vary substantially by destination. Our understanding of the consequences of climate change for destinations is often limited to a single dimension of these impacts (e.g. the impact on ski operations at a mountain destination). Rarely have multiple impacts been assessed for a single destination (e.g. reduced ski season, closure of major nearby ski-area competitors, mitigation policy increasing transport costs from some markets but not others and lengthened seasons for warm-weather tourism activities at a mountain destination). Assessments are also rare of the impact of climate change for multiple competing destinations (a series of regional or global competing destinations) in order to attempt to estimate aggregate impact on supply for certain types of tourism products (e.g. ski or coastal tourism resorts) and implications for marketplace competition. Furthermore, how the impacts of climate change will interact with other major long-term drivers of the tourism sector remains highly uncertain (Scott 2006a; Hall and Lew 2009; Hall 2010c), although some dimensions (fuel prices, increasing travel safety and health concerns, increased environmental and cultural awareness, environmental limitations such as water supply) are beginning to be considered in newer climate change studies. A number of authors (e.g. Gössling and Hall 2006a, 2006b; Scott 2006a; Hall 2008b; Scott *et al.* 2008a; Hall and Lew 2009; Dawson and Scott 2010) have repeatedly identified the need for broader systems-based approaches that would be capable of capturing these multiple impacts as well as changes in demand and socio-economic conditions. Improved understanding of this complexity is required in order to estimate more accurately the aggregate impacts of climate change for tourism destinations (see Box 5.1).

Second, as several authors have observed (Scott *et al.* 2003; Gössling and Hall 2006a; Scott *et al.* 2008a; Simpson *et al.* 2008; Hall 2009b), consideration of adaptation, innovation and resilience has been inadequate in climate change assessments in the tourism sector. Because so few studies of climate change impacts in tourism have incorporated adaptive responses, most of the impacts discussed in this chapter should be considered 'potential climate change impacts' rather than 'residual climate change impacts' (see Box 5.1). The social experiments offered by climate change analogue events represent particularly valuable learning opportunities to understand adaptation (both successful and failures) and an opportunity to assess the effectiveness of current climate adaptations against future climate conditions. Climate change analogues unfortunately continue to remain under-utilized in the tourism sector. With the increased significance of climate change adaptation in the scientific community, governments and more recently the business community, tourism researchers need to advance their capacity to understand adaptation in the tourism sector (see chapter 6) and incorporate it into more holistic climate change assessments.

Third, the conclusion of studies that have explored the impacts associated with a range of climate change scenarios (from high and low emission scenarios or using the full range of GCMs and emission scenarios) has been that the projected impacts are much greater under high emission scenarios (the trajectory global emissions are currently on or exceed) (see chapter 1). In several cases, the impacts of the high emission scenarios in the 2050s exceeded the impacts of the low emission scenarios in the 2080s. This illustrates very clearly the importance of GHG emissions and reinforces the conclusion of academics (e.g. Gössling and Hall 2006a; Scott *et al.* 2008a) as well as governmental (UNWTO *et al.* 2008) and tourism industry reports (WTTC 2009) that climate change mitigation is in the self-interest of the tourism sector worldwide. A +4°C world is not conducive to the environmental conditions, economic growth and social stability essential for future tourism development.

Fourth, in order to provide increasingly relevant information to governments and tourism industry decision-makers, destination impact assessments will need to better incorporate impact indicators that are meaningful to the tourism sector and the communities that rely on tourism. Even the most sophisticated modelling of impact indicators that are irrelevant to tourism operations (e.g. snow depth on the first of April only) is of no use. Researchers should also endeavour to adopt common indicators in order to facilitate comparisons of impacts across a broader range of tourism destinations. A key objective for the climate change and tourism research community is to identify where the greatest vulnerability exists at destination or community level (clusters of operators at risk), because it is here that implications for employment and livelihoods, and thus social conditions, are the most significant. These social impacts leading from tourism industry impacts have not been analysed for any community and are an important potential contribution for scholars with expertise on socio-cultural aspects of tourism. One of the most meaningful impact indicators for decision-makers is the costs of climate change-associated damages and adaptation costs (both market and non-market costs – see Box 5.1). This information is central to influencing policy, and this is a very important task for tourism economists and broader collaborations with economists and the tourism sector.

So although Scott (2011) argued that climate change and tourism researchers are increasingly able to provide information on climate change impacts that is relevant to decision-making by governments or the tourism industry, this improving information base remains restricted in scope. Therefore innovative new approaches are needed to overcome the important limitations that are common to the destination impacts literature and advance our understanding of the vulnerability of tourism destinations and the economies and livelihoods of the communities and people who rely on the tourism industry.

Finally, although progress has been made with respect to broadening the geographical coverage of climate change impact studies for tourism destinations, there remain some very notable regional gaps. There is a virtual absence of any destination-specific empirical studies on Africa, South America and the Middle East, and very little information available for most of Asia and the Pacific Islands (Hall 2008a). As some of these destinations are highly dependent on tourism, or feature tourism as a major future economic development strategy, addressing these regional knowledge gaps and building research capacity in these regions should be a primary objective of organizations such as the UNWTO and WTTC, their national governments, and tourism and climate change researchers.

FURTHER READING AND WEBSITES

As this chapter outlines, there is a wealth of information on different types of destination impact. The latest overviews of actual and potential climate change impact are included in the publications of organizations such as IPCC, UNDP and UNEP, while tourism-specific information can be found on the website of the UNWTO. At a national level, please refer to government environment agency websites as well as the websites of national science academies.

UN Development Programme (Environment and Energy): www.beta.undp.org/content/undp/en/home/ourwork/environmentandenergy/overview.html.

UN Environment Programme: www.unep.org (refer to sections focusing on climate change, disasters and conflicts, environmental governance and resource efficiency).

Government, industry and destination adaptation to climate change

6

The imperative for adaptation to climate change has advanced substantially within the scientific community and moved higher up the UNFCCC and national policy agendas over the past decade. This is largely due to the limited progress on emissions reductions. As outlined in chapter 1, several recent assessments have concluded that if current global GHG emission trends continue, or even if the emission reduction commitments currently made by countries (at the time of writing) are successfully achieved, temperatures would exceed +2°C average global warming by 2100 (Meinshausen *et al.* 2009; Parry *et al.* 2009a; Anderson and Bows 2011), the level considered by many scientists and the Parties to the UNFCCC to represent 'dangerous interference with the climate system'. Consequently, it is now widely recognized that all societies and economic sectors inevitability will need to adapt to unavoidable climate change in the decades ahead. Indeed, recent studies have suggested that society should be preparing to adapt to +4°C global warming (Meinshausen *et al.* 2009; Parry *et al.* 2009a).

Chapter 5 examines the far-reaching consequences of climate change for tourism businesses and destinations and emphasizes that, regardless of whether they are potential 'winners' or 'losers' from climate change, all will need to adapt to climate change in order to minimize associated risks and capitalize on new opportunities in an economically, socially and environmentally sustainable manner. This chapter provides a brief overview of concepts related to climate change adaptation, including its different forms and characteristics. The adaptive capacity of tourism stakeholders is then considered. The second part of the chapter focuses on climate change adaptation research within the tourism sector. This research is organized into four main areas: studies that examine the effectiveness of key adaptation strategies to reduce climate change vulnerability; destination-scale case studies that identify and evaluate adaptation options and explore tourism stakeholders' perceptions of climate change and determinants of adaptation action (including key barriers); the development of conceptual frameworks for adaptation processes specific to the tourism sector; and finally a review of climate change adaptation policy by major international organizations and national governments.

> 'It is meaningless to study the consequences of climate change without considering the ranges of adaptive responses.'
>
> Adger and Kelly (1999)

CONCEPTS AND CHARACTERISTICS OF CLIMATE CHANGE ADAPTATION

As adaptation to climate change became the subject of greater scientific inquiry and policy development, the need to define more clearly the specific types of adaptation applications and to distinguish its characteristics has increased (Smit *et al.* 2000). The concept of adaptation within climate change and allied human dimensions of global environmental change literatures has evolved over time, so that a number of definitions exist. In the IPCC *Fourth Assessment Report* (IPCC 2007c), adaptation is referred to as

> an adjustment in natural or human systems to a new or changing environment. Adaptation to climate change refers to adjustment in natural or human systems in response to actual or expected climatic stimuli or their effects, which moderates harm or exploits beneficial opportunities.

Key aspects of this definition of adaptation are considered below.

In this chapter we are concerned with adaptation of socio-economic systems related to tourism, which can be by societies, institutions, individuals or governments, and can be motivated by economic, social or environmental drivers through many mechanisms, for example social activities, market activities, local or global interventions (Adger *et al.* 2007).

Adaptation, as defined by both the IPCC and UNFCCC, focuses on adjustments in socio-economic systems in response to *actual or expected climate stimuli* and associated impacts (negative or positive). It does *not* include adjustments to climate mitigation policies, which have often been included in discussions of adaptation with tourism stakeholders (largely because of concerns of the tourism industry regarding travel costs and tourist mobility). Adaptation can be to contemporary climate variability and extremes or to future climate change (including changes in means and changes in magnitude and frequency of extremes). Importantly, while adaptation to contemporary climate variability may include strategies that are well suited to future climate change, that is not necessarily the case, and some adaptations to current climate conditions may even be maladaptations to future conditions (see Box 6.1). Evidence suggests, however, that relying on past experience is not likely to be adequate in most destinations. Adaptation to current climate cannot be interpreted as adaptation to future climate change, as has been mistakenly done in some discussions of climate change and tourism. This is not simply academic semantics, for equating past or ongoing climate adaptations as purposeful climate change adaptation incorrectly attributes the intent and suggests these initiatives are capable of dealing effectively with the range of projected future climate regimes projected by climate models (that is, that tourism operators have evaluated the design capabilities of climate adaptations to cope with future climates). The following quote from a ski tourism operator in Tyrol, Austria illustrates this difference very clearly:

> It would be presumptuous to say that the ski areas have taken precautions against climate change because we haven't, we are not prepared. ... Snow production may be a main strategy but a strategy against a lack of snow, not against climate change. We have not taken any real precautionary measures against a marked warming trend yet.
>
> (Trawöger 2010)

Adaptation includes actions taken before impacts are observed (anticipatory) and after impacts have occurred (reactive). Both anticipatory and reactive adaptations can be planned (a deliberate decision). Reactive adaptation can also occur spontaneously.

Adaptations can take many forms: technical, policy, planning, legal, economic, institutional and behavioural. Table 6.1 illustrates a diverse range of technological, managerial, policy and behavioural adaptations that are currently used to deal with climate variability at the business and destination levels. The availability of these adaptation options will be context-specific and will vary according to environmental characteristics, government jurisdiction and tourism marketplace or business model. Climate adaptations are rarely undertaken in isolation as a single, discrete action, but commonly involve multiple adaptations that are very specific to the destination climate and its tourism products (Scott *et al.* 2009). The location-specific nature of climate adaptation creates a complex mix of adaptations within the global tourism sector.

A thorough description of adaptation therefore needs to specify the system being examined (who or what adapts), the climate-related stimulus (adaptation to what), and the form of adaptation (how adaptation occurs) (Smit *et al.* 2000).

With the recognition that adaptation to climate change will become increasingly necessary in the decades ahead has come the realization that the consequences of adaptation need to be better understood, including its benefits and costs (Box 6.1). Not all climate adaptations are environmentally sustainable or socially equitable, and some can exacerbate the climate change problem they are designed to adapt to. These are considered maladaptations (see Box 6.1). Barnett and O'Neill (2010) identified five criteria by which adaptations should be considered maladaptive: if they increase GHG emissions; disproportionately burden the most vulnerable; bear high opportunity costs relative to alternatives; reduce incentives to adapt; or foster path dependency through development trajectories that are difficult to change in the future.

A related concept is adaptive capacity, which the IPCC (2001c: 982) defined as 'The ability of a system to adjust to climate change (including climate variability and extremes), to moderate potential damages, to take advantage of opportunities, or to cope with the consequences.' When applied to social systems, adaptive capacity is considered the ability of an individual, organization or society to develop adaptation. The complex mix of social, economic, technological, biophysical and political conditions within which climate adaptation takes place determines the capacity of individual and social systems to adapt. The IPCC (Metz *et al.* 2007) has identified several factors that are thought to determine adaptive capacity within countries, organizations and communities: economic wealth, technology, information and skills, infrastructure, institutions and equity.

The dynamic nature of the tourism industry and its ability to cope with a range of recent shocks, including SARS, terrorism attacks in a number of nations, the Asian tsunami and the global financial crisis in 2008/09, suggests a relatively high adaptive capacity within the tourism sector overall (Scott *et al.* 2008a; Hall 2010d). The capacity to adapt to climate change is nonetheless thought to vary between the components of the tourism system (tourists, tourism service suppliers, destination communities and tour operators) (Elsasser

Table 6.1 *Portfolio of climate adaptations utilized by tourism stakeholders*

Type of adaptation	Tourism operators/ businesses	Tourism industry associations	Governments and communities	Financial sector (investors/ insurance)
Technical	• Snowmaking • Shore-protection structures, beach nourishment • Rainwater-collection and water-recycling systems • Cyclone-proof building design and structure	• Pilot-test structural adaptations • Develop websites with practical information on adaptation measures	• Reservoirs and desalination plants • Shore-protection structures, beach nourishment • Weather forecasting and early warning systems	• Develop and test advanced building design or material (fire-hurricane resistant) standards for insurance • Provide information to customers
Managerial	• Water conservation plans • Low-season closures • Product and market diversification • Regional diversification in business operations • Redirect clients away from impacted destinations	• Snow condition reports through media • Use of short-term seasonal forecasts to plan marketing activities • Training programmes on climate change adaptation • Encourage environmental management with firms (e.g. via certification)	• Fee structures for water • Demand-side management programmes (water, energy) • 'Coral bleaching response plan' • Convention/ event interruption insurance • Business subsidies (e.g. insurance or energy costs)	• Require use of advanced building standards • Adjust insurance premiums or not renew insurance policies • Restrict lending to high-risk business operations • Include social cost of carbon in financing and credit risk assessments

Policy	• Hurricane interruption guarantees • Comply with or exceed regulations (e.g. building code)	• Coordinated political lobbying for GHG emission reductions and adaptation mainstreaming • Seek funding to implement adaptation	• Coastal management plans and setback requirements • Building design standards (e.g. for hurricane-force winds)	• Consideration of climate change in financing and credit risk assessments
Research	• Physical risk analysis for properties	• Assess awareness of businesses and tourists, and knowledge gaps	• Monitoring programmes (e.g. bleaching or avalanche risk, beach water quality)	• Extreme event risk exposure
Education	• Water and energy conservation education for employees and guests	• Public education campaign (e.g. 'Keep Winter Cool')	• Water conservation campaigns • Campaigns on dangers of UV radiation	• Educate/inform potential and existing customers
Behavioural	• Real-time webcams of snow conditions • GHG emission offset programmes	• GHG emission offset programmes • Water conservation initiatives	• Extreme event recovery marketing	• Good practice in-house

Source: adapted from Scott *et al.* (2008a)

BOX 6.1 ADAPTATION CONCEPTS AND DEFINITIONS

Adaptation benefits: avoided damage costs or accrued benefits following the adoption and implementation of adaptation measures (IPCC 2007c).

Adaptation costs: costs of planning, preparing for, facilitating and implementing adaptation measures, including transition costs (IPCC 2007c).

Adaptation deficit: failure to adapt adequately to existing climate risks largely accounts for the adaptation deficit. Controlling and eliminating this deficit in the course of development is a necessary, but not sufficient, step in the longer-term project of adapting to climate change. Development decisions that do not properly consider current climate risks add to the costs and increase the deficit. As climate change accelerates, the adaptation deficit has the potential to rise much higher unless a serious adaptation programme is implemented (World Bank 2011a).

Low-regret adaptation: options where moderate levels of investment increase the capacity to cope with future climate risks. Typically, these involve over-specifying components in new-build or refurbishment projects. For instance, installing larger-diameter drains at the time of construction or refurbishment is likely to be a relatively low-cost option compared with having to increase specification at a later date due to increases in rainfall intensity (World Bank 2011a).

Mainstreaming adaptation: refers to the integration of adaptation objectives, strategies, policies, measures or operations such that they become part of the national and regional development policies, processes and budgets at all levels and stages (Lim and Spanger-Siegfried 2005).

Maladaptation: an action or process that increases vulnerability to climate change-related hazards. Maladaptive actions and processes often include planned development policies and measures that deliver short-term gains or economic benefits, but lead to exacerbated vulnerability in the medium to long term (Lim and Spanger-Siegfried 2005).

and Bürki 2002; Gössling and Hall 2006b; Scott and Jones 2006b, 2006c; Scott *et al.* 2008a). Figure 6.1 illustrates the relative adaptive capacity of major tourism stakeholders. Tourists have the greatest adaptive capacity (depending on three key resources: money, knowledge and time), with relative freedom to avoid destinations affected by climate change or to shift the timing of travel to avoid unfavourable climate conditions. The implications of their potential adaptations to climate change are discussed in chapter 7. Suppliers of tourism services and tourism operators at specific destinations have less adaptive capacity. Large tour operators, who do not own the infrastructure, are in a better position to adapt to changes at destinations because they can respond to clients' demands and provide

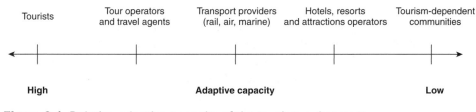

Figure 6.1 *Relative adaptive capacity of the tourism subsectors*
Source: adapted from Scott and Jones (2006b)

information to influence their travel choices. Destination communities and tourism operators with large investments in immobile capital assets (hotel, resort complex, marina,
casino) have the least adaptive capacity.

CLIMATE CHANGE ADAPTATION IN THE TOURISM SECTOR

Adaptation has figured less prominently in climate change research in tourism than in
some other economic sectors (e.g. agriculture), and remains an important knowledge gap,
particularly with respect to tourism-dependent destination communities (Scott *et al.*
2008a). In a review of the state of climate change adaptation in the tourism sector, Scott
et al. (2009) concluded that adaptation research and practice in tourism was about five to
seven years behind other leading sectors. This may be related to institutional arrangements
for tourism within governments (the lack of a national department or ministry for tourism
in some countries, and lesser research capacity to evaluate climate change risk than some
other sectors), and to the image sensitivity of the tourism industry, which has contributed
to the downplaying of climate change vulnerability.

Climate change adaptation research related to tourism has focused on four main areas. The
earliest studies examined the effectiveness of key adaptation strategies to reduce climate
change vulnerability (almost exclusively with respect to ski tourism). A number of destination-scale case studies have been conducted to identify climate change impacts (risks
and opportunities) and evaluate adaptation options. These case studies have also explored
tourism stakeholder awareness of climate change, perceived risk of climate change and
capacity to cope with climate change impacts (adaptive capacity), and determinants of
adaptation action (including key barriers). Building on the experience of early case studies
in tourism and the experience of adaptation practice in other sectors, conceptual frameworks have been developed for participatory adaptation processes specific to the tourism
sector. The final area of research has examined climate change adaptation policy by major
international organizations and national governments. Each of these themes is discussed
in this section.

Assessing the effectiveness of climate change adaptation strategies

As discussed in chapter 5, although ski tourism has repeatedly been identified as being
highly vulnerable to climate change, most climate change impact assessments of the ski

industry have not incorporated snow-making or the many other climate adaptations already utilized (see Figure 6.2). Snow-making is an integral climate adaptation that has been widespread in some ski tourism markets for over 20 years (eastern and central North America, Japan, Australia), and one that continues to increase in most other markets (the European Alps, western North America, New Zealand, China).

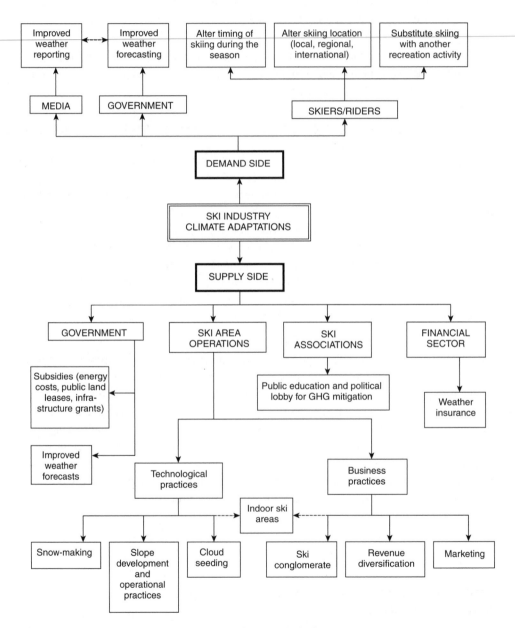

Figure 6.2 *Climate adaptation options in the ski industry*
Source: Scott and McBoyle (2007)

The earliest research on climate change adaptation in the tourism sector assessed the capacity of ski area operators to cope with projected climate change through the implantation of this key adaptation. The studies that fully incorporated snow-making (Scott *et al.* 2003, 2007a, 2008c, 2011; Hennessy *et al.* 2008; Steiger 2010, 2011a, 2011b; Steiger and Abegg 2012) have all found that the impact of climate change on ski areas (season length) was substantially lower than reported in previous studies that did not include snow-making.

With advanced snow-making in place, most ski area managers in North America, Australia and Austria are highly confident in their capacity to negate the impacts of future climate change (Scott *et al.* 2003, 2007a, 2008c; Scott and Jones 2005; Bicknell and McManus 2006; Wolfsegger *et al.* 2008). For example, almost half of Austrian ski area managers felt that with further adaptation of snow-making, their businesses would continue to be viable for at least another 75 years (Wolfsegger *et al.* 2008). Studies that have examined ski seasons and snow-making capacity under long-range scenarios suggest this confidence in the adaptive capacity provided by snow-making is overestimated (Steiger 2011a, 2011b; Steiger and Abegg 2012).

The projected large increases in snow-making requirements under climate change scenarios raise questions about the sustainability of this adaptation strategy. Communities and environmental organizations have expressed concern about the extensive water and energy use as well as chemical additives associated with snow-making. In regional ski markets, such as the European Alps or New England states, varied government regulations and support for snow-making (energy or water subsidies) may provide a competitive advantage to certain ski areas. For example, chemical–bacterial additives to facilitate snow-making at temperatures near freezing are allowed in Switzerland but banned in Germany, offering a competitive edge to Swiss ski areas. The broader social context of regional tourism activity must be considered when determining the net impact of additional snow-making activity on GHG emissions. If local ski areas are forced to close because of unreliable natural snow conditions, potentially thousands of skier visits may be displaced to other ski destinations or other forms of tourism. Alternative ski destinations may be nearby (less than 100 km away), but as was found in a study of potential ski area closures in the New England region of the USA (Scott *et al.* 2008c), in many cases ski tourism may be transferred 500 km or more, with attendant emissions from car, coach, rail or even air travel. If no alternative ski holiday is available regionally, skiers may opt for much more GHG-intensive travel options. For example, the poor ski conditions in the European Alps in the winter of 2006–07 generated a much higher than usual number of European skiers visiting ski resorts in the Rocky Mountains of North America. More comprehensive analysis of the GHG intensity of holiday options and the behavioural response of winter sports tourists is required to understand the net emissions impact of additional snow-making.

Product and market diversification is another common adaptation strategy embraced by ski resorts to cope with the business challenges of pronounced tourism seasonality. Over the past three decades, many ski areas in North America and Europe have diversified their operations beyond traditional ski activities to include the provision of skiing and snowboarding lessons, accommodation and retail sales. Many ski resorts have made substantial investments to provide alternative activities for non-skiing visitors (snowmobiling, skating, dog-sled rides, indoor pools, health and wellness spas, fitness centres, squash and tennis, games rooms, restaurants, retail stores) and have further diversified their business operations to become 'four-season resorts', offering activities such as golf, boating and white-water rafting, mountain biking, paragliding, horseback riding and other business lines. At many larger resorts, real-estate construction and management has also become a very important source of revenue. Many larger ski areas have profited from the sale of condominiums and other real estate as well as the management of these properties on behalf of their owners. The ability of diversification to overcome potential losses in ski tourism at the enterprise and community level has not been adequately assessed, due in part to the proprietary nature of business information. However, an analysis in two tourism regions of Quebec, Canada that compared projected revenue losses from reduced winter tourism demand (ski and snowmobile tourism) against revenue gains from warm-weather demand increases (park tourism, camping, golf, zoos, water parks) found the trade-off in seasons resulted in net loss of revenues (Scott *et al.* 2011). An expanded consideration of other summer tourism attractions that could not be included in the analysis because of lack of data (boating, festivals and events) may change the result, but for four-season resorts that primarily offer these warm-weather attractions, the result is likely to remain robust. The following statement from a ski tourism operator in Tyrol, Austria arrives at a similar conclusion.

> 'As far as increases in the summer season are concerned, there are limits. In summer we sell nature, silence and relaxation. If we get 30,000 hikers or mountain bikers in the area that's not manageable. For us summer can be a nice extra income but it can't – to no extent – alleviate big problems in winter. We have ideas how to cushion, but definitely no idea how to solve a real problem with ski tourism.'
>
> CEO, cable car company, quoted by Trawöger (2010)

Destination-scale adaptation case studies

Several studies have examined aspects of climate change adaptation with tourism stakeholders. Most of these multi-stakeholder case studies have focused on adaptation at the destination scale. These studies have had two main objectives. The first type engaged tourism stakeholders through various participatory approaches (workshops, focus groups, Delphi techniques) with the aim of identifying and prioritizing the types of impact anticipated from climate change, and of developing and sometimes evaluating adaptation options for future implementation. The place-specific adaptations that have emerged, and stakeholder evaluations of the feasibility and sustainability of adaptations, are not reviewed here – readers are referred to the original publications for portfolios of adaptations and the evaluation criteria utilized (Scott *et al.* 2005b; UNESCO 2006; Lemieux *et al.* 2008; Turton *et al.* 2010). A second group of studies has explored tourism stakeholders' awareness of climate change, their perceptions of the risks and opportunities posed by climate change, opinions on their capacity to cope with climate change impacts, and the determinants of adaptation action (including key barriers). Each of these themes is examined in this section.

A common aim of adaptation studies with tourism stakeholders has been to assess their awareness of climate change generally, and with specific reference to the risk it poses to their business or destination. The perspectives of stakeholders in a number of countries are summarized below.

In Canada, diverse tourism stakeholders have been involved in adaptation studies. Canada's parks are one of its primary tourism resources, attracting millions of domestic and international visitors each year. Awareness of climate change was found to be limited in the first study by Scott and Suffling (2000), but Parks Canada was committed to increasing its adaptive capacity through research and training, and became an early international innovator in terms of integrating adaptation into park management plans in the early 2000s (Welch 2005). A subsequent policy Delphi study with park managers identified 165 adaptation options across the primary management areas (tourism, conservation, public education/interpretive) (Lemieux et al. 2008). These options were then evaluated by an expert panel, which identified moderate to low capacity pertaining to policy formulation and management strategies. A separate panel of senior park agency decision-makers used a multiple criterion decision-facilitation matrix to evaluate further the institutional feasibility of the 56 most desirable adaptation options identified by the expert panel and prioritize them for consideration in a climate change action plan. Critically, only two of the 56 adaptation options were evaluated to be definitely implementable by senior decision-makers, due largely to fiscal and internal capacity limitations. A recent survey of all of Canada's federal, provincial and territorial park agencies found very high awareness and concern regarding climate change, as well as confirmed critical capacity issues (Lemieux et al. 2011). Although 94 per cent of park agencies indicated that climate change would substantially alter park policy and planning over the next 25 years, 91 per cent of the agencies conceded that they currently do not have the capacity necessary to respond effectively to climate change. Such capacity challenges are common in park agencies in developed countries and pervasive in developing countries, revealing the urgent need to increase adaptive capacity in the organizations that manage the world's most prominent nature-based tourism destinations. UNESCO (2006) reached a similar conclusion with respect to World Heritage Sites.

Studies with other government tourism agencies in Canada (Scott et al. 2005b, 2011; Scott and Jones 2005, 2006b) have revealed similar high levels of awareness and broad willingness to engage in climate change adaptation processes, which are often perceived by stakeholders as extensions of ongoing adaptation to contemporary climate. Tourism industry associations (national and provincial) and tourism operators are also well aware of climate change issues, but have been more cautious with respect to acknowledging impacts on the tourism sector out of concerns about negative press coverage, and understandably are hesitant to discuss climate change adaptation among competitors (Scott and Jones 2005; Jones et al. 2006). Dodds and Kuenzig's (2009) survey of destination-marketing organizations (DMOs) across Canada also found climate change was perceived as an important issue, although with clear geographical differences, as northern and coastal regions were more concerned about climate change impacts. Importantly, half of the respondents did not think their organization was sufficiently well informed to develop adaptation strategies.

Studies with tourism stakeholders conducted in several areas of Europe have found highly variable opinions on climate change and the need for adaptation. In Finland, interviews with a small sample of tourism operators found they were aware of climate change issues, but half did not believe in anthropogenic climate change (Sievänen *et al.* 2005; Saarinen and Tervo 2006). Their knowledge of how climate change could affect their business was much lower, and most were not concerned that climate change would have an impact on their operations. It was noted that just over half of the operators had contingency plans for climate variability, but operators refused to consider these as climate change adaptation. However, in follow-up interviews conducted two years later, the opinions of operators had changed, with the majority of respondents stating that they now believed in anthropogenic climate change (Tervo 2008). Half of the respondents believed that tourism entrepreneurs themselves could influence the intensity of climate change in the future, with climate change now being considered a more important factor for business future than competition for space and resources with other forms of land use such us forestry (Saarinen and Tervo 2010).

In England, studies in two regions with two different groups of tourism stakeholders revealed a clear dichotomy of opinion over climate change and tourism. In south-west England, Cheng's (2010) survey of tourism businesses found that 59 per cent agreed that '[tourism] businesses need to adapt to climate change', but that 47 per cent rated climate change as a low priority for their business. The study also found that the views of many tourism operators about climate change had been affected by the 'climategate' discussions in the media (see chapter 1). Several operators dismissed anthropogenic climate change and expressed opposition to the need for adaptation. In contrast, in north-west England, academics and government officials working in areas related to tourism management had high levels of climate change awareness, agreed that climate change posed a significant threat to key natural assets for tourism in the region, and rejected a 'do-nothing' approach to climate change adaptation (McEvoy *et al.* 2008).

In a survey of tourism experts from European universities about climate change implications for tourism in the Mediterranean region, Valls and Sarda (2009) found that 78 per cent of respondents felt that anthropogenic climate change was already under way, and a similar proportion agreed that the tourism industry still had time to adapt fully to the effects of climate change. Interestingly, the main consequences of climate change for Mediterranean tourism were perceived to be natural catastrophes and coastal damage. Shifts in traditional seasons and a decline in tourism demand were seen as two of the least important threats. This perspective stands in sharp contrast to studies that project declining demand and shifting tourism seasons as a result of temperature changes in the region (see chapter 7) and speculation by others that the region will be 'too hot' for summer tourism (see chapter 5). While uncertainties were expressed, there were no serious reservations over whether the tourism industry would be able to adapt to climate change, mainly through technological means, as increased tourism growth in the region was seen as perfectly compatible with tourism operations in an era of climate change (Valls and Sarda 2009). As Valls and Sarda (2009) note, the survey findings appear much more positive than the literature about the future of Mediterranean tourism. Similarly, optimism was found in a survey of small hotel owners in Grenada, Spain, who showed general awareness of climate change, but the majority (over 70 per cent) did not think climate change impacts would significantly affect their business (Jarvis and Pulido Ortega 2010).

Ski tourism operators in Austria and Switzerland have shown similar optimism with respect to their capacity to cope with future climate change through adaptation. In a survey of Austrian ski area managers, Wolfsegger *et al.* (2008) found most ski area managers were highly aware of the risk posed by climate change, but also highly optimistic about their capacity to negate the impacts of future climate change. With further adaptation, primarily snow-making, but including other business and technical options, 24 per cent believed that they could operate their businesses for another 30–45 years, while 44 per cent felt their businesses would be viable for at least another 75 years. Only two ski area managers believed that, even with further adaptation, their ski operations would no longer be viable in 15 years as a result of climate change. Hoffman *et al.* (2009) examined the determinants of climate adaptation at the business level among Swiss ski-lift operators and found that awareness of possible climate change impacts had a positive influence on the scope of corporate climate adaptation, but that the operators anticipated to be most vulnerable were not more likely to engage in climate adaptations. It was uncertain whether this represented a fatalistic perspective that further adaptation was not able to cope with climate change and thus seen as a wasted investment. If so, that would be a very different perspective than that of low-elevation operators in Austria. These studies and dialogue at several conferences indicate that the ski industry holds a very different perspective on the 'unavoidable physical collapse' of the ski industry, as often reported by the media (see chapter 5).

In Australia, Bicknell and McManus (2006) found that ski-resort managers acknowledged climate change, which differed from the blatant disregard that König (1998) reported a decade previously, although several did not fully accept anthropogenic climate change. While they were very aware of the studies that projected highly negative impacts on their industry, managers were sceptical of the findings and expressed frustration over the associated media coverage. Overall, they maintained a strong sense of optimism about their capacity to cope with future climate change, mainly through the use of snow-making, but also via other ongoing business strategies (e.g. business diversification). The potential impact of climate change on real estate sales and property values was cautiously acknowledged, but downplayed on the basis that real estate investment timelines consider ten years into the future, not to mid-century, when the impacts of climate change were expected. In another study in the Australian Alps, Roman *et al.* (2010) found that tourism stakeholders considered climate change as an issue to be 'aware of in principle' but of low priority due to its long timeframes. Specific operators' concerns varied substantially among those operating ski resorts and those in other areas, suggesting that a regional approach to climate change adaptation would be a substantial challenge. Based on workshops with diverse tourism stakeholders in five destinations across Australia, Turton *et al.* (2010) also concluded that the tourism sector was reasonably well informed of potential climate change impacts, but expressed varied opinions with respect to the need for adaptation and the capacity to adapt.

Buckley's (2008) analysis of adaptation by local government in Byron Bay, Australia came to a decidedly different conclusion with respect to how well informed adaptation decision-making was. This case study examined how misunderstandings and misrepresentations of the likely effects of sea-level rise were used by local government to 'plac[e] severe and irrational restrictions on development of residential and holiday accommodation' and 'prevent beachfront landowners from protecting their own properties against erosion' (Buckley 2008: 71). In this situation, it is argued that climate change science was misinterpreted for political reasons and needlessly damaged the destination's tourism industry by driving investment to nearby communities that did not adopt such restrictions on coastal properties.

In New Zealand, Hall (2006b) found that climate change was regarded by business as potentially significant in the future, but in the short term ranked well below other concerns. Significantly, where enterprises had been affected by extreme weather events, attitudes and behaviours towards climate change differed markedly from those in unaffected areas, although opposition to regulatory approaches with respect to climate change continued. Overall, given the perceived business costs and returns, adaptation rather than mitigation appeared to be the preferred strategies of tourism businesses.

Far fewer studies are available from developing nations, and specifically from tourism regions that are anticipated to be among the most vulnerable to climate change. A survey of tourism resort owners in Fiji found that 36 per cent had been affected by one or more climate-related impacts, and a similar proportion were aware of some potential impacts of climate change (e.g. cyclones, loss of coral reefs) (Becken 2005). With respect to adapting to climate-related risks, operators in Fiji focused on the high-risk, immediately relevant impacts from extreme events. A very small sample of tourism stakeholders in Jamaica were found to be well aware of climate change, with 75 per cent stating that it has already had an impact on tourism in the nation (Hall and Clayton 2009). Concerns over the capacity of Caribbean governments to lead climate change adaptation were emphasized.

Su *et al.* (2011) surveyed Taiwanese tourist hotels with respect to their awareness of climate change and their responses. Nearly all respondents (95 per cent) agreed that climate change exists. The media were their main sources of information on climate change. Over 65 per cent of respondents expected more negative impacts on the national tourism industry, hotel's region and hotel's business in the next five years, which is ten percentage points higher than that of the previous five years at the time of answering the survey. However, while hotels generally agreed with the statement that their hotel had a responsibility to respond to environmental and climate change impacts, respondents disagreed with the statement that their hotel contributed to climate change.

Together, these studies portray a positive situation, where climate change awareness in the tourism sector has been improving, even though, as noted, studies in England, Finland and Australia identified scepticism about anthropogenic climate change among some tourism stakeholders. In other cases, tourism stakeholders noted that public scepticism within their communities would pose a barrier to adaptation in the near term (Turton *et al.* 2010).

Some common findings that emerge from these studies in very diverse tourism regions include a general recognition of the need to adapt to climate change, and substantial optimism about the capacity of adaptation to overcome the challenges of climate change. While tourism stakeholders generally acknowledge the need to adapt to climate change, there remains little evidence of the development of adaptation plans, and very limited investment in adaptation by tourism operators or governments. Several key barriers to adaptation have been identified (scientific uncertainty, incompatibilities between business and climate change impact timelines, limited financial and technical capacity, the need for government leadership).

Chapter 5 outlines the many knowledge gaps with respect to our understanding of the potential impacts of climate change on key tourism assets and tourism destinations. These knowledge gaps have been recognized in several studies with tourism stakeholders (Scott and Jones 2005; Sievänen *et al.* 2005; Bicknell and McManus 2006; UNESCO 2007; Wolfsegger *et al.* 2008; Cheng 2010; Jarvis and Pulido Ortega 2010; Turton *et al.* 2010). A number of tourism stakeholders expressed low confidence in the current understanding of how climate change would affect tourism, and the need to improve this information base if adaptation was to proceed. UNESCO (2007) and Turton *et al.* (2010) have summarized the information gaps identified by tourism stakeholders, which serve as a useful guide for future destination focused studies. Positively, strong interest was observed in obtaining additional information on climate change impacts on destinations. For example, 69 per cent of tourism operators in south-west England were interested in finding out more about the implications of climate change for their region (Cheng 2010).

Tourism operators in several regions have expressed concern about media coverage of climate change implications for tourism and their destinations specifically. Negative media stories about the prospects of climate change may be more damaging to a destination's reputation and tourist demand than the actual climate change impacts, particularly in the short term (Hall 2008c, 2009c). The tourism literature on managing disasters and post-disaster communication and marketing have much to offer tourism destinations affected by climate-related events and those preparing proactively for future climate change impacts (even slow-onset impacts) (Hall 2010d).

> 'We are a "good news" industry and when you are talking about climate change, you talk about loss of beach, you talk about all the things we don't want to go to market with, and you say that we are going to have challenges. If you start from there, there is going to be absolute resistance from people against the idea of infusing it into their operations.'
>
> President of Jamaica Hotel and Tourist Association,
> quoted by Hall and Clayton (2009: 270)

The tourism industry is very image-sensitive and is therefore very cautious about even acknowledging concerns about climate change risks for fear of adversely affecting their reputation as a destination or sustainable business. For some tourism operators, acknowledging the potential impacts of climate change would invite negative reactions from investors, insurers or regulators, so that resistance to engaging in public adaptation processes is expected to continue in some destinations. The need to counter speculation and misinformation about climate change's impacts on tourism and its potential to cause reputational damage for destinations, or to bring about maladaptive investment or regulatory decisions, is yet another rationale for the tourism community to invest in climate change research (Scott 2011).

In several studies, tourism stakeholders noted that the timeframes of climate change scenarios and associated impacts were incompatible with conventional tourism business planning (Scott and Jones 2005; Sievänen *et al.* 2005; Bicknell and McManus 2006; Hall 2006b; Jarvis and Pulido Ortega 2010; Roman *et al.* 2010; Turton *et al.* 2010). The very long timelines of climate change impacts make it a low business management priority. In a number of cases, tourism stakeholders expressed a desire to have more details about climate change expected in the next 10–20 years and were less interested in long-range (mid-century) scenarios when climate is expected to be substantially different (Scott *et al.* 2005b; Scott and Jones 2005; Cheng 2010; Turton *et al.* 2010; Scott *et al.* 2011).

> '… climate change is what I term a luxury issue. When we have time, we'll move forward and deal with it.'
>
> Representative of Canadian destination-marketing organization
> quoted by Dodds and Kuenzig (2009: 3)

The perception of the tourism industry generally is that climate change adaptation is primarily the responsibility of governments. Governments need to provide leadership in terms of assessing the potential impacts of climate change on tourism and destinations and direct support for technological adaptations (e.g. water supply or coastal protection infrastructure) as well as indirect support to facilitate adaptation by tourism operators (e.g. subsidies, planning coordination and, where necessary, regulatory frameworks). Opinion over which level of government should lead adaptation differed. In Canada, because many stakeholders were from national and provincial government agencies, leadership was expected from the national government. In the Caribbean, stakeholders expressed reservations over the capacity of governments in the region to lead climate change adaptation (Hall and Clayton 2009). They also stressed the need for coordinated adaptation, with the probable need for government regulation, because voluntary initiatives would not be sufficient. Stakeholders in Australasia and some European studies were resistant to regulatory approaches. Turton *et al.* (2010) and Jarvis and Pulido Ortega (2010) argued that local government should act as a catalyst for destination climate change adaptation. However, local governments tend to have lower technical and financial capacity to support climate change adaptation. Buckley's (2008) case study provides further caution with respect to the capacity of local authorities to interpret climate change science and translate it into appropriate community-level land-use and development plans. The lone dissenting view about the primary responsibility of government for adaptation came from tourism experts in Europe, who rated companies as the most responsible for climate change adaptation, followed by state authorities, regional authorities, international organizations and, finally, local authorities (Valls and Sarda 2009).

Inadequate technical, human resource and financial capacities were recorded as critical barriers to climate change adaptation in all case studies. The very limited capacity of SMEs to engage in adaptation was emphasized in several studies (Becken 2005; Sievänen *et al.* 2005; Scott and McBoyle 2007; Hall and Clayton 2009; Cheng 2010; Jarvis and Pulido Ortega 2010; Roman *et al.* 2010; Turton *et al.* 2010). SMEs represent a high proportion of tourism businesses, and there is some evidence to suggest that they are more vulnerable to climate change (Scott and McBoyle 2007; Dawson *et al.* 2009; Steiger 2011a). SMEs have limited capital available for investment in infrastructure and little capacity to integrate climate risks into business planning. As Turton *et al.* (2010) point out, few SMEs can plan on timeframes longer than a couple of years, so that investments in adaptation, even those with payback periods of ten years, are not something they are willing to consider without financial support of government.

Tourism marketers are another group with limited capacity to respond to climate change (Hall 2008c, 2009c; Buzinde *et al.* 2010), particularly when environmental resources fundamental to destination image are adversely affected by climate change. Their response is critical to overcoming potential reputational damage from media coverage of real or potential impacts of climate change. Yet Dodds and Kuenzig (2009) found that DMOs expressed concerns that they did not have the appropriate skills and training to educate client members or devise adaptation strategies.

In noting the above, it is important to recognize that adaptation to climate change is very much place-dependent with respect to important factors such as the legislative capacity of different levels of government to act and the level of information available. Furthermore, many of the most important government influences on business and destination response will probably be non-tourism-specific: general regulations on energy efficiency and design, or financial programmes that support the adoption of energy conservation measures. In practice, developing an adaptation strategy with appropriate measures; a suitable mix of material, planning and regulatory response and market-based tools; and timely action requires overcoming numerous technical and institutional barriers (Bedsworth and Hanak 2010). Table 6.2 explores five such barriers observed by Bedsworth and Hanak (2010) in the California context:

- uncertain information on the extent and nature of climate-related impacts;
- conflicting goals and trade-offs between interests;
- backward-looking regulatory regimes in which existing regulatory frameworks are built around historical data rather than future scenarios or forecasts;
- coordination failures between actors and institutions;
- limits on current legal and institutional authority at different levels of governance.

These have been applied to four areas relevant to tourism: renewable electricity, coastal resources, air quality and ecosystem resources. Focusing on these areas highlights some of the policy-making difficulties faced by different levels of government as well as the need to recognize the implications of the limits to knowledge, authority and regulation in formulating climate adaptation strategies, including the need for new legislative and regulatory authority. However, it is important to note that, even if the climate were stable,

Table 6.2 *Barriers to adapting to climate change in California, 2010*

Area	Uncertain information	Trade-offs	Backward-looking regulation	Coordination	Limits to authority
Renewable electricity	Hydropower facilities will be less useful if the future is drier	Renewable energy sources are often initially costlier to users than carbon-based energy sources	Current reservoir-operation rules look backward and use historical data (relevant for hydropower)	Local government restrictions on renewable energy installations are common (even if federal and state approval has been granted and local utilities are supportive)	Federal and state approval for renewable energy installations does not compel local government to accept renewable energy proposals
Coastal resources	Early investments in coastal protection may have high costs if the rate of sea-level rise is on the low end of projections	Coastal protection, such as sea walls, affects the environment and public use; relocation of housing and infrastructure is costly	Federal flood-risk maps for coastal zones look backward and use historical data	Coastline-adaptation policies require intra-regional coordination (e.g. which areas to protect); local governments are uneven in their response	States cannot require local governments to update local coastal plans to take sea-level rise into account
Air quality	Uncertainty about local and regional climate impacts makes it difficult to develop forward-looking regulations	GHG emission-reduction measures could conflict with air-quality and public health goals	Federal regulations look backward and use historical data for setting standards, and ignore extreme events	State, regional and federal actors may have different agendas	State and regional authority to adjust emissions targets is limited
Ecosystem resources	Uncertainty about species response can make it harder to justify habitat conservation	Focusing on current at-risk species may limit availability of resources to protect those at risk in the future	Federal and state endangered species laws are backward-looking	Habitat conservation is often piecemeal, with broader inter-jurisdictional plans needed	Federal Endangered Species Act limits states' authority to develop forward-looking strategies

Source: adapted from Bedsworth and Hanak (2010)

these areas pose significant management challenges with respect to meeting the demands of interests and stakeholders. In many cases, climate change has only exacerbated such difficulties.

Adaptation frameworks for the tourism sector

A large literature exists on climate change adaptation practice, including sector-specific efforts, national government reports in many nations, and guidelines and tools from several international organizations and the United Nations (e.g. Lim and Spanger-Siegfried 2005; AIACC 2007). While these studies and implementation tools are not specific to tourism, they offer important guidance for tourism. For example, the four principles of the UNDP *Adaptation Policy Frameworks for Climate Change* (Lim and Spanger-Siegfried 2005) are highly relevant for tourism, and served as guidelines for some of the early destination-specific tourism adaptation processes identified in the previous section (Scott and Jones 2005; Scott *et al.* 2005b).

- Place adaptation in a development context. The impacts of climate change could negatively affect a country's sustainable development in diverse ways, including water resources, energy, health, agriculture, coastlines and biodiversity – all of which can influence the tourism sector. Consequently, the process of adaptation in the tourism sector cannot be undertaken in isolation and needs to be placed within the broader context of a country's sustainable development policies and strategies and to consider impacts and adaptations in other sectors.
- Build on current adaptive experience to cope with future climate variability. The tourism sector has tremendous experience coping with climate variability (see chapter 2), and much more assessment of its coping range is required as a starting point for adaptation. A wide range of tourism stakeholders need to be involved in the adaptation process to take full advantage of their diverse experience and expertise with adapting to current climate variability. Framing adaptation to recent climate anomalies (climate change analogues) assists tourism stakeholders to associate with their impacts and past responses (successful or failed).
- Recognize that adaptation occurs at different levels, in particular at the local level. Adaptation can be undertaken strategically at the national level, but implementation often takes place at the local destination, business or project level. Climate change is not just a challenge for governments and in the tourism sector – involvement of the tourism industry is critical, as their operations are and will be principally affected.
- Recognize that adaptation is an ongoing process. Adaptation will be an iterative process of implementing and evaluating strategies as climate conditions continue to evolve and the information base improves over the course of this century.

Climate change adaptation frameworks for application in the tourism sector have been developed by Becken and Hay (2007) and Simpson *et al.* (2008). Jopp *et al.* (2010) subsequently criticized these frameworks for not having an explicit tourist behaviour component, although presumably the demand-side perspective is provided by tourism operators and marketers who are well aware of market preferences or through the assessment of potential impacts (see Scott and Jones 2005; Scott *et al.* 2005b); and the difficulty of utilizing the risk management-based framework of Becken and Hay (2007) because of the large degree of uncertainty in quantifying climate change risks for tourism and the inadequate consideration of opportunities made possible by climate change. They propose a third framework designed specifically for regional tourism application.

It is not the objective of this book to recommend one approach over another, as none of these frameworks has been applied within the tourism sector (although the framework of Jopp *et al.* 2010 is being tested in Australia) and each has comparative strengths and may be more appropriate for use in specific circumstances. Instead, Box 6.2 identifies six common elements derived from a review of several well cited and tested adaptation frameworks by international organizations that have been involved in the practice of climate change adaptation since its integration within the UNFCCC (Lim and Spanger-Siegfried 2005; AIACC 2007). Further information sources on adaptation frameworks are available at the end of this chapter.

BOX 6.2 KEY ELEMENTS FOR A CLIMATE CHANGE ADAPTATION PROCESS

While all these elements are considered necessary to the process of adaptation, the degree to which they are emphasized depends on the situational context and the stakeholders involved.

STEP 1 – GETTING THE RIGHT PEOPLE INVOLVED IN A PARTICIPATORY PROCESS

A vital aspect in determining the eventual success of the adaptation process is to get the right people involved and to involve them in a participatory process. The purpose of multi-stakeholder processes is to promote better decision-making through an inclusive and transparent process that creates trust and a sense of ownership among stakeholders. Tourism is a highly diverse economic sector, and the perspectives of many local, national and, where applicable, international stakeholders should be sought – both those directly involved in the tourism sector or whose livelihoods are affected by tourism (government ministries, local government, tourism industry representatives, tourism labour representatives, local businesses and communities); and those in other sectors that might be affected by tourism adaptations (e.g. transportation, energy, agriculture), whose adaptations might affect tourism (e.g. insurance industry, health sector), or that have other relevant expertise (e.g. universities, NGOs).

STEP 2 – SCREENING FOR VULNERABILITY: IDENTIFYING CURRENT AND POTENTIAL RISKS

The next step is to understand how climate change may affect a region and what risks this would pose for the tourism sector. Understanding climate impacts is an essential early step, and the assessment should include examination of physical risks to tourism resources (e.g. biodiversity, water supply) and infrastructure (e.g. coastal resorts), business and regulatory risks (e.g. changes in insurance coverage), or market risks (e.g. changes in international competitiveness through transportation costs). AIACC (2007) specified that both current future risks (e.g. extreme climatic events, both sudden and slow onset) and potential future risks (e.g. changing climate means and variability) be assessed. Synthesizing information from existing national or regional climate change assessments may prove valuable at this stage to understand recent and projected climate changes and the implications for natural and human systems that are highly relevant to tourism. Because tourism has not been adequately considered in many previous climate change assessments, a scoping assessment of the range of tourism specific risks may also be needed to supplement existing information and to ensure knowledge gaps are addressed. Where little information is available, interviewing stakeholders about how recent climatic extremes have affected aspects of tourism operations (e.g. warmer tourism seasons, prolonged dry periods, extreme events that can serve as analogues for conditions expected under climate change) is a good place to begin. What do these analogue experiences reveal about existing climate sensitivities, and what analyses have been undertaken to better cope with future incidences?

STEP 3 – IDENTIFYING ADAPTATION OPTIONS

Work with tourism stakeholders to compile a list of alternative technologies, management practices or policies that may enable them to cope better with the anticipated impacts of climate change. This adaptation portfolio-building stage should include both preparatory and participatory activities. Preparatory activities should begin by identifying current adaptation strategies and policies in place to address current climate-related risks. Reviewing recent climate change reports from other communities and regions expected to face similar risks may be valuable for identifying additional adaptations utilized successfully in other tourism destinations. Participatory activities may include holding workshops or smaller focus group meetings with stakeholders. Where it is difficult or overly costly to bring a wide range of stakeholders to a workshop, field interviews with stakeholders by an adaptation team or Delphi techniques with key stakeholders and potential implementing partners can also be used to identify adaptation options. National and international experts in climate change risk assessment and adaptation should also be consulted to share information and experience from other nations and to help identify any potential gaps in the stakeholder-generated adaptation portfolio.

STEP 4 – EVALUATE ADAPTATION OPTIONS AND SELECT COURSE OF ACTION

The adaptation portfolio-building stage is likely to identify a long list of potential adaptations that may be difficult to analyse fully with limited time-frames and budgets. It is recommended that a second round of stakeholder consultation be done to present the full initial list of stakeholder-identified adaptations, and to determine criteria by which to evaluate adaptations and refine the portfolio of adaptations to be considered for implementation. A range of criteria can be used to evaluate adaptation strategies. Criteria applied in various AIACC (2007) projects include: net economic benefit, timing of benefits, distribution of benefits, consistency with development objectives, consistency with other government policies, cost, environmental impacts, spill-over effects, capacity to implement and social–economic–technological barriers. Some criteria may require an additional detailed analysis of each adaptation to be undertaken.

STEP 5 – IMPLEMENTATION

Implementation of the adaptation options selected in step 4 requires that the roles of implementing stakeholders, resource requirements and timelines be specifically defined. An implementation plan should be developed with the following components: strategic plan outlining actions and timelines of involved stakeholders; capacity-building needs assessment and training plan; financial/business plan covering expenditure needs and revenue sources; communication plan; sustainability plan; and plan for monitoring the perfor-mance of adaptations. Adaptation plans cannot stand alone and must relate to other existing planning processes and policies (i.e. 'mainstreaming' adaptation).

STEP 6 – MONITORING AND EVALUATION ADAPTATIONS

Climate change adaptation represents a long-term investment of human and financial resources. To ensure the optimal realization of this ongoing investment, the final step in this process is continuously to evaluate the effectiveness of the implemented adaptations. Again, several evaluation criteria are possible (e.g. cost, ease of implementation, delivered intended benefits, adverse impacts). The evaluation criteria and related indicators should be selected by stakeholders in step 5 as part of the monitoring and performance plan. Complete evaluation may prove difficult for some time, however, as the long-term risks posed by climate change that required the adaptation may not be realized for many years (even decades). As evaluation of the implemented adaptation strategies becomes possible, this continues the iterative process of adaptation by informing how the initial strategy will need to be refined.

(Developed by the authors; presented in revised form in Simpson *et al.* 2008)

A review of adaptation policy in the tourism sector

In 2010, the OECD and UNEP secretariats jointly conducted a survey on 'Climate Change and Tourism' to understand the extent to which countries are preparing to deal with the climate change challenge for tourism (OECD and UNEP 2011). As of January 2011, the secretariats had received replies from 18 countries: Australia, Austria, Egypt, Estonia, Germany, Greece, Hungary, Ireland, Israel, Japan, Mexico, the Netherlands, New Zealand, Poland, Portugal, Slovak Republic, Slovenia and South Africa.

Actions reported by each country are summarized in Table 6.3. In most countries, no strategic reviews exist of vulnerabilities and potential adaptation strategies for tourism. Only about a quarter of the countries (12 out of 44) have considered adaptation strategies for tourism, with two stating that no adaptation is necessary. Australia has been a leader in climate change adaptation in tourism, establishing a Tourism and Climate Change Taskforce in 2007 to undertake research (e.g. Turton *et al.* 2010) and to guide the development of an action plan. With two exceptions, none of the countries has planned or implemented adaptation policies for the sector. As outlined by Haas *et al.* (2008) for Austria, this may be a result of prevailing key research gaps that need to be addressed before adaptive measures can be implemented. Germany and Israel report that they have implemented policies with regard to flooding and water use, but in neither case are these specific to climate change or tourism. However, the definition of adaptation goals and the implementation of policies still appear in a development stage in virtually all countries.

'The inescapable conclusion is that current [national tourism] policy [to adapt to climate change], with few exceptions, is inadequate to the scale of the challenge, both on mitigation and on adaptation.'

OECD and UNEP (2011: 4)

The analysis identified potential reasons for inaction on climate change adaptation, including low awareness of the tourism sector's climate change adaptation needs and a lack of knowledge and research regarding the complexities of tourism–climate interrelationships (OECD and UNEP 2011). Individual governments could clearly do more to understand

Table 6.3 *Reported tourism-specific adaptation actions in selected countries*

Country	Adaptation strategies identified	Policy in place	Threats and opportunities
Australia	K (online web mapping tool); EI, AC, C	Under development	Increased costs
Austria	R	–	Effects on winter, summer, lake and city tourism, increased risk in Alps
Estonia	None existing	None existing	Unidentified
France	R, K	–	Declining climate comfort in summer, winter sports affected
Germany	R, C, ACT	Water Framework Directive, Floods Directive (EC)	Changing travel flows, lost winter tourism opportunities, water availability and quality
Greece	R	–	Droughts, more intense rainfall, floods and soil erosion, accelerating desertification, sea-level rise, salt-water intrusion in coastal aquifers, declining freshwater resources and water quality
Hungary	K, RISK	–	Summer cultural, activity and ecotourism vulnerable, increasing costs for heating, air conditioning, irrigation
Ireland	ACT, K,	Urban Wastewater Directive, Water Framework Directive, Award Schemes	Changing tourist flows, physical impacts moderate, drier and warmer summers, altered habitats and biodiversities, flooding, inundation of beaches, invasive species, water pollution

Continued

Table 6.3 *Continued*

Country	Adaptation strategies identified	Policy in place	Threats and opportunities
Israel	–	Water tariffs increased	Water availability, droughts, threatened destinations (Dead Sea)
Japan	None existing	None existing	–
Mexico	R	–	Beaches and shorelines affected; lack of awareness
New Zealand	R	–	–
Norway	R	–	–
Portugal	K, C, ACT	National Adaptation Plan	Coastlines affected (sea-level rise), risk of heavy rainfall, desertification
South Africa	AC	–	Biodiversity, coastal zones and beaches affected, rainfall, flooding, diseases

K: creating knowledge and awareness among stakeholders; EI: research into economic impact of climate change on key tourism activities; AC: analysis of adaptive capacity of tourism industry; C: communicate government policy and relevant information; R: research on adaptation needs; ACT: adaptation action plan; RISK: identification of risks
Source: OECD and UNEP (2011)

how tourism and related sectors (transport, food) contribute to, and will be affected by, climate change. This improved understanding would help them determine the appropriate policy mix to reduce the impact of tourism on climate change, including how mitigation goals and adaptive capacities could be increased without adversely affecting the sector. To move forward, governments may seek to fund research that addresses key knowledge gaps, possibly also taking into consideration adaptation plans developed in other economic sectors, which have made greater progress on this front.

Countries such as Austria (Haas *et al.* 2008), Germany (Federal Government 2008), Ireland (Government of Ireland 2007; Fáilte Ireland and The Heritage Council 2009) and South Africa (Republic of South Africa 2010) have shown that it is possible to identify current and future impacts and adaptation needs, even though key scientific uncertainties remain with regard to tourists' perceptions and potentially altered travel behaviour (see chapter 7). Most countries have identified research on adaptation needs as a priority.

Risks incurred in a partial understanding of the consequences of climate change for tourism are exemplified by two quotes taken from national adaptation plans in Sweden and Germany.

There is considerable evidence suggesting that summer tourism in the Mediterranean will be badly affected by higher summer temperatures, as these will be increasing more than on global average, while freshwater access is expected to decline ... It is likely that tourist flows to the Mediterranean will decline in the warmest summer months to the advantage of the Baltic ... If only a small share of travellers to the Mediterranean countries travel to Scandinavia instead, this would lead to a significant increase in visitor numbers in Sweden. Calculating that 1 per cent of tourism to the Mediterranean moves to Sweden would increase the number of bed nights by 10 million, representing a doubling of the current number of bed nights over the whole year in all of Sweden. This would, based on today's income from accommodation, correspond to [additional turnover of] almost 30 billion crowns per year.

(Regeringskansliet 2007: 406; authors' translation)

... changed climatic conditions could also open new opportunities for the tourism industry, for example through increased numbers of visitors in what has hitherto been the off season, or as a result of tourist flows shifting from southern to northern regions. Many Germans have tended to spend their summer holidays in the Mediterranean countries. The total stream of holidaymakers from central and northern Europe to southern Europe, with around 116 million arrivals, is the biggest touristical migration worldwide, and accounts for 41 per cent of internal tourism within Europe. Since there is an increasing probability that southern Europe will experience daily maximum air temperatures of 40°C or more during the peak season, travellers can expected to suffer increased heat stress, which can have unfavourable effects on the well-being of older persons and children in particular. In Germany, by contrast, rising temperatures and lower rainfall in the summer could tend to favour tourism, for example because of an extension of the summer season. The Potsdam Institute for Climate Impact Research expects Germany to become more attractive to tourists. Estimates suggest that the number of tourists coming to Germany could increase by 25 to 30 per cent.

(Federal Government 2008: 39)

Both quotes reveal a focus on a limited number of parameters (essentially 'summer temperature'), while it remains unclear which time horizons the assessments consider. However, as discussed in chapter 7, it is clear that travel motives are complex and cannot be deduced by a single parameter, while complexities of how changes in temperature are actually perceived by tourists are insufficiently understood. Not only is the basis of these assessments scientifically questionable, but they become highly problematic when extended to become estimates of economic gains associated with changing travel flows.

Overall, results provided by OECD and UNEP (2011) on national adaptation policies indicate that countries are focusing on already pressing issues such as freshwater availability or reduced snow cover. Key vulnerabilities are linked to increases in weather extremes, flooding, drought and water shortages (also affecting water-related tourism activities), declining water quality, changes in snow reliability, and coastal erosion, as well as conditions less suitable for tourism in already warm countries (increasing heat stress). Even though some countries have started to identify impacts and to discuss adaptation, where recommendations are actually made, these remain generic in character (e.g. 'greater role for research', 'restructuring to non-tourism economic activities'). With regard to timescales, some adaptation assessments focus on impacts occurring more than half a century from now, and it is less clear that investors and other stakeholders are capable of considering such long-term future scenarios. One conclusion from the OECD and UNEP (2011) review is that climate change adaptation assessments need to integrate greater complexity and more relevant timescales, as well as to examine better the scale on which policies can be implemented (individual tourists, businesses, regions or country-wide). Policy initiatives of international tourism organizations and other international organizations with strong interests and major collaborations with the tourism sector were also included in this analysis (OECD and UNEP 2011). One example is Su *et al.*'s (2011) analysis of the level of implementation of UNWTO and UNEP (2008) recommended measures in the accommodation sector (Table 6.4), which revealed moderate to low levels of implementation across a wide range of environmental initiatives.

The UNWTO (2010c) states that the organization has been working to raise awareness on climate change issues in the tourism sector since 2003, when it organized the first International Conference on Climate Change and Tourism. The second International Conference on Climate Change and Tourism in Davos, Switzerland in 2007 resulted in the Davos declaration, which included 'firm recommendations and a clear commitment for action to respond to the climate change challenge including the urgent adoption of a range of sustainable tourism policies' (UNWTO 2010c: 4).

Table 6.4 *Level of implementation of UNWTO and UNEP (2008) recommended measures for the accommodation sector to respond to climate change*

Rank	Environmental practice	Number of hotels	Level of implementation (mean)*
1	Implement control system for heating/cooling/lighting facilities	44	4.02
2	Provide locally produced, seasonal food	44	3.90
3	Recycle waste, raise customers' awareness of waste	45	3.80
4	Frequently clean and maintain electricity facilities	43	3.79
5	Reduce and pre-treat chemical and hazardous wastes	43	3.70
6	Measure and monitor resource use and waste production	44	3.61
7	Reduce use of materials	44	3.57
8	Use energy-efficient appliances	43	3.53
9	Implement water-saving and reuse measures	42	3.38
10	Encourage guests/staff to use green vehicles/public transport	44	3.16
11	Purchase fair trade/green label products where possible	44	3.00
12	Reduce use of air conditioning	44	2.98
13	Initiate a hotel environmental policy	42	2.88
14	Adapt hotel's products, marketing and positioning	44	2.72

Continued

Table 6.4 Continued

Rank	Environmental practice	Number of hotels	Level of implementation (mean)*
15	Set up environmental targets and benchmarking	43	2.67
15	Adapt building design for energy saving	43	2.67
16	Volunteer for local conservation or community projects	44	2.57
17	Implement environmental management system	43	2.53
18	Use alternative fuels and renewable energy	43	2.49
19	Implement energy-saving education/incentive for staff/guests	42	2.38
20	Develop environmental code of ethics for supplier chain	41	2.29
21	Involve and comply with climate change policies and plans	42	2.17
22	Provide climate change and environmental education for customers and staffs	44	2.05
23	Get involved in national tourism programme on energy efficiency and renewable energy use	41	1.95
24	Achieve environmental certification	41	1.88
25	Designate a manager with specific responsibility for environmental management system and emission issues	41	1.71
26	Integrate emission management with supply chain	39	1.69
27	Provide carbon-offset projects for guests	40	1.65
28	Offer incentives for adaptation and mitigation measures	43	1.58
28	Develop links with international policies, mechanisms, cooperation and standards on climate change	40	1.58
29	Get involved in climate change network to promote activities proposed in UNWTO's Davos report and declaration	37	1.41
30	Locate new establishments in low climate-risk areas	37	1.40
Overall mean, all responses			2.65

*Based on a scale of 1 = low; 3 = moderate; 5 = high level of implementation
Source: Su *et al.* (2011)

The WTTC (2010) suggests identifying climate change hotspots and supporting the adaptation efforts of communities. Tourism is seen to have the potential to contribute to a low-carbon future, 'combining initiatives which promote energy efficiency with the need to secure clean water supplies while protecting biodiversity'. Moreover, 'travel and tourism could be one of the most coherent, non-extractive, economic activities for forest communities and can act as a major tool for Reducing Emissions from De-forestation and Degradation (REDD)' (WTTC 2010: 9). While adaptation is repeatedly mentioned in the WTTC (2010) report *Climate Change – A Joint Approach to Addressing the Challenge*, there are no specific examples of adaptation strategies the industry should promote, or indications of the need for broader sectoral collaboration on adaptation capacity-building (e.g. technical knowledge-sharing) or adaptation financing. As the above examples illustrate, the focus of WTTC is largely climate change mitigation.

The OECD has been working on climate change economics and policy since the late 1980s. The OECD works closely with governments to help them to identify and implement least-cost policies to reduce GHG emissions in order to limit climate change, as well as to integrate adaptation to climate change into all relevant sectors and policy areas (Abegg *et al.* 2007). As OECD countries are the major international donors, OECD has a critical role in tracking climate finance, and in examining how public finance can be scaled-up and best targeted to help leverage private financial flows. In the wake of the economic crisis, the OECD is also looking at how measures that governments are taking to spur economic growth can best be formulated so that they support – and do not work against – the objectives of moving towards a green, low-carbon economy. Given the global nature of the climate change challenge, and its widespread economic, social and environmental impacts, the OECD is in a unique position to help countries put climate policy on a solid economic footing consistent with frameworks for development. Work on climate change is under way across the OECD, engaging government representatives from a range of ministries.

UNEP considers climate change in several divisions as well as through collaborations with other international bodies. Within the Global Partnership for Sustainable Tourism, recognizing the need to raise awareness on the topic of climate change among tourism industry actors and consumers, UNEP has developed a series of capacity-building and communications materials. These are designed to facilitate adaptation to the existing and expected effects of climate change, and to help anticipate and mitigate future impacts. Thus, in relation to the tourism industry, UNEP developed from 2007 and 2009 several seminars on climate change adaptation and mitigation in the sector, with the University of Oxford as academic partner and UNWTO as co-host, on frameworks, policies and practices targeting government and industry experts. Additionally, in 2009 UNEP, in partnership with the Caribbean Alliance for Sustainable Tourism, produced a handbook to support climate change adaptation efforts in tourism destinations and communities. This handbook aims to reduce the impacts of natural disasters on local coastal tourism communities. The handbook's approach builds on UNEP's Awareness and Preparedness for Emergencies at the Local Level (APELL) process, designed to create public awareness of hazards and to ensure that communities and emergency services are adequately trained and prepared to respond.

The Commission of the European Communities (CEC 2010: 5) mentions climate change twice as an issue in its policy framework document for tourism development in Europe. It states, as a challenge for tourism policy, that 'the supply of tourism services must in future take into account constraints linked to climate change, the scarcity of water resources, pressure on biodiversity and the risks to the cultural heritage posed by mass tourism'. It also outlines, as planned action, the need to 'facilitate identification by the European tourism industry of risks linked to climate change in order to avoid loss-making investments, and explore opportunities for developing and supplying alternative tourism services' (CEC 2010: 11).

CONCLUSION

A comparison of the state of climate change adaptation in the tourism sector with the study of Scott *et al.* (2008a) reveals some progress. Positively, the growing number of studies with tourism stakeholders suggests the level of climate change awareness within the sector has improved over the latter half of the 2000s. Nonetheless, these same studies and the review of national adaptation policies indicate that climate change adaptation still remains less developed than in other economic sectors.

Based on the ability of the tourism sector to recover from recent major shocks (disease outbreaks, terrorism, natural disasters, high oil prices and economic recession), it is thought that the sector possesses relatively high adaptive capacity. Evidence suggests that the tourism industry clearly shares this view and is highly confident in its ability to cope with the impacts of climate change through adaptation. However, with the exception of ski tourism operations, knowledge of the capacity of current climate adaptations utilized by the tourism sector to cope successfully with future climate change remains rudimentary. In an era of global climate change, it will no longer be sufficient to rely on past experience, and there is very little evidence that tourism businesses have assessed the impact of projected climate conditions on their operations. We cannot assume that past climate adaptations will be successful in the future (Scott *et al.* 2008a). As a result, tourism operators may be overestimating their adaptive capacity. Similarly, assessments that assume adaptation will be effective, and that fail to examine carefully the unique characteristics of tourism environments and market preferences, may also overestimate adaptive capacity in the sector. For example, many macro-scale studies of sea-level rise assume that structural coastal protection will occur because of the cost–benefit ratio. While this may be accurate for cities and critical coastal infrastructure throughout developed nations, the capacity of developing nations to afford the cost of coastal protection remains uncertain. This assumption of the economics of coastal protection has been extended to tourism; however, the economics for clusters of coastal resorts, let alone isolated resorts, remains questionable. Furthermore, while coastal protection may save resort infrastructure, without beach nourishment, coastal squeeze will eliminate the critical beach asset, degrading competitiveness and financial sustainability. That the tourism sector has not engaged better in assessing its vulnerability and adaptation capacity may be detrimental to its long-term sustainability.

Tourism stakeholders predominantly identified governments as responsible for leadership on climate change adaptation. The impact of national government support for initiating climate change adaptation is clearly visible in the tourism sector. Several of the countries where the majority of climate change adaptation studies have been conducted have bene-fited from government programmes supporting sectoral adaptation initiatives (e.g. the Climate Change Action Fund in Canada, the National Climate Change Adaptation Framework in Australia, and FinAdapt in Finland). Government agencies at all levels indicated that they did not have internal resources to dedicate to long-term issues such as climate change, and have relied on additional resources from temporary programmes to undertake adaptation-planning initiatives. This reliance on external resources suggests progress on adaptation is likely to remain sporadic for the foreseeable future. Tourism organizations in developing nations will typically have further resource constraints, restricting progress on adaptation in some of the countries where the tourism sector is anticipated to be most vulnerable unless international development agencies provide support in the future.

The review of adaptation policy development for tourism by national governments reveals that mainstreaming adaptation remains in its earliest stages. Considering that other tourism stakeholders look to national governments for leadership on climate change adaptation, tourism adaptation practice is likely to be further limited in lower levels of government. As the case study of local government involvement in climate change adaptation presented by Buckley (2008) highlights, where limited information or misinformation exists, there is the potential for climate change adaptation to become a political mechanism for tourism planning and development, where information on the impacts of climate change can be ignored or misused by governments as well as business and environmental lobbies.

While the business community has for some time been aware of the risks and opportunities associated with climate mitigation policies, and is taking into account the impacts of potential national and international regulations, shareholder perceptions and changes in consumer and supplier markets, fewer businesses are incorporating the risks and opportu-nities associated with the physical effects of climate change in their business planning (Sussman and Freed 2008). Corporate assessments of climate change risk differ from the interests of banks and insurers (Climate Service Center 2009). This situation is paralleled in the tourism sector, where several destination case studies of stakeholders' perceptions of climate change and adaptation reveal that the impact of mitigation policy on travel costs and tourism demand is a top concern and reducing GHG is perceived as a priority 'adapta-tion' in order to be able to market a destination as 'green'. As indicated, responses to climate policy are not considered climate change adaptation, but this perspective illustrates that the direct physical risks and opportunities of climate change are often of secondary importance, or not considered by tourism operators.

Unlike climate change mitigation efforts (chapter 4), where the benefits of reducing GHG emissions accrue at the global scale (with the exception of marketing advantages for com-panies), costs incurred in adaptation provide direct benefits to those who made the invest-ment, typically at the organization or local scale. Because of this direct benefit from investment in adaptation, Weaver (2011: 11) argues that 'adaptation often constitutes a rational "capitalism-compatible" response by individual businesses to actual or high-probability threats'. Presumably, therefore, there should be little opposition to adaptation from tourism operators and destinations. Interestingly, the experience of case studies in multiple countries has demonstrated this is not always the case, and there is resistance to engaging in public forms of climate change adaptation because of the fear of reputational damage and reduced visitation resulting from acknowledging concerns about climate change risks.

Furthermore, as Scott (2011) suggests, for strategic business reasons we should not expect to hear much from tourism operators about climate change adaptation. Where tourism companies see risk to their business or properties, they will not broadcast this vulnerability to guests, investors or insurers, as the perception of long-term vulnerability could have detrimental effects on the business in the short term through reduced visitation, reduced real estate values, increased insurance premiums, higher costs or greater difficulty accessing credit, and a lower sale price if attempting to sell the company. Because there is no business advantage to recognizing or making known climate vulnerability, companies are likely to downplay publicly any climate risk, and quietly adapt or divest high-risk assets. Similarly, where a company sees potential competitive advantage, again it will not broadcast this information to competitors, but will use it, like any other strategic business information, to improve its position in the marketplace.

Similar risks exist at the destination level, where there is a real risk for early adopters that development regulations (e.g. greater coastal set-backs, higher building standards) or other policy and planning adaptations (e.g. bans on snow-making additives, restricted water availability for golf courses) could direct tourism investment to other destinations that are not pursuing climate change adaptation. Like many other forms of environmental and planning regulations, in the absence of national adaptation planning frameworks, what we might expect in terms of climate change adaptation by destinations in the foreseeable future is the 'lowest common denominator' adaptation.

Communities with unique tourism products or with a history of progressive leadership are likely to emerge as innovators of climate change adaptation. These destinations should serve as learning sites for the collaboration between government, industry and academics needed for successful adaptation. The literature on understanding organizational change and innovation in business and government, for both tourism and other sectors (Hall and Williams 2008; Hall 2009b; Hjalager 2010), has much potential to provide important insights into assisting destinations to embrace climate change adaptation.

There is a need to further incorporate adaptation planning into decision-making throughout the tourism sector (government, industry, NGOs). There is also an important need for effective communication between the climate change science community and tourism stakeholders (tourism operators, communities, industry associations, national-level governments) regarding credible climate change information and informed interpretation of impacts from a tourism perspective (that is, meaningful tourism indicators).

Although the incompatibility of timeframes between normal business planning and projected climate change impact causes many tourism stakeholders to consider climate change adaptation a task for the distant future, the information requirements, policy changes, siting and regulatory approvals, and financing that are required for some types of adaptation by tourism destinations (e.g. coastal protection works, major water-supply projects) will require decades in some cases. Therefore the process of adaptation needs to commence in the near future in destinations anticipated to be affected by mid-century. Young and future tourism professionals will face most of the impacts discussed in chapter 5. There is an important role for tourism education and professional programmes to provide training on climate change issues as part of a broader environmental management or sustainable tourism curriculum in order to equip these future professionals with the skills to adapt to future challenges.

FURTHER READING AND WEBSITES

The following are recommended resources for those wanting to improve their understanding of the science, concepts, practice and policy pertaining to climate change adaptation:

Pelling, M. (2011) *Adaptation to Climate Change: From Resilience to Transformation.* London: Routledge.

Schipper, E. and Burton, I. (eds) (2008) *The Earthscan Reader on Adaptation to Climate Change.* London: Earthscan.

The broader business case for climate change adaptation is made by the following:

Huddleston, M. and Eggen, B. (2007) *Climate Change Adaptation for UK Businesses.* London: Met Office Consulting.

KPMG (2008) *Climate Changes Your Business.* www.kpmg.com/Global/en/IssuesAndInsights/ArticlesPublications/Documents/Climate-changes-your-business.pdf.

Network for Business Sustainability (2009) *Business Adaptation to Climate Change.* Ottawa: Network for Business Sustainability.

Sussman, F. and Freed, J. (2008) *Adapting to Climate Change: A Business Approach.* Arlington, VA: Pew Center on Global Climate Change.

World Business Council for Sustainable Development (2008) *Adaptation: An Issue Brief for Business.* Washington, DC: World Business Council for Sustainable Development.

Further guidance on adaptation frameworks, where they have been applied, and training requirements, is available from the UNFCCC Secretariat Compendium of Decision Tools to Evaluate Strategies for Adaptation to Climate Change:

www.aiaccproject.org/resources/ele_lib_docs/adaptation_decision_tools.pdf.

Consumer behaviour and tourism demand responses to climate change

7

The influence of climate change on patterns of tourism demand will be shaped by tourists' responses to the complexity of mitigation policy and its impacts on transportation systems (chapter 4); the wide range of climate change impacts on destinations (chapter 5); as well as broader impacts on society and economic development. As indicated in chapter 6, tourists have the largest adaptive capacity among the major stakeholders that comprise the tourism system because of the tremendous flexibility leisure tourists have to substitute the place, timing and type of holiday; even at very short notice. It is the responses of tourists that will govern the net impact of climate change on the global tourism system. Consequently, understanding tourists' perceptions and reactions to the impacts of climate change is essential to anticipating the potential geographical and seasonal shifts in tourism demand, as well as the decline or increase of specific tourism markets.

This chapter begins by providing a conceptual framework through which to examine tourists' responses to the varied impacts of climate change on other parts of the tourism system. Potential influences of climate change on travel motivations and important dimensions in destination-choice decisions are discussed, including the complexities and uncertainties associated with understanding how climate change could affect tourists' perceptions of destinations and thus travel behaviours. The chapter goes on to review the extant literature on potential impacts of climate change on tourism demand. This literature has become increasingly diverse, with studies that focus on individual tourist behaviour, through to aggregate tourism demand at the destination, to global tourism-system scales. Studies of tourists' behaviour have examined tourists' awareness and perceptions of climate change impacts on tourism and potential influences on travel behaviours, the role of weather and climate in tourists' decision-making processes, identifying climatic or environmental thresholds that could trigger tourists' behaviour change, understanding the role of the media coverage on climate change in influencing travel decisions, and the differential response of market segments to the impacts of climate change. Other studies have utilized bottom-up (individual operator to destination scale) or top-down (national to global scale) models of aggregate tourism demand to project changes (geographical, seasonal, overall demand) resulting from climatic change and other factors (demographic change, economic growth). Recent studies have also begun to explore how demand for tourism could be influenced by climate change impacts on other economic sectors and, similarly, how climate change impacts on tourism could affect the broader economy. To facilitate cross-connections with the considerations of impacts on tourism destinations (chapter 5), the wide range of tourism behaviour and demand-focused literature in this section has been organized according to the same four major impact pathways: direct impacts of climatic change, indirect climate-induced environmental change, indirect socio-economic change, and mitigation policy.

A CONCEPTUAL FRAMEWORK FOR ANALYSIS OF TOURISTS' RESPONSES TO CLIMATE CHANGE

The following sections provide an overview of uncertainties associated with tourists' responses to climate change, including a conceptual framework for analysis and a discussion of complexities with regard to perceptions, based on Gössling *et al.* (2011a). The motivations for leisure travel can be divided into a limited number of categories (Pearce 1993):

- relaxation needs;
- safety/security needs;
- relationship needs;
- self-esteem and development needs;
- self-actualization and fulfilment needs.

These motivations are linked to Maslow's (1970) hierarchy of needs, though Pearce (2005) outlines that these should not be understood as a hierarchy in a strict sense, rather they are a complex set of interacting needs. To capture fully all motives for travel, a sixth category, 'work', may have to be added, as business travel is part of tourism and in some cases unrelated to, or even contrary to, personal motivations (Hall 2005). Such an observation is also important as some motivations to travel are tied up with special interest travel motivations that are so significant for individual travel behaviour that they have been termed 'serious leisure' (Stebbins 1982). As Stebbins (1982: 253) observed:

> If leisure is … an improvement over work as a way of finding personal fulfillment, identity enhancement, self-expression, and the like, then people must be careful to adopt those forms with the greatest payoff. The theme here is that we reach this goal through engaging in serious rather than casual or unserious leisure.

The commitment to serious leisure is such that amateurs, hobbyists and careerists who engage in serious leisure can be described as having careers (Stebbins 1982). Although related to the motivations and travel careers of leisure tourists as identified by Pearce (2005), serious travellers' commitment to the pursuit of their interests may be such that some of the impacts of climate change on transportation systems and, particularly, at destinations may be less likely to affect them. Indeed, in some cases environmental change may even provide a motivation to engage in volunteering or other activities (Curtin 2010). Similarly, visiting friends and relations (VFR) tourism is also often marked by a strong social commitment that may mean the tourist will travel to a destination for family or relationship reasons, despite the effects of climate change (Gössling and Hall 2006b). As Forsyth *et al.* (2007: 2) noted with respect to the impacts of climate change on Australia's tourism markets: 'Not all tourism will be affected. Business and Visiting Friends and Relatives (VFR) tourism may be more or less unaffected by climate change.'

Pearce and Lee (2005) distinguish a subset of motivation factors and motive items (listed in Table 7.1) to provide an overview of the full range of motivational aspects that may stimulate travel. While climate change is but one aspect affecting travel motivation, its varied influence on travel motivation has not been considered in the literature previously. In Table 7.1 we consider the generalized impact of direct and indirect climate change, broader socio-economic change, as well as changes in transport costs on travel motives across a broad spectrum of travellers (impacts on individuals will vary substantially). Some motive items will be unaffected by climate change, while others will be positively or negatively affected. Notably, there is the potential for climate change to negatively affect motivational factors such as novelty, nature and isolation. The extensive negative impacts of climate change on a wide range of ecosystems and nature-based tourism destinations is outlined in chapter 5, and in some cases the availability and quality of some nature experiences will decline because of climate change, negatively affecting the ability of tourists to experience the novelty of some natural and cultural heritage, or to get close to or be harmonious with certain kinds of nature. Chapter 5 also indicates that for some tourism markets, the supply of operators and destinations (ski or coastal resorts, expansive beaches, healthy coral reefs) will contract under climate change, leading to increased visitation and congestion at the remaining destinations and thus reducing the ability of many tourists to experience isolation. 'Travellers' guilt' associated with knowledge of the environmental impacts associated with travel may also reduce the ability of some travellers to relax during holidays. In contrast, factors such as stimulation could be positively affected across a broad spectrum of tourists. Climate change will lead to rapidly changing landscapes and, in some destinations, more risky tourism experiences that could be construed as positive adventures, at least for the segment of tourists seeking novelty and stimulation. For the majority of travel motivations, the influence of climate change is anticipated to be negligible or mixed (positive and negative). In this context it needs to be considered, however, that not all motivational factors for travel are equally important, with factors relating to 'novelty', 'escape/relax' and 'relationship' having greater importance than others (Pearce 2005). Consequently, climate change should have greater relevance for perceptions associated with these factors.

Motives for travel are interlinked with destination attributes. Clearly, the destination or site chosen for a given holiday or leisure activity has to meet motivational demands. However, destinations will also have to appeal to tourists for other reasons, including uniqueness in terms of culture, cultural heritage or landscape, landscape elements or biodiversity resources; the existence of other resources such as climate; travel time and travel cost; perceived safety and security; existing facilities, services, access and socio-political stability; hospitality and hosts (Hall 2005). The respective combination of destination attributes and travel motives results in a destination's specific attractiveness (Gössling *et al.* 2011a; see Figure 7.1, points 1 and 4). While all of these destination attributes will be influenced

Table 7.1 *Travel motivation factors and their relevance under climate change scenarios*

Factors	Motivation factors	Potential climate change impact*
Novelty	Having fun	0
	Experiencing something different	+/−
	Feeling the special atmosphere of the destination	−
	Visiting places related to my personal interests	−/−
Escape/relaxation	Resting and relaxing	0
	Getting away from everyday psychological stress/ pressure	0
	Being away from daily routine	0
	Getting away from the usual demands of life	0
	Giving my mind a rest	0
	Not worrying about time	0
	Getting away from everyday physical stress/pressure	0
Relationship (strengthening)	Doing something with my companion(s)	0
	Doing something with my family/friend(s)	0
	Being with others who enjoy the same things as I do	0
	Strengthening relationships with my companion(s)	0
	Strengthening relationships with my family/friend(s)	0
	Contacting family/friend(s) who live elsewhere	0
Autonomy	Being independent	0
	Being obligated to no-one	0
	Doing things my own way	−
Nature	Viewing the scenery	−
	Being close to nature	−
	Getting a better appreciation of nature	−/+
	Being harmonious with nature	−/−
Self-development (host-site involvement)	Learning new things	+
	Experiencing different cultures	0
	Meeting new and varied people	0
	Developing my knowledge of the area	0
	Meeting the locals	0
	Observing other people in the area	0
	Following current events	0
Stimulation	Exploring the unknown	+
	Feeling excitement	++
	Having unpredictable experiences	+
	Being spontaneous	0
	Having daring/adventurous experiences	+
	Experiencing thrills	0
	Experiencing the risk involved	+

Continued

Table 7.1 *Continued*

Factors	Motivation factors	Potential climate change impact*
Self-development (personal development)	Developing my personal interests	0
	Knowing what I am capable of	+
	Gaining a sense of accomplishment	+/−
	Gaining a sense of self-confidence	+/−
	Developing my skills and abilities	+/−
	Using my skills and talents	+
Relationship (security)	Feeling personally safe and secure	−
	Being with respectful people	0
	Meeting people with similar values/interests	0
	Being near considerate people	0
	Being with others if I need them	0
	Feeling that I belong	0
Self-actualization	Gaining a new perspective on life	+
	Feeling inner harmony/peace	−
	Understanding more about myself	+
	Being creative	+
	Working on my personal/spiritual values	+
Isolation	Experiencing peace and calm	+/−
	Avoiding interpersonal stress and pressure	−
	Experiencing the open space	−
	Being away from crowds of people	−
	Enjoying isolation	−
Nostalgia	Thinking about good times I've had in the past	+/−
	Reflecting on past memories	+/−
Romance	Having romantic relationships	0
	Being with people of the opposite sex	0
Recognition	Sharing skill and knowledge with others	0
	Showing others I can do it	0
	Being recognized by other people	0
	Leading others	0
	Having others know that I have been there	++

*++, Strong positive impact; +, some positive impact; 0, no impact or impact difficult to predict; −, some negative impact; −/−, strong negative impact
Source: motivation factors adapted from Pearce and Lee (2005); potential climate change impact added by authors

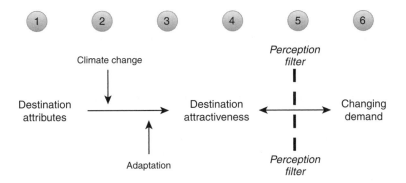

Figure 7.1 *Role of perceptions in defining destination attractiveness (see text for explanation of numbers)*
Source: Gössling *et al.* (2011)

by climate change (see chapter 5), some will be more affected than others (Figure 7.1, point 2).

There are fundamentally different timelines as to when the impacts of climate change will become relevant for tourism transport networks and destinations. As the headlines in Figure 2.1 illustrate, impacts resulting from extreme climate events or climate-sensitive environmental changes (e.g. coral reef bleaching, algal blooms), made more likely by global climate change, can occur at any time. These most immediate impacts linked to extremes are very difficult to project (Gössling and Hall 2006a). Increasing transport costs as a result of global and national climate policy are not likely to become significant in most parts of the world for years to come (OECD and UNEP 2011; see also chapter 4). Increasing costs of oil, and issues related to the overall state of the economy in tourism-generating regions, are likely to have a far more significant impact on travel behaviour than costs that arise from climate change *per se* for the foreseeable future (Gössling and Hall 2006a; Scott *et al.* 2008a; Hall 2010f). Even less relevant to the immediate future of tourist behaviour are long-term changes in climate parameters or climate-induced environmental change (Scott *et al.* 2008a). With regard to longer-term impacts, destinations may be able to cope with these changes to some degree through both anticipatory and reactive adaptation (point 3). For instance, changed price structures may offset increasing transport costs, and new products and increased marketing may seek to attract new customer groups – also tapping into new seasonal demands. The outcome of all these processes would then represent destination attractiveness at a given point in the future and constitute the basis for changing demand (point 6).

The greatest uncertainty in this model is represented by travellers' perceptions of change (point 5). Perceptions play a major role in tourist decision-making, representing an important intermediary stage of information processing, and are consequently highly important in influencing the actual outcome of the individual traveller's subjective negotiation of reported or experienced change. While there appears to be consensus on the significant role of perceptions (Gössling and Hall 2006a, 2006b; Scott *et al.* 2007b, 2008a; Moreno and Becken 2009), these are insufficiently understood and represent one of the key research gaps in the tourism and climate change literature. The following section discusses these issues of complexity and uncertainty in more detail.

Complexities and uncertainties regarding tourists' perceptions of climate change

Perception is the 'process of receiving and interpreting "information" through all senses' (Gössling *et al.* 2006: 423; see also Decrop 2006), and thus includes visual, audio, olfactory, haptic or sensual personal experiences, as well written, audio or visual accounts provided by third parties. There is consensus that a wide range of aspects shape tourist perceptions (for a discussion see Tasci *et al.* 2007). Preliminary insights with regard to the complexities of tourists' perceptions in the context of climate change are summarized as follows:

- Perceptions vary by holiday type and role.
- Perceptions change with age, culture and other socio-demographic variables.
- There are considerable differences in individual preferences, values and personalities.
- Perceptions evolve over travel careers and with the degree of specialization.
- Perception of climate change impacts is comparative.
- There are significant differences between *ex situ* and *in situ* perceptions.
- Perceptions are heavily influenced by media.
- The media will increase interest in 'last chance' tourism.
- Single events can have wide-ranging consequences for perceptions.
- Perceptions are complex, adaptive and hierarchical.
- Perceptions are likely to be context-dependent.
- Accuracy of understanding of weather parameters is insufficiently understood.
- Adaptive behaviour is insufficiently understood.
- Public perceptions of climate change can be ill-informed and highly polarized.

Perceptions vary by holiday type and role

Different forms of holiday, such as day trips, short trips, a main annual holiday or a once-in-a-lifetime trip, are characterized by varying planning horizons, as well as different expectations depending on the socio-cultural function of the trip and travel motives (Hall 2005). Consequently, it can be assumed that climate change will have a different impact on a holiday trip with a high degree of commitment, planned several months prior to travel, than on a spontaneous day or weekend trip. The temporal sequence of climate-related information is one important difference, where longer planning horizons would generally be focused on longer-term weather averages (climate), while shorter day- or week-long trips will be heavily influenced by weather forecasts and on-site weather (Scott and Lemieux 2009; see also chapter 2). Depending on the type of holiday or trip, there may also be varying degrees of resilience to climatic conditions (the degree to which climate change, specifically extremes, may be tolerated; Scott and Lemieux 2009). Rutty and Scott (2010), for instance, found that travellers to the Mediterranean perceive temperatures to be unacceptably warm differently, depending on whether their destination was a coastal beach resort or an urban sightseeing tour. They also found that media stories about heat-waves at their intended Mediterranean destination had a different impact on travel decisions depending on the level of commitment to the trip – more influence if still planning a trip than if travel and accommodations are already booked. Similarly, Ceron *et al.* (2012) and Hewer and Scott (2011) show that French and Canadian campers accept higher maximum temperatures than tourists staying in other accommodation, and that temperatures perceived as unacceptably warm depend on activities. Different resilience to climate change-induced environmental changes was also found among market segments visiting national parks in Canada's Rocky Mountains, where long-haul tourists, who travelled specifically to see certain attractions, were much less willing to visit the parks if these key features were affected by climate change than tourists from the region (defined as six hours' travel time) (Scott *et al.* 2007a, 2008d). With regard to changing transport costs, many segments of air travel are price-inelastic, that is, higher prices are tolerated because of the relative importance ascribed to air travel. Price elasticities are not identical for short-haul and long-haul travel, nor for business, VFR and leisure travel (Brons *et al.* 2002; Mayor and Tol 2007). Consequently, there is an important situational context in which perceptions of climate change and possible tourist responses need to be studied.

Perceptions change with age, culture and other socio-demographic variables

Perceptions of climate and climate change impacts also appear to differ to some extent among travellers from different cultural and climate contexts, as well as with other socio-demographic variables such as age or family status. For instance, as discussed in chapter 2, Scott *et al.* (2008b) and Rutty and Scott (2010) show that there are differences in preferred beach temperatures among young adult travellers from different countries. Moreno (2010), Wirth (2010) and Hewer and Scott (2011) found differences in climate preferences among young adult and senior travellers in Germany, the Netherlands and Canada. Research on weather perceptions of French travellers by Ceron *et al.* (2012) also showed that temperature preferences vary regionally within France, and between age cohorts, with older people (+60 years) being more heat-sensitive than younger people (18–24 years). Perceptions of suitable weather and concerns about weather risks during travel were found to differ among British travellers with varied family status (e.g. single professionals were far more resilient to weather than families with children) (Limb and Spellman 2001). Similarly, Buzinde *et al.* (2010) found that families interpreted efforts to correct beach erosion at Mexican coastal resorts, including the use of large geotubes filled with sand to prevent further erosion and trap sand, differently than some other market segments that preferred pristine natural beach conditions, because children enjoyed climbing and diving off these objects and they reduced wave action on the beach, creating safer conditions for small children.

Culture plays an extremely important role in travel behaviour, and what may be perceived as unattractive in one culture may be attractive in another. For example, the Indian Ministry of Tourism has sponsored campaigns in several Indian states to promote 'monsoon tourism', the wet season of torrential downpours that most travellers to India avoid. However, in the Indian context, the monsoon is a time of refreshment and renewal. A number of states, such as Goa and Kerala, have focused on promoting monsoon tourism to domestic visitors as well as to international tourists from the Middle East (Dhanesh 2010), for whom the cultural value of heavy rain is different. Denstadli *et al.* (2011) also found inter-cultural differences in summertime weather preferences among tourists in northern Scandinavia. Although it is recognized that there are different values attached to different dimensions of weather, such as rain, storms and snow, and seasonality in different cultures (Thornes and McGregor 2003), and that such forms of knowledge are important for understanding individual responses to climate change (Hulme 2008), there is a major knowledge gap as to the role that such culturally based perceptions play in tourists' decision-making and responses to climate change. As Scott *et al.* (2008b) have observed, most of our understanding of tourists' climate preferences has come from Europe, North America, New Zealand-Australia, and improved understanding of cross-cultural differences is an important area for future research.

Differences in individual preferences, values and personalities

It is generally accepted that there are considerable differences between individual holiday preferences and value systems (Gilg *et al.* 2005). For instance, the studies referred to in the previous two points have found that personal differences exist in climate preferences for holidays and interpretations of climate-induced environmental change, even among similar market segments and demographic cohorts, while Gössling *et al.* (2009d) indicate varying degrees of responsibility taken by air travellers for emissions, as well as their willingness to act on these through offsetting (see also McKercher *et al.* 2010; Wells *et al.* 2011). These results would support the importance of value systems, which can be differentiated on continuums of altruistic–egoistic, biocentric–anthropocentric and ecocentric–technocentric (Gilg *et al.* 2005). Depending on the belief systems constructed out of these value dimensions, it would, for instance, be possible for travellers in the 'egoistic–technocentric' domain to discard the notion of climate change and to see weather extremes as 'natural' hazards one has to live with and for which technical solutions exist, forming norms that do not translate into changes in behaviour. Overall, these examples would indicate that perceptions are value-dependent and thus not homogeneous between tourists. Likewise, travellers' personalities are not identical and may allow for differing responses to climate change, for instance depending on the degree of loyalty to certain locations (see also Dawson *et al.* 2011a).

Perceptions evolve over travel careers and with the degree of specialization

Pearce and Caltabiano (1983) argued that travel experience changes motivation, showing that the importance of 'self-development' and 'nature-seeking' increases with travel experience (see also Pearce and Lee 2005). As indicated in Table 7.1, climate change is not likely to influence motivation factors evenly, and consequently should have different consequences for experienced and specialized travellers depending on their interests.

For instance, research by Gössling *et al.* (2007b) and Dearden and Manopawitr (2011) indicates that novice divers are less affected by climate change, as they have a more limited knowledge of what constitutes healthy coral reefs, and of what to expect in terms of reef structure or species diversity. Moreover, diving may not be their primary holiday motivation. *Vice versa*, highly specialized divers may be very aware, not least through special-interest media, of the status of coral reefs in specific destinations and the loss of the very attractions for which they would visit particular destinations. Similar differences were found among the perceptions of risk posed by climate change to ski tourism and the potential behavioural responses of expert specialized skiers versus novices (König 1998; Behringer *et al.* 2000; Dawson and Scott 2010; Pickering *et al.* 2010; Dawson *et al.* 2011a). Overall, climate change will thus be perceived differently depending on the individual traveller's career and degree of specialization.

Perception of climate change impacts is comparative

Tourists comparing destinations will also perceive climate change impacts relatively. For instance, in a situation of warmer winters and reduced natural snow, an increasing number of destinations will become snow-unreliable (Abegg *et al.* 2007; Scott *et al.* 2008c; Steiger 2011b; see chapter 5 in this volume). However, less favourable conditions for winter sports in one location are likely also to be compared with other locations, with the possible outcome that identically 'marginal' conditions in several destinations within a given distance may lessen the perceived impact. Similar comparative situations are likely to emerge with respect to beach extent, health of coral reefs, and other environmental assets affected by climate change. If these tourism attractions are degraded across a large proportion of destinations, degraded conditions might take on the new normal status. As Scott *et al.* (2007b) point out, this could particularly be the case for new generations of travellers who did not know previous conditions, and therefore do not have the same perception of loss or degraded experience – for example, if they never saw a glacier in a certain destination and thus have no expectation to see it or perception of loss if they don't see it.

Differences between *ex situ* and *in situ* perceptions

It can be assumed that there are considerable differences in perception depending on how information about climate change is derived. For this purpose, it is useful to distinguish *ex situ* and *in situ* perceptions. In an *ex situ* situation, a travel decision is made without previous direct knowledge of a destination. The understanding of the destination is consequently building on images, text and verbal communication by third parties, including, for instance, advertisements, guide books and other media, and recommendations by friends or relatives or travel agents. For such externally derived information, credibility plays an important role. Advertisement is generally understood as 'controlled' information, and perceived as less credible than 'uncontrolled' information such as newspaper articles or word-of-mouth information, including travel commentary websites (such as *TripAdvisor*) and the wide range of emerging social networking technologies (e.g. Hall 2002). Importantly, in *ex situ* situations all information about a destination is derived from 'outside'.

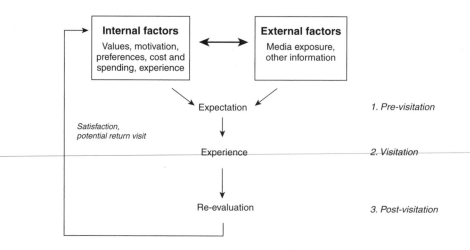

Figure 7.2 *Tourists' evolving understanding of destinations*

This situation is more complex where visits to a destination have been made previously, in which case perceptions are at least partially derived from *in situ* experiences. *In situ* perceptions will be shaped in three phases (pre-visitation, visitation and post-visitation; Figure 7.2). In the pre-visitation phase, tourists are likely to rely on a mix of previously acquired *in situ* knowledge (personal experiences) and third-party information. The understanding of a given destination or site is then shaped during the stay, where the experience of particular favourable or unfavourable conditions will affect perceptions. Depending on travel motivation, socio-demographic variables, geographical origin, costs and other variables, such as experience resulting in attachment to a place or familiarity with a given destination, deviation in personal experiences from expectations and preferences would then be interpreted and reflected upon in the post-visitation phase, leading to a re-evaluation of the suitability of the destination in terms of satisfaction for continued holiday-making. *In situ* perceptions are consequently more complex.

Perceptions are heavily influenced by media

Climate change has featured prominently in the media over the past decade, and there have been a number of critiques of this coverage in terms of the accuracy of the climate science; the equal voice and credibility given to climate experts, non-experts and political pundits in the name of journalistic balance; the often sensationalist portrayal of potential impacts; and most recently the media frenzy over 'climategate' and how this has only contributed to public confusion over climate change (Boykoff and Boykoff 2007; Boykoff and Roberts 2007; Billett 2010; Gavin 2010).

Media headlines about the impacts of climate change on the tourism sector have suffered from similar speculative and sensationalist tendencies (Gössling and Hall 2006a; Scott *et al.* 2008a; Scott and Becken 2010; Scott 2011). A number of headlines declaring the Mediterranean to be 'too hot' for summer tourism and the 'collapse' of ski tourism in the

European Alps or Rocky Mountains of the USA are contrasted with scientific information in chapter 5. Other common storylines suggest that the Baltic will become the new Mediterranean (*Metro* 2006) and that it is ethically despicable to participate in long-haul journeys because of GHG emissions (*Sunday Times* 2006). Where media speculation is convenient, tourism stakeholders may be quick to pick up on headlines, which can then turn into truth systems despite a lack of credible scientific evidence. An example is the 'Baltic as the new Mediterranean', which is now considered a 'fact' in climate change adaptation plans in Sweden and Germany (see statements in chapter 6 by Regeringskansliet 2007 and Federal Government 2008), even though scientifically this is highly disputable, given considerable differences in air and water temperatures, as well as cultural heritage including buildings and food (Scott *et al.* 2008b; Rutty and Scott 2010) (see also chapter 6). In the absence of credible information and recurrent exposure to such messages, consumers may eventually accept these speculative impacts as reality, particularly when consumers have only limited geographical knowledge of destinations (Selby 2004) and adjust their travel behaviours accordingly. Where this occurs, the reputational damage to destinations caused by disinformation could have a greater near-term impact on tourism visitation than the actual impacts of climate change (Gössling and Hall 2006a; Scott *et al.* 2008a; Scott 2011).

Various examples show that perceptions of climate and climate change impacts in a region are heavily influenced by media coverage and the geographical knowledge levels of consumers (see chapters 2 and 5). For example, market surveys found that media coverage of the three hurricanes that hit parts of Florida in August and September 2004 had created an impression that the entire state was heavily damaged. Reservation cancellations went up sharply, and one in four tourists were less likely to plan to visit Florida between July and September in the following years (Pack 2004). Tourists were unable to distinguish between areas that had been damaged and those that were unaffected by the hurricanes. The drought and resulting wildfires in the state of Colorado (USA) in 2002 provides a similar example of perceptions based on media coverage being as damaging for tourism as actual impacts of the climate event. Media coverage of major fires in some parts of the state, and a misstatement by a senior government official who said that 'it felt like the whole state was on fire', had a significant impact on summer tourism, with visitor numbers 40 per cent lower in some areas of the state, even well away from fire-affected areas (Butler 2002). In both Florida and Colorado, the state governments and tourism industry spent millions of dollars on marketing campaigns to inform consumers that the climate impacts were isolated. Similar effects have been observed as a result of wildfires in Portugal, Spain, France, Australia and New Zealand in recent years (see chapters 2 and 5).

In an attempt to gain insight into how climate-related media stories might influence decisions to travel to the Mediterranean, Rutty and Scott (2010) used excerpts from a 'heatwave' story in a popular UK newspaper to evaluate the influence on respondents' travel plans to the region. While the largest proportion of respondents (39 per cent) were unsure how the media story would influence their Mediterranean travel plans, 32 per cent stated that such stories would have a strong or very strong influence on their plans, and only a small percentage (12 per cent) were not influenced by this type of media story. Of those who were planning a Mediterranean holiday but had not yet booked their travel reservations, 52 per cent stated that they would change their travel plans in some manner (28 per cent would still book a Mediterranean holiday in a location that was not experiencing the heatwave; 19 per cent would change the dates of their holiday, and 5 per cent would go to another region). If their holiday reservations had already been booked, fewer respondents would change their plans, with the majority (58 per cent) stating they would still go forward with their Mediterranean holiday reservations as originally booked.

Even though Hall (2002: 44) cautions that 'unless a crisis [or risk] continues for a substantial length of time then it is extremely unlikely to have permanent impacts on destinations' [or activity] perceptions', indicating that media impacts may be short-lived, it can be assumed that where information is repetitive and fitting an already existing understanding, the media can shape perceptions fundamentally. This is, for instance, exemplified by Cohen and Higham's (2010) finding that media coverage on the impacts of air travel has led to generally raised awareness of emissions, although not necessarily leading to behavioural change, rather indicating complex responses (see also Gössling *et al.* 2009d; McKercher *et al.* 2010). Another example is the travel phenomenon variously called 'doom tourism' or 'last-chance tourism', where it is argued that coverage of climate change impacts in the media and travel magazines is increasing visitation to some destinations as tourists are inspired or advised to visit a place or an attraction 'before it's gone'. Box 7.1 considers whether 'last-chance tourism' is a reality.

BOX 7.1 'DISAPPEARING DESTINATIONS' AND 'LAST-CHANCE TOURISM': ARE THEY REAL?

The wide range of potential climate change impacts on tourism destinations has prompted a number of travel writers to speculate on which destinations are the most at-risk places that tourists should see 'before it's too late'. Beginning with *Condé Nast Traveller*'s list of climate change-threatened destinations in 2004 (Rogers 2004), a search revealed a total of 23 lists of climate change-vulnerable destinations published by travel-focused magazines, websites and travel sections of newspapers, as well as several travel books that included climate change in a broader range of threats to endangered destinations (Addison 2008; Hughs 2008; Lisagor and Hansen 2008; Drew 2011). A number of terms have been used to describe tourism related to seeking out destinations at risk from climate change, including: 'endangered destinations', 'tourism of doom', 'imperilled places', 'last-chance tourism', 'disappearing destinations', 'endangered vacations', 'climate change voyeurism', 'climate sightseeing' and 'climate disaster tourism' (Eijgelaar *et al.* 2010; Dawson *et al.* 2010; Lemelin *et al.* 2010).

This travel phenomenon is of growing interest to tourism researchers. Lemelin *et al.* (2010: 478) review the media coverage and academic discussions related to this phenomenon, and define last-chance tourism as 'where tourists explicitly seek vanishing landscapes or seascapes, and/or disappearing natural and/or social heritage'. Several other studies discuss this potential travel motivation and examine the paradox between GHG emissions associated with visiting destinations highly vulnerable to climate change (Burns and Bibbings 2009; Randles and Mander 2009; Dawson *et al.* 2010, 2011b; Eijgelaar *et al.* 2010; Farbotko 2010). Two academic books, *Disappearing Destinations* (Jones and Phillips 2011) and *Last Chance Tourism* (Lemelin *et al.* 2011) have also adopted these terms.

Discussion over emerging concepts and new terminology is common in tourism studies, and indeed all academic disciplines. Here we ask two central questions with respect to this travel phenomenon and the associated terms: are the terms accurate, and what evidence is there for this trend?

First, are these terms accurate? Two of the most frequently used terms are disappearing destinations and last-chance tourism. However, with the exception of some very low-lying island destinations (e.g. the Maldives), tourism destinations will not physically 'disappear' or 'vanish' as a result of climate change. Rather, climate change will alter the characteristics of some destinations, resulting in the degradation and potential loss of some, but not necessarily all, tourist attractions (e.g. a beach, glacier or specific species). Changes in the environmental characteristics of tourist destinations, or even the degradation/loss of cultural assets, is nothing new to tourism. Climate change will continue to become an increasingly important driver of the evolution of tourist destinations, just as many other factors identified in the tourism area-change literature (Hall 2005; Butler 2009; Hall and Lew 2009). As research on the product life-cycles of destinations suggests (Butler 2006), destinations heavily affected by climate change will enter a period of 're-orientation', and if key tourism assets can be restored or replaced by other attractions and the destination is able to successfully reinvent itself, it will enter a 'rejuvenation' stage, but where tourism assets are lost and cannot be replaced, the destination will 'decline' over time. As analogue situations tell us (e.g. past loss of beaches to hurricanes, rapidly retreating glaciers, reefs degraded by pollution, species lost to habitat loss or over-exploitation, heritage buildings lost to earthquakes, coastal areas severely damaged by tsunami), few places are likely to 'disappear' altogether as a destination (and thus to stop attracting any tourists).

The term 'last-chance tourism' is equally problematic. While a specific tourist attraction may be lost to climate change at a specific location (e.g. the often-cited example of polar bears at the community of Churchill, Canada; Lemelin *et al.* 2010), this is the only scale at which this term is meaningful. When a broader tourism-system scale is considered, there is currently no evidence to support the concept of 'last-chance tourism'. The few broad climate change impact assessments that have examined many tourism operators across multiple destinations – ski tourism in the European Alps (Abegg *et al.* 2007) and New England in the USA (Scott *et al.* 2008c), coastal tourism in the Caribbean (Scott *et al.* 2012) – have consistently found that, even without consideration of a full range of adaptation options, at least one-third to one-half of tourism operators at multiple destinations remained viable under climate change scenarios through the mid- to late-twenty-first century. In other words, even among those tourism destinations anticipated to be among the most vulnerable (winter sports tourism and coastal tourism), the type of tourism experience sought at these destinations is usually still available in many locations. The exceptions are attractions that are based on specific species or geomorphological attractions that have an extremely narrow range (e.g. some higher-altitude plant species in South Africa or south-western Australia). However, in most cases tourists may need to go to a different destination to obtain the sought-after attraction or experience, and it may be more difficult or more expensive to obtain – but it will be available. The literature that has examined tourists' perceptions of climate-induced environmental change (chapter 6) has found that tourists are very willing to substitute one location for another that still has the attraction or experience sought (Richardson and Loomis 2004; Uyarra *et al.* 2005; Scott *et al.* 2007b; Dawson *et al.* 2010). For the locations where the tourism experience is available, there will be the potential for new or expanded tourism opportunities.

Throughout this book, we have stressed the importance of spatial and temporal scales for understanding the consequences of climate change for tourism. When considering these terms, spatial scale is imperative. From a tourism systems perspective, there is currently little evidence to suggest that any specific tourism experience will be eliminated exclusively by climate change (at least not on time scales relevant to contemporary tourism development and planning), so 'last-chance tourism' therefore remains more connected to marketing hyperbole than actual fact.

Second, what evidence exists for this tourism trend? Lemelin *et al.* (2010: 477) also argue that 'this trend is embodied by a surge in the number of travellers to the Galapagos Islands and the polar regions'. Anecdotal evidence from travel writers is offered as evidence that many tourists were motivated to visit the polar regions to see wildlife species before they 'are gone forever' (Lemelin *et al.* 2010). It is not clear how Antarctic coastal wildlife is threatened by climate change in a way that would be observable by tourists and, while the disintegration of ice shelves is an attraction for tourists, this process will continue for centuries. Lemelin *et al.* (2010: 478) also suggest that travellers to Churchill, Canada for the purpose of viewing polar bears were 'strongly motivated by the stated vulnerability of the species and indicated that they wanted to see the bears before they disappear forever'. The study cited to support this contention interviewed a very small sample of tourists and found that four out of 18 (22 per cent) indicated they had 'last-chance' motives. Hall and Saarinen (2010a) suggest that polar tourism has grown in great part because climate change has made many locations more accessible, not because of last-chance tourism *per se*. Although they do note that climate change has become part of cruise promotion, they suggest that the new changing landscapes are as attractive to tourists as the role of iconic species or geomorphological features (Hall and Saarinen 2010b). Indeed, such is the rate of landscape change that in some cases the attraction is more of a first chance to see, rather than the last chance.

Can the increase in visitors to these regions be equally explained as forms of ecotourism or frontier tourism? Studies of polar tourists have revealed wild-life, curiosity (interest in learning), scenery, and remoteness as their primary motivations (Hall and Saarinen 2010c), which are similar to those of ecotourists more broadly. Similarly, 'last-chance-to-see' tourism has been promoted in the context of species facing extinction since the publication of British novelist Douglas Adams' book *Last Chance to See* in 1990. New frontiers have always been popular with tourists once travel became safe and affordable, whether it be isolated mountain peaks, tropical forests, deserts, underwater ecosystems, or more recently polar regions and even space tourism. Following the paths of explorers to Antarctica and through the Northwest Passage has the same mythical appeal as following explorers to the highest peaks and densest jungles.

It is also argued that some tourism operators are adopting last-chance tourism messages as part of their marketing to attract tourists. The only known study that attempts to identify the extent to which such marketing is occurring found that only 8 per cent of tourism operators operating near three destinations identified as vulnerable to climate change (Great Barrier Reef, Greenland and Mount Kilimanjaro) used such techniques to promote their products (Frew 2008). Other studies that examine tourism industry views on climate change have consistently found strong resistance to identifying their tourism product or destination as vulnerable to climate change (Wolfsegger *et al.* 2008; Hoffman *et al.* 2009; Valls and Sarda 2009; Cheng 2010; Jarvis and Pulido Ortega 2010; Roman *et al.* 2010; Turton *et al.* 2010; WTTC 2010). It is our experience that the tourism industry is generally very image-sensitive with respect to climate change. It would be very short-sighted for tourism operators to declare their tourism product to be highly vulnerable to climate change; adversely impacting investors' and insurers' perspectives on the sustainability of their business.

The cumulative evidence for this climate change-related tourism trend remains scarce at present and largely isolated to polar regions. Further work is required to document if this is a meaningful trend at diverse vulnerable destinations, and to understand how these travellers' motivations may differ from those of ecotourists or other types of traveller.

Overall, further research is needed to specify how the media are shaping perceptions of tourism under various climate change scenarios, and the consequences for destination choice and travel behaviour (see also Scott *et al.* 2008a; Buzinde *et al.* 2010). Furthermore, while the effectiveness of public relations and marketing campaigns to correct consumers' misperceptions about climate impacts on a destination is uncertain, this is likely to be an indispensable adaptation strategy for the tourism industry and governments alike (Scott 2006a; Scott *et al.* 2008a), and would benefit from improved understanding of the influence of media stories on travel decisions.

Single events can have wide-ranging consequences for perceptions

It has been hypothesized that negative perceptions can result in abrupt changes in travel behaviour, as well as longer-term behavioural modification (Gössling and Hall 2006a). Importantly, negative perceptions can arise out of single events. For instance, the European heatwave in summer 2003 led to fundamentally changed travel patterns in 2004, when many citizens in central Europe appear to have expected similar warm conditions, with the somewhat unexpected result that many decided to stay at home rather than go on holiday (Gössling and Hall 2006a). However, when it turned out that the 2004 summer would remain cold and rainy, there was a rush on last-minute travel to warm destinations in August 2004. Other extreme events may have similar consequences, with for instance Gössling and Nilsson (2010) reporting that a share of tourists to the Swedish island of Öland were reported seeking out other destinations after algal blooms in 2005 made coastal activities impossible in large parts of the island. Likewise, Denstadli *et al.* (2011) reported that a small share (5 per cent) of visitors left the holiday region in Vesterålen, Norway sooner than planned because of weather conditions perceived as unfavourable (rain and low visibility). Further research is needed in particular to understand the impacts of extreme weather and environmental events on tourists' behaviour in both the short and longer term.

Perceptions are complex, adaptive and hierarchical

Perceptions are complex, in the sense that they are always influenced by various parameters (for example, temperature alone never represents 'climate' or 'weather') (Gössling *et al.* 2006; Scott *et al.* 2008b), while considerations by individual travellers may result in unexpected outcomes. For instance, it has been noted above that perceptions are shaped by media reports, with uncontrolled information (newspaper articles, word-of-mouth, social media) having greater weight in influencing individuals. However, where information does not fit already existing beliefs, the result may be cognitive dissonance, where 'deviating' information is ignored (Tasci and Gartner 2007). A general uncertainty is also whether there are trade-offs, in the sense that changes perceived as 'negative' are weighted against 'positive' changes or other factors perceived positively. For example, it has long been recognized that consumers trade off desired travel experiences against factors such as time and cost as part of satisfying behaviour (Verhallen and van Raaij 1986).

Likewise, there is a general assumption that increasing temperatures will be positive for northern tourist destinations in Europe. This perception does not consider the impact of other (potentially negative) environmental changes in the region (e.g. Hall 2008c; Nilsson and Gössling 2012) or that tourists will still want to travel to the Mediterranean or to other regions where climatic resources are anticipated to degrade for other reasons (e.g. culture, food). For example, Moreno (2010) found 72 per cent of respondents from Belgium and the Netherlands would still travel to the Mediterranean for holidays even if their self-defined preferred climatic conditions were available in northern Europe. Further studies are needed to understand better the complex nuances of tourists' responses to changing climate and environmental conditions.

Tourists' potential acceptance of environmental change is also related to the expectations being created by tourism promotion and marketing, as well as the package purchased. For example, in a survey of Christmas tourists to Rovaniemi, 'the official home of Santa Claus' in Finland, Santa Claus was the most important reason to choose Rovaniemi as a destination (Tervo 2009). Other very important factors included snow, the image of 'real winter', the Christmas season, activities, reindeer and the image of a child-friendly environment. Respondents were provided with a number of different future climate change-related scenarios for Christmas tourism in Rovaniemi. Even given the presence of Santa, more than half the tourists reported that they would have not travelled to Rovaniemi if there was poor snow reliability. Fewer than a quarter of respondents stated that they would have considered Rovaniemi as a destination if there were no snow. In a scenario of poor snow reliability but with the availability of machine-made snow, approximately half of respondents indicated that they would not select Rovaniemi as a destination. Interestingly, almost 60 per cent of respondents indicated they were willing to pay more for their trip if it would help ensure the availability of snow-based activities (Tervo 2009). Buzinde et al. (2010) discuss how tourism industry representations of 'stable and pristine' beach landscapes are increasingly being challenged by tourists using online social networking sites, raising interesting questions regarding marketing ethics and threshold conditions, with some visitors feeling 'swindled' by the discrepancy between promotional materials and reality, and consumers' willingness to access worse environmental conditions for price discounts. Such results only serve to reinforce that the adaptive behaviour of tourists is insufficiently understood.

With regard to hierarchies, it is equally unclear how single negative experiences may dominate holiday experiences and affect holiday satisfaction. Anecdotal evidence suggests that travellers place great weight on single positive or negative events when recollecting holiday experiences, although such weight may change over time (Andressen and Hall 1988/89). Hierarchies for some aspects related to climate change have been established, however. For instance, there is consensus that selected weather parameters dominate perceptions – rain, for instance, will dominate experiences of summer beach and urban holidays (e.g. Gössling et al. 2006; Scott et al. 2008b). Likewise, weather extremes can be assumed to dominate over other holiday experiences. The range of parameters involved, as well as their interaction, is however still insufficiently understood.

Perceptions are likely to be context-dependent

Depending on the research situation, it can also be assumed that perceptions may vary. For instance, in a prolonged cold winter, a warm climate may be perceived as more desirable than in a warm summer; on a rainy, cold day, snow may become more desirable for a skier than on a 'perfect' snow day. This is of importance for research methodologies, as various *ex situ* surveys assessing the role of different weather parameters for holiday-making do not account for seasonality – the climatic conditions at the time of study (Scott *et al.* 2008b; Moreno 2010; Rutty and Scott 2010). Further research is warranted to quantify this influence.

Accuracy of understanding of weather parameters is insufficiently understood

A wide range of publications have sought to assess the consequences of direct climate change for tourism, based on the assessment of changes in weather parameters (see chapter 5 and discussion later in this chapter). It remains unclear, however, to what degree tourists are able to assess temperatures and other weather parameters accurately in either *in situ* or *ex situ* situations (see chapter 2). This regards both single parameters such as temperature, where it remains to be tested whether respondents can accurately identify observed temperatures *in situ*, and whether they can also distinguish the effect of other parameters, such as humidity and wind, in influencing felt temperatures. Similarly, with regard to transport and related environmental impacts, Higham and Cohen (2011) show that long-haul air travel is perceived as being less climatically harmful than short-haul air travel, even though the contrary is true. Consequently, it would be essential to carry out further studies to understand better the role of such complexities and the accuracy of existing studies in this regard.

Adaptive behaviour is insufficiently understood

As indicated, within the tourism sector tourists are considered to have the greatest capacity to adapt to the risks and opportunities posed by climate change. However, the adaptive capacity of tourists remains largely unexplored. For instance, tourists may learn to accept a 'green' winter environment in the absence of snow, to avoid hot periods of the day by staying in air-conditioned rooms or in areas with coastal breezes, to acclimatize physically over time with warmer average temperatures at home; to adjust their perception of acceptable or preferred environmental conditions (a reduced beach or a drastically diminished glacier becomes the norm); or to focus on a different set of activities supported by prevailing environmental conditions. There remains much scope to understand adaptive capacity better by assessing climate change analogue events and utilizing research techniques from fields such as acceptable limits to change.

Short-term versus longer-term changes in travel behaviour

The impact of many events, such as storms, drought and wildfires, on tourist perceptions of a destination is likely to be short-lived, but some impacts are likely to be more enduring in the minds of consumers and may over time widely alter the perceived attractiveness of a destination. The potential expansion of geographical areas susceptible to the transmission of vector-borne diseases such as malaria and dengue (Hall 2006c; Scott 2006a) to popular tourism destinations where these diseases are not currently prevalent is one such example. How would travellers respond if required to take malaria medication or other preventive procedures in order to go to the Azores, South Africa, Cuba or Mexico in the future? Travellers' responses to media coverage of regional outbreaks and perceived changes in disease risk could also have important implications for travel patterns, and remains an important area for further research.

Various short-term reactions to climate and climate-induced environmental conditions have been reported, including last-minute booking or spontaneous change of destinations (Scott and Lemieux 2009; Nilsson and Gössling 2012). However, many of the changes associated with climate change will occur in the medium- to long-term future. Consequently, the stated behavioural response to climate change impacts in the extant literature usually refers to unknown, hypothetical futures. Some exceptions, such as the study of Dawson and Scott (2010), utilize change scenarios based on recent historical events that tourists may have experienced or read about. Moreover, in longer-term scenarios, destinations, tour operators and activity providers may be able to adapt, possibly in many situations still offering viable and attractive experiences. It is thus questionable whether studies suggesting long-term changes in travel behaviour will prove to predict longer-term changes in travel patterns correctly.

Public perceptions of climate change can be ill-informed and highly polarized

General climate change perceptions of some members of the general public have been found to be ill-informed and highly polarized (compared with terms such as 'political correctness' and 'liberal' in the USA), such that persistent patterns of environmental ignorance and the emotional response to climate change are thought to represent important barriers to behavioural change (Hoffman 2010). Public scepticism toward climate change has implications for how climate-induced changes in the environment are perceived and the willingness of tourists to engage in behavioural changes to reduce the carbon footprint of their holidays (see related discussion by Weaver 2011 and Scott 2011).

> 'At their core, both environmental problems and environmental solutions are organizationally and culturally rooted. While technological and economic activity may be the direct cause of environmentally destructive behavior, individual beliefs, cultural norms and societal institutions guide the development of that activity. ... Unfortunately, the present reality is that we tend to overlook the social dimensions of environmental issues and focus strictly on their technological and economic aspects. ... Pricing alone ignores the critical social context.'
>
> Hoffman (2010: 295)

As these complexities reveal, advancing our understanding of tourists' responses to the various impacts of climate change remains a highly challenging, but fundamental, research area if accurate projections of changes in geographical and temporal patterns of tourism demand are to be possible.

The next section summarizes the available literature on potential tourist responses (at individual and aggregate scales) to the four major climate change impact pathways: direct impacts of climatic change, indirect climate-induced environmental change, indirect socio-economic change, and mitigation policy. The limitations of these studies with respect to the aforementioned complexities are noted where applicable.

TOURIST RESPONSES (ADAPTATION) TO CLIMATE CHANGE

Climate, the natural environment, income and discretionary wealth, personal safety and travel costs influence travel motivations and destination choice. Because all these influential factors are anticipated to be affected by climate change (see chapters 4 and 5), the implications for tourists' behaviour and patterns of demand at the local, national and international scales could be profound. Because the impacts of climate change on tourism operations and destinations are closely intertwined with tourists' behaviour, the following sections examine the potential influence of four major types of climate change impacts on tourism demand: direct impacts of a changed climate; indirect impacts of environmental change; mitigation policy and tourist mobility; and societal change (economic growth and social–political stability).

Tourists' responses to a changing climate

Chapter 2 set out a number of lines of evidence that demonstrated the intrinsic importance of weather and climate for tourist decision-making (motivations, destination choices and timing of travel) and the vacation experience. Consequently, changes in the spatial and temporal distribution of climate resources are anticipated to have important consequences for tourism demand from the global to destination scale, and projecting these changes has been a salient area of research since the early 2000s.

Changes in global demand patterns

Top-down simulation models (Hamilton *et al.* 2005; Berrittella *et al.* 2006) have been used to explore the potential impact of climatic change, in conjunction with a range of other macro-scale factors, such as population growth, per capita income, and other variables included in the IPCC *SRES* scenarios, on aggregate international tourism demand and the potential geographical redistribution of tourist departures and arrivals.

The general geographical patterns of projected changes in tourism flows as projected by Hamilton *et al.* (2005a, 2005b) for 2025 and Berrittella *et al.* (2006) for 2050 are compared in Table 7.2. Broadly, the anticipated impacts include a gradual shift in international tourism demand to higher-latitude countries. Tourists from temperate nations that currently dominate international travel (e.g. northern Europe, Japan, northern USA and Canada) are

Table 7.2 *Regional changes in international arrivals and departures*

Region*	Hamilton et al. (2005): 2025[†]		Berrittella et al. (2006): 2050[†]		
	Arrivals	Departures	Arrivals	Departures	Macro-economic impact
USA	Increase (+)	Decline (+)	Decline (−)	Decline (+/+)	Increase (+)
EU	Varied (+/−)	Varied (+/−)	Decline (−/−)	Decline (+/+)	Decline (−)
EEFSU	Increase (+ to ++)	Decline (+/+)	Increase (+)	Decline (+/+)	Increase (+/+)
JPN	Increase (+)	Decline (+)	Decline (−)	Decline (+)	Increase (+)
RoA1	Varied (− to ++)	Varied (− to ++)	Increase (+/+)	Decline (+/+)	Increase (+/+)
EEx	Decline (− to −/−)	Increase (−)	Decline (−/−)	Increase (−/−)	Decline (−/−)
CHIND	Varied (− to +)	Varied (− to +)	Decline (−)	Decline (+)	Increase (+)
RoW	Decline (− to −/−)	Increase (− to −/−)	Decline (−/−)	Increase (−/−)	Decline (−/−)

*Regional classifications are from Berrittella et al. (2006); country results from Hamilton et al. (2005) have been approximated to these groupings of nations
EU, European Union; EEFSU, Eastern Europe and Former Soviet Union; JPN, Japan; RoA1, rest of Annex 1 nations, i.e. developed nations including Canada, Australia, New Zealand; EEx, energy exporters; CHIND, China and India; RoW, rest of world, i.e. developing nations including Africa, Caribbean, South America, Pacific Islands
[†]Results for approximately 1°C global warming scenario are shown

projected to spend more holidays in their home country or nearby regions, adapting their travel patterns to take advantage of new climatic opportunities closer to home. Demand for international travel to subtropical and tropical nations is projected to decline, with fewer arrivals from temperate nations and increased outbound travel from these nations. These aggregated regional results mask some of the subregional differences observable in Hamilton *et al.* (2005a, 2005b), such as the projected positive impact for Scandinavian nations versus the negative impact for Mediterranean nations of Europe; and the much more positive impacts projected for countries such as Canada and Russia compared with other nations in their groups (e.g. Australia, Eastern Europe).

Berrittella *et al.* (2006) assessed the macro-economic effects of the expected changes in geographical patterns of tourism demand. A net negative macro-economic impact was found for three regions: the European Union; the energy-exporting nations group, which includes many Arab nations; and the 'rest of the world' category of nations, which includes the Caribbean. China and India were expected to be largely unaffected, although with slightly positive changes in tourist numbers. Net gains were projected for North

America, Australasia, Japan, Eastern Europe and the former Soviet Union. It was estimated that the global economic impact of climatic change-induced change in tourism demand could become 'non-negligible' loss under higher emission scenarios (Berrittella *et al.* 2006).

The most recent iteration of these simulation models, by Bigano *et al.* (2007), incorporated indicators of domestic tourism to examine potential trade-offs between domestic and international holidays, and ran model outputs into the later decades of the twenty-first century, where climate change becomes a greater influence on spatial redistribution of tourism demand, but where socio-economic scenarios and environmental impacts of climate change become much more uncertain. Overall, domestic tourism globally is not affected by climatic change because positives and negatives in individual countries cancel each other in the average; however, aggregate international tourism declines slightly as more tourists are projected to stay in their home country. Broad geographical patterns of demand change were similar to those in Table 7.2, with increased tourism projected in currently cooler countries and reduction in warmer countries. Specifically, as a result of climatic change by 2100, domestic tourism may double in colder countries and fall by 20 per cent in warmer countries versus the baseline without climate change. The projected range of impacts from climatic change for international tourism is even greater, where for individual countries, international arrivals may fall by up to 60 per cent of the base value or increase by over 200 per cent.

While the specific projections of models of any social system 50–75 years in the future are highly uncertain, if even remotely accurate, then the implications of climatic change for geographical redistribution of tourism demand are indeed significant. As statements from policy-makers in Sweden and Germany that the Baltic will become the new Mediterranean indicate (see chapter 6), some destinations perceive the climate opportunities for future tourism development to be very large, and raise important questions regarding supply capacity (and carrying capacity) in benefiting destinations as well as over-capacity in destinations adversely affected.

Global-scale simulation models of tourism demand are necessarily highly simplified and have important limitations that are specified by the authors (Hamilton *et al.* 2005a, 2005b; Berrittella *et al.* 2006) and other researchers (Gössling and Hall 2006b; Bigano *et al.* 2006a; Eugenio-Martin and Campos-Soria 2010; Weaver 2011). These studies are well cited in the climate change and tourism literature, the IPCC AR4 and other governmental reports. In virtually all cases, a discussion of the limitations set out by the authors and others has unfortunately been omitted. These limitations and the contrasting results from other studies are important for scholars and policy-makers to be aware of, for they identify key knowledge gaps for further inquiry and provide vital caveats for policy-makers that, in some cases, are anticipating either large gains (e.g. that the Baltic will become the new Mediterranean – *Metro* 2006; Regeringskansliet 2007; Federal Government 2008) or substantial losses in tourism demand (e.g. that the Mediterranean will become too hot for summer tourism, with attendant decline in tourism demand – Parry *et al.* 2007; Ciscar *et al.* 2011). Box 7.2 outlines the specified limitations of these global-scale simulation models, a number of which also apply to country-scale modelling studies that are summarized next.

BOX 7.2 THE CHALLENGES AND UNCERTAINTIES OF GLOBAL-SCALE TOURISM-DEMAND MODELLING

The limitations of global-scale simulation models of tourism demand under climate change have been set out by the authors (see Hamilton and Tol 2007 for a good summary) and further elaborated on in review papers (Gössling and Hall 2006b; Bigano *et al.* 2006a, 2006b; Weaver 2011) as well as other studies that provide insights into some of the model specifications. The main limitations are as follows.

TOURISM DATA

The models utilized the best available tourism data; however, despite considerable progress made in recent decades, international tourism statistics are often not uniform (domestic tourism even more so, as statistical definitions of a tourist can vary by jurisdiction within nations) because definitions and methods of data collection tend to differ from country to country. Databases of international tourists include only total arrivals and departures. Types of tourist (business or leisure travellers) and travel purpose (beach holiday, urban sightseeing–shopping trip, mountain trekking, visiting friends and relatives), which have very different influencing factors (including different climatic requirements), are not differentiated in the data sets and thus must be treated as a singular generic tourist in the demand model. As Lyons *et al.* (2009) conclude from an analysis of micro-trip data, there is no single 'representative tourist' from Ireland available for this type of demand modelling.

MODEL RESOLUTION

Temporally, these models operate on annual timeframes. This has important implications for how variables are specified. Annual climate (proxied as annual mean temperature) has no meaning for tourists, with the possible exception of equatorial destinations, and therefore there is no direct connection between this variable and tourism decision-making. With an annual temporal resolution, an important drawback is that seasonal shifts in demand cannot be modelled. As Maddison (2001) projected, the outcome of climate change for tourism demand in destinations such as Greece was a seasonal shift away from the peak summer season to the spring and autumn months, when temperatures were more consistent with those preferred by the majority of tourists.

Spatially, countries are the unit of analysis, and therefore remarkable variations that exist within large countries (as both push and pull factors for tourism) are masked. Therefore climate (typically proxied as temperature) is averaged for each country in the model (or has used capital cities to represent the climate of the nation). While this may introduce little error for small countries with relatively uniform temperatures (e.g. Barbados and many other small islands), the representation of climate becomes problematic for large, climatically diverse countries and may lead to inaccurate results in subsequent analyses (Scott *et al.* 2008a; Eugenio-Martin and Campos-Soria 2010). For example, in the USA, which has ten distinct Köppen climatic regions within its borders, a single annual temperature regime is representative of the subtropical coastal state of Florida and the mountainous, temperate states such as Colorado. Micro-climates that may be so very important to coastal tourism in warm countries (e.g. the Mediterranean basin) cannot be accounted for at this coarse spatial resolution.

PARAMETERIZATION OF DEMAND-INFLUENCING FACTORS

Statistics-based models used to explain or predict travel flows estimate the aggregate behaviour of tourists as a function of a given set of factors, such as climate, travel distance and income, that are thought to attract or deter international tourists (e.g. coastal features, security risks). These models can only attempt to include factors that influence travel decisions for which there are national data sets for a wide range of countries – for both current conditions and future scenarios. Unfortunately, data for some factors, such as disposable household income and available leisure time, are available only for some countries and there are no long-term scenarios. Therefore these aspects cannot be incorporated in the simulation models at present.

For other factors, proxy variables are used, but the adequacy with which they represent the influence on tourism decision-making remains uncertain. Because temperature has been found to be statistically significant in econometric studies of climate and tourism demand, and is always available at the required resolution in historical and climate change data sets, it has been used as the proxy variable for climate. As discussed in chapter 2 and acknowledged by the modelling teams, climate is of course more complex than just temperature, and tourists consider a range of meteorological variables in their decision-making. Depending on the type of tourism destination being considered, the relative importance of weather variables has been found to differ (Scott *et al.* 2008b; Credoc 2009; Rutty and Scott 2010), and these findings contrast with the assumption that temperature is the most important weather parameter in these simulation models. Furthermore, while changes in average temperatures have been considered in the models, the consequences of changes in climate variability (e.g. temperature extremes such as heatwaves and cold waves) are not considered. Experiences with such phenomena in Europe in recent years suggest that extremes can have a substantial influence on travel patterns. The foundations of these models are based on the conclusion that '[p]eople from any country prefer the same climate for their holidays' (Bigano *et al.* 2006a: 399), but other statistical models of tourism demand (Eugenio-Martin and Campos-Soria 2010) as well as stated preference studies (Scott *et al.* 2008b; Denstadli *et al.* 2011) indicate that climatic preferences vary to some extent by nation and cultural group, and this has not been incorporated adequately into existing global tourism demand-simulation models. Resolving this apparent contradiction is an important area for future research.

Coastal features are another example. Coasts are undoubtedly important to many tourists and length of coastlines is used to represent this aspect. As Bigano *et al.* (2006a: 393) note, 'coastline is not a proxy for beach length'. The geomorphological characteristics of the coastline, e.g. the proportion of sandy beaches versus rocky cliffs, and the climatic and water temperatures have highly significant implications for whether or not length of coastline is a major tourism asset. For example, Russia and Canada have the longest coastlines in the world, but because of air and water temperature, these coasts do not have nearly the same value for tourism as the coastlines of even some of the smallest tropical islands, such as Barbados or the Seychelles. Coastlines are also projected to change substantially under climate change as a result of coastal inundation and erosion from sea-level rise (SLR; see chapter 5). The impact of SLR on coastal resources is not represented in these simulation models, nor is any other form of climate change-induced environmental change expected to affect tourism destinations (see chapter 5).

In other cases, the range of scenarios for key variables has been fully explored. Mobility is a precondition for tourism, and scenarios of future costs of transportation (particularly air travel, which is central to international tourism) vary widely under different assumptions of the future price of oil and costs associated with the stringent mitigation policies needed to avoid 'dangerous' climate change (see chapter 4). The range of mobility futures for scenarios run through to the 2050s has not been considered adequately in tourism demand simulations thus far (Gössling and Hall 2006a).

Finally, there are some factors that will have a major impact on tourism in some nations, but are fundamentally unpredictable (Gössling and Hall 2006a). The impact of these random events (e.g. acts of terrorism, natural disasters such as earthquakes and tsunami, political revolutions) cannot be modelled easily. Bigano *et al.* (2007: 28) suggest that 'Such events may wreck short-term predictions, but their effect on long-term trends is much smaller.'

THE ROLE OF TOURISTS' PERCEPTIONS

Bigano *et al.* (2006a: 393) recognize that, while these models utilize objective explanatory variables, 'what matters is the perception of the tourist. As tourist perceptions are not regularly measured, we assume that the perceived status of the destination is close to the "real" status.' As we set out at the beginning of this chapter, there are important uncertainties with respect to the role of perceptions in tourist decision-making and whether the assumption of 'perfect information' is valid for some factors. These complexities are not captured in existing models. Shaw and Loomis (2008: 262) point out that all rigorous macro-economic models must be informed by what individuals do, as they represent aggregate tourist behaviour, and therefore 'be consistent with microeconomic model predictions of what individuals do in response to climate change'. Yet currently there appear to be a number of important contrasts between behavioural studies of tourists (perceptions of preferred climatic conditions, reactions to beach erosion and other environmental change) and the assumptions of macro-scale models. Research is needed to better support or refine the postulated relationships in the simulation models with the richness of finer-resolution behavioural or micro-economic studies.

RELIANCE ON OTHER MODELS

Global tourism-demand models that explore the impacts of climate change rely on other models for data inputs or other model specifications, including the FUND, IMAGE and COSMIC models developed for climate change integrated assessments. The data sets on economic growth projections for specific countries will need frequent updating to account for economic events, such as those that have unfolded in some Eurozone nations (e.g. Ireland, Iceland, Greece) as well as political events, such as the revolutionary wave that has taken place in many Arabic nations. Integrated assessment models are subject to a range of critiques regarding data, variable parameterization and model assumptions (Ackerman *et al.* 2009; Stanton *et al.* 2009; Ackerman and Munitz 2011; Cullenward *et al.* 2011; van Vuuren *et al.* 2009, 2011).

Changes in national demand patterns

The possible response of tourists to changing climate conditions and the implications for demand have also been assessed at the regional–national scale and for a range of specific tourism subsectors. These analyses overcome some of the limitations of global-scale studies set out in Box 7.2 (monthly temporal resolution, incorporation of multiple climate variables). Maddison (2001) used a pooled travel-cost model to examine the implications of climatic change for the flow of British tourists (with the assumption the British climate was unchanged) in the 2030s, and found total visits to Greece remained largely unchanged, but the seasonality of visits was altered substantially (greater in spring and autumn shoulders, declines in summer). Results were similar for Spain, although with greater potential for increased visits in all seasons except the summer months.

National-scale tourism demand-simulation models based on the global-scale models discussed previously have been implemented for the UK, Ireland and Germany (Hamilton and Tol 2007). It is concluded that tourists from all three nations would spend more holidays in their home country, increasing domestic tourism. There is some support for this pattern from anomalously warm summers in the UK (Giles and Perry 1998; Agnew and Palutikof 2006), where domestic tourism increased while outbound tourism decreased. Tourism in the northern regions of the UK and Ireland are projected to benefit more extensively from greater demand, while in Germany the interior regions benefit more as they warm more quickly than coastal regions. International arrivals into these nations are projected to decrease at first, as tourists from other Western European nations also stay closer to home, but then to increase with more arrivals from 'increasingly rich tropical nations' (Hamilton and Tol 2007: 161). Some have questioned whether greater numbers of international tourists from subtropical and tropical regions will emerge, given the projected broad impacts of climate change on their national economies and on the tourism sector in particular.

A similar approach was used by Moore (2010) to examine the implications of climatic change on travel demand to the Caribbean region. The regional impact was projected to be relatively minor (–1 to +3 per cent) even under climate scenarios for the later decades of the twenty-first century. Results for individual countries were more varied, but none exceeded ±10 per cent, suggesting that changes in temperature and precipitation (as modelled) would be relatively inconsequential to decisions to visit the region.

Other studies have used different techniques to explore the impacts of climatic change for tourism demand to specific countries. Hein *et al.* (2009) utilized a relationship between the tourism climatic index and tourist arrivals to Spain, and estimated that climate change would decrease annual tourist flows 5–14 per cent under scenarios for the 2060s. In Switzerland, Serquet and Rebetez (2011) found positive correlations between hot summer temperatures at lower elevations (cities in mountain valleys) and the number of overnight stays at mountain resorts at higher elevations. Mountain resorts in closest proximity to major urban centres experienced greater effects. They concluded that, if tourists already respond on a short-term basis to hot summer temperatures, then as the frequency of higher temperatures increases under climate change, tourists are likely to travel to alpine resorts more frequently and for longer periods.

Changes in subsector demand patterns: park, ski and beach tourism

Subnational, destination-specific studies have also been conducted to establish the influence of climate and climate change on specific tourism market segments. Illustrative of the expanded opportunities associated with an improved and expanded warm-weather tourism season in temperate regions is projected nature-based tourism in Canada's system of national parks. Tourism in many of the parks in Canada and the northern USA is constrained by winter conditions, except where skiing operations exist. Assuming other socio-economic factors that influence park visitation (e.g. desire to interact with nature, travel costs, amenity requirements) remain relatively constant, Canada's national parks are projected to have higher visitation under a warmer climate. With the assumption that there would be little change in other factors that influence park visitation patterns, total annual visits to the national parks analysed by Jones and Scott (2006) were projected to increase by approximately 6–8 per cent by 2035. Annual visits by 2050 could increase between 9 and 29 per cent. Although visitation to Canada's national parks is projected to increase system-wide under climate change, important regional and seasonal differences were projected. Parks on the coast of the Atlantic Ocean were projected to experience the largest increases, with visitation levels potentially doubling under the warmest scenarios. Most of the parks are projected to experience the largest increases in visitation during the spring (April to June) and autumn (September to November) shoulder seasons, with minimal increase during the traditional peak months of July and August. Golf tourism in and around some of these national parks was projected to increase even more with similar geographical and seasonal patterns, for example, 27 to 61 per cent in the Great Lakes region; 48 to 74 per cent in the Atlantic Coast region (Scott and Jones 2007). The projected increase in visitation (an additional 1–4 million visitors annually by mid-century) would generate substantial growth in park revenues and the economies of nearby communities. Increased visitation could also exacerbate visitor-related ecological pressures and crowding issues at popular park attractions in high-visitation parks. Under a changed climate, more intensive visitor-management strategies may be required to support sustainable tourism in these parks and prevent the degradation of visitor experience.

A similar study of potential visitation changes to US national parks, representing 12 different climate regimes across the country, found much more diverse impacts (Hyslop 2007). The number of visits to parks in the northern regions was projected to increase annually, with the majority of increases occurring in the spring and autumn shoulder seasons. In contrast, parks in the south are projected to experience decreased annual visitation as temperatures become uncomfortably hot, particularly under high emissions scenarios. Small changes in visitation (up to 10 per cent annually) were projected under low emission scenarios, while larger visitation changes (up to 47 per cent annually) resulted out of higher emission scenarios.

As discussed in chapter 5, winter sports tourists are faced with the prospect of less natural snow fall and shorter, more variable ski seasons in the future. A critical question for the future of the ski industry and closely associated tourism operators is how tourists might respond to these changed conditions and what will happen to overall demand in the ski tourism marketplace. Insights into these questions come from studies using a range of techniques (tourist surveys, statistical analyses, climate change analogues) in several nations.

Marketplace surveys have been conducted with skiers in several countries to examine how skiers might change their behaviour if climate change conditions were realized. König (1998) conducted the first ski tourist survey in Australian ski resorts. Faced with a scenario described as 'the next five winters would have very little natural snow', 25 per cent of respondents indicated they would continue to ski as often in Australia, while 31 per cent said they would ski less often, but still in Australia. An even greater portion of skiers (38 per cent) indicated they would substitute destinations and ski overseas (mainly in

New Zealand and Canada), and a further 6 per cent would quit skiing altogether (König 1998). Pickering *et al.* (2010) repeated König's (1998) survey in 2007 to examine if skiers' perceptions of climate change had changed. In the repeat survey, 90 per cent of skiers stated that they would ski less often in Australian resorts if the next five years had low natural snow (up from 75 per cent of skiers surveyed in 1996), 69 per cent would ski less often, 5 per cent would give up, and 16 per cent would ski at the same level but overseas. Nearly all skiers thought that climate change would affect the ski industry (87 per cent, compared with 78 per cent in 1996), and that this would occur sooner than the perception of respondents in the 1996 survey. Fifty-nine per cent of respondents believed this would take effect before 2030, 25 per cent before 2060, and 29 per cent between 2060 and 2100 (Pickering *et al.* 2010).

A survey using König's (1998) hypothetical future scenario of skiing conditions conducted at resorts in Switzerland found that 30 per cent of respondents would not change their skiing behaviour, 11 per cent would ski at the same location but less often, 28 per cent would ski at a more snow-reliable resort at the same frequency, 21 per cent would ski at a higher resort less often, and 4 per cent would give up skiing altogether (Behringer *et al.* 2000). A survey of skiers in Austria (Unbehaun *et al.* 2008) found that, if there were 'several consecutive years of snow deficiency', 68 per cent of respondents were willing to travel further to a snow-secure destination and 25 per cent would no longer ski. When asked to trade off additional costs and additional travel distances for a snow-secure destination, the majority of winter sports tourists are willing to incur some additional cost, but the majority reach thresholds at about 10 per cent additional cost and two hours' additional driving. Snow-independent activities were accepted as temporary substitutes, but not for the entire winter holiday.

Three notable limitations of these initial attempts to examine climate change vulnerability of skier demand include the descriptive scenario of future ski seasons and snow condi-tions, and whether the stated adaptive behaviours of skiers to future conditions are any different from responses to historical conditions. The climate change scenarios used by König (1998) and Behringer *et al.* (2000) asked, 'Assuming that the next five winters had very little natural snow: please circle one option as to where you would ski/snowboard.' How is 'very little snow' to be interpreted? Does that mean snow fall has been so deficient that ski areas did not open at all, or are open only half as long as usual, or are open an average length of time but have poor conditions (e.g. bare patches), or that ski areas are open an average length of time but had to rely heavily on snow-making? Depending on how the respondent perceived this scenario, the response could be very different. These studies did not examine how individuals have responded to marginal snow conditions in the past. Because of this omission, we do not have a reasonable understanding of whether a net difference in behaviour change should be expected in the future. If a high proportion of respondents are already engaging in certain substitution behaviours, and a similar pro-portion indicate that they will engage in the same type of substitution behaviour in the future, then demand impacts would be similar to those seen during marginal seasons of the past, and climate change analogue seasons would provide important insights into future demand changes. None of these surveys differentiated the stated behavioural responses of various market segments, which is one of the strengths offered by survey techniques over statistical analyses of visitation data. Scott *et al.* (2008a) and Dawson and Scott (2010) have suggested that available tourism and recreation concepts that could greatly enhance our understanding of climate-induced behaviour change (e.g. substitution, specialization and destination loyalty) have not been explored adequately.

Surveys in the New England region of the USA have sought to address these limitations. Scott and Vivian (unpublished data) presented skiers with a scenario that resembled the record warm winter of 2001–02 that many would have experienced, as half of the respondents had been skiing for more than 20 years. The scenario provided quantitative description of natural snow fall, the length of the season and conditions during key holiday periods:

> Average snow fall at ski resorts in New England is 40% lower than normal, with little snow fall until after Christmas, except at high elevations. The ski season in the region starts 3 weeks late and is only 118 days long (an average winter is 132 days). A number of resorts have only some of their runs and chairlifts open over Christmas–New Year and Spring Break holidays.

When asked if another such winter were to occur, 87 per cent would ski their usual frequency, with fewer total days over the winter; 11 per cent would ski more often than normal to make up for a shorter season; 1 per cent would stop skiing for the entire winter; and none of the respondents indicated they would stop skiing altogether. When asked where they would ski, 60 per cent stated that they would wait until their usual ski area(s) in New England were open; 23 would travel further to find better snow conditions within the region; 9 per cent would take a ski holiday outside New England instead; and 3 per cent said they would do something else instead of skiing. These results appear to differ qualitatively from earlier surveys and are more consistent with observed demand responses in the region during climate change analogue winters (see below).

A second climate scenario was based on ski operations modelling using the worst-case scenario for the 2050s high emissions model (from Scott *et al.* 2008c) and quantitatively described the additional change anticipated to ski season length, and where in the region ski areas would be operating:

> Average snow fall at ski resorts in New England is 65% lower than normal, with little snow fall until after Christmas, except at the highest elevations. The ski season in the region starts 4 weeks late and is only 99 days long (an average winter is 132 days). Many resorts in the region are closed over Christmas–New Year and Spring Break holidays. Only ski/snowboard areas in Vermont, Northern New Hampshire, and Northern Maine have good snow conditions.

Under this scenario, behavioural adaptation increased, but not markedly: 84 per cent would ski their usual frequency and fewer total number of days; 11 per cent would ski more often than normal to make up for a shorter season; 4 per cent would stop skiing for the entire winter; and 1 per cent would stop skiing altogether. Destination substitution also increased slightly, with 30 per cent stating they would travel further to find better snow conditions within New England and 11 per cent taking their ski holiday outside of the region. The limited variation between responses during past conditions and future seasons indicates that behavioural adaptation to future climate change may be remarkably similar to what has been observed in recent analogously warm winter seasons (Dawson *et al.* 2009; Scott and Vivian unpublished data).

This survey, and also that by Dawson *et al.* (2011a), found significant differences in the behavioural adaptation of skier market segments. Those highly committed to skiing were more likely to ski more often than normal to make up for a shorter winter season. Skiers with high destination loyalty were more willing to wait until their usual resort in New England opened to go skiing, and less likely to substitute place. Understandably, those with significant capital investment in ownership of a recreational property in the area were less likely to engage in place substitution and much more likely to wait until their usual ski resorts opened. If the average ski-season length was reduced by one-third over a normal winter (similar to the record warm winter of 2001–02), 5 per cent of recreation property owners would use the property more often, 57 per cent would use the property about the same amount as they do on average, 30 per cent would use the property less often, 7 per cent a lot less often, and 1 per cent would sell the property. If the ski area that their property was near closed as a result of consecutive poor winter conditions (some ski resort closures are projected in the region by Scott *et al.* 2008c and Dawson and Scott 2010): 28 per cent would use the property about the same amount as they do now, 12 per cent would use the property a little less often, 33 per cent a lot less often, and 21 per cent would sell the property. The implications of reduced ski operations or closures for the value of recreational properties in the region is clear and probably can be generalized to other ski regions or coastal regions that could see beach extent and quality decline.

A survey of tourists to New Zealand ski areas in 2009 indicated that visitors were very aware that climate change would affect their participation in winter sports activities (Prince 2010). Respondents mostly agreed that the climate in New Zealand and overseas is changing and that the effects on snow conditions would be negative over time ('less snow', 'shorter snow season', 'worsening snow conditions', 'higher snowline'). Very few visitors (2 per cent) thought it would have no effect(s) on snow conditions at New Zealand ski fields. When questioned about the effects of skiing under conditions with very little natural snow at the New Zealand ski fields, most respondents indicated that they were willing to continue. New Zealand visitors were more willing still to ski than international visitors (58 versus 38 per cent). Arguably, this is in part because of the increased financial costs in travelling overseas, especially because many respondents had indicated that they were already opposed to any increase in ticket prices for the use of more snow-making. This could signal they are more likely to be flexible in time (e.g. ski only when there is good snow coverage) and space (e.g. between different local ski fields) as a form of behavioural adaptation to changes in climate, weather and snow conditions.

Climate change analogues have also been used to explore the impact on ski tourism demand of marginal conditions and shortened ski seasons. In the Canadian provinces of Ontario and Quebec, the record warm winter of 2001–02 resulted in reductions in skier visits of between 7 and 15 per cent compared with a climatically normal winter (for the 1961–90 period) (Scott 2006b; Scott *et al.* 2011). In the nearby New England region of the USA, skier visits were 11 per cent lower during the same record warm winter (Dawson *et al.* 2009). Steiger (2011b) examined the impact of the extraordinarily record warm winter of 2006–07 in the Tyrol region of Austria, and found ski-lift transports were 11 per cent lower than the average of the three previous years. This finding stands in sharp contrast to the stated responses found in skier surveys in the region (Behringer *et al.* 2000; Unbehaun *et al.* 2008), and suggests the need for further investigation into this discrepancy. While the impact of this analogue winter was relatively minor across the region, major differences were observed in the impact on demand at different ski operators. Ski-lift transports were 44 per cent lower at small ski areas and 33 per cent lower at low-lying ski areas, while conversely the impact on extra-large ski areas and high-altitude ski areas (–6 and –2 per cent, respectively) was negligible.

Statistical approaches have also been used to examine the impact of climate change on ski tourism demand. Fukushima *et al.* (2003) utilized the relationships between snow fall and skier visits from three ski areas in Japan to estimate that reduced snow fall resulting from a 3°C warming scenario would reduce overall skier visits in the country by 30 per cent. Ski areas in southern regions were considered the most vulnerable, with skier visits falling by 50 per cent. Conversely, the impact of climate change on skier demand was projected to be negligible in northern high-altitude ski areas.

Adaptive responses have been examined in other winter sports tourism markets as well. In the multi-billion-dollar North American snowmobile industry, sales of equipment has been used as a proxy for future demand (Scott 2006b). Recent market trends in new snowmobile and all-terrain vehicles (ATVs) sales are starting to reflect adaptation to shorter and more variable winter trail seasons. Since the 1970s, new snowmobile sales have exceeded ATV sales in Canada, as recently as 1996 by a 2.5:1 margin (Suthey Holler Associates 2003). In 1999–2000, ATVs outsold snowmobiles for the first time, and in 2004 ATV sales approximately doubled those of snowmobiles. Snowmobile and ATV sales in the northern United States could be considered a spatial analogue for Canada under a warmer climate. With a shorter winter season in this region, ATV sales exceeded snowmobiles by over 4:1 in 2002, and by over 8:1 across the United States (Suthey Holler Associates 2003).

Landauer and Pröbstl (2008) postulate, for instance, that cross-country skiers have a strong preference for natural winter landscape experiences, ruling out the use of indoor skiing or alternative skiing regions. In contrast, snow-making is accepted by the broad majority of respondents, which could possibly be seen as an adaptation over time, as snow-making has been considered a technical solution to insufficient snow in some regions since the 1980s (see chapter 5). However, should winter landscapes change from white to green, and precipitation mostly fall as rain rather than snow, it is unclear whether there may be further acceptance of so-far unwarranted technical adaptation solutions.

The final destination type for which there is more limited evidence of how direct climatic changes may influence tourism demand is coastal beach tourism. Analyses of tourist visitation and spending patterns during the 2003 European heatwave indicated that demand for beach tourism was stimulated as travellers sought relief from the heat of large cities on the coasts (Létard *et al.* 2004; Martinez Ibarra 2011). Similar patterns have been observed in other nations during exceptionally warm periods (Bigano *et al.* 2005; McEvoy *et al.* 2008; Scott *et al.* 2008a).

Tourists' responses to climate-induced environmental change

Although climatic conditions are important in affecting tourist demand and behaviour, it is important to emphasize that it is the holistic impact of climate change on tourism environments that tourists will respond to. Tourists' perceptions of environmental change brought about by global climate change will be particularly important for destinations where nature-based tourism is a primary tourism segment and where ecosystems are highly sensitive to climatic change (Gössling and Hall 2006a; Scott 2006a; Scott *et al.* 2008a; Hall and Lew 2009). As Prideaux *et al.* (2010) note, climate change will force a transformation of how tourists perceive landscapes. This section focuses on three sensitive destination types – alpine landscapes, beaches and coral reefs – as well as specific changes in biodiversity.

Alpine landscapes

The perceived quality of the alpine environment is an important attraction for tourism in mountain regions, and there have been repeated warnings about the potential negative effects of environmental change in reducing the attractiveness of mountain landscapes, to the extent that there may be an adverse impact on tourism (see chapter 5). There are only a few studies that explicitly focus on the perception and assessment of the temporal change of mountain landscapes, which is somewhat surprising, given the role of alpine landscape features, and glaciers in particular, as a motif of climate change (Brönnimann 2002). Nevertheless, there is growing recognition that landscape management in order to meet the needs of a range of stakeholders, including tourists, is going to be increasingly needed in order to respond to the impacts of climate change (Höchtl *et al.* 2005).

The limited knowledge of how tourists might respond to changes in sensitive mountain landscapes comes mainly from the Rocky Mountain region of North America. When presented with hypothetical scenarios of how the climate and landscape (including physical and biological changes) might be affected by climate change over the next 30–50 years, the responses of tourists in three national parks (Rocky Mountain National Park, USA; Waterton-Glacier International Peace Park, USA and Canada; Banff National Park, Canada) were remarkably consistent. When asked how the specified changes in the mountain environments anticipated to occur in the early to mid-twenty-first century would have affected their intention to visit that park, the majority of visitors indicated that they would not change their visitation patterns. In Rocky Mountain National Park, visitor response to the environmental change scenarios for the 2020s and 2030s resulted in a 10–14 per cent increase in annual visitation under scenarios of moderate warming, while the 'extreme heat' scenario would cause a 9 per cent decline in visitation (Richardson and Loomis 2004). The environmental change scenarios constructed for similar timeframes were also found to have minimal influence on the intention to visit Waterton-Glacier International Peace Park or Banff National Park, with almost all visitors still intending to visit the parks, and 10 per cent indicating they would visit them more often, presumably due to improved climatic conditions (Scott and Jones 2005; Scott *et al.* 2007b). There is also the potential that media coverage of melting glaciers might motivate more people to visit these parks over the next 20–30 years to personally see the changes to these landscapes (see Box 7.1) or show children the impacts of climate change on the landscape.

In the studies that examined the potential impacts of greater environmental change, an important threshold was reached for many visitors to Waterton-Glacier International Peace Park and Banff National Park in scenarios that might occur by the end of the twenty-first century (Scott *et al.* 2008d). A substantial number of tourists (19 per cent in Waterton-Glacier and 31 per cent in Banff) indicated they would not visit the parks if the specified environmental changes occurred. The projected loss of glaciers in the region was noted as a significant heritage loss and the most important reason cited for not intending to visit the park in the future. Another 36–38 per cent of tourists indicated they would plan to visit less often. Visitors most likely to be negatively affected by climate-induced environmental change were long-haul tourists and ecotourists, motivated by the opportunity to view pristine mountain landscapes and wildlife. The impact of environmental change was more pronounced in Banff National Park, which has a much greater proportion of international tourists.

Similar results have been recorded in other mountain regions. A survey that specifically investigated the impact of a retreating glacier that is a main attraction in Lijiang, China showed that 20–43 per cent of domestic tourists would not visit the destination in the absence of the glacier (Yuan *et al.* 2006). In a completely different tropical mountain rainforest landscape in Queensland, Australia, Prideaux *et al.* (2010) used the substantial damage from a recent category 5 tropical cyclone as a proxy to explore tourists' perceptions of landscape change under climate change (rather than scenario-based approaches used in the North America studies). In the survey three months after the cyclone, 54 per cent of tourists reported noticing substantial damage to the landscape, but by 15 months only 6 per cent reported noticing impacts. They concluded that to most tourists (largely from urban areas), scenic value would not change greatly once obvious physical damage to the forest is obscured by new growth and the impression of a vibrant forest is restored. Therefore the impact of cyclones, and by extension climate change, on tourist demand was anticipated to be minimal. The use of similar climate change analogues, as long as the impacts are similar to what is anticipated under climate change, deserves greater attention by researchers in other mountain regions of the world.

If realized, such reductions in visitation would have major impacts on local economies. However, Scott *et al.* (2007b, 2008d) included an important caveat in their discussion of the potential impacts on tourism demand of long-range climate-induced environmental changes. They note that the perception of some contemporary visitors that the landscape has been degraded from some former state if the specified environmental changes occurred may not be shared by a visitor born in the 2040s, who has no experience of the former condition. The twentieth century offers some historical analogues for such visitor perceptions of changing mountain landscapes. For example, ice caves in Glacier National Park, the US portion of Waterton-Glacier International Peace Park, were an important tourist attraction in the 1930s, but melted decades ago, so that contemporary tourists have no experience of these ice caves or any perception that the park landscape is less attractive than it was 80 years ago. More generally, glaciers have been melting and vegetation responding to warmer temperatures throughout mountain destinations over the latter half of the twentieth century, and these environmental changes have had no known impact on visitation levels to these regions. Contemporary visitors still value these mountain landscapes and the recreation opportunities they provide, even though they are different from those of previous decades (Scott *et al.* 2007b, 2008d). It seems, therefore, that despite concerns that climate-induced environmental change may have an adverse impact on the aesthetics and hazards in mountain tourism destinations (see chapter 5), the temporal scale of these changes is such that, with the exception of a few high-profile tourism attractions (e.g. well known glaciers or snow-capped peaks such as Mount Kilimanjaro), or some key tourist market segments (e.g. expert mountaineers), the eventual impact on visitation to mountain destinations may actually be minor as the frame of reference of mountain landscapes evolves for future generations of tourists.

This interpretation is, of course, debatable, as others might argue that landscape values and preferences are relatively stable between contiguous generations. For the next generation at least, there is likely to be substantive transfer of knowledge of alpine landscape change from previous generations and through interpretive programmes, for example, markers on the ground that track the terminus of glaciers each year, or historical photographs in interpretive centres and visitor brochures. It therefore remains uncertain to what extent the stated behavioural intentions of contemporary visitors would translate similarly to visitors a generation or two from now.

Another dimension of landscape change in alpine destinations that has not been examined in the literature is the potential impact of abandoned ski areas and landscape change in alpine areas. As mentioned in chapter 5, income from winter ski tourism is an important form of livelihood diversification for many alpine farms in Italy, Switzerland and Austria, and the decline in ski tourism would have secondary impacts on the sustainability of these farms in an era of declining agricultural subsidies. These cultural alpine landscapes have developed over centuries and are increasingly appreciated as heritage and tourism commodities. However, alpine agriculture in these countries is in transition, raising some concern about landscape aesthetics as farmlands are abandoned and allowed to revert back to forest (Höchtl *et al.* 2005; Soliva *et al.* 2010). Using photo manipulations of an alpine region depicting different land-use conditions, Soliva *et al.* (2010) found that Swiss respondents visually preferred low-intensity land use over reforested landscapes, and that spontaneous (natural) reforestation was less liked at higher elevations. The broader literature on alpine landscape aesthetics and land-use change has revealed differing perceptions among nationalities and generations (e.g. Höchtl *et al.* 2005; Soliva *et al.* 2010). These studies have focused almost exclusively on local residents and those from alpine regions, and have not yet examined international tourist perceptions or any potential impact on intention to visit, so the implications of climate change for ecological change and the secondary impacts of abandoned ski areas/pastures for mountain destinations remain uncertain.

Coral reefs

As discussed in chapter 5, coral reefs are an ecosystem that is extremely vulnerable to global environmental change, mainly through increasing acidification and higher sea temperatures, but also through SLR, changes in UV exposure, and increased extreme events – all as a result of CO_2 emissions and associated climate change – causing coral bleaching and spread of disease (Hoegh-Guldberg *et al.* 2007; Baker *et al.* 2008). These climate change-related pressures are often exacerbated by other human impacts on the coastal and marine environments within which reefs are situated (e.g. pollution, overfishing, damage by tourists). Some 20 per cent of the world's coral reefs have already been lost, and another 50 per cent are at immediate risk of extinction (Riegl *et al.* 2009).

The impact of climate change-related reef degradation and loss on dive tourism and tourism destinations remains uncertain (Gössling *et al.* 2007b), but because reef health is thought to be important to the experience and satisfaction of dive tourists (Fenton *et al.* 1998; Roman *et al.* 2007) it has been suggested that the impacts of coral reef bleaching on dive tourism could be highly significant. Unlike mountain landscapes, where environmental change has only begun and will need to continue for decades in some locations before it is highly noticable to tourists, the major coral-bleaching events that have occurred in the late 1990s and 2000s have provided scholars with important opportunities to study the impact of real events on tourist behaviour (see also Box 6.2 on climate change analogues). These studies have focused on determining the level of awareness of dive tourists, the level of impact on their satisfaction and willingness to return to the destination if similar events occurred, and how various market segments were affected differently.

A number of studies have examined the impacts of the 1998 bleaching event, which was the largest and most severe recorded worldwide. In El Nido, the Philippines and nearby islands, severe coral bleaching in 1998 led to 30–50 per cent coral mortality, and a typhoon that same year caused further damage to local reefs. Whether divers or not, most tourists (95 per cent) coming to El Nido have at least some interest in the local marine environment (Cesar 2000). However, general awareness of coral bleaching among tourists was limited (44 per cent). The bleaching event did not affect budget tourist arrivals, but fewer budget tourists went diving during their stay. The impact at resorts, some of which cater to the high-end dive tourism market, was much worse. The annual economic losses were estimated to be US$6–8.4 million over the next ten years, concurrent with the anticipated coral recovery timeframe. If, however, there is no significant coral recovery, or a bleaching event re-occurs, the economic losses were projected to increased to between US$15–27 million (Cesar 2000).

Similar studies in 1999 for the destinations of Zanzibar, Tanzania and Mombasa, Kenya also found that awareness of bleaching among tourists was low in Zanzibar (28 per cent) and limited in Mombasa (45 per cent) (Westmacott *et al.* 2000). Of those aware of the bleaching, 80 per cent stated that knowledge of bleaching in the area would affect their decision to visit the region or to dive there. The availability of a previous valuation study of the coral reefs for Zanzibar tourism by Andersson (1997) just prior to the bleaching event provided an excellent 'before-and-after' comparison to examine if consumers' perceptions of the value of their holiday to Zanzibar had changed as a result of the bleaching event. Analysis found that the level of reef use was comparable before (1996/97) and after (1999) the bleaching event, and the consumer surplus remained largely unchanged (Westmacott *et al.* 2000). Interestingly, a second survey of tourists in Zanzibar in 2001 found much higher level of awareness of coral bleaching (72 per cent) and that tourists would have been willing to pay US$84 to visit reefs in the area that were not bleached, suggesting a loss of economic value between US$1.6 and 4.8 million (Ngazy *et al.* 2002). Andersson's (2007) comparison of tourism perceptions at the East African islands of Zanzibar and Mafia found that serious divers who visit Mafia were more aware of bleaching than the inexperienced recreational divers who dominate Zanzibar (62 versus 29 per cent), again highlighting how market segments will be affected differently by the impacts of climate change. When asked if they would be willing to dive on a bleached reef, a non-negligible minority indicated they would (40 per cent at Zanzibar and 33 per cent at Mafia), suggesting there may still be a market for degraded reefs for some time.

Analysis of the impacts of the same bleaching event in the Maldives produced comparable findings. Surveys of European travellers on their way to the Maldives showed low to moderate awareness of recent bleaching damage (awareness ranged from 15 per cent of Dutch travellers to 50 per cent of German travellers) (Westmacott *et al.* 2000). No decline in tourism arrivals to the Maldives was observed in the year after the bleaching event, although the growth was not as strong as expected, suggesting there may have been up to a 2 per cent decline in arrivals due to the bleaching (up to US$63 million over two years). When asked about the most disappointing aspect of their holiday to the Maldives, dead corals was the most frequent response (48 per cent; see quote below). In an effort to establish the perceived lost economic value of coral bleaching, tourists were also asked how much extra they would have been willing to pay to go to areas of the Maldives that were not affected by coral reef bleaching. Divers were willing to pay an extra US$319, while non-divers were willing to pay US$261, for an estimated consumer welfare loss of over US$19 million as a result of the bleaching.

> 'This is a warning to anyone, like me, who loves to snorkel on coral. THERE IS NO LIVE CORAL IN THE MALDIVES. Those photographs showing vivid gardens of scarlet, purple and blue coral were taken before El Niño raised the water temperature and bleached all the coral. What is left is dead looking, grey, brown or yellow. This means that all the interesting little reef fish that used to inhabit the coral are also gone. Yes, there are a few brightly coloured fish around but that is not what I came to see. Apparently, there is some live coral below 20 metres but that is only accessible to divers.'
>
> 'Lack of live coral in the Maldives', posting on Tripadvisor.com
> (6 December 2006)

In other coastal destinations, the impact of climate change was also projected to adversely affect tourist preferences for these destinations. For the Great Barrier Reef (GBR), Prideaux *et al.* (2009) conducted a survey of visitors at a nearby domestic airport and asked if they would consider visiting the region if coral bleaching occurred. A small proportion (13 per cent) stated that they would not revisit the region, while a much larger proportion (41 per cent) indicated they were not sure. Similarly, in a survey of divers and snorkellers in North Queensland, Kragt *et al.* (2009) determined that the number of annual trips undertaken by these visitors would fall by 80 per cent in the event of 'reef quality decline' due to bleaching (a scenario described as 80 per cent decrease in coral cover, 30 per cent decrease in coral diversity, and 70 per cent decrease in fish diversity). As a result of such studies, the Great Barrier Reef Foundation (2009) suggests that at least 50 per cent of international and domestic overnight tourists who visited the GBR coral sites in the course of their trip would be unlikely to have made their trip if the GBR sites had suffered permanent bleaching.

In their case study of Mauritius, Gössling *et al.* (2007b) found that there may be large differences between dive destinations. They found that the state of coral reefs was largely irrelevant to dive tourists and snorkellers, as long as a certain threshold level was not exceeded. This threshold level was defined by visibility, abundance and variety of species, and the occurrence of algae or physically damaged corals, and was not exceeded in the case study despite the fact that considerable damage had already occurred. This is consistent with the findings of Main and Dearden (2007) that 85 per cent of recreational divers failed to perceive any damage to reefs in Phuket, Thailand after the 2004 Indian Ocean tsunami.

In the Caribbean, 76 per cent of tourists at the diving destination of Bonaire (where 99 per cent of respondents took at least one dive during their trip) indicated they would be unwilling to return for the same holiday price in the event that corals suffered 'severe bleaching and mortality' (Uyarra *et al.* 2005). The impact on the 3S destination of Barbados was much lower, where only 26 per cent indicated they would be unwilling to return for the same price. Importantly, many respondents indicated they would return to both islands for a lower price, suggesting a willingness to substitute price for reef conditions. These findings further indicate that there may be a market for climate change-degraded reefs in the future, but at a discounted price, which will have economic implications even if arrival levels can be maintained. Pricing and alternative tourist attractions will be an important component of destination adaptation.

A survey of dive tourism operators from across the Caribbean found the widespread 2005 bleaching event and the more restricted 2010 event had limited impact on their operations and dive tourists (Sealy-Baker 2011). The majority of dive tour operators (60 per cent) indicated recent bleaching events had no impact on their business, while 14 per cent suffered some financial losses or adjusted the number of dives each day to reduced demand. When asked what divers' reactions to coral bleaching were, operators agreed that divers were aware and concerned, while as many agreed and disagreed that divers were asking for information about coral bleaching and, as a group, disagreed that divers were coming to see bleaching. As the following quote illustrates, dive tour operators generally felt that many divers were aware of bleaching and wanted to know more about it, but did not change their travel patterns because of it:

> Well [divers] ask if there's bleaching because they read about it, they are asking how it happens and how it goes and will it recover and so on. So the few who notice that there is something different from how it was are definitely asking questions and want to know more about it. They are not saying oh, I'm not coming next year because you have dead corals here and going back to Bali or wherever else, but they are definitely asking questions and want to know about it.
>
> (quoted in Sealy-Baker 2011: 67)

This literature reveals a number of similarities with demand impacts in ski tourism and alpine landscapes. Experienced divers with specialized motivations and expectations in terms of reef conditions make conscious decisions for certain dive sites, and will be more affected by degraded reefs than beginner or recreational divers who are focused on building their skills or participating in a dive as part of a broad range of tourism activities. Similarly, as more reefs are degraded by climate change, specialized divers will concentrate in reef areas that remain relatively unaffected, which could result in increased crowding and decreased satisfaction. Although evidence suggests there will remain a market for dive tourism on degraded reefs, the market will be comprised of beginner and recreational divers, who have been found to be lower-yield tourists, and thus may affect the financial viability of dive tour operators and other tourism businesses at destinations that specialize in dive tourism. Like the above discussion of how generational cohorts may perceive changes in alpine landscapes, Dearden and Manopawitr (2011) question whether future generations of divers will perceive bleaching and degraded reef conditions in the same way as contemporary divers, if they have no frame of reference of previous, more pristine conditions. They compare this potential generational change in perceptions with the 'shifting baselines syndrome' in fisheries, described by Pauly (1995). This is an excellent example of how lessons in other natural resource sectors may help advance understanding of how tourists will respond to environmental changes brought about by climate change.

Beach loss to sea-level rise

Although it is well established that beach tourism is one of the largest tourism segments worldwide, and that a large proportion of the world's monitored beaches are eroding and will be increasingly vulnerable to SLR (chapter 5), relatively few studies have examined the potential impact of beach loss on tourism demand.

In Europe, an early study by Braun *et al.* (1999) examined the combined scenarios of temperature and precipitation changes with SLR and beach loss on the likelihood of travelling to the Baltic and North Sea coasts of Germany. With increased temperatures, the likelihood of choosing the north German coast for a holiday was only slightly higher than the scenario where climate change impacts on the Mediterranean were very limited. In scenarios where there were some potential negative impacts on the German coasts (e.g. beach erosion, increased frequency of rain), the likelihood of visiting was substantially lower, even when possible adaptations were included (e.g. greater setbacks of tourism infrastructure, more diversified indoor activities). Similar scenarios of climatic and environmental change on the East Anglian coast of the UK by Coombes *et al.* (2009) resulted in minor reductions in visitation (less than 1 per cent) from beach loss (although beach loss was presented as 13–16 m loss to beaches with widths of over 250 m), whereas there were large gains in visitation through warmer temperatures (24–26 per cent).

In the Caribbean region, a survey of tourists to the islands of Barbados and Bonaire found that under a severe scenario, where 'beaches largely disappeared', 77 per cent would be unwilling to return to the 3S destination of Barbados for the same price, while 43 per cent would be unwilling to return to Bonaire, which is known as a dive tourism destination and where beaches are not the primary tourism asset (Uyarra *et al.* 2005). Buzinde *et al.* (2010) use qualitative approaches (tourist interviews and online tourist comments about destinations and trip experiences) to examine tourists' perceptions of beach erosion and attempts to restore the beach in the Playcar, Mexico area. They classify tourists' responses into three categories: positive, negative and reconciliatory. Those who adopted a positive view of the situation focused largely on the additional recreational opportunities provided by the erosion-control structures in the water (large sand-filled geotextile bags described as 'beach whales' and used for climbing and diving platforms). Those with negative reactions were often unaware of the severely eroded beaches and had expected the state of beaches as reflected in tourism marketing images of the destinations. This discrepancy between promoted images and reality left them feeling deceived and swindled. Like the negative-response group, the reconciliatory group viewed the beach erosion-control measures as aesthetically unpleasant, but seemed to understand they were necessary to protect the beach and were part of ongoing work to restore it through beach nourishment. Some of these tourists appeared aware of the degraded condition of the beach before arrival and therefore did not have a sense of misrepresentation by tourism operators. Interestingly, the study also found that some tourists associated these degraded beach conditions with climate change and expected this to become more common in the future. It is unclear what proportion each group represented, which is important in order to understand the potential impact on destination reputation and future demand. It is also unclear what role prior knowledge of the degraded state of the beaches or price discounts played in the varied interpretations of these three groups or their respective intentions to return to this destination at the same price or discounted prices. These remain areas for future research.

The only study that has attempted to translate potential demand changes from SLR and lost beach areas into economic losses for a destination was conducted by Bin *et al.* (2007) in the US state of North Carolina. They estimated that with a 46 cm rise in sea level by 2080 (although note that many current scientific projections exceed 1 m SLR), there would be substantial losses in recreation benefits from reduced opportunities for beach trips (US$3.5 billion discounted at a 2 per cent rate) and fishing trips (US$430 million using a 2 per cent discount rate). The study also examined the impact of SLR on coastal properties, and estimated the same SLR would put US$1.2 billion of residential property at risk in four counties alone. A large proportion of these properties are recreational homes and with their loss, tourism spending would decline further than what was estimated.

Other forms of environmental change

The pollution of water bodies and outbreaks of species perceived as harmful or aesthetically unpleasant are likely to affect tourist perceptions (Nilsson and Gössling 2012). Insects including beetles, ticks and mosquitoes can be a nuisance to recreationists, and also, as in the case of parasites acting as vectors for disease or mosquitoes, represent a potential health hazard (for the role of ticks as vectors see e.g. Jaenson *et al.* 1994; Gustafson and Lindgren 2001). Rising populations of harmful species such as jellyfish may prevent bathing and swimming activities, or negatively affect experiences of these activities. For example, in Hawaii and the Gulf of Mexico, accumulation of jellyfish is understood to have affected tourist destinations, although the link is unsubstantiated (e.g. Gershwin *et al.* 2010).

In the cases of algal blooms, water quality can be negatively affected, which can be perceived as an aesthetic problem or a health hazard. Although the reasons for outbreaks of marine pests are complex, anthropogenic environmental change in the form of pollution, eutrophication, ocean acidification and global warming is seen as an underlying factor (Schernewski and Schiewer 2002; Heisler *et al.* 2008; Jöhnk *et al.* 2008; Paerl *et al.* 2011). Reference to the consequences of algal blooms for tourism has been largely speculative (e.g. Hall and Stoffels 2006; Englebert *et al.* 2008; Gershwin *et al.* 2010; Galil *et al.* 2010). Except for basic statistical reviews (e.g. Becheri 1991), the only study to date that explicitly investigates the consequences of algal blooms for tourism, although without linking these to climate change, is an assessment of the effects of the severe algal blooms observed in the Adriatic Riviera in 1989 (Gasperoni and Dall'Aglio 1991). Based on a survey of tourists, all of whom had some knowledge of algal blooms before travelling to the coast, Gasperoni and Dall'Aglio (1991) found that 17 per cent were not influenced by algal blooms at all, but the broad majority reported 'substantial reductions in the amount of time dedicated to all seawater-related activities, especially sea-swimming' (Gasperoni and Dall'Aglio 1991: 264). Consequently, 73 per cent reported a negative influence of the algae on their holidays, with foreigners in particular being sensitive to environmental change. As the occurrence of algal blooms was known, this possibly excludes a share of more sensitive tourists who cancelled their holidays in the Adriatic because of the blooms.

During the past two decades, the North Sea and the Baltic Sea have also been exposed to major outbreaks of algal blooms, with some destinations reporting significant impacts on tourism (Barometern 2007). In a study of visitors to southern Sweden, Nilsson and Gössling (2012) found that tourists paid considerable attention to extreme environmental events. In this study of perceptions of algal blooms in the Baltic Sea, almost 98 per cent of respondents stated that they were aware of algal blooms, and 64 per cent had had personal experiences with algal blooms. Results also indicate that the overall perception of blooms is shaped by media reports that discursively link algae with health risks. This has repercussions for risk perceptions, with a general understanding that algae are a potential health threat and aesthetically problematic. Notably, the understanding of 'aesthetically problematic' includes visual, olfactory and haptic dimensions. The study showed that only a small share of travellers (0.6 per cent) perceived algal blooms as entirely unproblematic, while 2 per cent shortened their stay by going home; 4 per cent stayed, but changed bathing locations; 4 per cent shortened their holiday in the destination, moving to another area; and 10 per cent stayed, but changed their activities. Including those 19 per cent of travellers reporting not to have been able to take a bath as often as they would have liked, Nilsson and Gössling (2012) concluded that 39 per cent of respondents were affected by algal blooms. For 11 per cent, experiences with algae had such an impact that they chose not to return to the destination in subsequent years.

In other cases, it is the reduction of wildlife abundance or loss of species that is anticipated to affect tourism visitation. The multi-billion-dollar freshwater sport fishery of North

America and the associated tourism market would be affected by climate change through altered geographical distribution of species and abundance of sought-after sport fish (particularly cold-water species). A US EPA (1995) study estimated annual economic damage in the US sportfishing industry at US$320 million in the 2050s. Notably, this study found that, when alternative modelling assumptions were used, the estimated damages increased substantially, suggesting the need for further research to narrow the range of uncertainty. Other regional studies similarly project large changes in fish habitats, but the implications for tourism have not been assessed, nor has there been a rigorous analysis of potential adaptation strategies such as lake-stocking strategies or anglers' choice of species.

Dawson *et al.* (2010) examined the implications of declining polar bear populations in Churchill, Canada. Their survey of tourists engaged in polar bear watching found many were willing to visit Churchill to view the bears under varying scenarios for the polar bear population. Over 80 per cent of respondents indicated that, even if environmental conditions altered the population dynamics of polar bears so much that they were able 'to view only a quarter of the number of bears they actually saw on a 2007 visit', they would still visit Churchill in the future. If, in the future, polar bear populations were to 'appear unhealthy' (very skinny), which is already beginning to occur and is expected to continue, over 60 per cent of visitors would still visit the destination. If visitors were 'not guaranteed to see any bears' (they may or may not see them), only 50 per cent of visitors indicated they would still visit. The majority of visitors indicated that if they 'could not view polar bears in Churchill' (they were locally extirpated), 72 per cent of visitors 'would be willing to travel somewhere else so they could view polar bears'.

The implications of climate change for the many other forms of wildlife viewing, whether it be organized tourism events associated with avitourism (bird watching) or simply part of the attraction of a nature-based destination, remains uncertain. As discussed in chapter 5, much will depend on the capacity of managers of protected areas to maintain suitable habitat or assist species in their adaptation to climate and ecosystem changes.

Demand implications of mitigation policies and tourism mobility

The GHG emissions from tourism, and the mitigation policy frameworks (national, international) that determine the extent to which components of the tourism system must reduce their emissions, are discussed in chapters 3 and 4. These national, international and sectoral mitigation policies have an impact on tourists' behaviour in two ways. First, policies, especially market-based instruments, may lead to an increase in the costs of travel. Second, climate change mitigation policies can increase awareness of the climate impacts of travel and foster environmental attitudes that lead tourists to change their travel patterns.

Analysis of how mitigation policies could affect the price of travel, and thereby tourism demand, have focused entirely on air travel. A number of studies have analysed the price sensitivity of air passengers, and the results vary considerably (Brons *et al.* 2002).

Overall, it appears that leisure travellers are more price-sensitive than business travellers, and short-haul travellers are more sensitive to price increases than those on long-haul trips. The reason for this is that there are more choices and possibilities for substitution for shorter trips compared with long ones (Brons *et al.* 2002), while the commitments of business travellers make them more likely to pay higher ticket prices so as to provide both flexibility and certainty in their travel arrangements (Hall 2009d). Also, tourists who can afford long-distance holidays are likely to be wealthier than average (Hall 2005). Research also found that tourists are more likely to adjust their behaviour (travel less by air) in response to higher prices in the longer term, rather than immediately. This means that an increase in airfares may not have an immediate behavioural effect (tourists cannot change their plans quickly), but over time, tourists may learn to avoid the pricier option of air travel and select alternative transport options such as trains, or less costly destinations.

Studies that have examined the potential impact of aviation sector mitigation policies on international tourism demand have consistently found that policies as currently proposed would have little impact on overall tourism demand. Although all studies project a small decrease in the growth of international tourist arrivals versus a scenario with no emission reduction policies, in all cases, demand for air travel and international tourism continues to increase (Mayor and Tol 2007, 2010a, 2010b; Gössling *et al.* 2008; Pentelow and Scott 2010, 2011). Consequently, there is no evidence to suggest that mitigation policies for international aviation would have even a moderate impact on tourism demand through 2020. As discussed in chapter 6, the impact of mitigation policies is not on tourism demand *per se*, but rather on the geographical distribution of travel and potentially even on length of stay, so that the impacts for destinations are projected to vary from country to country depending on proximity to primary markets.

Tourists' perceptions of the carbon footprint of travel, and in particular air travel, is probably of similar consequence to tourism demand as prices change due to mitigation policies. There has been substantial media coverage on this topic (Scott *et al.* 2008a; Burns and Bibbings 2009), as well as a growing awareness of the potential environmental impacts of air travel as reflected in public opinion polls and traveller surveys (Becken 2007; Brouwer *et al.* 2008; Gössling *et al.* 2009d; McKercher *et al.* 2010; Higham and Cohen 2011). However, some studies have found a 'psychology of denial' in terms of awareness of the contributions of air travel to climate change (Gössling *et al.* 2009d; Hares *et al.* 2010; Cohen *et al.* 2011).

This dichotomy is reflected in the varied level of support that tourist surveys have identified with respect to 'willingness to pay' for carbon offsets or to change holiday travel patterns in order to reduce the environmental consequences of air travel. The aforementioned traveller surveys have generally found a willingness to pay some additional charge to reflect environmental costs (in the form of a levy or carbon offset). Brouwer *et al.* (2008) estimate that, based on the stated willingness to pay to offset GHG emissions, funds in the order of €23 billion could be generated annually from air travellers. However, the proportion of travellers who self-declare they have purchased offsets or always purchase offsets is typically less than 5 per cent (Gössling *et al.* 2009d; McKercher *et al.* 2010; Higham and Cohen 2011). Hall (2009c) found awareness and willingness to pay for offsets to be even lower in a survey of New Zealand business travellers. The inclination of tourists voluntarily to change their travel behaviour to reduce environmental impacts is equally low, as the personal benefits of travel trump the social costs of climate change (Becken 2007; Hall 2009c; Randles and Mander 2009; McKercher *et al.* 2010; Higham and Cohen 2011).

While these contrasting results suggest that customers' attitudes toward flying are unlikely to have a meaningful impact on tourism demand in the near term, long-haul destinations have expressed concerned that 'growing guilt over the impact of jet flights on global warming' could adversely impact the tourism economy (Bartlett 2007; Boyd 2007) (see Box 4.1). A number of nations are seeking to position themselves as 'carbon-neutral' destinations in order to address growing concerns over air travel (Gössling 2009), although the lack of rigour and credible plans to achieve 'carbon neutrality' (Gössling and Schumacher 2010) has attracted criticism in the media (e.g. Goodall 2009) and tourism scholars alike (e.g. Weaver 2011).

Demand implications of climate-induced societal change

Long-range scenarios for global tourism development do not exist (the UNWTO's 2001 *Tourism 2020 Vision* is the longest projection available). Consequently, it has not been possible for tourism scholars to examine the implications of climate change for the socio-economic conditions that are the foundation of long-range tourism scenarios. When the long-range IPCC *SRES* scenarios for the size and distribution of global population and economic activity are considered (Table 7.3), the importance of interpreting the varied socio-economic indicators associated with fundamentally different development pathways for the future of tourism is unmistakable.

Similarly, while it is well established that climate change poses a potential risk to future economic development and political stability in some countries (Parry *et al.* 2007; UNDP 2007), and that international tourism is highly sensitive to economic conditions, the economic scenarios of the Stern Review (Stern 2007) and competing interpretations of

Table 7.3 Socio-economic characteristics of the IPCC scenarios *(Special Report on Emission Scenarios)*

Scenario indicators	1990 baseline	2100 scenarios*			
		A1FI	A2	B1	B2
Global population (billions)	5.3	7.1	15.1	7.0	10.4
Global GDP (US$ trillions)	23	525	243	328	235
Income ratio between developed and developing countries	16:1	1.5:1	4.2:1	1.8:1	3:1

*See chapter 1 for explanation of scenarios
Source: Nakicenovic *et al.* (2000)

the global economic impact of climate change (Heal 2009) have also not been systematically interpreted for the tourism sector. Any reduction of global or regional GDP resulting from climate change would reduce the discretionary wealth available to consumers for tourism and have negative repercussions for anticipated strong future growth in tourism demand. When scenarios of climatic change and socio-economic change from the *SRES* scenarios were utilized in the models of Hamilton *et al.* (2005) and Berrittella *et al.* (2006), the socio-economic impacts had a far greater effect on overall tourism demand than did climate.

Interpreting the implications of the technological and economic conditions that are the foundation for GHG stabilization scenarios (see chapters 1 and 4) is another area of research that is largely unexplored. Setting carbon emission-reduction goals of 50 per cent by 2035 (as the WTTC 2009 has), or reducing emissions by a factor of three or four from tourism in developed countries by 2050 (Dubois and Ceron 2006), inevitably leads to significant changes in tourism patterns and demand. Examining the demand-side implications of reaching these targets has been insufficient.

Attempts to assess the impacts of climate change on tourism demand within models of climate change impacts on the broader economy have been conducted in Europe and Australia. Ciscar *et al.* (2011) integrated climate scenarios and physical-impact models with a multi-sectoral economic model to examine the potential consequences of climate change on agriculture, river floods, coastal areas, tourism and human health in Europe. They concluded that, if the climate of the 2080s occurred in the present EU economy, the annual damage due to climate change in terms of GDP loss would range between 0.2 and 1 per cent (€20–65 billion) for the 2.5°C scenario with 49 cm SLR and the 5.4°C scenario with 58 cm SLR. Tourism was one of the least vulnerable sectors and, given full flexibility to respond to climate change, tourism was estimated to have an EU-wide growth in bed nights of 1–7 per cent (€4–18 billion in spending) for the low and high temperature-change scenarios. These aggregated economic impacts mask important regional variations. For tourism, central Europe, the British Isles and northern Europe were all projected to see increased tourism demand (17, 18 and 25 per cent, respectively), while a 4 per cent reduction was projected for southern Europe. Like the aforementioned economic models of the impact of climate change on tourism demand, this study examined only the impact of climatic change on tourism and did not consider climate-induced environmental change, even though SLR impacts were included in the analysis with a coastal systems model.

The Garnaut Climate Change Review (Garnaut 2008) was a major effort to examine the implications of climate change for the Australian economy. Although tourism is an important part of the Australian economy, and many of its natural assets for tourism have been identified as at risk from climate change, tourism was mentioned only briefly in this national economic analysis. Pham *et al.* (2010) utilized a regional economic model and input on the impact of climate change on tourism demand from tourism industry stakeholder workshops (described in Turton *et al.* 2010) to assess the impact of climate change on domestic and international tourism at five major destinations, using economy-wide conditions from the Garnaut Review as boundary conditions. Although industry-based, the stakeholder scenarios of how climate change would affect tourism demand were arbitrary (10, 40 and 60 per cent reductions in visitor nights by 2020, 2050 and 2070, respectively). Although the impact on national GDP was estimated to be largely insignificant, individual tourism-reliant destinations were severely affected by the assumptions of how tourism demand would change, with gross regional product (GRP) exceeding 30 per cent in the Victorian Alps region (ski tourism) and 17 per cent in the Tropical North Queensland region (GBR) by mid-century.

Both modelling efforts acknowledge that the use of recent historical data to calibrate economic parameters, which are unlikely to accurately reflect aspects of economic behaviour and linkages between tourism and other sectors into the distant future as technology, social and policy frameworks evolve, is an important limitation. Improving understanding of the impacts of climate change on tourism demand was identified as critical to advance economic analyses of the implications of climate change for tourism and local–regional economies.

CONCLUSION

With their capacity to adapt to the effects of climate change by substituting the place, timing and type of holiday in their travel decisions, tourists will play a pivotal role in the eventual impacts of climate change on the tourism industry and destinations. The evidence available suggests that the geographical and seasonal redistribution of tourism demand resulting from changes in climate may be very large for some individual destinations and countries by mid-century, although some tourist segments show resilience to many forms of climate change impact.

The direct effect of climate change might be significant enough to alter major intra-regional tourism flows where climate is of paramount importance, including northern Europe to the Mediterranean, North America to the Caribbean, and to a lesser extent Northeast Asia to Southeast Asia, although a greater understanding is needed of the place of climate in tourists' decision-making processes and of climatic thresholds of key market segments before this can be confirmed. However, the net effect of a change in climate on tourism demand at the global scale may be limited, as there is no evidence to suggest that a change in climate will lead directly to a significant reduction of the global volume of tourism, only to its redistribution geographically and seasonally. Similarly, the many types of climate-induced environmental change are anticipated to have important impacts on tourism demand for specific destinations. At the global scale, the impact will be one of redistribution of tourism demand, but no aggregate decline in demand is anticipated. Further research is required to identify critical threshold levels that trigger behavioural changes in tourists. Additional multi-destination analyses are required to better understand destination-substitution behaviour.

There is no evidence that climate change mitigation policies, as currently proposed, would result in a net decline in global tourism demand; rather they would only slow anticipated growth in tourism arrivals to certain nations. As the highest projected impact of mitigation policies on tourism arrivals are in low to medium developed countries, the economic impacts for these countries are significant in some cases and raise important questions about the equity of international climate change policy.

The implications of climate change-related societal change for tourism demand is the least understood impact pathway. If, as some economic analyses indicate (Stern 2007), global economic growth were to be adversely affected by climatic change, then a parallel negative impact on tourism demand would be highly likely. There are two notable knowledge gaps with respect to this area of climate change and tourism research. First, there is limited understanding of how climate change impacts will interact with other, longer-term social and market trends influencing tourism demand, including ageing populations in industrialized countries, increasing travel safety and health concerns, increased environmental and cultural awareness, advances in information and transportation technology, as well as shifts toward shorter and more frequent holidays. Second, to date no study has attempted to assess the synergistic effects of multiple impact pathways on tourism demand for a specific destination. This is particularly important for long-haul destinations and tourism-dependent regions where substantial impacts on major tourism assets are expected. The development of methodological frameworks to examine these questions is an important area of future inquiry.

This chapter provides an extensive discussion of the complexities of tourists' perceptions of the impacts of climate change on tourism destinations. If long-range shifts in tourism demand are to be projected more accurately, improved understanding of tourists' climate preferences and key climatic and environmental thresholds (perceptions of coral bleaching, diminished or lost glaciers, degraded coastlines, reduced biodiversity or wildlife prevalence) are essential. There are substantial methodological and conceptual challenges associated with this research, but there are many tools available in the tourism studies literature and other disciplines that have potential to contribute to advancing this key area of inquiry. Over a decade ago, Braun *et al.* (1999) noted the challenge of making destination scenarios of climate, environmental and social change 'feel real' for tourists. The use of climate change analogues, landscape visualization and other approaches used to assess limits of acceptable change deserve greater attention by climate change and tourism researchers. As the literature cited in this chapter demonstrates, there is a range of concepts from tourism studies and other behavioural disciplines (e.g. environmental psychology, behavioural geography, marketing) that would assist in understanding better the varied responses of tourist market segments to climate change-related impacts. The demand response to climate change is a spectrum, as some tourists with low thresholds for change will act early, while others possess greater tolerance for change. This spectrum of tourist responses will translate into varied impacts for destinations that cater to different tourism markets, and understanding this spectrum is a central task for the climate change and tourism research community.

FURTHER READING AND WEBSITES

For a broader discussion on the interactions of tourism and other forms of global environmental change, the following book is recommended:

Gössling, S. and Hall, C.M. (eds) (2006) *Tourism and Global Environmental Change*. London: Routledge.

Overviews of conceptual frameworks and research themes related to tourism behaviour are available from:

Pearce, P.L. (2005) *Tourist Behaviour. Themes and Conceptual Schemes.* Clevedon: Channelview.

Reisinger, Y. (2009) *International Tourism: Cultures and Behavior.* Oxford: Butterworth-Heinemann.

Further information on long-range scenario development and future studies for the tourism sector is available from:

Yeoman, I. and McMahon-Beattie, U. (2005) 'Developing a scenario planning process using a blank piece of paper', *Tourism and Hospitality Research,* 5(3): 273–285.

Varum, C., Melo, C., Alvarenga, A. and de Carvalho, P. (2011) 'Scenarios and possible futures for hospitality and tourism', *Foresight,* 13(1): 19–35.

Hall, C.M. and Lew, A. (2009) *Understanding and Managing Tourism Impacts: An Integrated Approach.* London: Routledge, chapter 7 deals with futures for tourism and provides an overview of forecasting in tourism.

8 | Conclusion

'What's the use of having developed a science well enough to make predictions ... if, in the end, all we're willing to do is stand around and wait for them to come true?'

Sherwood Rowland, co-winner of the 1995 Nobel Prize for his work for his research into the effects of chlorofluorocarbon (CFC) gases on the ozone layer (Brooks 2011)

As this book indicates, there is a substantial amount of information and knowledge with respect to tourism and climate change. However, the understanding of this relationship is uneven with respect to destinations, topics, and the institutions that influence tourism and climate change policy and decision-making. This final chapter identifies where some of the major knowledge gaps are, and highlights significant future issues with respect to tourism and climate research.

KNOWLEDGE GAPS AND THE CONTEXT OF KNOWLEDGE

Climate change knowledge is highly contested and, at times, extremely controversial – for example, when Sir David King, the former Chief Scientific Adviser to the British Government, warned in 2004 that global warming posed 'a bigger threat than terrorism', the warning so incensed then US President George W. Bush that he phoned British Prime Minister Tony Blair to ask him to gag the scientist (Harvey 2011). Hansen (2009) details efforts by the same administration to censor US Government climate science reports and to prevent him from speaking publicly on climate change as Director of the NASA Goddard Institute for Space Studies. Just before the UN Climate Change Summit in Copenhagen in December 2009, substantial controversy emerged over a set of over 1000 private emails and many other documents that were stolen from the University of East Anglia's Climatic Research Unit and then selectively leaked to the media in what became referred to as 'climategate'. The selected contents of the emails were used by some to suggest that climate scientists had deliberately manipulated or hidden data. Such claims by climate sceptics, including the well known US environmental expert Sarah Palin (Pearce 2010), were comprehensively rejected in a series of independent parliamentary, scientific and university inquiries (see Scott 2011 for further details and discussion in a tourism context). However, the reviews did highlight that scientists had difficulties in knowing how to deal with freedom of information requests and with the overall openness of their work. It has even been suggested that freedom of information laws have been used to harass some climate scientists in targeted, organized campaigns of requests for data and other research materials in order to intimidate them and slow down research (Jha 2011). Some climate scientists who monitored the media coverage of the controversy found examples of fabricated quotes and misrepresentation of science. In some instances, the errors by the media were so egregious that complaints were filed with press-oversight bodies, and in some cases retractions were later published (e.g. in *The Sunday Times* and *Frankfurter Rundschau*). In other cases, misinformation and defamation of internationally leading scientists went unanswered. While retractions are helpful, the damage to the public's understanding of the science of climate change by such distortions is not undone.

All this raises the question as to why climate change research is so controversial, especially given its potential economic, environmental and social effects. Despite the portrayal of scientific controversy over climate change in some media, there is actually virtually no controversy among the climate and wider science community over whether or not anthropogenic climate change is real. Anderegg *et al.* (2010) reviewed the publication and citation data of 1372 climate researchers and concluded that 97–98 per cent of the most research-active climate scientists support the tenets of anthropogenic climate change outlined by the IPCC, and that the relative expertise and scientific prominence of scientists unconvinced of anthropogenic change are substantially below that of the convinced researchers. However, the paper received some criticism from those claiming that the approach used by Anderegg *et al.* (2010) only confirms the notion that the peer-review process is reinforcing the consensus view on anthropogenic climate change (Kintisch 2010). Such criticisms raise broader issues regarding scientific evaluation and 'normal science'. Similar overwhelming levels of agreement were found in separate surveys of publishing climate researchers by Doran and Zimmerman (2009) and Rosenberg *et al.* (2010), as well as by a joint declaration of 32 national science academies confirming anthropogenic climate change (*Science* 2001).

'We rely on various experts on a daily basis (from lawyers to doctors to pilots) and demand thorough assessments of their skills and credibility as experts. When one needs heart surgery, does one seek out an auto mechanic? Yet, with respect to matters of climate science, far too many in the media and some politicians are willing to give equal voice to non-experts with no credibility in the field (i.e. non-scientists … and scientists that have never contributed to the body of peer-reviewed literature they so readily critique).'

Scott (2011: 24–25)

A consensus that anthropogenic climate change exists certainly does not mean there is consensus over how the problem should be framed, evaluated and 'solved'. Indeed, it is also of note that different disciplines appear to have different degrees of agreement with the concept of anthropogenic climate change, as well as different approaches. In a survey of earth scientists, Doran and Zimmerman (2009) found a very high degree of support for anthropogenic climate change among climatologists, but much lower support among resource geologists. However, overall, 90 per cent of respondents agreed that temperatures have risen compared with pre-1800 levels, and 82 per cent agreed that humanity has significantly influenced the global temperature. Similarly, Heal (2009: 276) contends:

> [t]here is an amazing disjunction between economists and natural scientists on this issue: most natural scientists take it as completely self-evident that the consequences of climate change justify significant actions to mitigate the build up of GHG, whereas there is a range of opinions on this matter among economists, with the conventional wisdom being until very recently very different from that in the scientific community.

Knowledge of something also does not necessarily mean that action will be taken, at least in policy and decision-making terms. Hulme (2009: xxii) notes that '[e]ven when scientists, politicians and publics agree on the basic principles and most robust findings of climate science, there is still plenty of room for disagreement about what the implications of that science are for action'. The World Bank (2010: 19) further commented that

> climate-smart development policy must tackle the inertia in the behaviour of individuals and organizations. Domestic perception of climate change will also determine the success of a global deal—its adoption but also its implementation. And while many of the answers to the climate and development problem will be national or even local, a global deal is needed to generate new instruments and new resources for action.

As chapter 7 discusses, for reactive policies to be framed and developed, it is essential that there is public support, especially in light of opposition from vested interests that may be affected by climate change mitigation (Giddens 2009). Yet individuals' willingness to respond to climate change differs across countries and does not always translate into either personal or political concrete actions. Although awareness of environmental problems normally increases with wealth, concern about climate change does not (Sandvik 2008). 'Individuals (and nations) with higher incomes (and higher carbon dioxide emissions) may disregard global warming as a way to avoid incurring the potential costs of solutions associated with lower levels of consumption and lifestyle changes' (World Bank 2010: 9).

There is a strong need to communicate climate change to a wider audience and, as chapter 2 notes, to understand better the role of climate information in both tourism-sector and individual tourist decision-making, in order to reduce climate risk. Obviously, this book is geared towards a specialist audience. Nevertheless, tourism organizations, government agencies, scientific bodies and interest groups do need to be aware of the principles by which they may best communicate climate change information to specific audiences (see Box 8.1).

BOX 8.1 COMMUNICATING CLIMATE CHANGE

Rather than an aberration, denial needs to be considered a coping strategy deployed by individuals and communities facing unmanageable and uncomfortable events. Resistance to change is never simply the result of ignorance—it derives from individual perceptions, needs, and wants based on material and cultural values.

(World Bank 2010: 8)

The way issues are framed – the words, metaphors, stories, and images used to communicate information – greatly influences actions. According to Lorenzoni *et al*. (2007), frames trigger deeply held assumptions, cultural and institutional models, and worldviews in judging the climate change message and in accepting or rejecting it accordingly. In many cases, if the facts do not fit the frames, it is the facts that are rejected, not the frame (World Bank 2010). Furthermore, individuals tend to rank climate change lower than other environmental and economic issues that are 'closer to home' or 'day-to-day' in character (Hall 2006b; Moser and Dilling 2007). Humans 'are "myopic decision-makers" who strongly discount future events and assign higher priorities to problems closer in space and time' (World Bank 2010: 7). As Hall (2006b: 236) concluded in his study of New Zealand rural entrepreneurs' perceptions of climate change,

> people rank risks not only on scientific reports of the probability of harm, but also on how well processes are understood, their visibility, the frequency and extent of catastrophic events, the relative distribution of risk and vulnerability, and how well individuals can control their exposure.

Based on such understandings, it can then be decided whether climate change communication is best served by repeating or breaking the dominant discourse, or by reframing an issue using different concepts, languages and images to evoke a different way of thinking and facilitate alternative choices (World Bank 2010). Either way, climate change discourses need to be situated in people's locality in order to increase their saliency and trustworthiness (Lorenzoni and Pidgeon 2006). According to Lorenzoni *et al*. (2007) and World Bank (2010), applying this approach to communications on climate change could take many forms:

- where appropriate, place the climate change issue in the context of higher values, such as responsibility, stewardship, competence, vision and ingenuity;
- characterize mitigation as being about new thinking, new technologies, planning ahead, smartness, farsightedness, balance, efficiency and prudent caring;

- simplify the model, analogy or metaphor to help make it more understandable for the public, and establish a conceptual hook to make sense of information and set up appropriate reasoning;
- refocus communications to emphasize the human causes of the problem and the solutions that exist to address it, suggesting that humans can and should act to prevent the problem now;
- evoke the existence and effectiveness of solutions up front;
- above all, make the communication relevant to the audience and their concerns.

Aside from our alarm about how this economic calculus of costs and benefits has come to be the only legitimate basis of appeal for action to slow the onset of climate change ... we are concerned that the conservative nature of the prevailing climate change damage assessment has not been sufficiently appreciated. Some simplification is inevitable and, in fact, both necessary and desirable, if specialist knowledge is to be communicated to wider audiences. The challenge is to remain open about the often tacit assumptions and social commitments built into the technical details of scientific knowledges.

Demeritt and Rothman (1999: 405)

SCIENTIFIC WORLDVIEWS ON CLIMATE CHANGE

Differences in worldview do not exist just between climate change researchers and the general public. They also exist within climate change research. There is a fundamental difference in worldviews or ontologies between the natural sciences and social sciences in terms of how climate change questions may be framed. In his book *Why We Disagree About Climate Change*, Hulme (2009) argues that the discourses on the mutating idea of climate change vary among the sciences, social sciences and humanities, as well as the disciplines that comprise them. When climate change is examined '... from these different vantage points, we see that – depending on who one is and where one stands – the idea of climate change carries quite different meanings and seems to imply quite different courses of action' (Hulme 2009: xxvi). The natural sciences are distinguished academically, philosophically and, to an extent, methodologically from the formal, behavioural and social sciences as well as from the humanities (Weyl 2009). The term 'natural sciences' generally refers to those areas of organized knowledge that utilize a naturalistic approach and are concerned with the material aspects of existence. The formal sciences, including mathematics, statistics, computer science and logic, are not necessarily as concerned with the validity of theories based on observations in the real world as are the natural sciences. Nevertheless, formal science methods are extremely important for the methods and frameworks used in the natural sciences. In contrast, 'social science' refers to a range of fields, usually also including tourism studies, that are primarily concerned with the study of society and the individuals, institutions and social structures within it. The natural sciences have had considerable influence within the social sciences with respect to research philosophies, particularly positivism, as well as method, primarily quantitative. However, the social sciences are also characterized by an extremely significant interpretive tradition that tends to be qualitative, or at least a combination of qualitative and quantitative methodologies, and an understanding of society and the environment as being socially constructed (Hall 2012).

Ontologies condition the ways of seeing, creating and understanding not only different forms of knowledge, but also their acceptability. Different fields of knowledge have different ontologies. Ontologies are therefore not just academic concerns, but also determine how problems are defined and how they should be understood. Such a situation is directly related to how the environment is defined and understood; the difficulties that can exist in legal and institutional decision-making systems in recognizing the validity and standing of different environmental knowledge claims; the ways in which different institutions, interest groups, individuals and researchers define environmental management problems and issues; differences in standards of what can be regarded as 'proof'; and the difficulties that exist in getting different groups of researchers to work together on environmental problems (Jerneck *et al.* 2011; Hall 2012).

Bhaskar (2008) has suggested that there are three main ontological traditions within science: classical empiricism, transcendental idealism, and transcendental realism. Classical empiricism recognizes 'the ultimate objects of knowledge' as 'atomistic events in which knowledge and the world may be viewed as surfaces whose points are in correspondence or actually fused' (Bhaskar 2008: 14–15). The positivist account, which is usually associated with the scientific method of the natural sciences, presupposes an ontology of empirical realism, whereby the world consists of 'experience and atomistic events constantly conjoined' (Bhaskar 2008: 221–222), in which there is a dichotomous and oppositional division between humans (or the individual) and the environment. Such an approach has a strong relationship to the role of reductionism and mechanism as distinguishing features of western natural science, ontological reduction being the thesis 'that the properties of any entity may be understood by knowing the properties of its parts, because nothing can be explained about the entity without reference to its parts' (Keller and Golley 2000: 172).

The classical empiricist approach has been extremely significant in teaching and research on tourism, but the framework it provides is often taken for granted without consideration of the assumptions that organize and circumscribe the field of analysis (Castree 2002). For example, consider the use of the metaphor of tourism or tourist impact on the environment that has become strongly embedded in tourism and wider discourse (Hall and Lew 2009), so much so that 'the metaphor of human impacts has come to frame our thinking and circumscribe debate about what constitutes explanation' (Head 2008: 374). This metaphor is derived from the material realist ontology of classical empiricism, and has several features (Hall 2012).

- The emphasis on *the moment(s) of collision between two separate entities* (e.g. the 'impact' between tourism and the environment) has favoured explanations and methods that depend on correlation in time and space (Weyl 2009), to the detriment of the search for mechanisms of connection and causation rather than simple correlation (Head 2008).

- The emphasis on the moment(s) of impact *assumes a stable natural, social or economic baseline* (Hall and Lew 2009), *and an experimental method in which only one variable is changed* (Head 2008). This approach is inappropriate for understanding complex and dynamic socio-environmental systems (Head 2008; Hall and Lew 2009).

- Perhaps most profoundly influential (Head 2008) is the way the terms 'tourism impacts' or 'tourist impacts' ontologically *position tourism and tourists as 'outside' the system under analysis*, as outside of nature (or whatever it is that is being impacted) (Hall and Lew 2009). This is ironic, given that research on global climate and environmental change demonstrates just how deeply entangled tourism is in environmental systems, yet the metaphor remains in widespread use.

- Placing a significant explanatory divide between humans and nature requires the *conflation of bundles of variable processes* under headings like 'human', 'climate', 'environment' and 'nature' (Head 2008).

- Dichotomous explanations are characterized by their *veneer of simplicity and elegance*. Yet 'the view that causality is simple takes many more assumptions than the view that it is complex' (Head 2008: 374).

In contrast, transcendental idealism suggests that scientific knowledge, such as models, involves artificial constructs and is not independent of human activity in general. From this perspective, 'the natural world becomes a construction of the human mind, or, in its modern versions, of the scientific community' (Bhaskar 2008: 15). The third position, which Bhaskar himself supports, is that of transcendental realism, which regards the objects of knowledge as the structures and mechanisms that generate phenomena, and the knowledge that is produced as a result of the social activity of science. According to Bhaskar (2008: 15), 'these objects are neither phenomenon (empiricism) nor human constructs imposed upon the phenomena (idealism), but real structures which endure and operate independently of our knowledge, our experience and the conditions which allow us access to them'. Both transcendental realism and transcendental idealism reject the empiricist notion of science. Although they both agree that there could be no knowledge without the social activity of science, they differ over how nature is held (Hall 2012).

There are clearly different ontological positions with respect to understanding the environment. These exist between and within sciences, and are extremely important in terms if how we try and understand the interrelationships between tourism, climate change and the environment at different scales. For example, for ecology, and correspondingly for studies of tourism and environmental change, a key question is: 'Can an ecological entity be understood through an analysis of its biotic and abiotic components (reductionism), or must any ecological entity be explained by treating it as a unitary entry with unique characteristics (holism)?' (Keller and Golley 2000: 171). Such a question is extremely important for addressing how research on tourism and climate change at various scales of analysis, in both space and time, can be integrated into a satisfactory explanatory whole, that is, the extent to which the results of global research can be scaled down to a local scale, or *vice versa*. In the case of ecology, a reductionist ontology asserts that the essence of an entity, such as an ecosystem, community or even the biosphere, is a function of the sum of its parts, and therefore knowledge of the parts is adequate for understanding the whole. In contrast, holism asserts that some entities have emergent properties that are not properties of the parts, but rather of the whole, and therefore knowledge is neither necessary nor sufficient to understand the whole. In between the reductionist and holist positions lies emergentism, in which knowledge of the parts and their relations is a necessary but not sufficient condition to understand the whole (Keller and Golley 2000). As noted in chapter 2 with respect to macro-scale models of climate change and the capacity to transfer results between scales, such ontological differences raise fundamental questions about how future trends and even desired futures can actually be understood and predicted (Hall 2012). However, the situation becomes even more complicated when also trying to incorporate social science perspectives.

Geuss (1981) held that critical theories were fundamentally different from theories in the natural sciences, because while the natural sciences claim to be objective, critical theories are reflective. For example, Habermas (1978) placed great theoretical emphasis on what he termed cognitive or knowledge-constitutive interests in order to explain the connections between knowledge and action. Habermas' critique of the relationship between theory and practice in modern science has been especially influential given that he believed that science had divorced itself from the means of understanding its social context (Unwin 1992). Yet, despite these criticisms, the hegemony of the positivist outlook, with its strongly quantitative and instrumentalist approach, has fashioned the popular image and understanding of science (Bhaskar 2008).

Integrating social and biophysical perspectives and the construction of climate change

The need to integrate social and biophysical perspectives in climate and environmental change research is widely recognized (Füssel and Klein 2006; Conrad 2009; Hulme 2009; O'Brien 2011), including with respect to the need for ontological shifts (Turnpenny *et al.* 2011) and the production of local knowledge (Slocum 2010). However, given the institutional nature of knowledge, some types of social science assessment will be more acceptable than others in institutional processes. For example, in the expert review of the first draft of the IPCC *Working Group II Fourth Assessment Report on Climate Change Impacts, Adaptation and Vulnerability*, Karen O'Brien of the University of Oslo commented, 'Methodologies that incorporate social science perspectives such as institutions and social networks, or that include human behavior, such as agent based modelling and actor network theory, are excluded from this assessment of assessment methodologies' (IPCC 2005: 15). In response, the writing team noted,

> Addressed – we can understand the reviewer's perspective, and have tried to address those concerns, though they seem to reflect a very structured view of what different methods are and are not, which is not very amenable to the treatment that we have developed here.
>
> (IPCC 2005: 15)

Demeritt's (2001a, 2001b) examination of the construction of climate change and the politics of science is particularly informative (see also the response of Schneider 2001).

Demeritt retraces the history of climate modelling and associated climate science to identify tacit social and epistemic communities that are characterized by the technocratic and
reductionist inclinations of climate change science. It is important to emphasize, however,
that Demeritt was not denying the existence of climate change (and nor are the present
authors) – but highlighting the way in which climate change science is constructed and
how this leads into issues of the politics of the dominant, natural science-led formulation
of the climate change problem, as well as issues of public trust and climate change communication (Hall 2012). Significantly, Demeritt's critique well anticipated the issues
associated with media representation of 'climategate' and climate change sceptics as well
as the broader politics of climate change policy. As Demeritt (2001a: 309) notes,

> public representations of science seldom acknowledge the irreducibly social dimension of
> scientific knowledge and practice. As a result, disclosure of the social relations through which
> scientific knowledge is constructed and conceived has become grounds for discrediting both
> that knowledge and any public policy decisions based upon it.

Demeritt (2001a, 2001b, 2006) argues that climate change has been constructed in
narrowly technical and reductionist scientific terms by the IPCC and other international
and national scientific bodies on climate change, and that this promotes certain kinds of
knowledge at the expense of others. 'For the most part, climate change model projections
have been driven by highly simplistic business-as-usual scenarios of human population
growth, resource consumption, and GHG emissions at highly aggregated geographic
scales' (Demeritt 2001a: 312), that operate at a global scale, rather than framing the problem in terms of alternative, no less relevant forms such as the structural imperatives of the
capitalist economy that drives emissions (Wainwright 2010); the north–south divide in
terms of emissions; or a focus on issues of poverty and deprivation (Hall 2012). As
Demeritt (2001a: 316) argues, 'by treating the objective physical properties of [greenhouse
gases] in isolation from the surrounding social relations serves to conceal, normalize, and
thereby reproduce those unequal social relations'. Similarly, Wainwright (2010) suggests
that although climate scientists engage in debates about the meaning of their results, they
rarely reopen the 'black boxes' that are taken for granted in their research, and provides
an example of how carbon may be considered by physical and social scientists:

> Two physical scientists might engage in heady debate about the precise role of CO_2 or CH_4 in
> forcing a certain atmospheric process, but it is hard to imagine that carbon's basic qualities – its
> atomic number or weight, chemical properties, and so on – would be called into question. By
> contrast, two social scientists discussing, say, the hegemony of carbon emissions markets in
> climate policy discourse … would need to agree on the meaning of hegemony, markets, climate
> policy, discourse, and so on. This turns out to be no mean feat, because distinct interpretations of
> these and related concepts reflect different conceptions of the world … there is no metalanguage
> that lies outside of social life with which to objectively calibrate these concepts. Consequently,
> debates over the meaning of the building-block concepts for social thought are, by necessity,
> complex and interminable.
>
> (Wainwright 2010: 984)

The dominant scientific positions within the IPCC process have also led to the physical
reductionism of simulation modelling becoming the most authoritative method for studying the climate system (Demeritt 2001a, 2001b). Yet the appeals of formal quantitative

evaluation methods are social and political as much as technical and scientific (Hall 2012). Just as significantly, it also makes them more credible from a public perspective of natural science, 'insofar as adherence to rigidly uniform and impersonal and in that sense "procedurally objective" … rules limits the scope for individual bias or discretion and thereby guarantees the vigorous (self-)denial of personal perspective necessary to make knowledge seem universal, trustworthy and true' (Demeritt 2001a: 324).

Issues surrounding how climate change research is constructed scientifically are significant for understanding tourism's relationship to global change, and therefore the future of tourism (and the planet), for a number of reasons (Hall 2012). First, it helps explain why anthropogenic climate change has been defined primarily in environmental rather than political terms, or terms that require a more fundamental reframing of the contemporary capitalist economic system. Second, even though, as noted above, there has been a call for more social science information to be brought into the climate change assessment process, this has been assessed primarily in terms of neoclassical economic contributions (e.g. Stern 2007), which have themselves been greatly influenced by the ontology of natural science, and which are also dominated by formal modelling (Dietz and Stern 2008). Moreover, the neoclassical economic belief that the social and economic value of things can be expressed in terms of aggregate individual willingness to pay, or in monetary terms at all, is open to substantial critique (Demeritt and Rothman 1999: 404) because:

- it is anthropocentric and ignores intrinsic value (Leopold 1966);
- the world cannot be broken down into discrete and alienable entities to which monetary values might meaningfully be attached (Norgaard 1985);
- it confuses values and preferences (Sagoff 1988);
- its narrow decisionist framework artificially abstracts information about human values and preferences from an ongoing and multidimensional social process of (re-)expressing them (Wynne 1997);
- money conceals a profound asymmetry in the apparent equality of the exchange relation (Harvey 1996).

Indeed, Demeritt and Rothman (1999: 406) argue that one of the major difficulties in managing the effects of climate change is that 'to speak of the "science" of economic damage assessment as if it were a separate domain exchanging independently generated ideas with policymakers is to conceal the shared commitments that define them as part of a single cultural and political order'. The choice of valuation procedure, like the definition of value itself, as used by governments and supranational authorities such as the UNWTO, WTTC and WEF in relation to tourism and climate change, is part of a mediated social construction of knowledge and is ultimately personal and political rather than objective and rational (Demeritt and Rothman 1999; Hall 2012). Demeritt (2001a: 309) suggests that, instead of accepting an idealized vision of scientific truth and denying the socially situated and contingent nature of scientific knowledge, 'the proper response to it is to develop a more reflexive understanding of science as a situated and ongoing social practice, as the basis for a more balanced assessment'. Indeed, one of the problems of the portrayal of 'climategate' and the results of climate science being suddenly politicized is that it is fundamentally misleading. Science has always been political, in the sense that its outputs and outcomes, if not the process itself, lead to winners and losers. As Demeritt and Rothman (1999: 405–406) observe,

> Politics do not begin (or end) with explicit policy recommendations. The very process of valuing impacts, in monetary or any other terms, is inevitably value laden and politically saturated. But monetary valuation is by no means the only way in which particular values and social commitments enter into the practice of climate change damage assessment.

'Given the immensely difficult negotiations involved with an international climate change treaty, it is enormously tempting for politicians to argue that policy must be based upon an objective scientific assessment of the economic costs and benefits, thereby absolving themselves of any responsibility to exercise discretion and leadership. Because this division of labor enhances their power and prestige, it is also attractive to the IPCC participants. But this is a dangerous strategy. It provides neither a very democratic nor an especially effective basis for crafting a political response to global climate change. It enshrines in apparent scientific objectivity the particular values embodied by the IPCC assessment. When, as is almost inevitable, these political presumptions are publicly exposed and deconstructed, there is the danger that the resulting acrimony will tend to harden positions rather than make negotiation and mutual understanding easier.'

Demeritt and Rothman (1999: 406)

The arguments of Demeritt and Rothman were extremely prescient given the difficulties in obtaining a post-Kyoto agreement and a consensus on climate change action at the Copenhagen COP in 2009. It is important to re-emphasize that those who write of the social construction of climate change are not arguing that it does not exist; rather they are focusing on how the issue is constructed, developed and communicated, and how this may meet some interests and worldviews but not others. How a problem is defined clearly delimits its possible solutions. The significance of such a perspective is recognized by the World Bank, which, when discussing the barriers to behavioural change in relation to climate change, stated:

> People naturally tend to resist and deny information that contradicts their cultural values or ideological beliefs. This includes information that challenges notions of belonging and identity as well as of rights to freedom and consumption. Notions of needs and the priorities deriving from them are socially and culturally constructed.

(World Bank 2010: 8)

The point here is that the social construction of climate change adaptation and mitigation is not limited to public behaviour, but is also essential to understanding scientific and political behaviour.

KNOWLEDGE GAPS IN CLIMATE CHANGE AND TOURISM

As the above discussion has emphasized, knowledge of climate change needs to be seen in an appropriate context. But this does not mean we cannot identify where there are significant gaps and where they need to be overcome. An obvious general starting point, which applies to both climate and environmental change research overall as well as tourism-related studies, is that there is a real need to try and encourage mutual understandings of different worldviews and disciplinary constructions of climate change. This is something that can be done in the longer term through better education, and in the short term via the development of multi-disciplinary and trans-disciplinary research groups and meetings in which common understandings and shared language of inquiry can be developed (Coles *et al.* 2006). However, there are also other ways in which we can identify knowledge gaps in relation to tourism and climate change, and these are discussed below.

Variability in coverage

Despite tourism's undoubted economic significance and its role as an economic justification for biodiversity conservation, the place of tourism within major general climate change assessments is relatively weak (Parry *et al.* 2009b). For example, as chapter 5 notes, there was only one tourism-related reference in the 2007 Stern Review. In contrast, tourism has been given much greater emphasis in the IPCC climate change reports and the IPCC Working Group on Impacts, Adaptation and Vulnerability in particular (Amelung *et al.* 2008; Hall 2008a). Table 8.1 compares reference to tourism and cognate terms in comparable chapters of the 2001 (McCarthy *et al.* 2001) and 2007 (Parry *et al.* 2007) reports of the IPCC Working Group II. It is noticeable that, although there are similar total numbers of references to tourism, tourist, recreation and cognate terms in the regional chapters of the two reports, there is substantial variation between chapters.

Changes in word counts may reflect aspects of the issue ecology of content in IPCC reports, and are therefore a crude surrogate measure of importance and levels of knowledge on particular subjects. This is also partly a reflection of the authors of the various

Table 8.1 *References to tourism, tourist, recreation and cognate terms in regional chapters of IPCC Working Group II 2001 and 2007 reports on Impacts, Adaptation and Vulnerability*

Regional chapter	Citations*	
	Third report (2001)	Fourth report (2007)
Africa	3	28
Asia	9	7
Australia and New Zealand	22	31
Europe	34	40
Latin America	10	10
North America	86	27
Polar regions	11	7
Small island states	26	47
Regional chapter totals	201	197

*References in text, figures, tables and headings with respect to tourism, tourist, recreation and cognate terms (references to terms in bibliographic information are not included in tallies of key words)

chapters, as well as their levels of knowledge of tourism, and the selection of references they cite within the range of academic publications on tourism and climate change (Hall 2008a; Scott 2008). Table 8.2 provides a more detailed analysis of tourism within the Working Group II report by outlining the frequency of substantive comments on tourism issues (a sentence or more) in Parry *et al.* (2007).

Wilbanks *et al.*'s (2007) chapter on industry, settlement and society probably provides the most substantial assessment of the impact of climate change on tourism within the IPCC context. Tourism was identified as a 'climate-sensitive human activity' with the chapter, concluding that vulnerabilities of industries to climate change are 'generally greater in certain high-risk locations, particularly coastal and riverine areas, and areas whose economies are closely linked with climate sensitive resources, such as ... tourism; these vulnerabilities tend to be localized but are often large and growing' (Wilbanks *et al.* 2007: 359). However, as noted throughout this book, one of the greatest problems with assessing the impacts of climate change on tourism is that both direct and indirect effects will vary greatly with location. The IPCC chapters, as well as the literature discussed in this book, highlight island and coastal tourism systems, along with skiing/winter tourism destinations, as major focal points of research given their vulnerability as well as economic significance.

Rosenzweig *et al.*'s (2007: 111) claim that

> as a result of the complex nature of the interactions that exist between tourism, the climate system, the environment and society, it is difficult to isolate the direct observed impacts of climate change upon tourism activity. There is sparse literature about this relationship at any scale

probably did not apply in 2007, and certainly does not apply to the same degree at present, as the literature in the various chapters in this book indicates. However, it should be noted that comments with respect to a sparse literature on the tourism and climate change relationship also potentially highlight the failure of climate change researchers to expand the range of journals and books that they use in any review to include the broader social scientific literature as well as what appears in climate and environmental change journals. Nevertheless, it must also be recognized that the interrelationship between tourism and climate change is far more complex than deterministic models of temperature change and travel patterns (see chapters 2 and 5) suggest, and there is an acknowledged need to understand tourists' behaviour better in a climate change context (see chapter 7).

The substantial research on tourism emissions is not given due coverage in the IPCC literature, although it is noted in some regional and national government studies (Garnaut 2008; Ciscar *et al.* 2011). Probably one of the main reasons for this is that tourism does not fit easily into many of the standard integrated assessment models used for developing climate change forecasts and scenarios, as tourism is not a standard industry classification that is used in economic modelling. Although sufficient data exist so that international flows of tourists can be broadly modelled, it needs to be emphasized that international travel accounts for only a little over 15 per cent of all tourism. Therefore there are immense gaps in knowledge with respect to understanding domestic flows and the trade-offs between domestic and international travel. The incorporation of improved international and domestic travel data into climate change modelling, along with better economic and

Table 8.2 *Dimensions of the tourism and climate change relationships covered in the IPCC Working Group II 2007 reports on Impacts, Adaptation and Vulnerability*

Dimension	Specified locations	Number of chapters	Chapter sources
Effects on winter tourism	Asia, Europe, Bolivia	8	Adger et al. (2007): 721, 722, 734; Alcamo et al. (2007): 543–4, 561, 565; Cruz et al. (2007): 489; Field et al. (2007): 634; Hennessy et al. (2007): 523; IPCC (2007b): 14, 18; Magrin et al. (2007): 589; Rosenzweig et al. (2007): 117; Wilbanks et al. (2007): 363
Sensitivity of tourism to climate change	Africa, Asia, Australia and New Zealand, Europe, North America, tropical destinations, small islands	7	Alcamo et al. (2007): 543–4; Boko et al. (2007): 450, 459; Cruz et al. (2007): 489; Hennessy et al. (2007): 523; Mimura et al. (2007): 689, 697; Schneider et al. (2007): 790; Wilbanks et al. (2007): 363, 368, 375, 380
Effects on coastal tourism	Africa, Americas, Caribbean, Mediterranean, Florida, Thailand, Maldives, small islands	6	Alcamo et al. (2007): 543–4; Boko et al. (2007): 440, 449; Field et al. (2007): 634; Magrin et al. (2007); 584, 599, 600; Mimura et al. (2007): 689, 696, 698, 701, 703; Nicholls et al. (2007): 335–6, 337
Skiing	Asia, Australia and New Zealand, Europe	6	Adger et al. (2007): 721, 722, 734; Alcamo et al. (2007): 557, 561; Field et al. (2007): 634; Hennessy et al. (2007): 523; Magrin et al. (2007): 589; Rosenzweig et al. (2007): 89, 111

Degradation of coral reef, coral bleaching	Africa, Australia, small islands	5	Boko et al. (2007): 439; Fischlin et al. (2007): 235; Hennessy et al. (2007): 523; IPCC (2007b): 13, 15; Mimura et al. (2007): 689, 696; Nicholls et al. (2007): 320
Effects on wild faunal diversity/nature-based tourism	Marine ecosystems, Mediterranean-type ecosystems, small islands, southern Africa, tropical savannah systems	4	Boko et al. (2007): 435, 459; Field et al. (2007): 634; Fischlin et al. (2007): 225, 226, 234; Mimura et al. (2007): 696
Issues in tourism's adaptation to climate change	Small islands	4	Alcamo et al. (2007): 561; Anisimov et al. (2007): 673, 676; Mimura et al. (2007): 705; Wilbanks et al. (2007): 380
Effects of extreme events	Mexico, North America, small islands	3	Field et al. (2007): 626; Magrin et al. (2007): 585; Mimura et al. (2007): 693, 702
Effects on mountain tourism	Europe	2	Fischlin et al. (2007): 223; Rosenzweig et al. (2007): 88, 89
Effects on summer tourism	Europe, especially Mediterranean	1	Alcamo et al. (2007): 565
Reservation economies	North America	1	Field et al. (2007): 625

business data, would be a major boon for identifying the adaptive capacities of destinations along with a better understanding of their vulnerability.

As emphasized by Hall (2008a) and Scott *et al.* (2008a), major geographical gaps remain in the climate change and tourism literature, including South America, Asia, the Middle East and Africa, although notable improvements in the level of information available for the Caribbean region have occurred since these publications. Some highly significant tourism environments are also significantly under-researched (e.g. urban areas and World Heritage sites). For example, in the case of Africa, Boko *et al.* (2007: 450) stress that 'very few assessments of projected impacts on tourism and climate change are available' and later note: 'There is a need to enhance practical research regarding the vulnerability and impacts of climate change on tourism, as tourism is one of the most important and highly promising economic activities in Africa. Large gaps appear to exist in research on the impacts of climate variability and change on tourism and related matters' (Boko *et al.* 2007: 459). On the basis of a review of the then extant literature (particularly Gössling and Hall 2006a and Scott *et al.* 2008a), and the IPCC and Stern reports, Hall (2008a) provides a basic overview of the relative level of tourism-specific climate change knowledge and estimated impact of climate change on tourism by regions used in the IPCC reports. Table 8.3 presents an updated version of Hall (2008a) that takes into account subsequent research and findings, including with respect to potential adaptive capacity.

However, the approach used in Table 8.3 is based on our collective experience, and is included here more for purposes of promoting discussion to at least alert readers to the research needs of some major tourism regions, particularly in developing countries. It is apparent that there are major gaps in research at a destination level, and while national and even international research may provide a valuable context for destination research, and even suggest potential impacts of climate change as well as responses, it is no substitute for detailed destination-scale analysis on vulnerability and adaptive capacity. It should also be noted that much research has focused on those areas that appear most immediately vulnerable to climate change, rather than on the most significant areas for tourism at a destination or economic scale. For example, there is a major gap in understanding the effects of climate change in urban areas, which, in most parts of the world, is where tourism is often concentrated.

Knowledge of policy-making processes

A policy problem is a sub-issue, issue or suite of issues perceived to require resolution in some way, thus posing the challenge of choosing and framing the optimum policy response(s) (Table 8.4). Climate change is a meta-policy problem that has led to new institutional arrangements and policy settings at international, national and local scales (see chapters 1 and 4 for a discussion of some of these arrangements). Improvements in climate policy are not just a function of better models or more data. It is essential that policy-makers understand the assumptions that go into forecasts and scenarios, and their relative advantages and disadvantages. There is therefore a need for improved communication of existing knowledge and knowledge gaps to policy-makers and all tourism stakeholders, and assistance in the capacity to respond to knowledge gaps so that they do not

Table 8.3 Relative level of tourism-specific climate change knowledge and estimated impact of climate change on tourism, by IPCC region

Region	Estimated impact of climate change on existing tourism destinations under B2 scenario	Relative level of tourism-specific climate change knowledge
Africa	Moderately–strongly negative	Extremely poor
Asia	Weakly–moderately negative (coastal areas most vulnerable)	Extremely poor (China and Japan are judged poor)
Australia and New Zealand	Moderately–strongly negative (Great Barrier Reef, Australian alps and coastal areas most vulnerable)	Poor–moderate (highest in Great Barrier Reef, rapidly increasing in Australian alps and coastal areas)
Europe	Weakly positive–moderately negative (Mediterranean coast most vulnerable)	Moderate (highest in alpine areas)
Latin America	Weakly–moderately negative	Extremely poor
North America	Weakly negative–weakly positive (coastal areas most vulnerable)	Moderate (highest in Canada)
Polar regions	Weakly negative–weakly positive	Poor–moderate (highest in Antarctic peninsula and Nordic countries)
Small islands	Strongly negative	Poor–moderate (highest in tropical and subtropical islands)

become an excuse for inaction. For example, there is some evidence that elected officials may hold different perceptions of a problem and its solution than resource managers and planners (Wardell 2010). Varied views of the risks posed by climate change and response options are clearly evident among tourism stakeholders in chapter 6. In addition, more attention needs to be given to the nature of the actual decision-making process itself that is used by politicians and interest groups in determining policy and the political risk-taking involved.

In a wider vein, there is also a dearth of knowledge of the relative influence of tourism bodies on climate policy at international and national levels, and the manner in which economic and climate concerns may be traded off. There is therefore a need for better understanding of the institutional arrangements that surround climate change and the extent to which organizations have the capacities to adapt and to learn new approaches to solving complex climate-related policy problems (Hall 2011a).

Table 8.4 *Policy framing and response attributes with respect to sustainability problems*

Attributes	Descriptor
Policy framing	1 Spatial scale of cause or effect
	2 Magnitude of possible impacts on natural systems and/or human systems
	3 Temporal scale of possible impacts: timing and/or longevity
	4 Reversibility
	5 Measurability of factors and processes
	6 Degree of complexity and connectivity
Response framing	7 Nature of cause(s): discrete, fundamental, systemic
	8 Perceived relevance to the polity
	9 Tractability: availability and acceptability of means
	10 Level and basis of public concern
	11 Existence of policy goals

Source: after Dovers (1996); Hall and Lew (2009); Hall (2011a)

In order to assist tourism-related climate policy-making, there is an overall need for improved consistent international and domestic travel data, updated on a regular basis. Such information is essential for calculating emissions and identifying travel patterns. Although consumer-provided tourism information is relatively easy to find, there is a major issue with respect to supply-side information. Although business sensitivity is recognized, there is a need for greater researcher access to information with respect to tourism business consumption patterns and emissions. Without such data, it becomes much harder for researchers to respond to sectoral needs to improve efficiency and lower emissions.

From a more critical policy perspective, the degree of policy failure with respect to conservation of natural capital is considerable, but it has not yet been matched by an accompanying conceptual policy change that removes the focus on economic growth and the market (Czech 2008). The national and global institutional arrangements that surround sustainable development, tourism and climate change remain wedded to assumptions based on the compatibility between the environment and economic growth and acceptance of market forces. While this is the case, the implementation of alternative steady-state approaches remains problematic (Gareau 2008). Far too much attention has been given to the assumption that a well designed institution is 'good' because it facilitates cooperation and network development (Becken and Clapcott 2011), rather than a focus on norms and institutionalization, as first and necessary steps in the assessment of what kind of changes institutional arrangements are promoting and their potential outcomes. Such an approach, which is widespread in tourism studies (Hall 2011b), has reinforced limited and incremental change rather than conceptual policy learning and paradigm change.

Behavioural knowledge

There is a need for a much greater understanding of knowledge of individual and organizational behaviour with respect to climate change. Importantly, this needs to be approached not only at a global comparative scale, but also within the context of situated national, cultural and local knowledge. At the individual level, there are many knowledge gaps with respect to its place in the complex psychological process of travel planning and *in situ* tourist decisions, making this an important area for further research. Such research should be concerned with not only the overall relationship between weather and tourist satisfaction, but also the gaps that exist with regard to the entry points and decisiveness of weather and climate in tourists' decision-making processes overall. Undertaking studies in this area would greatly enhance understanding of the longer-term capacities of tourists to respond to climate change. Work is similarly needed to improve understanding of climate change-induced environmental change and where limits of acceptable change begin to influence tourist satisfaction.

In terms of behaviour, there is also a need for a better understanding of how environmental and climate change concerns are translated into action. For example, although there appears to be substantial willingness to pay for carbon offsets, there is a considerable gap with regard to actual payments (Gössling *et al.* 2009d). Furthermore, there is a need to gain greater insights into consumer, business and destination responses to new regulatory and taxation regimes that may be introduced as a result of climate change, particularly to identify compliance levels and behavioural effects.

Behavioural studies also need to be conducted on and within business and tourism organizations with respect to gaining a better understanding of their climate change perceptions and how they are translated into actions. This will require focusing particularly on physical and financial risk perceptions as well as the capacity for business innovation. In terms of assisting business decision-making, researchers also need a better understanding of business information requirements.

Scales and interactions

Integrating insights from local-scale case studies with global-scale models is a major issue in climate change research. As Wilbanks and Kates (1999: 601) argued, '[i]mproving the understanding of linkages between macroscale and microscale phenomena and process is one of the great overarching intellectual challenges of our age in a wide range of sciences'. Much climate change research has focused on the macro-scale, with top-down interpretation of global climate change for local places. 'Focusing exclusively on a larger scale can lead to ready generalizations that are just that – much too general' (Wilbanks and Kates 1999: 608). With the limitations of the top-down paradigm increasingly recognized, an increased emphasis on bottom-up, local-scale studies is recommended, as well as efforts to develop conceptual frameworks that provide greater compatibility and integration between top-down and bottom-up approaches.

As we stress throughout this volume, subnational destination characteristics and market dynamics matter tremendously for tourism. With the development of global-scale models of the implications of climate change for international tourism flows and greater place-specific details emerging from subnational case studies during the 2000s, a greater focus on integrating knowledge from different scales remains an essential task for the climate change and tourism research community. Understanding the interaction of scales has always been an interest of geographers (Hall and Lew 2009), but is also a key focus of scholars with interests in complexity and system dynamics, and these disciplines have much to contribute to this task.

Knowledge of adaptation and vulnerability

As noted above, knowledge of the implications of climate change for tourism is extremely uneven. This situation is also reflected in the state of knowledge on adaptation costs and benefits both within and between sectors (Agrawala and Fankhauser 2008; Parry *et al.* 2009b).

It is important to recognize that that the adaptive capacities of the developed world and the less-developed economies are markedly different. That a destination is physically vulnerable to climate change does not automatically mean it will be unable to adapt, although it is acknowledged that some destinations, such as the Maldives and other low-lying island states, would clearly have difficulties in responding to SLR. Nevertheless, as Table 8.5 illustrates, those economies that are highly dependent on tourism tend to be island states, which are also some of the most vulnerable to the effects of climate change (McMullen and Jabbour 2009). However, other factors that contribute greatly to destination vulnerability include tourism being concentrated in coastal regions and estuarine systems; having environmentally based activities such as safaris and ecotourism; and having a substantial proportion of its market based on long-haul travel (more than six hours) (Gössling and Hall 2006a; Gössling *et al.* 2008, 2009a; Scott *et al.* 2008a).

Destinations should seek to assess their dependency and vulnerability regarding energy-intense tourism. Destinations would seem well advised to restructure their tourism products towards low-carbon and/or high-value tourism. Many models now exist to strategically reduce the energy intensity of tourism markets with a focus on maintaining or increasing yield. For instance, Gössling *et al.* (2005) have used eco-efficiency as an integrated indicator combining ecological and economic information. Knowledge of the energy intensity of various markets and activities as well as life-cycle analysis of infrastructure can thus help to make decisions in favour of low-carbon tourism that generates high revenue (Gössling 2010).

In the case of developing countries, current levels of investment are already considered far from adequate, and this leads to high current vulnerability to climate, including its variability and extremes – what Burton (2004) refers to as an 'adaptation deficit'. This is one of the main reasons why developing countries are so vulnerable to climate change. Achievement of the Millennium Development Goals would go a long way towards reducing the adaptation deficit. Sachs and McArthur (2005) costed this at about US$200 billion by 2015, while Parry *et al.* (2009b) suggested that it would be equivalent to official development assistance of approximately 0.7 per cent of GDP of OECD countries.

Table 8.5 *Developing economies with visitor expenditure greater than 5 per cent of GDP*

Developing economies	UNCTAD economic groupings*	Nominal GDP 2006 (US$ million)	Total expenditure of visitors 2006 (US$ million)	Visitor expenditure as percentage of GDP 2006	Percentage area in low-elevation coastal zones <10 million¶	Renewable resources (percentage of total primary energy supply)
China, Macao SAR	A	14,293	9,337	65.3	Island	–
Palau	A, F	156	90	57.6	Island	–
Anguilla	A	201	†107	53.2	Island	–
Cook Islands	B	177	90	50.8	Island	–
Maldives	B, F, G	907	†434	47.9	100	–
Seychelles	A, F	707	323	45.7	Island	–
Aruba	A	2,380	1,076	45.2	Island	–
Turks and Caicos Islands	A	648	†292	45.1	Island	–
St Lucia	A, F	933	†347	37.2	4.1	–
Vanuatu	B, F, G	361	109	35.6	7.4	–
Antigua and Barbuda	A, F	ᵉ1,002	†347	34.6	Island	–
Bahamas	A, F	ᵉ6,175	2,079	33.7	93.2	–
Barbados	A, F	3,446	†978	28.4	Island	–
Cape Verde	B, F	1,116	286	25.6	Island	–
Lebanon	A	22,064	5,491	24.9	1.8	5
St Kitts and Nevis	A, F	487	†116	23.8	Island	–
Saint Vincent and Grenadines	B, F	ᵉ480	†113	23.5	Island	–
Grenada	A, F	ᵉ413	†93	22.5	6.5	–
Dominica	B, F	316	†68	21.5	4.5	–
Samoa	B, F, G	ᵉ429	91	21.2	8.4	–
Belize	B	1,217	†253	20.8	15.6	–
Cayman Islands	A	2,447	509	20.8	Island	–
Fiji	B, F	3,103	636	20.5	10.6	–
Jamaica	B, F	10,316	2,094	20.3	6.9	10
Mauritius	B, E	6,413	1,302	20.3	6.1	–
Zimbabwe	C, E	1,765	338	19.2	0.0	70
Sao Tome and Principe	C, D, F, G	74	††14	18.9	Island	–
Montserrat	A	46	†8	17.4	Island	–
Cambodia	C, G	ᵉ6,648	1,080	16.2	7.4	71
Jordan	B	14,336	2,008	14.0	0.0	2
French Polynesia	A	5,643	785	13.9	Island	–
Gambia	C, D, G	511	69	13.5	–	–

Continued

Table 8.5 Continued

Developing economies	UNCTAD economic groupings*	Nominal GDP 2006 (US$ million)	Total expenditure of visitors 2006 (US$ million)	Visitor expenditure as percentage of GDP 2006	Percentage area in low-elevation coastal zones <10 million¶	Renewable resources (percentage of total primary energy supply)
Dominican Republic	B	31,593	†3,792	12.0	4.7	–
Bahrain	A, H	ᵉ15,884	1,786	11.2	Island	0
Morocco	B	65,365	6,899	10.5	–	4
Tunisia	B	30,673	2,999	9.8	3.3	14
Mongolia	C, E	ᵉ3,041	261	8.6	0.0	3
Costa Rica	B	ᵉ22,333	1,890	8.5	3.5	53
Panama	B	ᵉ17,251	1,450	8.4	–	25
Malaysia	B, I, J, K	148,941	12,355	8.3	6.2	5
China, Hong Kong SAR	A, I, K	ᵉ188,381	15,311	8.1		0
Thailand	B, I, J, K	206,247	15,653	7.9	6.9	19
Egypt	B	110,075	8,133	7.4	–	4
Namibia	B	ᵉ6,402	473	7.4	–	21
Tanzania	C, D, G	13,228	950	7.2	0.3	90
Madagascar	C, D, G	5,506	386	7.0	2.7	–
Micronesia (Federated States of)	B, F	245	†17	6.9	Island	–
Tonga	B, F	232	†16	6.9	Island	–
Comoros	C, D, F, G	398	27	6.8	Island	–
Syrian Arab Republic	B, H	ᵉ31,320	2,113	6.7	–	2
El Salvador	B	ᵉ18,485	1,175	6.4	–	58
Botswana	B, E	8,836	539	6.1	–	23
Suriname	B	1,820	109	6.0	–	
Vietnam	C	57,983	3,200	5.5	20.2	49
Eritrea	C, D, G	ᵉ1,088	60	5.5	–	74
New Caledonia	A	4,743	†258	5.4	Island	
Singapore	A, I, J, K	132,155	7,069	5.3	Island	0
Honduras	C	9,301	490	5.3	5.8	45
Laos	C, E, G	ᵉ3,484	†173	5.0	0.0	–

*A–C, 2000 per capita current GDP = (A) >$4500 (high-income); (B) $1000–4500 (middle-income); (C) <$1000 (low-income); D, heavily indebted poor countries (HIPCs); E, landlocked developing countries (LLDCs); F, small island developing states (SIDS); G, least developed countries (LDCs); H, major petroleum exporters; I, major exporters of manufactured goods; J, emerging economies; K, newly industrialized economies
ᵉEstimate
†Visitor expenditure excluding transport
‡Most recent available figure
¶Where data are unavailable, island status is noted
SAR, Special Administrative Region
Source: derived from UNCTAD (2008); UNDP (2010); World Bank (2010) data

The expansion of international tourism has also been heavily promoted by the UNWTO as a form of development assistance, although the extent to which this is appropriate is something of a moot point, given not only the outflow of funds from developing countries but also the impacts of the emissions from such travel (Gössling *et al.* 2009a).

Another major research gap when considering adaptation to climate change is how much impact is being avoided by such modifications. Most impacts are projected to increase non-linearly with climate change, and adaptation costs are similar to those of impacts (Parry *et al.* 2007). Therefore it will probably be inexpensive to avoid some impacts, but prohibitively expensive to avoid others; and some impacts, such as ocean acidification and coral reef bleaching (see chapter 5), probably cannot be avoided in the foreseeable future even if funds were unlimited because the technologies are not available (Parry *et al.* 2009b) (see Figures 8.1 and 8.2). Although the nature of the curve in Figures 8.1 and 8.2 will differ greatly between destinations (high- versus low-elevation ski resorts; inshore shallow reefs near pollution sources versus offshore, deeper, well managed reefs), it is likely in most cases that 'adaptation to (say) the first 10 per cent of damage will be disproportionately cheaper than for 90 per cent of damage' (Parry *et al.* 2009b: 12). There is therefore an urgent need to assess the relative cost–benefits of different adaptive measures at destinations; and, as part of this, to identify the likely residual damage – the damage that will not be adapted to over the longer term because it is either unfeasible or too expensive; as well as to try and assess how adaptation costs themselves will change over time. Depending on its nature, such residual damage may have significant effects on tourism. Future research on adaptation will need to be conducted at various scales and to utilize a wide range of case studies conducted in different environments, and in destinations with different emphases on markets and activities. This is especially significant because destination and business decisions about adaptive measures will affect market perceptions as

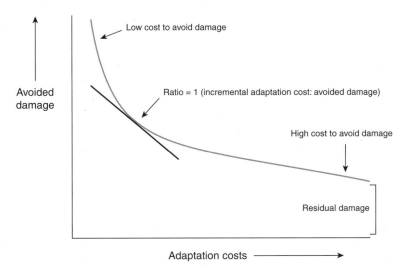

Figure 8.1 *Adaptation costs, avoided damage and residual damage compared over time*

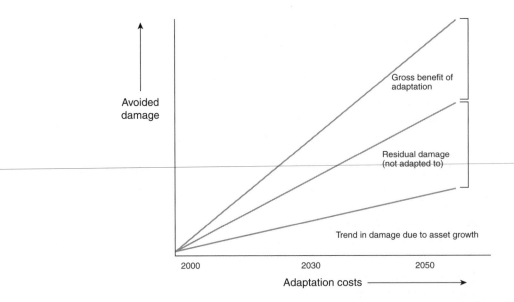

Figure 8.2 *Adaptation costs, avoided damage and residual damage, 2000–50*

well as the capacity to engage in various activities. Different adaptations will affect markets differentially. Such research would need to specify the time period clearly, and expected climate changes and results for multiple timeframes would be useful. We would also support Parry *et al.*'s (2009b: 14) recommendations that

> Non-climate trends need careful portrayal, especially the future levels of non-climate investment. Costs of adapting to varying amounts of impact should be analysed, thus providing a choice range for preparedness to pay; and there needs to be some analysis of the residual impact that adaptation is not likely to avoid, and the resulting damage costs that we need to anticipate.

As has been emphasized throughout the book, climate change can be adapted to, but such responses do cost, and destinations, industry and government will need to consider their investment decisions carefully in order to ensure they are investing in the most appropriate manner. But, of course, this also assumes that the effects of climate change at a global scale are not such that the level of economic and financial insecurity, along with resource and environmental insecurity, is such that the international tourism industry has been affected severely. A number of authors warn of grave consequences of +4°C global warming for ecosystems, the global economy and society (see Hansen 2009; New *et al.* 2011) – a world in which concerns over tourism development would largely be considered inconsequential.

As indicated, the vulnerability of the tourism sector is not just a matter of the direct physical impacts of climate change on the environment, knowledge of which, as a number of chapters in this book indicate, is reasonably strong in general terms for a number of tourism assets and activities. The relative vulnerability of a destination is also related to a number of other socio-economic dimensions, such as the overall level of economic development, infrastructure capacity, education and literacy levels, food and energy security, and the proportion of the population below the poverty line (which together greatly determine the adaptive capacity of a region). The relative economic dependence on tourism is also significant, especially the extent to which it is dependent on long-haul markets. In the broader context, the relative vulnerability of major competitors and the relative capacity to transfer economic resources from tourism to other sectors are also important factors in determining destination vulnerability.

Table 8.6 provides a summary of various global assessments of the relative vulnerability of tourism to climate change by region. Most are expert-based assessments, with some based on comprehensive evaluation of the literature available at the time (Gössling and Hall 2006b; Hall 2008a; Scott *et al.* 2008a) and others based on undisclosed data sets and index weighting criteria (Deutsche Bank Research 2008). Hamilton *et al.* (2005) is based entirely on a macro-scale international tourism demand model, which, as discussed in chapter 7, does not account for climate change-induced environmental changes at destinations or changes in transportation costs. Other ratings of regional tourism vulnerability are available, but some examine only a single type of tourism destination (e.g. beach–coastal tourism – Perch-Nielsen 2010), or are so speculative and with known errors (see e.g. comments in chapter 5 with respect to the Halifax Insurance report of 2006) that they are not included in Table 8.6. Despite the different approaches, some commonalities emerge, particularly with respect to SIDS that are highly economically dependent on tourism, and nations in Africa and Asia that are looking to tourism as an important component of future economic development. The confluence of large anticipated impacts on natural tourism assets, heightened security risks and potential destabilization of governments, relatively lower adaptive capacity, and distance to major markets (and thus potential sensitivity to mitigation policies) put tourism in these developing regions at the greatest risk. However, as noted above, and by Gössling and Hall (2006b) and Scott *et al.* (2008a), until the major geographical gaps in climate change and tourism research are addressed and more systematic region-level assessments are conducted, definitive statements on the net impacts on the tourism sector and the potential for future tourism development will not be possible.

While such vulnerability rankings are sometimes considered academic exercises, they should not be trivialized. With limits on international funds to support climate change adaptation, some nations and NGOs have suggested that adaptation funding through the UNFCCC and other mechanisms should be prioritized on the basis of the vulnerability of countries, and vulnerability indices have been identified as one approach to determine which countries are particularly vulnerable (Füssel 2009; Klein 2009). In addition to being a potential tool for guiding international adaptation funds and official development assistance, such vulnerability rankings could also be used to prioritize sectoral support for adaptation. The third action item of WTTC's (2009: 15) climate change strategy states that the tourism industry '… will identify climate change hotspots [most vulnerable destinations] where we operate and develop strategies to help local communities increase their adaptive capacity to cope with the impacts of climate change'. A tourism-specific vulnerability index could be used to identify hotspots in need of priority assistance. There is also the danger that such rankings could be used by investors to identify countries and destinations that pose a greater financial risk, deterring tourism development, in some cases perhaps unnecessarily. With increased interest in climate change vulnerability

Table 8.6 *Estimated climate change vulnerability of tourism by region*

Timeframe assessed	Vulnerability categories	Developing regions	Developed regions	Study
2030	Negatively affected (slightly or strongly)[*]	South America, Caribbean/ Mexico, Southeast Asia (including China and India), Middle East, Africa[†]	Mediterranean Europe and Balkans, Australia	Deutsche Bank Research (2008)
Mid-21st century	Negatively affected (moderately to strongly)[‡]	Africa, Asia (especially coastal China), Latin America, small island nations	Mediterranean, Florida and south-west USA, Queensland	Gössling and Hall (2006b); Hall (2008a)
2025, +4°C warming scenario	Negative impact on tourist arrivals[¶]	Caribbean/ Mexico, South America (except Chile), Africa (except Zambia and Zimbabwe), Middle East, Southeast Asia (except China)	Southern Europe/ Mediterranean, Australia and New Zealand	Hamilton et al. (2005)
Mid-21st century	Vulnerability hotspots[§]	Caribbean, Indian and Pacific Ocean small island nations[**]	Mediterranean, Australia and New Zealand	Scott et al. (2008a)

[*]Impact criteria considered: climate changes, regulatory burdens, substitution effects, adaptation capacities (data sources and indicators not identified)
[†]Many nations in the Middle East and Africa were 'not examined' but no rationale was provided regarding availability of information for selected nations
[‡]Impact criteria considered: land and marine biodiversity loss, urbanization, water security, sea-level rise, regime change, fuel costs, temperature changes, disease potential
[¶]Impact criteria considered: change in annual average temperature
[§]Impact criteria considered: summer and winter climate change, increases in extreme events, sea-level rise, land and marine biodiversity loss, water scarcity, political destabilization, health impacts/disease potential, transportation costs, relative importance of tourism to the economy (also utilizes results of Gössling and Hall 2006b)
[**]South America, Africa, Middle East, Southeast Asia also identified as potentially vulnerable but not listed as hotspots due to insufficient information on magnitude of potential impacts

indicators and indices, the development of more robust indicators relevant to the tourism sector is a useful area for future research.

CONCLUSION

> 'Our most important responsibility to the future is not to coerce it but to attend to it. Collectively, [such actions] might be called "future preservation", just as an analogous activity carried out in the present is called historical preservation.'
>
> Kevin Lynch (1972: 115)

This chapter outlines some of the major gaps and issues with respect to knowledge on tourism and climate change and future research. It highlights that climate change research could be framed in different ways, with quite substantial implications for the role of social science research on climate change in particular. It also suggests that the manner in which climate change has predominantly been framed with respect to 'objective' science has also been one of the reasons why the politics of climate change has become so dramatic. It therefore suggests that acknowledgement of the political implications of climate research actually needs to become more overt, and to be embraced as part of an improved communication capacity for climate change science. It also suggests that there is a significant research gap with respect to the policy-making process overall for tourism and climate change.

A number of other research gaps are identified with respect to both demand and organizational behaviour with respect to climate change. Also of major importance is the need to connect research conducted at various scales of analysis, as well as the substantial variability with respect to regional and destination analysis. In a number of regions, and less developed countries in particular, some of those places that may be the most vulnerable to climate change have the least amount of information available for decision-making.

Tourists have the greatest capacity to adapt to the impacts of climate change, with relative freedom to avoid destinations affected by climate change. Climate, the natural environment, personal safety and travel cost are four primary factors in destination choice, and global climate change is anticipated to have significant impacts on all these factors at the regional level. It is tourists' responses to the complexity of destination impacts that will reshape consumer demand patterns and play a pivotal role in the eventual impacts on destinations. The perceptions of future impacts of climate change at destinations will be central to the decision-making of tourists, tourism investors, governments and development agencies alike, since perceptions of climate conditions or environmental changes are just as important to consumer choices as the actual conditions. Perceptions of climate change impacts in a region are often heavily influenced by the nature of media coverage. It is therefore critical to avoid the type of media speculation and misinformation that is likely to be more damaging to a tourism destination in the near term than the actual impacts of climate (see discussion in chapters 5 and 7).

All tourism businesses and destinations will need to adapt to climate change in order to minimize associated risks and capitalize on new opportunities in an economically, socially and environmentally sustainable manner. However, knowledge of the capability of current climate adaptations to adapt successfully to future change remains rudimentary, and therefore the future residual damage of climate change at destinations remains unknown. As chapters 4, 6 and 7 discuss, climate change is entering slowly into the decision-making processes of a range of tourism stakeholders – investors, insurance companies, tourism enterprises, governments, development organizations, national park and conservation agencies, and tourists. However, studies that have examined the climate change risk appraisal of local tourism officials and operators have consistently found relatively low levels of concern and little evidence of long-term strategic planning in anticipation of future changes in climate. Considering that the large information requirements, policy changes and investments required for effective adaptation by tourism destinations will require decades to implement in some cases, the process of adaptation must commence now for destinations anticipated to be among those affected by mid-century. Development organizations in particular will need to enhance their understanding of not only the implications of climate change for the sustainability of tourism products and supporting services at the destination (e.g. coastal zones, coral reefs, water supply, heritage assets) and of tourism development projects, but also the implications of emerging climate policy regimes for relative costs and accessibility. Most importantly, a more critical perspective is required, not only with respect to policy and the framing of climate change research, but also as to whether tourism will be the best development alternative. As Le Blanc (2009) notes regarding the relationship between sustainable development and climate change, in order to be effective, efforts to combat climate change will have to be integrated into the broader context of social and economic development. Therefore broader debates and issues that occur with respect to the sustainability of tourism are going to have a direct effect on tourism's contribution to, and relationships with, climate change as well as broader environmental change (Gössling 2002; Gössling and Hall 2006a; Scott 2006a, 2011; Hall 2011a; Weaver 2011).

Is tourism ready? Undoubtedly, the information base for climate change decision-making is steadily improving. As Scott (2011: 19) notes, '[t]he field is now developing a critical mass of knowledge and research techniques better suited to deliver relevant knowledge for government and the private sector'. Nonetheless, this book identifies a large number of remaining knowledge gaps, and limited preparedness among a range of tourism stakeholders. A multi-sectoral comparison of climate change preparedness by the international business advisory firm KPMG (2008) similarly concluded that both tourism and aviation sectors were in the 'danger zone', based on relatively low awareness of climate change risks and little evidence of strategic planning to cope with the associated physical and regulatory challenges. The task ahead for the tourism community to collaborate with scholars from a broad range of disciplines and professionals from transport, financial services, international development, coastal and protected areas management, and other sectors to develop climate change response strategies remains enormous and should not be underestimated.

FURTHER READING AND WEBSITES

For a debate over the place of climate change in sustainable tourism, see:

Weaver, D. (2011) 'Can sustainable tourism survive climate change?', *Journal of Sustainable Tourism*, 19(1): 5–15; and the response by:

Scott, D. (2011) 'Why sustainable tourism must address climate change', *Journal of Sustainable Tourism*, 19: 17–34.

For those interested in various dimensions of the relationship between tourism, sustainable development and climate change in policy terms, see:

Gössling, S., Hall, C.M. and Weaver, D. (2009c) *Sustainable Tourism Futures: Perspectives on Systems, Restructuring and Innovations*. New York: Routledge.

Hall, C.M. (2008b) *Tourism Planning*. Harlow, UK: Pearson.

On the issue of how climate change issues are framed, see:

Demeritt, D. (2001a) 'The construction of global warming and the politics of science', *Annals of the Association of American Geographers*, 91: 307–337; and the response by:

Schneider, S. (2001) 'A constructive deconstruction of deconstructionists: a response to Demeritt', *Annals of the Association of American Geographers*, 91: 338–344; and the further response by:

Demeritt, D. (2001b) 'Science and the understanding of science: a reply to Schneider', *Annals of the Association of American Geographers*, 91: 345–348; also

Hulme, M. (2009) *Why We Disagree About Climate Change: Understanding Controversy, Inaction and Opportunity*. Cambridge: Cambridge University Press.

The Guardian provides a good overview of 'climategate' and the issues associated with it:

'Climate wars. Guardian special investigation': www.guardian.co.uk/environment/series/climate-wars-hacked-emails.

'Q&A: Climategate': www.guardian.co.uk/environment/2010/jul/07/climate-emails-question-answer.

Bibliography

Aall, C. (2011) 'Energy use and leisure consumption in Norway: an analysis and reduction strategy', *Journal of Sustainable Tourism*, 19(6): 729–745.

Abegg, B., Konig, U., Buerki, R. and Elsasser, H. (1998) 'Climate impact assessment in tourism', *Applied Geography and Development*, 51: 81–93.

Abegg, B., Agrawala, S., Crick, F. and De Montfalcon, A. (2007) 'Climate change impacts and adaptation in winter tourism', in *Climate Change in the European Alps: Adapting Winter Tourism and Natural Hazards Management*. Paris: Organisation for Economic Cooperation and Development, pp. 25–58.

Access Economics (2007) *Measuring the Economic and Financial Value of the Great Barrier Reef Marine Park: 2005–2006*, a report to the Great Barrier Reef Marine Park Authority. Canberra: Access Economics.

ACIA (2005) *Arctic Climate Impacts Assessment*. Cambridge: Cambridge University Press.

Ackerman, F. and Munitz, C. (2011) *Climate Damages in the FUND Model: A Disaggregated Analysis*, Stockholm Environment Institute Working Paper WP-US-1105. Washington, DC: Stockholm Environment Institute.

Ackerman, F., DeCanio, S., Howarth, R. and Sheeran, K. (2009) 'Limitations of integrated assessment models of climate change', *Climatic Change*, 95: 297–315.

Adams, R. (1973) 'Uncertainty in nature, cognitive dissonance, and the perceptual distortion of environmental information: weather forecasts and New England beach trip decisions', *Economic Geography*, 49: 297–307.

Addison, A. (2008) *Disappearing World: 101 of the Earth's Most Extraordinary and Endangered Places*. New York: HarperCollins.

Adger, N. and Kelly, M. (1999) 'Social vulnerability to climate change and the architecture of entitlements', *Mitigation and Adaptation Strategies for Global Change*, 4(3): 253–266.

Adger, W.N., Agrawala, S., Mirza, M.M.Q., Conde, C., O'Brien, K., Pulhin, J., Pulwarty, R., Smit, B. and Takahashi, K. (2007) 'Assessment of adaptation practices, options, constraints and capacity', in M.L. Parry, O.F. Canziani, J.P. Palutikof, P.J. van der Linden and C.E. Hanson (eds) *Climate Change 2007: Impacts, Adaptation and Vulnerability. Contribution of Working Group II to the Fourth Assessment Report of the Intergovernmental Panel on Climate Change*. Cambridge: Cambridge University Press.

Adler, J. (2006) 'Climate change: prediction perils', *Newsweek*, 23 October.

Agnew, M. (1995) 'Tourism', in J. Palutikof, S. Subak and M. Agnew (eds) *Economic Impacts of the Hot Summer and Unusually Warm Year of 1995*. Norwich: Department of the Environment.

Agnew, M. and Palutikof, J. (2006) 'Impacts of short-term climate variability in the UK on demand for domestic and international tourism', *Climate Research*, 31: 109–120.

Agrawala, S. and Fankhauser, S. (eds) (2008) *Economic Aspects of Adaptation to Climate Change. Costs, Benefits and Policy Instruments*. Paris: Organisation for Economic Co-operation and Development.

AHDR (2004) *Arctic Human Development Report*. Akureyri: Stefansson Arctic Institute.

Ahn, S., De Steiguer, J., Palmquest, R. and Holmes, T. (2000) 'Economic analysis of the potential impact of climate change on recreational trout fishing in the southern Appalachians: an application of a nested multinomial logit model', *Climatic Change*, 45: 493–509.

AIACC (2007) *Climate Change Vulnerability and Adaptation in Developing Country Regions*. Draft Final Report of the AIACC [Assessments of Impacts and Adaptations to Climate Change] Project. www.aiaccproject.org/Final%20Reports/final_reports.html

Air New Zealand (2008) *World-First Biofuel Test Flight*. www.airnewzealand.co.nz/aboutus/biofuel-test/default.htm

Airbus (2011) *Airbus Global Market Forecast 2010–2029*. www.airbus.com/company/market/gmf2010

Alcamo, J., Moreno, J.M., Nováky, B., Bindi, M., Corobov, R., Devoy, R.J.N. Giannakopoulos, C., Martin, E., Olesen, J.E. and Shvidenko, A. (2007) 'Europe', in M.L. Parry, O.F. Canziani, J.P. Palutikof, P.J. van der Linden and C.E. Hanson (eds) *Climate Change 2007: Impacts, Adaptation and Vulnerability. Contribution of Working Group II to the Fourth Assessment Report of the Intergovernmental Panel on Climate Change*. Cambridge: Cambridge University Press.

Allianz Group and World Wildlife Fund (2006) *Climate Change and Insurance: An Agenda for Action in the United States*. New York: Allianz Group.

Alongi, D.M. (2008) 'Mangrove forests: resilience, protection from tsunamis, and responses to global climate change', *Estuarine Coastal and Shelf Science*, 76: 1–13.

Altalo, M. and Hale, M. (2002) *Requirements of the US Recreation and Tourism Industry for Climate, Weather and Ocean Information*. Consultant's report to National Oceanic and Atmospheric Administration.

Amelung, B. and Lamen, M. (2007) 'Estimating the greenhouse gas emission from Antarctic tourism', *Tourism in Marine Environments*, 4(2–3): 121–133.

Amelung, B. and Viner, D. (2006) 'Mediterranean tourism: exploring the future with the tourism climate index', *Journal of Sustainable Tourism*, 14(4): 349–366.

Amelung, B., Nicholls, S. and Viner, D. (2007) 'Implications of global climate change for tourism flows and seasonality', *Journal of Travel Research*, 45(2): 85–296.

Amelung, B., Moreno, A. and Scott, D. (2008) 'The place of tourism in the IPCC Fourth Assessment Report: a review', *Tourism Review International*, 12(1): 5–12.

American Meteorological Society (2011) *Glossary of Meteorology*. amsglossary.allenpress.com/glossary/search?id=applied-climatology

Amsterdam Climate Office (2008) *New Climate Amsterdam*. Amsterdam: Amsterdam Climate Office.

Anderegg, W.R.L., Prall, J.W., Harold, J. and Schneider, S.H. (2010) 'Expert credibility in climate change', *Proceedings of the National Academy of Sciences, USA*, 107(27): 12107–12109.

Anderson, K. and Bows, A. (2011) 'Beyond "dangerous" climate change: emission scenarios for a new world', *Philosophical Transactions of the Royal Society*, 369: 20–44.

Andersson, J. (1997) 'The value of coral reef for the current and potential tourism industry on Unguja Island, Zanzibar', in R. Johnstone, J. Francis and C. Mohando (eds) *Coral Reefs: Values, Threats and Solutions*, Proceedings of the National Conference on Coral Reefs, Zanzibar, Tanzania.

Andersson, J. (2007) 'The recreational costs of coral bleaching – a stated and revealed preference study of international tourists', *Ecological Economics*, 62: 704–715.

Andressen, B. and Hall, M. (1988/89) 'The importance of intense negative outdoor experiences', *Recreation Australia*, 8(1): 6–8.

Anisimov, O.A., Vaughan, D.G., Callaghan, T.V., Furgal, H., Marchant, H., Prowse, T.D., Vilhjálmsson, H. and Walsh, J.E. (2007) 'Polar regions (Arctic and Antarctic)', in M.L. Parry, O.F. Canziani, J.P. Palutikof, P.J. van der Linden and C.E. Hanson (eds) *Climate Change 2007: Impacts, Adaptation and Vulnerability. Contribution of Working Group II to the Fourth Assessment Report of the Intergovernmental Panel on Climate Change*. Cambridge: Cambridge University Press.

Arora, V., Scinocca, J., Boer, G., Christian, J., Denman, K., Flato, G., Kharin, V., Lee, W. and Merryfield, W. (2011) 'Carbon emission limits required to satisfy future representative concentration pathways of greenhouse gases', *Geophysical Research Letters*, 38(5): L05805.

Arrhenius, S. (1896) 'On the influence of carbonic acid in the air upon the temperature of the ground', *London, Edinburgh, and Dublin Philosophical Magazine and Journal of Science*, 41(5): 237–275.

Artal Tur, A., Garcá Sánchez, A. and Sánchez García, J.F. (2008) 'The length of stay determinants for sun-and-sand tourism: an application for the Region of Murcia', *XVI Jornadas ASEPUMA – IV Encuentro Internacional*, 16(1): 801. www.uv.es/asepuma/XVI/801.pdf

Associated Press (2002a) 'Royal Gorge tourism hurt by fires, drought', 3 September 2002.

Associated Press (2002b) 'Rough year for rafters', 3 September 2002.

Association of British Insurers (2004) *A Changing Climate for Insurance: A Summary Report for Chief Executives and Policymakers*. www.climate-insurance.org/upload/pdf/Dlugolecki_Catovsky2004_A_Changing_Climate_for_Insurance.pdf

ATAG (2011) *Powering the Future of Flight. The Six Easy Steps to Growing a Viable Aviation Biofuels Industry*. Geneva: Air Transport Action Group.

Atmosfair (2010) *Offset your Cruise*. www.atmosfair.de/en/act-now/contribute-now-compact/

Atmosfair (2011) *Atmosfair Airline Index 2011*. cdn.atmosfair.de/atmosfair_Airline_Index_2011_en.pdf

Aviation Global Deal Group (2009) *A Sectoral Approach to Addressing International Aviation Emissions*, Discussion Note 2.0, 9 June 2009. www.agdgroup.org/pdfs/090609_AGD_Discussion_Note_2.0.pdf

Bach, W. and Gössling, S. (1996) 'Klimaökologische Auswirkungen des Flugverkehrs', *Geographische Rundschau*, 48: 54–59.

Baker, A.C., Glynn, P.W. and Riegl, B. (2008) 'Climate change and coral reef bleaching: an ecological assessment of long-term impacts, recovery trends and future outlook', *Estuarine, Coastal and Shelf Science*, 80(4): 435–471.

Baker, B.B. and Moseley, R.K. (2007) 'Advancing treeline and retreating glaciers: implications for conservation in Yunnan, PR China', *Arctic Antarctic and Alpine Research*, 39: 200–209.

Bakos, G.C. and Soursos, M. (2003) 'Techno-economic assessment of a stand-alone PV/hybrid installation for low-cost electrification of a tourist resort in Greece', *Applied Energy*, 73:183–193.

Barkham, P. (2008) 'Waves of destruction', *The Guardian*, 17 April. www.guardian.co.uk/environment/2008/apr/17/flooding.climatechange?intcmp=239

Barnaby, W. (2009) 'Do nations go to war over water?', *Nature*, 458: 282–283.

Barnes, B. (2002) 'The heat wave vacation', *The Wall Street Journal*, 2 August 2002.

Barnett, J. and Adger, N. (2007) 'Climate change, human security and violent conflict', *Political Geography*, 26(6): 639–655.

Barnett, J. and O'Neill, S. (2010) 'Maladaptation', *Global Environmental Change*, 20(2): 211–213.

Barometern (2007) 'Turistchef vädjar om att ha kvar Fröken alg'. www.barometern.se/nyheter/lanet_old/turistchef-vadjar-ombr-att-ha-kvar-froken-alg(65867)gm

Barr, S., Shaw, G., Coles, T. and Prillwitz, J. (2010) '"A holiday is a holiday": practicing sustainability, home and away', *Journal of Transport Geography*, 18: 474–481.

Bartlett, L. (2007) *Australia Fears Jet Flight Guilt Could Hit Tourism*. www.spacemart.com/reports/Australia_Fears_Jet_Flight_Guilt_Could_Hit_Tourism_999.html

Baum, T. and Lundtorp, S. (eds) (2001) *Seasonality in Tourism*. Amsterdam: Pergamon.

BBC (2006) 'Package holidays "will be history"', *BBC News Europe*, 26 August. news.bbc.co.uk/2/hi/science/nature/5288092.stm

BBC (2010) 'Sunshine PR success for tourism', *BBC Jersey*, 30 July. news.bbc.co.uk/local/jersey/hi/people_and_places/arts_and_culture/newsid_8872000/8872807.stm

BBC (2011a) 'Jersey Tourism's weather campaign wins marketing award', *BBC News Jersey*, 25 March. www.bbc.co.uk/news/world-europe-jersey-12863736

BBC (2011b) 'Spain speed limit cut over high oil prices', *BBC News Europe*, 7 March. www.bbc.co.uk/news/world-europe-12663092

BBC (2011c) 'EU Commission tightens rules for biofuel use', *BBC News Europe*, 19 July. www.bbc.co.uk/news/world-europe-14205848

Beccali, M., La Gennusa, M., Lo Coco, L. and Rizzo, G. (2009) 'An empirical approach for ranking environmental and energy saving measures in the hotel sector', *Renewable Energy*, 34: 82–90.

Becheri, E. (1991) 'Rimini and Co – the end of a legend? Dealing with the algae effect', *Tourism Management*, 12(3): 229–235.

Bechrakis, D.A., McKeogh, E.J. and Gallagher, P.D. (2006) 'Simulation and operational assessment for a small autonomous wind–hydrogen energy system', *Energy Conversion and Management*, 47: 46–59.

BeCitizen (2010) *The Maldives' 2009 Carbon Audit*. Paris: BeCitizen.

Beckage, B., Osborne, B., Gavin, G.G., Pucko, C., Siccama, T. and Perkins, T. (2008) 'A rapid upward shift of a forest ecotone during 40 years of warming in the Green Mountains of Vermont', *Proceedings of the National Academy of Sciences, USA*, 105: 4197–4202.

Becken, S. (2002) 'Analysing international tourist flows to estimate energy use associated with air travel', *Journal of Sustainable Tourism*, 10: 114–131.

Becken, S. (2004) 'How tourists and tourism experts perceived climate change and carbon-offsetting schemes', *Journal of Sustainable Tourism*, 12: 332–345.

Becken, S. (2005) 'Harmonizing climate change adaptation and mitigation: the case of tourist resorts in Fiji', *Global Environmental Change*, 15: 381–393.

Becken, S. (2007) 'Tourists' perception of international air travel's impact on the global climate and potential climate change policies', *Journal of Sustainable Tourism*, 15: 351–368.

Becken, S. (2008) 'Developing indicators for managing tourism in the face of peak oil', *Tourism Management*, 29: 695–705.

Becken, S. and Clapcott, R. (2011) 'National tourism policy for climate change', *Journal of Policy Research in Tourism, Leisure & Events*, 3(1): 1–17.

Becken S. and Hay J. (2007) *Tourism and Climate Change – Risks and Opportunities*. Clevedon: Channel View Publications.

Becken, S. and Lennox, J. (2012) 'Implications of a long-term increase in oil prices for tourism', *Tourism Management*, 33(1): 133–142.

Becken, S. and Patterson, M. (2006) 'Measuring national carbon dioxide emissions from tourism as a key step towards achieving sustainable tourism', *Journal of Sustainable Tourism*, 14: 323–338.

Becken, S. and Wilson, J. (2007) 'Trip planning and decision making of rental vehicle tourists – a quasi-experiment', *Journal of Travel and Tourism Marketing*, 20(3/4): 47–63.

Becken, S., Wilson, J. and Reisinger, A. (2010) *Weather, Climate and Tourism: A New Zealand Perspective. Land Environment and People – Technical Report 20*. Lincoln, NZ: Lincoln University.

Becker, S. (2000) 'Bioclimatological rating of cities and resorts in South Africa according to the climate index', *International Journal of Climatology*, 20: 1403–1414.

Bedsworth, L.W. and Hanak, E. (2010) 'Adaptation to climate change', *Journal of the American Planning Association*, 76: 477–495.

Behringer, J., Buerki, R. and Fuhrer, J. (2000) 'Participatory integrated assessment of adaptation to climate change in alpine tourism and mountain agriculture', *Integrated Assessment*, 1: 331–338.

Bélisle, F.J. (1984) 'Tourism and food imports: the case of Jamaica', *Economic Development and Cultural Change*, 32(4): 819–842.

Beniston, M. (2004) 'The 2003 heat wave in Europe: a shape of things to come? An analysis based on Swiss climatological data and model simulations', *Geophysical Research Letters*, 31: L02202.

Beniston, M. (2008) 'Sustainability of the landscape of a UNESCO World Heritage Site in the Lake Geneva region in a greenhouse climate', *International Journal of Climatology*, 28: 1519–1524.

Berkelmans, R., De'ath, G., Kininmonth, S. and Skirving, W.J. (2004) 'A comparison of the 1998 and 2002 coral bleaching events of the Great Barrier Reef: spatial correlation, patterns and predictions', *Coral Reefs*, 23: 74–83.

Bermudez-Contreras, A., Thomson, M. and Infield, D.G. (2008) 'Renewable energy powered desalination in Baja California Sur, Mexico', *Desalination*, 220: 431–440.

Berrittella, M., Bigano, A., Roson, R. and Tol, R. (2006) *General Equilibrium Analysis of Climate Change Impacts on Tourism*, Working Paper No. 17. Ecological and Environmental Economics (EEE) Programme.

Besancenot, J. (1991) *Clima y Turismo*. Barcelona: Masson.

Besancenot, J.P., Mouiner, J. and De Lavenne, F. (1978) 'Les conditions climatiques du tourisme, littoral', *Norois*, 99: 357–382.

Bhaskar, R. (2008) *A Realist Theory of Science*, 2nd edn. Abingdon: Routledge.

Bhatnagar, P. (2005) 'Uncertainty and fear to rebuild: Mississippi officials say casino companies are hesitant to rebuild, state could lose billions', *CNNMoney*, http://money.cnn.com/2005/09/05/news/economy/katrina_business

Bibbings, L. and Burns, P. (2009) 'The end of tourism? Climate change and societal challenges', *21st Century Society*, 4(1): 31–51.

Bicknell, S. and McManus, P. (2006) 'The canary in the coalmine: Australian ski resorts and their response to climate change', *Geographical Research*, 44: 386–400.

Bigano, A., Hamilton, J. and Tol, R. (2005) 'The effect of climate change and extreme weather events on tourism in Italy', in A. Lanza, A. Markandya and F. Pigiaru (eds) *The Economics of Tourism and Sustainable Development*. Cheltenham: Edward Elgar.

Bigano, A., Hamilton, J.M. and Tol, R.S.J. (2006a) 'The impact of climate on holiday destination choice', *Climatic Change*, 76: 389–406.

Bigano, A., Hamilton, J.M., Maddison, D.J. and Tol, R.S.J. (2006b) 'Predicting tourism flows under climate change: an editorial comment on Gössling and Hall (2006)', *Climatic Change*, 79(3/4): 175–180.

Bigano, A., Hamilton, J. and Tol, R. (2007) 'The impact of climate change on domestic and international tourism: a simulation study', *Integrated Assessment Journal*, 7(1): 25–49.

Bigano, A., Bosello, F., Roson, R. and Tol, R. (2008) 'Economy-wide impacts of climate change: a joint analysis for sea level rise and tourism', *Mitigation and Adaptation Strategies for Global Change*, 13(8): 765–791.

Billett, S. (2010) 'Dividing climate change: global warming in the Indian mass media', *Climatic Change*, 99(1/2): 1–16.

Bin, O., Dumas, C., Poulter, B. and Whitehead, J. (2007) *Measuring the Impacts of Climate Change on North Carolina Coastal Resources*. Washington, DC: National Commission on Energy Policy.

Bird, E. (1985) *Coastline Changes*. New York: Wiley & Sons.

Board of the Millennium Ecosystem Assessment (2005) *Living Beyond Our Means: Natural Assets and Human Well-being, Statement from the Board*. Washington, DC and Nairobi: World Resources Institute and United Nations Environment Programme.

Boeing (2011) *Current Market Outlook 2010–2029*. www.boeing.com/commercial/cmo

Bohdanowicz, P. (2002) 'Thermal comfort and energy savings in the hotel industry', in *Proceedings of the 16th Congress of the International Society of Biometeorology*, Kansas City, Missouri. Boston: American Meteorological Society, 396–400.

Bohdanowicz, P. (2009) 'Theory and practice of environmental management and monitoring in hotel chains', in S. Gössling, C.M. Hall and D. Weaver (eds) *Sustainable Tourism Futures: Perspectives on Systems, Restructuring and Innovations*. London: Routledge.

Bohdanowicz, P. and Martinac, I. (2007) 'Determinants and benchmarking of resource consumption in hotels – case study of Hilton International and Scandic in Europe', *Energy and Building*, 39: 82–95.

Boko, M., Niang, I., Nyong, A., Vogel, C., Githeko, A., Medany, M., Osman-Elasha, B., Tabo, R. and Yanda, P. (2007) 'Africa', in M.L. Parry, O.F. Canziani, J.P. Palutikof, P.J. van der Linden and C.E. Hanson (eds) *Climate Change 2007: Impacts, Adaptation and Vulnerability. Contribution of Working Group II to the Fourth Assessment Report of the Intergovernmental Panel on Climate Change*. Cambridge: Cambridge University Press.

Boletín official del Estado (2011) 'Disposiciones Generales', *Ministerio de la Presidencia*, 55(I): 25249. www.boe.es/boe/dias/2011/03/05/pdfs/BOE-A-2011-4120.pdf

Bonazza, A., Messinaa, P., Sabbionia, C., Grossi, C.M. and Brimblecombe, P. (2009a) 'Mapping the impact of climate change on surface recession of carbonate buildings in Europe', *Science of the Total Environment*, 407(6): 2039–2050.

Bonazza, A., Sabbioni, C., Messina, P., Guaraldi, C. and De Nuntiis, P. (2009b) 'Climate change impact: mapping thermal stress on Carrara marble in Europe', *Science of the Total Environment*, 407(15): 4506–4512.

Boodhoo, Y. (2005) 'Climate services for sustainable development', *WMO Bulletin*, 54(1): 8–11.

Borger, J. (2008) 'US biofuel subsidies under attack at food summit', *guardian.co.uk*, 3 June 2008. www.guardian.co.uk/environment/2008/jun/03/biofuels.energy

Botkin, D.B., Saxe, H., Araújo, M.B., Betts, R., Bradshaw, R.H.W., Cedhagen, T., Chesson, P., Dawson, T.P., Etterson, J.R., Faith, D.P., Ferrier, S., Guisan, A., Hansen, A.S., Hilbert, D.W., Loehle, C., Margules, C., New, M., Sobel, M.J. and Stockwell, D.R.B. (2007) 'Forecasting the effects of global warming on biodiversity', *Bioscience*, 57: 227–236.

Boyd, A. (2007) 'Carbon tax threatens to ground Asia tourism', *Asian Times Online*. www.atimes.com/atimes/Asian_Economy/ID19Dk01.html

Boyd, D.R. (2001) *Canada vs. the OECD*. Victoria: Eco-Chair of Environmental Lay and Policy.

Boykoff, M. and Mansfield, M. (2008) '"Ye olde hot aire": reporting on human contributions to climate change in the UK tabloid press', *Environmental Research Letters*, 3(2): 1–8.

Boykoff, M.T. and Roberts, J.T. (2007) *Media Coverage of Climate Change: Current Trends, Strengths, Weaknesses*, Human Development Report Office Occasional Paper 2007/3. Geneva: United Nations Development Program.

Boykoff, T. and Boykoff, J. (2007) 'Climate change and journalistic norms: a case-study of US mass-media coverage', *Geoforum*, 28(6): 1190–1204.

Brander, L.M., Rehdanz, K., Tol, R.S.J. and van Beukering, P.J.H. (2009) *The Economic Impact of Ocean Acidification on Coral Reefs*, Working Paper No. 282. Dublin: ESRI.

Braun, O., Lohmann, M., Maksimovic, O. Meyer, M., Merkovic, A., Messerschmidt, E., Riedel, A. and Turner, M. (1999) 'Potential impact of climate change effects on preference for tourism destinations: a psychological pilot study', *Climate Research*, 11: 247–254.

Breiling, M. and Charamaza, P. (1999) 'The impact of global warming on winter tourism and skiing: a regionalized model of Austrian snow conditions', *Regional Environmental Change*, 1(1): 4–14.

Broderick, J. (2009) 'Voluntary carbon offsets: a contribution to sustainable tourism?', in S. Gössling, C.M. Hall and D. Weaver (eds) *Sustainable Tourism Futures. Perspectives on Systems, Restructuring and Innovations*. London: Routledge.

Brönnimann, S. (2002) 'Picturing climate change', *Climate Research*, 22: 87–95.

Brons, M., Pels, A., Nijkamp, P., Rietveld, P. (2002) 'Price elasticities of demand for passenger air travel: a meta-analysis', *Journal of Air Transport Management*, 8(3): 165–175.

Brook, B.W. (2009) 'Kakadu: a climate change hotspot', in W. Steffen, A. Burbidge, L. Hughes, R. Kitching, D. Lindenmayer, W. Musgrave, M. Stafford Smith and P. Werner (eds) *Australia's Biodiversity and Climate Change: A Strategic Assessment of the Vulnerability of Australia's Biodiversity to Climate Change*, A report to the Natural Resource Management Ministerial Council commissioned by the Australian Government. Canberra: CSIRO.

Brook, B.W. and Whitehead, P.J. (2006) 'The fragile millions of the tropical north', *Australasian Science*, 27: 29–30.

Brook, B.W., Sodhi, N.S. and Bradshaw, C.J.A. (2008) 'Synergies among extinction drivers under global change', *Trends in Ecology & Evolution*, 23: 453–460.

Brooks, M. (2011) 'Let's step out of the lab as the climate changes around us', *New Statesman*, 3 June. www.newstatesman.com/environment/2011/05/climate-scientists-ozone

Brouwer, R., Brander, L. and Van Beukering, P. (2008) '"A convenient truth": air travel passengers' willingess to pay to offset their CO_2 emissions', *Climatic Change*, 90: 299–313.

Brovkin, V., Raddatz, T., Reick, C.H., Claussen, M. and Gayler, V. (2009) 'Global biogeophysical interactions between forest and climate', *Geophysical Research Letters*, 36: L07405.

Brown, M.A., Southworth, F. and Sarzynski, A. (2008) *Shrinking the Carbon Footprint of Metropolitan America*. Washington, DC: Metropolitan Policy Program, Brookings Institute.

Browne, D. (2009) 'Rise in UK departure tax could cut London's European gateway role', *ETN*, 8 July. www.eturbonews.com/10256/rise-uk-departure-tax-could-cut-londons-european-gateway-role

de Bruijn, K., Dirven, R., Eijgelaar, E. and Peeters, P. (2008) *Reizen op grote voet 2005. De milieube-lasting van vakanties van Nederlanders. Een pilot-project in samenwerking met NBTC–NIPO Research*. Breda, the Netherlands: NHTV Breda University of Applied Sciences.

de Bruijn, K., Dirven, R., Eijgelaar, E. and Peeters, P. (2010) *Travelling Large in 2008. The Carbon Footprint of Dutch Holidaymakers in 2008 and the Development since 2002*. Breda, the Netherlands: NHTV Breda University of Applied Sciences, NRIT Research and NBTC–NIPO Research.

Bruun, P. (1962) 'Sea-level rise as a cause of shore erosion', *Journal Waterways and Harbours Division*, 88(1–3): 117–130.

Buckley, R. (2008) 'Misperceptions of climate change damage on coastal tourism: case study of Byron Bay, Australia', *Tourism Review International*, 12: 71–88.

Buhaug, Ø., Corbett, J.J., Endresen, Ø., Eyring, V., Faber, J., Hanayama, S., Lee, D.S., Lee, D., Lindstad, H., Markowska, A.Z., Mjelde, A., Nelissen, D., Nilsen, J., Pålsson, C., Winebrake, J.J., Wu, W.-Q. and Yoshida, K. (2009) *Second IMO GHG Study 2009*. London: International Maritime Organization.

Buisson, L., Thuiller, W., Leks, S., Lim, P. and Grenouillet, G. (2008) 'Climate change hastens the turnover of stream fish assemblages', *Global Change Biology*, 14: 2232–2248.

Bundesregierung (2010) *Luftverkehrssteuer*, 1 September. www.bundesregierung.de/nn_774/Content/DE/Artikel/2010/09/2010-09-01-luftverkehrssteuer.html

Bundesregierung (2011) *Die Klimapolitik der Bundesregierung* (*Climate Policy of the German Government*). www.bundesregierung.de/Webs/Breg/un-klimakonferenz/DE/Klimapolitik-DerBundesregierung/klimapolitik-der-bundesregierung.html

Burkhardt, U. and Kärcher, B. (2011) 'Global radiative forcing from contrail cirrus', *Nature Climate Change*, 1: 54–58.

Bürki, R. (2000) 'Klimaaenderung und Tourismus im Alpenraum – Anpassungsprozesse von Touristen und Tourismusverantwortlichen in der Region Ob- und Nidwalden', thesis, Department of Geography, University of Zurich.

Bürki, R., Elsasser, H., Abegg, B. (2003) 'Climate change – impacts on the tourism industry in mountain areas', *1st International Conference on Climate Change and Tourism*, Djerba, April 2003. Madrid: UN World Tourism Organization. www.breiling.org/snow/djerba.pdf

Burns, P.M. and Bibbings, L.J. (2009) 'The end of tourism? Climate change and societal changes', *21st Century Society*, 4(1): 31–51.

Burton, I. (2004) 'Climate change and the adaptation deficit', in A. Fenech, D. MacIver, H. Auld, R. Bing Rong and Y. Yin, *Climate Change: Building the Adaptive Capacity*, Adaptation and Impacts Research Group Occasional Paper 4. Ottawa: Meteorological Service of Canada, Environment Canada.

Burton, I. and Lim, B. (2005) 'Achieving adequate adaptation in agriculture', *Climatic Change*, 70: 191–200.

Butler, A. (2002) 'Tourism burned: visits to parks down drastically, even away from flames', *Rocky Mountain News*, 15 July.

Butler, D. (2008) 'Architects of a low-energy future', *Nature*, 452(7187): 520–523.

Butler, R. (2001) 'Seasonality in tourism: issues and implications', in T. Baum and S. Lundtorp (eds) *Seasonality in Tourism*. London: Pergamon.

Butler, R. (2006) *The Tourism Life Cycle*. Clevedon: Channelview.

Butler, R. (2009) 'Tourism destination development: cycles and forces, myths and realities', *Tourism Recreation Research*, 34(3): 247–254.

Butler, R. and Jones, P. (2001) 'Conclusions – problems, challenges and solutions', in A. Lockwood and S. Medlik (eds) *Tourism and Hospitality in the 21st Century*. Oxford: Butterworth-Heinemann.

Buzinde, C.N., Manuel-Navarrete, D., Yoo, E. and Morais, D. (2010) 'Tourists' perceptions in a climate of change: eroding destinations', *Annals of Tourism Research*, 37: 333–354.

C40 Cities (2010) *Cities and Climate Change*. www.c40cities.org/climatechange.jsp

Campbell, A., Kapos, V., Scharlemann, J.P.W., Bubb, P., Chenery, A., Coad, L., Dickson, B., Doswald, N., Khan, M.S.I., Kershaw, F. and Rashid, M. (2009) *Review of the Literature on the Links between Biodiversity and Climate Change: Impacts, Adaptation and Mitigation*, Technical Series No. 42. Montreal: Secretariat of the Convention on Biological Diversity.

Canadell, J.G., Le Quéré, C., Raupach, M.R., Field, C.B., Buitenhuis, E.T., Ciais, P., Conway, T.J., Gillett, N.P., Houghton, R.A. and Marland, G. (2007) 'Contributions to accelerating atmospheric CO_2 growth from economic activity, carbon intensity, and efficiency of natural sinks', *Proceedings of the National Academy of Sciences, USA*, 104(47): 18866–18870.

Capaldo, K.P., Corbett, J.J., Kasibhatla, P., Fischbeck, P. and Pandis, S.N. (1999) 'Effects of ship emissions on sulphur cycling and radiative climate forcing over the ocean', *Nature*, 400: 743–746.

Carbon Disclosure Project (2010a) 'Carbon Disclosure Project reveals water constraints now a boardroom issue for global corporations', Media Release, 12 November. London: Carbon Disclosure Project.

Carbon Disclosure Project (2010b) *CDP Water Disclosure 2010 Global Report. On behalf of 137 Investors with Assets of US$16 trillion*, report written for Carbon Disclosure Project by ERM. London: Carbon Disclosure Project.

Carbon Trust (2010) *Hospitality. Saving Energy Without Compromising Service*. www.carbontrust. co.uk/publications/pages/publicationdetail.aspx?id=CTV013

Caribbean Tourism Organization (2003) *Blueprint for New Tourism*. London: World Travel and Tourism Council.

Carlson, A.E. (2003) 'Federalism, preemption, and greenhouse gas emissions', *University of California Davis Law Review*, 37(1): 281–319.

Carnival (2008) *2008 Annual Environmental Management Report*. http://phx.corporate-ir.net/pheonix.zhtml?c=140690&p=irol-sustainability-env

Casola, J., Kay, J., Snover, A., Norheim, R. and Binder, L. (2005) *Climate Impacts on Washington's Hydropower, Water Supply, Forests, Fish and Agriculture*. Seattle: University of Washington, Centre for Science and the Earth System.

Castree, N. (2002) 'False antitheses? Marxism, nature and actor networks', *Antipode*, 34:111–146.

CBC (2008) 'The beetle and the damage done', *Canadian Broadcasting Corporation News*, 23 April. www.cbc.ca/news/background/science/beetle.html

CBD and GIZ (2011) *Biodiversity and Livelihoods. REDD-plus Benefits*. Montreal and Eschborn: Secretariat of the Convention on Biological Diversity/Gesellschaft fur Internationale Zusammenarbeit. www.cbd.int/doc/publications/for-redd-en.pdf

CEC (2007) *Agenda for a Sustainable and Competitive European Tourism*. Brussels: Commission of the European Communities. http://ec.europa.eu/enterprise/sectors/tourism/documents/communications/commission-communication-2007/index_en.htm

CEC (2010) *Europe, the World's No 1 Tourist Destination – A New Political Framework for Tourism in Europe*. Brussels: Commission of the European Communities.

Centrec Consulting Group, LLC (2007) *An Investigation of the Economic and Social Value of Selected NOAA Data and Products for Geostationary Operational Environmental Satellites (GOES)*, report submitted to NOAA's National Climatic Data Center. Centrec Consulting Group. www.economics.noaa.gov/?file=bibliography#noaa.2004

Ceron, J.-P., Dubois, G., Gössling, S., Hall, C.M. and Scott, D. (2012) 'Climate perceptions and preferences of French tourists', *Climate Research*.

Cesar, H. (2000) *Impacts of the 1998 Coral Bleaching Event on Tourism in El Nido, Phillippines*, prepared for Coastal Resources Center Coral Bleaching Initiative, University of Rhode Island, Narragansett.

CHA–CTO (2007) *CHA–CTO Position Paper on Global Climate Change and the Caribbean Tourism Industry*. St Martin, Barbados: Caribbean Hotels Association and Caribbean Tourism Organization.

Chamber of Shipping, Australian Shipowners' Association, Royal Belgian Shipowners' Association, Norwegian Shipowners' Association, Swedish Shipowners' Association (2009) *A Global Cap and Trade System to Reduce Carbon Emissions from International Shipping*. London: Chamber of Shipping of the UK.

Chan, W.W. and Lam, J.C. (2003) 'Energy-saving supporting tourism: a case study of hotel swimming pool heat pump', *Journal of Sustainable Tourism*, 11(1): 74–83.

Chen, C., Hill, J.K., Ohlemüller, R., Roy, D.B. and Thomas, C.D. (2011) 'Rapid range shifts of species associated with high levels of climate warming', *Science*, 333(6045): 1024–1026.

Cheng, M. (2010) *Results of the SWCCIP Tourism Group's Tourim Business Survey – 2010*. Southwest Tourism. www.oursouthwest.com/climate/registry/100400-tourism-survey-report.pdf

Christ, C., Hilel, O., Matus, S. and Sweeting, J. (2003) *Tourism and Biodiversity: Mapping Tourism's Global Footprint*. Washington, DC: Conservation International.

Church, J.A. and White, N.J. (2011) 'Sea-level rise from the late 19th to the early 21st century', *Surveys in Geophysics*, March: 1–18.

CIA (2011) *The World Factbook*. https://www.cia.gov/library/publications/the-world-factbook

CIBSE (2006) *Guide A: Environmental Design. Category: Heating, Air Conditioning and Refrigeration*. London: Chartered Institution of Building Services Engineers.

Cioccio, L. and Michael, E.J. (2007) 'Hazard or disaster: tourism management for the inevitable in northeast Victoria', *Tourism Management*, 28: 1–11.

Ciscar, J.-C., Iglesias, A., Feyen, L., Szabó, L., Van Regemorter, D., Amelung, B., Nicholls, R., Watkiss, P., Christensen, O.B., Dankers, R., Garrote, L., Goodess, C.M., Hunt, A, Moreno, A., Richards, J. and Soria, A. (2011) 'Physical and economic consequences of climate change in Europe', *Proceedings of the National Academy of Sciences, USA*, 108(7): 2678–2683.

Clark, D. (2009) 'Maldives first to go carbon neutral', *The Observer*, 15 March. www.guardian. co.uk/environment/2009/mar/15/maldives-president-nasheed-carbon-neutral

CLIA (2009) *CLIA Cruise Market Overview: Statistical Cruise Industry Data through 2008.* Arlington, VA: Cruise Lines International Association.

Climate Service Center (2009) *Requirements for the Climate Service Center from the Perspective of the Financial Sector.* Hamburg: Climate Service Centre. www.cfi21.org/fileadmin/user_upload/CSC-Bericht_englisch_web.pdf

Cofala, J., Amann, M., Heyes, C., Wagner, F., Klimont, Z., Posch, M., Schöpp, W., Tarasson, L., Jonson, J.E., Whall, C. and Stavrakaki, A. (2007) *Analysis of Policy Measures to Reduce Ship Emissions in the Context of the Revision of the National Emissions Ceilings Directive*, submitted to the European Commission, DG Environment. Laxenburg, Austria: International Institute for Applied Systems Analysis.

Coghlan, A. and Prideaux, B. (2009) 'Welcome to the wet tropics: the importance of weather in reef tourism relisience', *Current Issues in Tourism*, 12(2): 89–104.

Cohen, E. and Neal, M. (2010) 'Coinciding crises and tourism in contemporary Thailand', *Current Issues in Tourism*, 13(5): 455–475.

Cohen, M.J. (2009) 'Sustainable mobility transitions and the challenge of countervailing trends: the case of personal aeromobility', *Technology Analysis & Strategic Management*, 21(2): 249–265.

Cohen, S.A. and Higham, J.E.S. (2011) 'Eyes wide shut? UK consumer perceptions on aviation climate impacts and travel decisions to New Zealand', *Current Issues in Tourism*, 14(4): 323–335.

Cohen, S., Higham, J., Cavaliere, C. (2011) 'Binge flying: behavioural addiction and climate change', *Annals of Tourism Research*, 38(3): 1070–1089.

Coles, T., Duval, D.T. and Hall, C.M. (2005) 'Tourism, mobility, and global communities: new approaches to theorising tourism and tourist spaces', in W.F. Theobald (ed.) *Global Tourism.* Amsterdam: Elsevier.

Coles, T., Hall, C.M. and Duval, D. (2006) 'Tourism and post-disciplinary inquiry', *Current Issues in Tourism*, 9(4/5): 293–319.

Colwell, R.K., Brehm, G., Cardelus, C.L., Gilman, A.C. and Longino, J.T. (2008) 'Global warming, elevational range shifts, and lowland biotic attrition in the wet tropics', *Science*, 322: 258–261.

Connell, J.F. and Williams, G. (2005) 'Passengers' perceptions of low cost airlines and full service carriers: a case study involving Ryanair, Aer Lingus, Air Asia and Malaysia Airlines', *Journal of Air Transport Management*, 11(4): 259–272.

Conrad, J. (2009) 'Climate research and climate change: reconsidering social science perspectives', *Nature and Culture* 4(2): 113–122.

Continental Airlines (2009) *Continental Airlines Flight Demonstrates Use of Sustainable Biofuels as Energy Source for Jet Travel.* www.continental.com, About Continental, News Releases.

Cook, K.H. and Vizy, E.K. (2008) 'Effects of twenty-first century climate change on the Amazon rain forest', *Journal of Climate*, 21: 542–560.

Coombes, E., Jones, A. and Sutherland, W. (2009) 'The implications of climate change on coastal visitor numbers: a regional analysis', *Journal of Coastal Research*, 25(4): 981–990.

Coppock, J. (1982) 'Geographical contribution to leisure', *Leisure Studies*, 1: 1–27.

Corbett, J.J. and Koehler, H.W. (2003) 'Updated emissions from ocean shipping', *Journal of Geophysical Research*, D: Atmosphere, 108(D20): 4650 (doi: 10.1029/2003JD003751).

Corbett, J.J., Wang, C., Winebrake, J.J. and Green, E. (2007a) *Allocation and Forecasting of Global Ship Emissions.* Boston: Clean Air Task Force and Friends of the Earth International.

Corbett, J.J., Winebrake, J.J., Green, E., Kasibhatla, P., Eyring, V. and Lauer, A. (2007b) 'Mortality from ship emissions: a global assessment', *Environment Science and Technology*, 41(24):8512–8518.

COTA (2009) 'COTA receives funding to develop tourism – mountain pine beetle strategy', Vancouver, BC: Council of Tourism Associations of British Columbia. www.cotabc.com/press/tourism_news_archive.aspx?year=2009

Council of the European Union (2010) *Report from the Commission to the European Parliament and the Council. Monitoring the CO_2 emissions from new passenger cars in the EU: Data for the year 2008*. http://register.consilium.europa.eu/pdf/en/10/st05/st05515.en10.pdf

Cowell, P.J., Thom, B.G., Jones, R.A., Everts, C.H. and Simanovic, D. (2006) 'Management of uncertainty in predicting climate-change impacts on beaches', *Journal of Coastal Research*, 22: 232–245.

Crapo, D. (1970) 'Recreational activity choice and weather: the significance of various weather perceptions in influencing preference for selected recreational activities', PhD thesis, Department of Resource Development, Michigan State University.

Credoc (2009) *Climat, météorologie et fréquentation touristique, rapport final*, 29 July. Paris: Ministère de l' Écologie, de l' Énergie, du Développement durable et de la Mer.

Crompton, J. (1979) 'Motivations of pleasure vacation', *Journal of Leisure Research*, 6:408–424.

Crowe, R.B., McKay, G.A. and Baker, W.M. (1973) *The Tourist and Out-door Recreation Climate of Ontario, Vol. 1. Objectives and Definitions of Season*. Report No. REC-1-73. Toronto: Atmospheric Environment Service, Environment Canada.

Cruz, R.V., Harasawa, H., Lal, M., Wu, S., Anokhin, Y., Punsalamaa, B., Honda, Y., Jafai, M., Li, C. and Huu Ninh, N. (2007) 'Asia', in M.L. Parry, O.F. Canziani, J.P. Palutikof, P.J. ven der Linden and C.E. Hanjon (eds), *Climate Change 2007: Impacts, Adaptation and Vulnerability. Contribution of Working Group II to the Fourth Assessment Report of the Intergovernmental Panel on Climate Change*. Cambridge: Cambridge University Press.

Cullenward, D., Schipper, L., Sudarshan, A. and Howarth, R. (2011) 'Psychohistory revisited: fundamental issues in forecasting climate futures', *Climatic Change* 104(3/4): 457–472.

Curtin, S. (2010) 'What makes for memorable wildlife encounters? Revelations from "serious" wildlife tourists', *Journal of Ecotourism*, 9(2): 149–168.

Curtis, S., Arrigo, J., Long, P. and Covington, R. (2009) *Climate, Weather, and Tourism: Bridging Science and Practice*. Greenville, NC: Centre for Sustainable Tourism, East Carolina University. www.research2.ecu.edu/Tourism/Documents/Summary%20Report.pdf

Czech, B. (2008) 'Prospects for reconciling the conflict between economic growth and biodiversity conservation with technological progress', *Conservation Biology*, 22: 1389–1398.

Dalton, G.J., Lockington, D.A. and Baldock, T.E. (2007) 'A survey of tourist operator attitudes to renewable energy supply in Queensland, Australia', *Renewable Energy*, 32: 567–586.

Dalton, G.J., Lockington, D.A. and Baldock, T.E. (2008) 'Feasibility analysis of stand-alone renewable energy supply options for a large hotel', *Renewable Energy*, 33: 1475–1490.

Dalton, G.J., Lockington, D.A. and Baldock, T.E. (2009) 'Case study feasibility analysis of renewable energy supply options for small to medium-sized tourist accommodations', *Renewable Energy*, 34: 1134–1144.

Daufresne, M. and Boet, P. (2007) 'Climate change impacts on structure and diversity of fish communities in rivers', *Global Change Biology*, 13: 2467–2478.

Davis, T. (2009) 'Climate-change damage may double cost of insurance', *The Sunday Times*, 22 March. business.timesonline.co.uk/tol/business/industry_sectors/banking_and_finance/article5949991.ece

Dawson, J. and Scott, D. (2010) 'Systems analysis of climate change vulnerability of the US Northeast ski sector', *Journal of Tourism and Hospitality Planning*, 7(3): 219–235.

Dawson, J., Scott, D. and McBoyle, G. (2009) 'Analogue analysis of climate change vulnerability in the US Northeast ski tourism', *Climate Research*, 39(1): 1–9.

Dawson, J., Stewart, E.J., Lemelin, H. and Scott, D. (2010) 'The carbon cost of polar bear viewing in Churchill, Canada', *Journal of Sustainable Tourism*, 18: 319–336.

Dawson, J., Havitz, M. and Scott, D. (2011a) 'The influence of ego involvement on climate-induced substitution and place loyalty among alpine skiers', *Journal of Travel Tourism and Marketing*, 28: 388–404.

Dawson, J., Johnston, M.E., Stewart, E.J., Lemieux, C.J., Lemelin, H. and Maher, P. (2011b) 'Ethical considerations of last chance tourism', *Journal of Ecotourism*, 10(3): 250–265.

Dearden, P. and Manopawitr, P. (2011) 'Climate change – coral reefs and dive tourism in South-east Asia', in A. Jones and M. Phillips (eds) *Disappearing Destinations*. Wallingford, UK: CABI Publishing.

DECC (2011) *Carbon Budgets*. London: Department of Energy & Climate Change. www.decc.gov. uk/en/content/cms/emissions/carbon_budgets/carbon_budgets.aspx

Decrop, A. (2006) *Vacation Decision Making*. Wallingford, UK: CABI Publishing.

Defra (2005) *The Validity of Food Miles as an Indicator of Sustainable Development*. Harwell, UK: AEA Technology Environment. http://archive.defra.gov.uk/evidence/economics/foodfarm/reports/documents/foodmile.pdf

Demeritt, D. (2001a) 'The construction of global warming and the politics of science', *Annals of the Association of American Geographers*, 91: 307–337.

Demeritt, D. (2001b) 'Science and the understanding of science: a reply to Schneider', *Annals of the Association of American Geographers*, 91: 345–348.

Demeritt, D. (2006) 'Science studies, climate change and the prospects for constructivist critique', *Economy and Society*, 35: 453–479.

Demeritt, D. and Rothman, D. (1999) 'Figuring the costs of climate change: an assessment and critique', *Environment and Planning A*, 31: 389–408.

Deng, S. and Burnett, J. (2000) 'A study of energy performance of hotel buildings in Hong Kong', *Energy and Buildings*, 31: 7–12.

Denstadli, J., Jacobsen, J. and Lohmann, M. (2011) 'Tourist perceptions of summer weather in Scandinavia', *Annals of Tourism Research*, 38(3): 920–940.

De Souza, M. (2011) 'Kent says Canada won't jump into cap-and-trade', *Ottawa Citizen*, 26 January.

Deutsche Bahn AG (2010) *Kennzahlen und Fakten zur Nachhaltigkeit 2009*. www.deutschebahn. com/site/shared/de/dateianhaenge/publikationen_broschueren/holding/nachhaltigkeitskennzahlen_2009.pdf

Deutsche Bank Research (2008) *Climate Change and Tourism: Where Will the Journey Take Us?* Berlin: Deutsche Bank Research.

Dhanesh, G. (2010) 'Kerala: God's own country', in D. Moss and M. Powell with B. DeSanto (eds) *Public Relations Cases: International Perspectives*, 2nd edn. London: Taylor & Francis.

Diamond, J.M. (2005) *Collapse: How Societies Choose to Fail or Survive*. New York: Viking.

Dickinson, J. and Lumsdon, L. (2010) *Slow Travel and Tourism*. London: Earthscan.

Dietz, S. and Stern, N. (2008) 'Why economic analysis supports strong action on climate change: a response to the Stern Review's critics', *Review of Environmental Economics and Policy*, 2:94–113.

Di Silvestro, R. (2009) 'Where have Yellowstone amphibians gone? New study suggests that global warming plays a role in decline of frogs, toads and salamanders', *National Wildlife*, 1 February. www.nwf.org/News-and-Magazines/National-Wildlife/Animals/Archives/2009/Where-Have-Yellowstone-Amphibians-Gone.aspx

Dodds, R. and Kuenzig, G. (2009) *Climate Change and Canada's Municipal Destination Marketing Organizations*. Toronto: Icarus Foundation. www.theicarusfoundation.com/pdf/ExecSummaryDMOandCC_Final.pdf

DOE/NETL (2009) *Affordable, Low-carbon Diesel Fuel from Domestic Coal and Biomass*. Washington, DC: US Department of Energy/National Energy Technology Laboratory. www.netl. doe.gov/energy-analyses/pubs/CBTL%20Final%20Report.pdf

Donner, S.D., Knutson, T.R. and Oppenheimer, M. (2007) 'Model-based assessment of the role of human-induced climate change in the 2005 Caribbean coral bleaching event', *Proceedings of the National Academy of Sciences, USA*, 104(13): 5483–5488.

Doran, P. and Zimmerman, M. (2009) 'Examining the scientific consensus on climate change', *Eos, Transactions, American Geophysical Union*, 90(3): 22–23.

Dovers, S. (1996) 'Sustainability: demands on policy', *Journal of Public Policy*, 16: 303–318.

Dowling, R.K. (2010) *Cruise Ship Tourism: Issues, Impacts, Cases*. Wallingford, UK: CABI Publishing.

Drew, P. (2011) *100 Places to Remember Before they Disappear*. New York: Harry N. Abrams Inc.

Dubois, G. and Ceron, J.P. (2006) *How Heavy Will the Burden Be? Using Scenario Analysis to Assess Future Tourism Greenhouse Gas Emissions*. Marseille, France: Tourism Transports Territoires Environnement Conseil. www.tec-conseil.com/IMG/pdf/AS_5_20_PROOF_01.pdf

Dubois, G. and Ceron, J.P. (2009) 'Carbon labelling and restructuring travel systems: involving travel agencies in climate change mitigation', in S. Gössling, C.M. Hall and D. Weaver (eds) *Sustainable Tourism Futures. Perspectives on Systems, Restructuring and Innovations*. London: Routledge.

Dwyer, L., Forsyth, P., Spurr, R. and Hoque, S. (2010) 'Estimating the carbon footprint of Australian tourism', *Journal of Sustainable Tourism*, 18: 355–376.

Dye, T. (1992) *Understanding Public Policy*, 7th edn. Englewood Cliffs, NJ: Prentice Hall.

Earth Policy Institute (2010) 'US feeds one quarter of its grain to cars while hunger is on the rise', press release. Washington, DC: Earth Policy Institute. www.earth-policy.org/data_highlights/2010/highlights6

EC (2009) *Energy Efficiency. Energy Labelling of Domestic Appliances*. http://ec.europa.eu/energy/efficiency/labelling/labelling_en.htm

ECRSAS (2005) *Statements on Oil*, 14 October. Stockholm: Energy Committee at the Royal Swedish Academy of Sciences.

EEA (2010) *Biodiversity, Climate Change and You. EEA Signals 2010*. Copenhagen: European Environment Agency.

EHAS (2010) *BioEthanol for Sustainable Transport. Results and Recommendations from the European BEST Project*. Stockholm: Environmental and Health Administration of Sweden. www.best-europe.org

ehotelier.com (2008) 'PATA questions UK government policy on air departure taxes', *ehotelier.com*, 28 November. ehotelier.com/hospitality-news/item.php?id=A15077_0_11_0_M

Eijgelaar, E., Thaper, C. and Peeters, P. (2010) 'Antarctic cruise tourism: the paradoxes of ambassadorship, "last chance tourism" and greenhouse gas emissions', *Journal of Sustainable Tourism*, 18: 337–354.

El-Raey, M., Dewindar, K. and El-Hattab, M. (1999) 'Adaptation to the impacts of sea level rise in Egypt', *Mitigation and Adaptation Strategies for Global Change*, 4: 343–361.

Elsasser, H. and Bürki, R. (2002) 'Climate change as a threat to tourism in the Alps', *Climate Research*, 20: 253–257.

Elsasser, H. and Messerli, P. (2001) 'The vulnerability of the snow industry in the Swiss Alps', *Mountain Research and Development*, 21(4): 335–339.

Emmanuel, K. and Spence, B. (2009) 'Climate change implications for water resource management in Barbados tourism', *Worldwide Hospitality and Tourism Themes*, 1(3): 252–268.

Emmons, H., Dean, B.V., Nunnikhoven, T.S., Rossiter, D.R., Rao, R.N., de Kluyver, C. and Richard, B. (1975) 'A market analysis for Cleveland Zoological Park', in S.P. Lodamy (ed.) *Management Science Applications to Leisure Time Operations*. New York: Elsevier.

Endler, C. and Matzarakis, A. (2011) 'Climatic potential for tourism in the Black Forest, Germany – winter season', *International Journal of Biometeorology*, 55: 339–351.

Endresen, O., Soergaard, E., Sundet, J.K., Dalsoeren, S.B., Isaksen, I.S.A., Berglen, T.F. and Gravir, G. (2003) 'Emission from international sea transportation and environmental impact', *Journal of Geophysical Research*, D: Atmosphere, 108(D17): 4560 (doi: 10.1029/2002JD002898).

Energy Star (2010) *About Energy Star*. Washington, DC: US Environmental Protection Agency and US Department of Energy. www.energystar.gov

Englebert, E.T., McDermott, C. and Kleinheinz, G.T. (2008) 'Effects of the nuisance algae, *Cladophora*, on *Escherichia coli* at recreational beaches in Wisconsin', *Science of the Total Environment*, 404: 10–17.

Environment Canada (2002) *Backgrounder on Sea Level Rise and Climate Change Impacts and Adaptation Needs on PEI (Canada)*. Ottawa: Environment Canada.

Environment Canada (2009) *Humidex Calculator – 2004*. www.weatheroffice.gc.ca/mainmenu/faq_e.html#weather6

ETC and UNWTO (2009) *ETC–UNWTO Symposium: Tourism & Travel in the Green Economy – A Contribution to the Davos Process on Climate Change and Tourism*, 14–15 September 2009, Gothenburg, Sweden, Delegates' Briefing Papers. European Travel Commission and UN World Tourism Organization. www.etc-corporate.org/resources/uploads/BriefingMaterial_ 20090911.pdf

Eugenio-Martin, J.L. and Campos-Soria, J.A. (2010) 'Climate in the region of origin and destination choice in outbound tourism demand', *Tourism Management*, 31: 744–753.

EUHOFA, IH&RA and UNEP (2001) *Sowing the Seeds of Change*, Environmental Teaching Pack for the Hospitality Industry. Paris: International Association of Hotel Schools, International Hotel and Restaurant Association and United Nations Environment Programme.

EurActiv (2009) 'EU carbon tax on new Commission's agenda early next year', *EurActiv.com*, 4 November. www.euractiv.com/en/climate-change/eu-carbon-tax-new-commission-agenda-early-year/article-187029

European Parliament and Council (2009) 'Directive 2008/101/EC of the European Parliament of the Council of 19 November 2008 amending Directive 2003/87/EC so as to include aviation activities in the scheme for greenhouse gas emission allowance trading within the Community', *Official Journal of the European Union*, 13 January.

Ewing, R., Bartholomew, K., Winkelman, S., Walters, J., Chen, D., McCann, B. and Goldberg, D. (2007) *Growing Cooler: The Evidence on Urban Development and Climate Change*. Washington, DC: Urban Land Institute.

Eyring, V., Köhler, H.W., van Aardenne, J. and Lauer A. (2005a) 'Emissions from international shipping: 1. The last 50 years', *Journal of Geophysical Research*, 110: D17305 (doi: 10.1029/2004JD005619).

Eyring, V., Köhler, H.W., Lauer, A. and Lemper, B. (2005b) 'Emissions from international shipping: 2. Impact of future technologies on scenarios until 2050', *Journal of Geophysical Research*, 110: D17306 (doi: 10.1029/2004JD005620).

Eyring, V., Isaksen, I.S.A., Berntsen, T., Collins, W.J., Corbett, J.J., Endresen, O., Grainger, R.G., Moldanova, J., Schlager, H. and Stevenson, D.S. (2009) 'Transport impacts on atmosphere and climate: shipping', *Atmosphere and Environment*, 44(37): 4735–4771 (doi: 10.1016/j.atmosenv., 04.059).

Fabian, P. (1974) *Residence Time of Aircraft Exhaust Contaminants in the Stratosphere*, CIAP Contract No. 05-30027. Washington, DC: US Department of Transportation.

Fabian, P. (1978) 'Ozone increase from Concorde operations?', *Nature*, 272: 306–307.

Fáilte Ireland and The Heritage Council (2009) *Climate Change, Heritage and Tourism – Implications for Ireland's Coast and Inland Waterways*. Kilkenny: The Heritage Council. www.heritagecouncil.ie/fileadmin/user_upload/Publications/Marine/ClimateReportWeb_version_june_09FINAL.pdf

FAO (2010) *Global Forest Resources Assessment 2010 Main Report*, FAO Forestry Paper 163. Rome: Food and Agricultural Organization of the United Nations.

Farajzadeh, H. and Matzarakis, A. (2009) 'Quantification of climate for tourism in northwest of Iran', *Meterological Applications*, 16(4): 545–555.

Farajzadeh, M. and Ahmad, A.A. (2010) 'Assessment and zoning of tourism climate of Iran using Tourism Climate Index (TCI)', *Physical Geography Research Quarterly*, 71: 31–42.

Farbotko, C. (2010) '"The global warming clock is ticking so see these places while you can": voyeuristic tourism and model environmental citizens on Tuvalu's disappearing islands', *Singapore Journal of Tropical Geography*, 31(2): 224–238.

Fargione, J., Hill, J., Tilman, D., Polasky, S. and Hawthorne, P. (2008) 'Land clearing and the biofuel carbon debt', *Science*, 319: 1235–1237.

Fawkes, S. (2007a) 'Space tourism and carbon dioxide emissions', *The Space Review*, 19 February. www.thespacereview.com/article/813/1

Fawkes, S. (2007b) 'Carbon dioxide emissions resulting from space tourism', *Journal of the British Interplanetary Society*, 60: 409–413.

FCFC (2009) *Food Carbon Footprint Calculator*. www.foodcarbon.co.uk

February, E.C., West, A.G. and Newton, R.J. (2007) 'The relationship between rainfall, water source and growth for an endangered tree', *Austral Ecology*, 32: 397–402.

Federal Government (2008) *German Strategy for Adaptation to Climate Change*. Berlin: Federal Government. www.bmu.de/files/english/pdf/application/pdf/das_gesamt_en_bf.pdf

Fenton, D., Young, M. and Johnson, V. (1998) 'Re-presenting the Great Barrier Reef to tourists: implications for tourist experience and evaluation of coral reef environments', *Leisure Sciences*, 20(3): 177–192.

Field, C.B., Mortsch, L.D., Brklacich, M., Forbes, D.L., Kovacs, P., Patz, J.A., Running, S.W. and Scott, M.J. (2007) 'North America', in M.L. Parry, O.F. Canziani, J.P. Palutikof, P.J. van der Linden and C.E. Hanson (eds) *Climate Change 2007: Impacts, Adaptation and Vulnerability. Contribution of Working Group II to the Fourth Assessment Report of the Intergovernmental Panel on Climate Change*. Cambridge: Cambridge University Press.

Fischlin, A., Midgley, G.F., Price, J.T., Leemans, R., Gopal, B., Turley, C., Rounsevell, M.D.A., Dube, O.P., Tarazona, J. and Velichko, A.A. (2007) 'Ecosystems, their properties, goods, and services', in M.L. Parry, O.F. Canziani, J.P. Palutikof, P.J. van der Linden and C.E. Hanson (eds) *Climate Change 2007: Impacts, Adaptation and Vulnerability. Contribution of Working Group II to the Fourth Assessment Report of the Intergovernmental Panel on Climate Change*. Cambridge: Cambridge University Press.

Font, X., Cochrane, J. and Tapper, R. (2004) *Tourism for Protected Area Financing: Understanding Tourism Revenues for Effective Management Plans*. Leeds: Leeds Metropolitan University.

Forbes, D.L. (ed.) (2011) *State of the Arctic Coast 2010: Scientific Review and Outlook*. International Arctic Science Committee, Land–Ocean Interactions in the Coastal Zone, Arctic Monitoring and Assessment Programme, International Permafrost Association. Geesthacht: Helmholtz-Zentrum.

Ford, J.D., Keskitalo, E.C.H., Smith, T., Pearce, T., Berrang-Ford, L., Duerden, F. and Smit, B. (2010) 'Case study and analogue methodologies in climate change vulnerability research', *WIREs Climate Change*, 1(3): 374–392.

Forster, P., Ramaswamy, V., Artaxo, P., Berntsen, T., Betts, R., Fahey, D.W., Haywood, J., Lean, J., Lowe, D.C., Myhre, G., Nganga, J., Prinn, R., Raga, G., Schulz, M. and Van Dorland, R. (2007) 'Changes in atmospheric constituents and in radiative forcing', in S. Solomon, D. Qin, M. Manning, Z. Chen, M. Marquis, K.B. Averyt, M. Tignor and H.L. Miller (eds) *Climate Change 2007: The Physical Science Basis. Contribution of Working Group I to the Fourth Assessment Report of the Intergovernmental Panel on Climate Change*. Cambridge and New York: Cambridge University Press.

Forsyth, P.J., Dwyer, L. and Spurr, R. (2007) *Climate Change Policies and Australian Tourism: Scoping Study of the Economic Aspects*. Brisbane: Sustainable Tourism CRC.

Fosgerau, M. (2005) 'Speed and income', *Journal of Transport Economics and Policy*, 39: 225–240.

Fowbert, J.A. and Smith, R. (1994) 'Rapid population increases in native vascular plants in the Argentine Islands, Antarctic Peninsula', *Arctic and Alpine Research*, 26: 290–296.

Francia, C. and Juhasz, F. (1993) *The Lagoon of Venice, Italy, Coastal Zone Management: Selected Case Studies*. Paris: Organisation for Economic Co-operation and Development, pp. 109–134.

Frändberg, L. (2008a) 'Paths in transnational time–space: representing mobility biographies of young Swedes', *Geografiska Annaler B*, 90(1): 17–28.

Frändberg, L. (2008b) 'How normal is travelling abroad? Differences in transnational mobility between groups of young Swedes', *Environment and Planning A*, 41(3): 649–667.

Fransen, T. (2009) 'The carbon cuts promised by developing countries at Copenhagen', *The Guardian*, 10 December. www.guardian.co.uk/environment/interactive/2009/dec/10/copenhagen-carbon-emissions-developing-countries?INTCMP=SRCH

de Freitas, C.R. (1990) 'Recreation climate assessment', *International Journal of Climatology*, 10: 89–103.

de Freitas, C.R. (2003) 'Tourism climatology: evaluating environmental information for decision making and business planning in the recreation and tourism sector', *International Journal of Biometeorology*, 48: 45–54.

de Freitas, C.R. (2005) 'The climate–tourism relationship and its relevance to climate change impact assessment', in C.M. Hall and J. Higham (eds) *Tourism, Recreation and Climate Change: International Perspectives*. Clevedon: Channelview Press.

de Freitas C.R., Scott, D. and McBoyle, G. (2008) 'A second generation climate index for tourism (CIT): specification and verification', *International Journal of Biometeorology*, 52: 399–407.

Frew, E.A. (2008) 'Climate change and doom tourism: advertising destinations "before they disappear"', in J. Fountain and K. Moore (eds) *Recreating Tourism: Proceedings of the New Zealand Tourism and Hospitality Research Conference 2008*, 3–5 December, Hanmer Springs, New Zealand. CD-Rom. Canterbury, New Zealand: Lincoln University.

Frost, W. and Hall, C.M. (eds) (2009) *Tourism and National Parks: International Perspectives on Development, Histories and Change*. London: Routledge.

Frumhoff, P.C., McCarthy, J.J., Melillo, J.M., Moser, S.C. and Wuebbles, D.J. (2007) *Confronting Climate Change in the U.S. Northeast: Science, Impacts, and Solutions*. Synthesis report of the Northeast Climate Impacts Assessment (NECIA). Cambridge, MA: Union of Concerned Scientists.

Fukushima, T., Kureha, M., Ozaki, N., Fujimori, Y. and Harasawa, H. (2003) 'Influences of air temperature change on leisure industries: case study on ski activities', *Mitigation and Adaptation Strategies for Climate Change*, 7: 173–189.

Füssel, H. (2009) 'Review and quantitative analysis of indices of climate change exposure, adaptive capacity, sensitivity, and impacts', background note to the *World Development Report 2010*. New York: World Bank.

Füssel, H.-M. and Klein, R.J.T. (2006) 'Climate change vulnerability assessments: an evolution of conceptual thinking', *Climatic Change*, 75: 301–329.

G8 (2009) *Declaration of the Leaders: The Major Economies Forum on Energy and Climate*. www.g8italia2009.it/static/G8_Allegato/MEF_Declarationl.pdf

Gable, F. (1987) 'Changing climate and Caribbean coastlines', *Oceanus*, 30(4): 53–56.

Gable, F. (1990) 'Caribbean coastal and marine tourism: coping with climate change and its associated effects', in M. Miller and J. Auyong (eds) *Proceedings of the 1990 Congress on Coastal and*

Marine Tourism, 25–31 May 1990, Honolulu, Hawaii, USA. Newport, Oregon: National Coastal Resources Research and Development Institute.

Gable, F. (1997) 'Climate change impacts on Caribbean coastal areas and tourism', *Journal of Coastal Research*, 27: 49–70.

Galil, B.S., Gershwin, L.-A., Douek, J. and Rinkevich, B. (2010) '*Marivagia stellata gen. et sp. nov.* (Scyphozoa: Rhizostomeae: Cepheidae), another alien jellyfish from the Mediterranean coast of Israel', *Aquatic Invasions*, 5(4): 331–340.

Gallastegui, I.G. and Spain, S. (2002) 'The use of eco-labels: a review of the literature', *European Environment*, 12(6): 316–331.

Galloway, R.W. (1988) 'The potential impact of climate change on Australian ski fields', in Pearman, G.I. (ed.) *Greenhouse: Planning for Climate Change*. Melbourne: CSIRO.

Gamble, D.W. and Leonard, L.A. (2005) *Coastal Climatology Products for Recreation and Tourism End Users in Southeastern North Carolina*. Charleston, SC: NOAA Coastal Services Center.

Gardiner, B., Blennow, K., Carnus, J.-M., Fleischer, P., Ingemarson, F., Landmann, G., Lindner, M., Marzano, M., Nicoll, B., Orazio, C., Peyron, J.L., Reviron, M.-P., Schelhaas, M.-J., Schuck, A., Spielmann, M. and Usbeck, T. (2010) *Destructive Storms in European Forests: Past and Forthcoming Impacts*, Report to the European Commission, Directorate-General for the Environment. Joensuu: European Forest Institute.

Gareau, B.J. (2008) 'Dangerous holes in global environmental governance: the roles of neoliberal discourse, science, and California agriculture in the Montreal Protocol', *Antipode*, 40: 120–130.

Garnaut, R. (2008) *The Garnaut Climate Change Review*. Canberra: Commonwealth of Australia. www.garnautreview.org.au/index.htm

Gartner, W. and Lime, D. (2000) *Trends in Outdoor Recreation, Leisure and Tourism*. Wallingford, UK: CABI Publishing.

Gasperoni, G. and Dall'Aglio, S. (1991) 'Tourism and environmental crises: the impact of algae on summer holidays along the Adriatic Riviera in 1989', *Marketing and Research Today*, 19: 260–270.

Gaston, K.J. (2003) *The Structure and Dynamics of Geographic Ranges*. Oxford: Oxford University Press.

Gavin, N. (2010) 'Pressure group direct action on climate change: the role of the media and the web in Britain – a case study', *British Journal of Politics and International Relations*, 12(3): 459–475.

Gerbens-Leenes, W., Hoekstra, A.Y. and van der Meer, T.H. (2009) 'The water footprint of bioenergy', *Proceedings of the National Academy of Sciences, USA*, 106(25): 10219–10223.

German Advisory Council on Global Change (2007) *Climate Change as a Security Risk*. Berlin: German Advisory Council on Global Change.

German Hotel and Restaurant Association (2009) *Energiekampagne, Gastgewerbe* (Energy Campaign, Restaurants). www.dehoga-bundesverband.de/home/page-sta_1433.html

Gershwin, L.A., de Nardi, M., Winkel, K.D. and Fenner, P.J. (2010) 'Marine stingers: review of an under-recognized global coastal management issue', *Coastal Management*, 38(1): 22–41.

Getz, D. (2002) 'Why festivals fail', *Event Management*, 7: 209–219.

Geuss, R. (1981) *The Idea of a Critical Theory: Habermas and the Frankfurt School*. Cambridge: Cambridge University Press.

Giddens, A. (2009) *The Politics of Climate Change*. Cambridge: Polity Press.

Gigerenzer, G., Hertwig, H., van den Broek, E., Fasolo, B. and Katsikopoulos, K. (2005) '"A 30% chance of rain tomorrow": how does the public understand probabilistic weather forecasts?', *Risk Analysis*, 25(3): 623–629.

Gilbert, R. and Perl, A. (2008) *Transport Revolutions. Moving People and Freight without Oil*. London: Earthscan.

Giles, A. and Perry, A. (1998) 'The use of a temporal analogue to investigate the possible impact of projected global warming on the UK tourist industry', *Tourism Management*, 19(1): 75–80.

Gilg, A., Barr, S. and Ford, N. (2005) 'Green consumption or sustainable lifestyles? Identifying the sustainable consumer', *Futures*, 37: 481–504.

Goetz, A.R. and Vowles, T.M. (2009) 'The good, the bad, and the ugly: 30 years of US airline deregulation', *Journal of Transport Geography*, 17: 251–263.

Goetz, S.J., Mack, M.C., Gurney, K.R., Randerson, J.T. and Houghton, R.A. (2007) 'Ecosystem responses to recent climate change and fire disturbance at northern high latitudes: observations and model results contrasting northern Eurasia and North America', *Environmental Resource Letters*, 2: 1–9.

Gómez-Martín, B. (2004) 'An evaluation of the tourist potential of the climate in Catalonia (Spain): a regional study', *Geografiska Annaler*, 86: 249–264.

Gómez-Martín, B. (2005) 'Weather, climate and tourism: a geographical perspective', *Annals of Tourism Research*, 32(3): 571–591.

Gómez-Martín, B. (2006) 'Climate potential and tourist demand in Catalonia (Spain) during the summer season', *Climate Research*, 32: 75–87.

Goodall, C. (2009) 'Maldives' carbon neutral plan is not greenwash, just imperfect progress', *guardian.co.uk*, 26 March. www.guardian.co.uk/environment/cif-green/2009/mar/26/maldives-carbon-neutral-greenwash

Gössling, S. (2002) 'Global environmental consequences of tourism', *Global Environmental Change*, 12: 283–302.

Gössling, S. (2006) 'Tourism and water', in S. Gössling and C.M. Hall (eds) *Tourism and Global Environmental Change. Ecological, Social, Economic and Political Interrelationships.* London: Routledge.

Gössling, S. (2009) 'Carbon neutral destinations: a conceptual analysis', *Journal of Sustainable Tourism*, 17(1): 17–37.

Gössling, S. (2010) *Carbon Management in Tourism: Mitigating the Impacts on Climate Change.* London: Routledge.

Gössling, S. and Hall, C.M. (eds) (2006a) *Tourism and Global Environmental Change.* London: Routledge.

Gössling, S. and Hall, C.M. (2006b) 'Uncertainties in predicting tourist travel flows based on models', *Climatic Change*, 79(3/4): 163–173.

Gössling, S. and Hall, C.M. (2006c) 'Uncertainties in predicting travel flows: common ground and research needs. A reply to Tol *et al.*', *Climatic Change*, 79(3/4): 181–183.

Gössling, S. and Hall, C.M. (2008) 'Swedish tourism and climate change mitigation: an emerging conflict?', *Scandinavian Journal of Hospitality and Tourism*, 8(2): 141–158.

Gössling, S. and Hickler, T. (2006) 'Tourism and forest ecosystems', in S. Gössling and C.M. Hall (eds) *Tourism and Global Environmental Change.* London: Routledge.

Gössling, S. and Nilsson, J.H. (2010) 'Frequent flyer programmes and the reproduction of mobility', *Environment and Planning A*, 42: 241–252.

Gössling, S. and Peeters, P. (2007) '"It does not harm the environment!" – An analysis of discourses on tourism, air travel and the environment', *Journal of Sustainable Tourism*, 15: 402–417.

Gössling, S. and Schumacher, K. (2010) 'Implementing carbon neutral destination policies: issues from the Seychelles', *Journal of Sustainable Tourism*, 18: 377–391.

Gössling, S. and Upham, P. (2009) 'Introduction: aviation and climate change in context', in S. Gössling and P. Upham (eds) *Climate Change and Aviation.* London: Earthscan.

Gössling, S., Peeters, P.M., Ceron, J.-P., Dubois, G., Patterson, T. and Richardson, R.B. (2005) 'The eco-efficiency of tourism', *Ecological Economics*, 54(4): 417–434.

Gössling, S., Bredberg, M., Randow, A., Svensson, P. and Swedlin, E. (2006) 'Tourist perceptions of climate change: a study of international tourists in Zanzibar', *Current Issues in Tourism*, 9(4/5): 419–435.

Gössling, S., Broderick, J., Upham, P., Peeters, P., Strasdas, W., Ceron, J.-P. and Dubois, G. (2007a) 'Voluntary carbon offsetting schemes for aviation: efficiency and credibility', *Journal of Sustainable Tourism*, 15: 223–248.

Gössling, S., Lindén, O., Helmersson, J., Liljenberg, J. and Quarm, S. (2007b) 'Diving and global environmental change: a Mauritius case study', in B. Garrod and S. Gössling (eds) *New Frontiers in Marine Tourism: Diving Experiences, Management and Sustainability*. Amsterdam: Elsevier.

Gössling, S., Peeters, P. and Scott, D. (2008) 'Consequences of climate policy for international tourist arrivals in developing countries', *Third World Quarterly*, 29(5): 873–901.

Gössling, S., Hall, C.M. and Scott, D. (2009a) 'The challenges of tourism as a development strategy in an era of global climate change', in E. Palosou (ed.) *Rethinking Development in a Carbon-Constrained World. Development Cooperation and Climate Change*. Helsinki: Ministry of Foreign Affairs.

Gössling, S., Ceron, J.-P., Dubois, G. and Hall, C.M. (2009b) 'Hypermobile travellers', in S. Gössling and P. Upham (eds) *Climate Change and Aviation*. London: Earthscan.

Gössling, S., Hall, C.M. and Weaver, D. (2009c) *Sustainable Tourism Futures: Perspectives on Systems, Restructuring and Innovations*. New York: Routledge.

Gössling, S., Hultman, J., Haglund, L, Källgren, H. and Revahl, M. (2009d) 'Voluntary carbon offsetting by Swedish air travellers: towards the co-creation of environmental value?', *Current Issues in Tourism*, 12: 1–19.

Gössling, S., Garrod, B., Aall, C., Hille, J. and Peeters, P. (2010a) 'Food management in tourism. Reducing tourism's carbon "foodprint"', *Tourism Management*, 32(3): 1–10.

Gössling, S., Hall, C.M., Peeters, P. and Scott, D. (2010b) 'The future of tourism: a climate change mitigation perspective', *Tourism Recreation Research*, 35(2): 119–130.

Gössling, S., Scott, D., Hall, C.M., Dubois, G. and Ceron, J.-P. (2011a) 'Consumer behaviour and demand response of tourists to climate change', *Annals of Tourism Research*, 39(1): 63–58.

Gössling, S., Lohmann, M., Peeters, P. and Eijgelaar, E. (2011b) *Das Reiseverhalten der Deutschen 2020: Wege zur Emissionsreduktion im Tourismus im Einklang mit der deutschen Klimapolitik*. Berlin: Atmosfair.

Gössling, S., Garrod, B., Aall, C., Hille, J. and Peeters, P. (2011c) 'Food management in tourism: reducing tourism's carbon "foodprint"', *Tourism Management*, 32(3): 534–543.

Gössling, S., Peeters, P., Hall, C.M., Ceron, J.-P., Dubois, G., Lehmann, L.V. and Scott, D. (2011d) 'Tourism and water use: supply, demand, and security – an international review', *Tourism Management*, 33(1): 1–15.

Government of France (2011) *Le bonus écologique, c'est facile et ça rapporte!* Paris: Ministère de L'Écologie, du Développement durable, des Transports et du Logement. www.developpement-durable.gouv.fr/Le-bonus-ecologique-c-est-facile.html

Government of Ireland (2007) *Ireland National Climate Change Strategy 2007–2012*. Dublin: Department of the Environment, Heritage and Local Government. www.environ.ie/en/Environment/Atmosphere/ClimateChange/NationalClimateChangeStrategy/Publications-Documents/FileDownLoad,1861,en.pdf

Government of the Maldives (2001) *First National Communication of the Republic of Maldives to the United Nations Framework Convention on Climate Change*. Malé: Government of Maldives.

Great Barrier Reef Foundation (2009) *The Reef and Climate Change: Frequently Asked Questions*. www.barrierreef.org/ResearchVision/WhyResearch/TheReefandclimatechange.aspx

Green, A.E., Hogarth T. and Shackleton, R.E. (1999) 'Longer distance commuting as a substitute for migration in Britain', *International Journal of Population Geography*, 5: 49–68.

The Guardian (2006) 'Climate change could bring tourists to UK', 28 July. www.guardian.co.uk/travel/2006/jul/28/travelnews.uknews.climatechange

Gude, V.G., Nirmalakhandan, N. and Deng, S. (2010) 'Renewable and sustainable approaches for desalination', *Renewable and Sustainable Energy Reviews*, 14: 2641–2654.

Gustafson, R. and Lindgren, E. (2001) 'Tick-borne encephalitis in Sweden', *The Lancet*, 358(9275): 16–18.

Gwynne, P. (1975) 'The cooling world', *Newsweek*, 28 April. www.denisdutton.com/cooling_world.htm

Haas, W., Weisz, U., Balas, M., McCallum, S., Lexer, W., Pazdernik, K., Prutsch, A., Radunsky, K. Formayer, H., Kromp-Kolb, H. and Schwarzl, I. (2008) *Identifikation von Handlungsempfehlungen zur Anpassung an den Klimawandel in Österreich: 1, Phase, 2008*. Vienna: Lebensministerium.

Habermas, J. (1978) *Knowledge and Human Interests*, 2nd edn. London: Heinemann.

Haites, E. (2009) 'Linking emissions trading schemes for international aviation and shipping emissions', *Climate Policy*, 9: 415–430.

Haites, E., Pntin, D., Attzs, M. and Bruce, J. (2002) 'Assessment of the impact of climate change on CARICOM countries', *Environmentally and Socially Sustainable Development – Latin America and Caribbean Region*. Washington, DC: World Bank. www.margaree.ca/reports/Climate-ChangeCARICOM.pdf

Halifax Travel Insurance (2006) *Holiday 2030*, press release. www.hbosplc.com/media/pressreleases/articles/halifax/2006-09-01-05.asp?section+Halifax

Hall, A. and Clayton, A. (2009) 'How will climate change impact the tourism industry?', *Worldwide Hospitality and Tourism Themes*, 1(3): 269–273.

Hall, C.M. (1998) 'Historical antecedents of sustainable development and ecotourism: new labels on old bottles?', in C.M. Hall and A. Lew (eds) *Sustainable Tourism Development: Geographical Perspectives*. London: Addison Wesley Longman.

Hall, C.M. (2001) 'Trends in ocean and coastal tourism: the end of the last frontier?', *Ocean and Coastal Management*, 44: 601–618.

Hall, C.M. (2002) 'Travel safety, terrorism and the media: the significance of the issue–attention cycle', *Current Issues in Tourism*, 5: 458–466.

Hall, C.M. (2003) 'Health and spa tourism', in S. Hudson (ed.) *International Sports & Adventure Tourism*. New York: Haworth Press.

Hall, C.M. (2004) 'Scale and the problems of assessing mobility in time and space', paper presented at the Swedish National Doctoral Student Course on Tourism, Mobility and Migration, hosted by Department of Social and Economic Geography, University of Umeå, Umeå, Sweden, October.

Hall, C.M. (2005) *Tourism: Rethinking the Social Science of Mobility*. Harlow: Prentice-Hall.

Hall, C.M. (2006a) 'Tourism, biodiversity and global environmental change', in S. Gössling and C.M. Hall (eds) *Tourism and Global Environmental Change*. London: Routledge.

Hall, C.M. (2006b) 'New Zealand tourism entrepreneur attitudes and behaviours with respect to climate change adaptation and mitigation', *International Journal of Innovation and Sustainable Development*, 1(3): 229–237.

Hall, C.M. (2006c) 'Tourism, disease and global environmental change: the fourth transition', in S. Gössling and C.M. Hall (eds) *Tourism and Global Environmental Change: Ecological, Economic, Social and Political Interrelationships*. London: Routledge.

Hall, C.M. (2007) 'Pro-poor tourism: do "tourism exchanges benefit primarily the countries of the South"?', *Current Issues in Tourism*, 10: 111–118.

Hall, C.M. (2008a) 'Tourism and climate change: knowledge gaps and issues', *Tourism Recreation Research*, 33: 339–350.

Hall, C.M. (2008b) *Tourism Planning*. Harlow, UK: Pearson.

Hall, C.M. (2008c) 'Santa Claus, place branding and competition', *Fennia*, 186(1): 59–67.

Hall, C.M. (2009a) 'Archetypal approaches to implementation and their implications for tourism policy', *Tourism Recreation Research*, 34(3): 235–245.

Hall, C.M. (2009b) 'Tourism firm innovation and sustainability', in S. Gössling, C.M. Hall and D.B. Weaver (eds) *Sustainable Tourism Futures: Perspectives on Systems, Restructuring and Innovations*. New York: Routledge.

Hall, C.M. (2009c) 'Changement climatique, authenticité et marketing des régions nordiques: conséquences sur le tourisme finlandais et la "plus grande marque au monde" ou "Les changements climatiques finiront-ils par tuer le père Noël?"', *Téoros*, 28(1): 69–79.

Hall, C.M. (2009d) 'International business travel by New Zealand firms: an exploratory study of climate change mitigation and adaptation practices', in J. Carlsen, M. Hughes, K. Holmes and R. Jones (eds) *See Change: Tourism & Hospitality in a Dynamic World*. Fremantle: Curtin University of Technology.

Hall, C.M. (2009e) 'Degrowing tourism: *décroissance*, sustainable consumption and steady-state tourism', *Anatolia: An International Journal of Tourism and Hospitality Research*, 20(1): 46–61.

Hall, C.M. (2010a) 'Tourism and environmental change in polar regions: impacts, climate change and biological invasion', in C.M. Hall and J. Saarinen (eds) *Tourism and Change in Polar Regions: Climate, Environments and Experiences*. London: Routledge.

Hall, C.M. (2010b) 'Tourism and the implementation of the Convention on Biological Diversity', *Journal of Heritage Tourism*, 5(4): 267–284.

Hall, C.M. (2010c) 'Tourism and biodiversity: more significant than climate change?', *Journal of Heritage Tourism*, 5(4): 253–266.

Hall, C.M. (2010d) 'Crisis events in tourism: subjects of crisis in tourism', *Current Issues in Tourism*, 13(5): 401–417.

Hall, C.M. (2010e) 'Changing paradigms and global change: from sustainable to steady-state tourism', *Tourism Recreation Research*, 35(2): 131–145.

Hall, C.M. (2010f) 'Equal access for all? Regulative mechanisms, inequality and tourism mobility', in S. Cole and N. Morgan (eds) *Tourism and Inequality: Problems and Prospects*. Wallingford, UK: CABI Publishing.

Hall, C.M. (2011a) 'Policy learning and policy failure in sustainable tourism governance: from first and second to third order change?', *Journal of Sustainable Tourism*, 19(4–5), 649–671.

Hall, C.M. (2011b) 'Framing governance theory: a typology of governance and its implications for tourism policy analysis', *Journal of Sustainable Tourism*, 19(4–5), 437–457.

Hall, C.M. (2011c) 'Consumerism, tourism and voluntary simplicity: we all have to consume, but do we really have to travel so much to be happy?', *Tourism Recreation Research*, 36(3): 298–303.

Hall, C.M. (2012) 'The natural science ontology of environment', in A. Holden and D. Fennell (eds) *A Handbook of Tourism and the Environment*. London: Routledge.

Hall, C.M. and Härkönen, T. (eds) (2006) *Lake Tourism: An Integrated Approach to Lacustrine Tourism Systems*. Clevedon, UK: Channel View Press.

Hall, C.M. and James, M. (2011) 'Medical tourism: emerging biosecurity and nosocomial issues', *Tourism Review*, 66(1/2): 118–126.

Hall, C.M. and Jenkins, J.M. (1995) *Tourism and Public Policy*. London: Routledge.

Hall, C.M. and Lew, A. (2009) *Understanding and Managing Tourism Impacts: An Integrated Approach*. London: Routledge.

Hall, C.M. and Mitchell, R. (2008) *Wine Marketing*. Oxford: Elsevier.

Hall, C.M. and Page, S.J. (2006) *The Geography of Tourism and Recreation: Environment, Place and Space*, 3rd edn. London: Routledge.

Hall, C.M. and Saarinen, J. (2010a) 'Polar tourism: definitions and dimensions', *Scandinavian Journal of Hospitality and Tourism*, 10(4): 448–467.

Hall, C.M. and Saarinen, J. (2010b) 'Geotourism and climate change: paradoxes and promises of geotourism in polar regions', *Téoros*, 29(2): 77–86.

Hall, C.M. and Saarinen, J. (2010c) *Tourism and Change in Polar Regions: Climate, Environments and Experiences*. London: Routledge.

Hall, C.M. and Saarinen, J. (2010d) 'Last chance to see? Future issues for polar tourism and change', in C.M. Hall and J. Saarinen (eds) *Tourism and Change in Polar Regions: Climate, Environments and Experiences*. London: Routledge.

Hall, C.M. and Sharples, L. (2008) 'Food events and the local food system: marketing, management and planning issues', in C.M. Hall and L. Sharples (eds) *Food and Wine Festivals and Events around the World: Development, Management and Markets*. Oxford: Butterworth-Heinemann.

Hall, C.M. and Stoffels, M. (2006) 'Lake tourism in New Zealand: sustainable management issues', in C.M. Hall and T. Härkönen (eds) *Lake Tourism: An Integrated Approach to Lacustrine Tourism Systems*. Clevedon, UK: Channelview Press, pp. 182–206.

Hall, C.M. and Williams, A. (2008) *Tourism and Innovation*. London: Routledge.

Hall, C.M. and Wilson, S. (2010) 'Tourism, conservation and visitor management in the sub-Antarctic islands', in C.M. Hall and J. Saarinen (eds) *Tourism and Change in Polar Regions: Climate, Environments and Experiences*. London: Routledge.

Hall, C.M., Sharples, E. and Smith, A. (2003) 'The experience of consumption or the consumption of experiences? Challenges and issues in food tourism', in C.M. Hall, E. Sharples, R. Mitchell, B. Cambourne and N. Macionis (eds) *Food Tourism Around the World: Development, Management and Markets*. Oxford: Butterworth-Heinemann.

Hall, C.M., Müller, D. and Saarinen, J. (2008) *Nordic Tourism*. Clevedon, UK: Channel View Press.

Hall, C.M., James, M. and Wilson, S. (2010) 'Biodiversity, biosecurity, and cruising in the Arctic and sub-Arctic', *Journal of Heritage Tourism*, 5(4): 351–364.

Hamilton, J.M. and Lau, M.A. (2005) 'The role of climate information in tourist destination choice decision-making', in S. Gössling and C.M. Hall (eds) *Tourism and Global Environmental Change*. London: Routledge.

Hamilton, J.M. and Tol, R.S.J. (2007) 'The impact of climate change on tourism in Germany, the UK and Ireland: a simulation study', *Regional Environmental Change*, 7: 161–172.

Hamilton, J.M., Maddison, D.J. and Tol, R.S.J. (2005a) 'Climate change and international tourism: a simulation study', *Global Environmental Change*, 15: 253–266.

Hamilton, J.M., Maddison, D.J. and Tol, R.S.J. (2005b) 'Effects of climate change on international tourism', *Climate Research*, 29(3): 245–254.

Hamilton, L., Rohall, D., Brown, B., Hayward, G.F. and Keim, B.D. (2003) 'Warming winters and New Hampshire's lost ski areas: an integrated case study', *International Journal of Sociology and Social Policy*, 23(10): 52–73.

Hansen, J. (2006) 'Climate change: on the edge', *The Independent*, 17 February. www.independent.co.uk/environment/climate-change-on-the-edge-466818.html

Hansen, J. (2007) 'Scientific reticence and sea level rise', *Environmental Research Letters* (doi: 10.1088/1748-9326/2/2/024002).

Hansen, J. (2009) *Storms of my Grandchildren*. New York: Bloomsbury.

Hares, A., Dickinson, J. and Wilkes, K. (2010) 'Climate change and the air travel decisions of UK tourists', *Journal of Transport Geography*, 18(3): 466–473.

Harlfinger, O. (1991) 'Holiday biometeorology: a study of Palma de Majorca, Spain', *GeoJournal*, 25: 377–381.

Harrabin, R. (2011) 'Biofuels targets are "unethical", says Nuffield report', *BBC News UK*, 13 April. www.bbc.co.uk/news/uk-13056862

Harrison, K. and Winter, C. (2005) 'Crocodiles and oil prices: Northern Territory tourism as the "canary in the coal mine" for global tourism', in *4th International Symposium on Aspects of*

Tourism: Mobility and Local–Global Connections, 23–24 June 2005. Eastbourne, UK: University of Brighton. http://eprints.usq.edu.au/425/

Harrison, R., Kinnaird, V., McBoyle, G., Quinlan C. and Wall, G. (1986) 'Climate change and downhill skiing in Ontario', *Ontario Geographer*, 28: 51–68.

Harrison, S.J., Winterbottom, S.J. and Sheppard, C. (1999) 'The potential effect of climate change on the Scottish tourist industry', *Tourism Management*, 20(2): 203–211.

Harrison, S.J., Winterbottom, S.J. and Johnson, R. (2001) *Climate Change and Changing Patterns of Snowfall in Scotland*. Edinburgh: Scottish Executive Central Research Unit. www.scotland.gov.uk/Resource/Doc/156666/0042099.pdf

Hartz, D.A., Brazel, A.J. and Heisler, G.M. (2006) 'A case study in resort climatology of Phoenix, Arizona, USA', *International Journal of Biometeorology*, 51: 73–83.

Harvey, D. (1996) *Justice, Nature, and the Geography of Difference*. Oxford: Blackwell.

Harvey, F. (2011) 'Climate change will increase threat of war, Chris Huhne to warn UK climate secretary to tell defence experts that conflict caused by climate change risks reversing the progress of civilisation', *The Guardian*, 6 July. www.guardian.co.uk/environment/2011/jul/06/climate-change-war-chris-huhne

Harvey, M. and Pilgrim, S. (2011) 'The new competition for land: food, energy, and climate change', *Food Policy*, 36: 40–51.

Hay, B. (1989) 'Tourism and Scottish weather', in S. Scotland, J. Harrison and K. Smith (eds) *Weather Sensitivity and Services*. Edinburgh: Scottish Academy Press.

Hayhoe, K., Cayan, D., Field, C., Frumhoff, P., Maurer, E., Miller, N., Moser, S., Schneider, S., Cahill, K., Cleland, E., Dale, L., Draper, R., Hanemann, R.M., Kalksteing, L., Lenihan, J., Lunch, C., Neilson, R., Sheridan, S. and Verville, J. (2004) 'Emissions pathways, climate change and impacts on California', *Proceedings of the National Academy of Sciences, USA*, 101(34): 12422–12427.

Head, L. (2008) 'Is the concept of human impacts past its use-by date?' *The Holocene*, 18: 373–377.

Heal, G. (2009) 'The economics of climate change: a post-Stern perspective', *Climatic Change*, 96: 275–297.

Hein, L., Metzger, M.J. and Moreno, A. (2009) 'Potential impacts of climate change on tourism: a case study for Spain', *Current Opinion in Environmental Sustainability*, 1(2): 170–178.

Heisler, J., Gilbert, P.M., Burkholder, J.M., Anderson, D.M., Cochlan, W., Dennison, W.C., Dortch, Q., Gobler, C.J., Heil, C.A., Humphries, E., Lewitus, A., Magnien, R., Marshall, H.G., Sellner, K., Stockwell, D.A., Stoecker, D.K. and Suddleson, M. (2008) 'Eutrophication and harmful algal blooms: a scientific consensus', *Harmful Algae*, 8: 3–13.

Hennessy, K., Fitzharris, B., Bates, B.C., Harvey, N., Howden, S.M., Hughes, L., Salinger, J. and Warrick, R. (2007) 'Australia and New Zealand', in M.L. Parry, O.F. Canziani, J.P. Palutikof, P.J. van der Linden and C.E. Hanson (eds) *Climate Change 2007: Impacts, Adaptation and Vulnerability. Contribution of Working Group II to the Fourth Assessment Report of the Intergovernmental Panel on Climate Change*. Cambridge: Cambridge University Press.

Hennessy, K.L., Whetton, P.H., Walsh, K., Smith, I.N., Bathols, J.M., Hutchinson, M. and Sharpies, J. (2008) 'Climate change effects on snow conditions in mainland Australia and adaptation at ski resorts through snow making', *Climate Research*, 35: 255–270.

Heo, I. and Lee, S. (2008) 'The impact of climate change on ski industries in South Korea', *Journal of the Korean Geographical Society*, 43(5): 715–727.

Hewer, M. and Scott, D. (2011) *Influence of Weather on Ontario Park Visitors*, Ontario Parks Technical Report. Toronto: Ontario Parks Agency.

Hickman, L. (2011) 'Isles of Scilly turn heat on Jersey over "warmest place in Britain" claim', *The Guardian*, 11 April. www.guardian.co.uk/uk/2011/apr/10/isles-of-scilly-jersey-warmest-britain?INTCMP=SRCH

Higham, J.E.S. and Cohen, S.A. (2011) 'Canary in the coalmine: Norwegian attitudes towards climate change and extreme long-haul air travel to Aotearoa/New Zealand', *Tourism Management*, 32: 98–105.

Hilbert, D.W., Graham, A. and Hopkins, M.S. (2007) 'Glacial and interglacial refugia within a long-term rainforest refugium: the Wet Tropics Bioregion of NE Queensland, Australia', *Palaeogeography, Palaeoclimatology, Palaeoecology*, 251: 104–118.

Hill, A. (2009) 'Holiday deals abroad vanish in rush to flee the rain', *The Observer*, 9 August.

Hille, J., Aall, C. and Grimstad Klepp, I. (2007) *Miljøbelastninger fra norsk fritidsforbruk – en kartlegging*. www.vestforsk.no/filearchive/rapport-1-07-fritidsbruk.pdf

Hirsch, R.L. (2007) *Peaking of World Oil Production: Recent Forecasts*, DOE/NETL-2007/1263, 5 February. National Energy Technology Laboratory.

Hirsch, R.L., Bezdek, R. and Wendling, R. (2005) *Peaking of World Oil Production: Impacts, Mitigation & Risk Management*, report prepared for the US Department of Energy.

Hirt, P., Gustafson, A. and Larson, K. (2008) 'The mirage in the valley of the sun', *Environmental History*, 13: 482–514.

Hjalager, A.-M. (2010) 'A review of innovation research in tourism', *Tourism Management*, 31: 1–12.

HM Revenue & Customs (2008) *Air Passenger Duty – Introduction*. London: HM Revenue & Customs. http://customs.hmrc.gov.uk/channelsPortalWebApp/downloadFile?contentID=HMCE_CL_001170

Höchtl, F., Lehringer, S. and Konold, W. (2005) '"Wilderness": what it means when it becomes a reality – a case study from the southwestern Alps', *Landscape and Urban Planning*, 70: 85–95.

Hodas, D.R. (2004) 'State law responses to global warming: is it constitutional to think globally and act locally?', *Pace Environmental Law Review*, 21: 53–81.

Hoegh-Guldberg, O. and Hoegh-Guldberg, H. (2008) *The Impact of Climate Change and Ocean Acidification on the Great Barrier Reef and its Tourist Industry*, paper commissioned for the Garnaut Climate Change Review. Cambridge: Cambridge University Press.

Hoegh-Guldberg, O., Mumby, P.J., Hooten, A.J., Steneck, R.S., Greenfield, P., Gomez, E., Harvell, C.D., Sale, P.F., Edwards, A.J., Caldeira, K., Knowlton, N., Eakin, C.M., Iglesias-Prieto, R., Muthiga, N., Bradbury, R.H., Dubi, A. and Hatziolos, M.E. (2007) 'Coral reefs under rapid climate change and ocean acidification', *Science*, 318: 1737–1742.

Hoffman, A. (2010) 'Climate change as a cultural and behavioral issue: addressing barriers and implementing solutions', *Organizational Dynamics*, 39(4): 295–305.

Hoffmann, V.H., Sprengel, D.C., Ziegler, A., Kolb, M. and Abegg, B. (2009) 'Determinants of corporate adaptation to climate change in winter tourism: an econometric analysis', *Global Environmental Change*, 19: 256–264.

Honey, M. and Krantz, D. (2007) *Global Trends in Coastal Tourism*. Stanford, CA: Stanford University, Center on Ecotourism and Sustainable Development. www.responsibletravel.org/resources/documents/reports/Global_Trends_in_Coastal_Tourism_by_CESD_Jan_08_LR.pdf

Höppe, P.R. and Seidl, H.A. (1991) 'Problems in the assessment of the bioclimate for vacationists at the seaside', *International Journal of Biometeorology*, 35: 107–110.

Horton, R., Herweijer, C., Rosenzweig, C., Liu, J., Gornitz, V. and Ruane, A. (2008) 'Sea level projections for current generation CGCMs based on semi-empirical method', *Geophysical Research Letters*, 35 (doi: 10.1029/2007GL032486).

Houston, J.R. (2002) 'The economic value of beaches: a 2002 update', *Shore and Beach*, 70:9–12.

Hovenden, M.J., Wills, K.E., Schoor, J.K.V., Williams, A.L. and Newton, P.C.D. (2008) 'Flowering phenology in a species-rich temperate grassland is sensitive to warming but not elevated CO_2', *New Phytologist*, 178: 815–822.

Howden, S. and Crimp, S. (2007) *Effect of Climate and Climate Change on Electricity Demand in Australia*, paper presented at MSSANZ International Congress on Modelling and Simulation, University of Canterbury, Christchurch, New Zealand, December. www.mssanz.org.au/MODSIM01/Vol%202/Howden.pdf

Howitt, O.J.A., Revol, V.G.N., Smith, I.J and Rodger, C.J. (2010) 'Carbon emissions from international cruise ship passengers' travel to and from New Zealand', *Energy Policy*, 38(5): 2252–2560.

Hu, Y. and Ritchie, J. (1993) 'Measuring destination attractiveness: a contextual approach', *Journal of Travel Research*, 32(20): 25–34.

Huddleston, M. and Eggen, B. (2007) *Climate Change Adaptation for UK Businesses*. London: Met Office Consulting.

Hughes, T.P., Graham, N.A.J., Jackson, J.B.C., Mumby, P.J. and Steneck, R.S. (2010) 'Rising to the challenge of sustaining coral reef resilience', *Trends in Ecology and Evolution*, 25:633–642.

Hughs, H. (2008) *Frommer's 500 Places to Visit before they Disappear*. Hoboken, NJ: Frommer's.

Hulme, M. (2008) 'Geographical work at the boundaries of climate change', *Transactions of the Institute of British Geographers*, 33(1): 5–11.

Hulme, M. (2009) *Why We Disagree About Climate Change: Understanding Controversy, Inaction and Opportunity*. Cambridge: Cambridge University Press.

Hulme, M., Conway, D. and Lu, X. (2003) *Climate Change: An Overview and its Impact on the Living Lakes*, report prepared for the 8th Living Lakes Conference, 'Climate Change and Governance: Managing Impacts on Lakes', Zuckerman Institute for Connective Environmental Research, University of East Anglia, Norwich, UK, 7–12 September. Norwich: Tyndall Centre for Climate Change Research, University of East Anglia.

Hyslop, K. (2007) 'Climate change impacts on visitation in national parks in the United States', Master's thesis, Department of Geography, University of Waterloo, Canada. uwspace.uwaterloo.ca/bitstream/10012/2660/1/MES%20Thesis%20-%20Revised%20Final.pdf

Hystad, P. and Keller, P. (2006) 'Disaster management: Kelowna tourism industry's preparedness, impact and response to a 2003 major forest fire', *Journal of Hospitality and Tourism Management*, 13(1): 44–58.

IAPAL (2008) *International Air Passenger Adaptation Levy: A Proposal by the Maldives on Behalf of the Group of Least Developed Countries (LDCs)*. International Air Passenger Adaptation Levy. unfccc.int/files/kyoto_protocol/application/pdf/maldivesadaptation131208.pdf

IATA (2008) *Jet Fuel Price Monitor*. Montreal/Geneva: International Air Transport Association. www.iata.org/whatwedo/economics/fuel_monitor/index.htm

IATA (2009) *The IATA Technology Roadmap Report*. Montreal/Geneva: International Air Transport Association. www.iata.org/SiteCollectionDocuments/Documents/Technology_Roadmap_May2009.pdf

IATA (2011) *Aviation and Climate Change. Pathway to Carbon-neutral Growth in 2020*. Montreal/Geneva: International Air Transport Association. www.iata.org/SiteCollectionDocuments/AviationClimateChange_PathwayTo2020_email.pdf

IBAC (2009) *Business Aviation Statement on Climate Change*. Montreal: International Business Aviation Council. www.ibac.org/business-aviation-and-the-environment/business-aviation-statement-on-climate-change

ICAO (2009) *Group on International Aviation and Climate Change (GIACC)*, Report, 1 June. International Civil Aviation Organization. www.icao.int/icao/en/atb/meetings/GIACC/GiaccReport_Final_en.pdf

ICLEI (2008) *Mitigation*. Bonn, Germany: ICLEI (International Council for Local Environmental Initiatives) – Local Governments for Sustainability. www.iclei.org/index.php?id=10828

ICLEI (2009) *World Mayors and Local Governments Climate Protection Agreement*. Bonn, Germany: ICLEI (International Council for Local Environmental Initiatives) – Local Governments for Sustainability. www.globalclimateagreement.org/index.php?id=10372

IEA (2009) *World Energy Outlook 2009*. Paris: International Energy Agency.

IEA (2010) *World Energy Outlook 2010*, press release. London: International Energy Agency. www. worldenergyoutlook.org/docs/weo2010/press_release.pdf

IEA and OECD (2009) 'Transport, energy and CO_2: moving toward sustainability', International Energy Agency and Organisation for Economic Co-operation and Development, *International Journal of Health Services*, 37(1): 193–198.

IMO (2009a) *Prevention of Air Pollution from Ships*, Second IMO GHG Study. London: International Maritime Organization. www.imo.org/includes/blastDataOnly.asp/data_id%3D26047/INF-10.pdf

IMO (2009b) *Climate Change: A Challenge for IMO Too!*, background paper. London: International Maritime Organization. www.imo.org/includes/blastDataOnly.asp/data_id%3D26316/backgroundE.pdf

IMO (2011) *Mandatory energy efficiency measures for international shipping adopted at IMO environment meeting*. Marine Environment Protection Committee (MEPC) – 62nd session: 11–15 July 2011. London: IMO. www.imo.org/MediaCentre/PressBriefings/Pages/42-mepc-ghg.aspx

Instanes, A., Anisimov, O., Brigham, L., Goering, D., Ladanyi, B., Larsen, J.O. and Khrustalev, L.N. (2005) 'Infrastructure: buildings, support systems, and industrial facilities', in *Arctic Climate Impact Assessment (ACIA)*. Cambridge: Cambridge University Press.

International Snowmobile Manufacturers Association (2006) *International Snowmobile Industry Facts and Figures*. Haslett, MI: International Snowmobile Manufacturers Association. www.snowmobile.org/pr_snowfacts.asp

International Tourism Partnership (2008) *Going Green: Minimum Standards toward a Sustainable Hotel*. London: International Tourism Partnership www.tourismpartnership.org/downloads/Going%20Green.pdf

IPCC (1996) 'The science of climate change', in J.T. Houghton, L.G. Meira Filho, B.A. Callander, N. Harris, A. Kattenberg and K. Maskell (eds) *Climate Change 1995*. Cambridge and New York: Cambridge University Press/Intergovernmental Panel on Climate Change.

IPCC (2001a) *Glossary of Terms*. Geneva: Intergovernmental Panel on Climate Change. www.ipcc.ch/pdf/glossary/tar-ipcc-terms-en.pdf

IPCC (2001b) 'The scientific basis', in J.T. Houghton, Y. Ding, D.J. Griggs, M. Noguer, P.J. van der Linden, X. Dai, K. Maskell and C.A. Johnson (eds) *Climate Change 2001: Contribution of Working Group I to the Third Assessment Report of the Intergovernmental Panel on Climate Change*. Cambridge: Cambridge University Press.

IPCC (2001c) 'Impacts, adaptation and vulnerability', in J.T. Houghton, Y. Ding, D.J. Griggs, M. Noguer, P.J. van der Linden, X. Dai, K. Maskell and C.A. Johnson (eds) *Climate Change 2001: Contribution of Working Group II to the Third Assessment Report of the Intergovernmental Panel on Climate Change*. Cambridge: Cambridge University Press.

IPCC (2005) *IPCC WGII Fourth Assessment Report, Climate Change Impacts, Adaptation and Vulnerability*, Expert Review of First Order Draft, Specific Comments, Chapter 2, December 5. www.ipcc-wg2.gov/AR4/FOD_COMMS/Ch02_FOD_comments.pdf

IPCC (2007a) 'Communication of uncertainty in the Working Group II Fourth Assessment', in M.L. Parry, O.F. Canziani, J.P. Palutikof, P.J. van der Linden and C.E. Hanson (eds) *Climate Change 2007: Impacts, Adaptation and Vulnerability. Contribution of Working Group II to the Fourth Assessment Report of the Intergovernmental Panel on Climate Change*. Cambridge: Cambridge University Press.

IPCC (2007b) 'Summary for policymakers', in S. Solomon, D. Qin, M. Manning, Z. Chen, M. Marquis, K.B. Avery, M. Tignor and H.L. Miller (eds) *Climate Change 2007: The Physical Science Basis. Contribution of Working Group I to the Fourth Assessment Report of the Intergovernmental Panel on Climate Change*. Cambridge: Cambridge University Press.

IPCC (2007c) *Glossary*. Geneva: Intergovernmental Panel on Climate Change. www.ipcc.ch/pdf/assessment-report/ar4/syr/ar4_syr_appendix.pdf

IUCN (2007) *Forest Fires in the Mediterranean – Background Information*. Gland, Switzerland: International Union for Conservation of Nature. www.uicnmed.org/web2007/documentos/Background_med_forest_fires.pdf

IUCN, UNEP and WWF (1991) *Caring for the Earth: A Strategy for Sustainable Living*. Gland, Switzerland: International Union for Conservation of Nature, United Nations Environment Programme and World Wide Fund for Nature.

Jacobson, M.Z. (2005) 'Studying ocean acidification with conservative, stable numerical schemes for nonequilibrium air–ocean exchange and ocean equilibrium chemistry', *Journal of Geophysical Research*, D: Atmospheres, 110: D07302.1–D07302.17 (doi: 10.1029/2004JD005220).

Jaenson, T.G.T., Tälleklint, L., Lundqvist, L., Olsen, B., Chirico, J. and Mejlon, H. (1994) 'Geographical distribution, host associations and vector roles of ticks (Acari: Ixodidae and Argasidae) in Sweden', *Journal of Medical Entomology*, 31: 240–256.

Jarvis, N. and Pulido Ortega, A. (2010) 'The impact of climate change on small hotels in Granada, Spain', *Tourism and Hospitality Planning and Development*, 7(3): 283–299.

JATO (2011) *Rich Nations Falling Behind Europe on Car CO_2 Emissions*. Troy, MI: JATO Dynamics, Inc. www.jato.com/PressReleases/Rich%20Nations%20Falling%20Behind%20Europe%20on%20Car%20CO2%20Emissions.pdf

Jennings, S. (2004) 'Coastal tourism and shoreline management', *Annals of Tourism Research*, 23(8): 503–508.

Jerneck, A., Olsson, L., Ness, B., Anderberg, S., Baier, M., Clark, E., Hickler, T., Hornborg, A., Kronsell, A., Lövbrand, E. and Persson, J. (2011) 'Structuring sustainability science', *Sustainability Science*, 6(1): 69–82.

Jha, A. (2011) 'Freedom of information laws are used to harass scientists, says Nobel laureate', *The Guardian*, 25 May. www.guardian.co.uk/politics/2011/may/25/freedom-information-laws-harass-scientists

Jodha, N. (1991) 'Mountain perspective and sustainability: a framework for development strategies', in M. Banskota, N. Jodha and U. Pratap (eds) *Sustainable Mountain Agriculture Perspectives and Issues*, vol. 1, pp. 41–82.

Jöhnk, K.D., Huisman, J., Sharples, J., Sommeijer, B., Visser, P.M. and Stroom, J.M. (2008) 'Summer heatwaves promote blooms of harmful cyanobacteria', *Global Change Biology*, 14: 495–512.

Johnson, L. (2008) 'Texas weather blasts nature tourism', *KUHF Houston Public Radio*, 29 October. app1.kuhf.org/houston_public_radio-news-display.php?articles_id=1225310536

Jones, A.L. and Phillips, M. (eds) (2011) *Disappearing Destinations: Climate Change and the Future Challenges of Coastal Tourism*. Wallingford, UK: CABI Publishing.

Jones, B. and Scott, D. (2006) 'Climate change, seasonality and visitation to Canada's national parks', *Journal of Parks and Recreation Administration*, 24(2): 42–62.

Jones, B., Scott, D. and Gössling, S. (2005) 'Lakes and streams', in S. Gössling and C.M. Hall (eds) *Tourism and Global Environmental Change*. London: Routledge.

Jones, B., Scott, D. and Abi Khaled, H. (2006) 'Implications of climate change for outdoor event planning: a case study of three special events in Canada's Capital Region', *International Journal of Event Management*, 10(1): 63–76.

Jones, G.V., White, M.A., Cooper, O.R. and Storchmann, K. (2005) 'Climate change and global wine quality', *Climatic Change*, 73(3): 319–343.

Jonsson, C. and Devonish, D. (2008) 'Does nationality, gender, and age affect travel motivation? A case of visitors to the Caribbean island of Barbados', *Journal of Travel and Tourism Marketing*, 23(3): 398–408.

Jopp, R., DeLacy, T. and Mair, J. (2010) 'Developing a framework for regional destination adaptation to climate change', *Current Issues in Tourism*, 13(6): 591–605.

Jopp, R., DeLacy, T., Fluker, M. and Mair, J. (2011) 'Using a regional tourism evaluation framework to determine climate change adaptation options for Victoria's Surf Coast', *Asia Pacific Journal of Tourism Research*, in press.

Jordbruksverket (2011) *Matkonsumtionens senaste utveckling – köttkonsumtionen stiger igen.* www.jordbruksverket.se/download/18.e01569712f24e2ca09800020/PM+konsumtion+5+april. pdf

Jorgensen, F. and Solvoll, G. (1996) 'Demand models for inclusive tour charter: the Norwegian case', *Tourism Management*, 17: 17–24.

Jotzo, F. (2010) *Comparing the Copenhagen Emissions Targets*, CCEP Working Paper, Canberra: Centre for Climate Economics & Policy, Crawford School of Economics and Government, The Australian National University.

Kågeson, P. (2007) *Linking CO$_2$ Emissions from International Shipping to the EU ETS*, report by Nature Associates. Dessau, Germany: Federal Environment Agency.

Kaswan, A. (2009) 'Climate change, consumption and cities,' *Fordham Urban Law Journal*, 36: 253–312.

Katz, M.E., Cramer, B.S., Mountain, G.S., Katz, S. and Miller, K.G. (2001) 'Uncorking the bottle: what triggered the Paleocene/Eocene thermal maximum methane release', *Paleoceanography*, 16(6): 549–562.

Katz, R.W. and Murphy, A.H. (2000) 'Economic value of weather and climate forecasts', *Climatic Change*, 45(3/4): 601–606.

Kaufman, D.S., Schneider, D.P., McKay, N.P., Ammann, C.M., Bradley, R.S., Briffa, K.R., Miller, G.H., Otto-Bliesner, B.L., Overpeck, J.T., Vinther, B.M. and Arctic Lakes 2k Project Members (2009) 'Recent warming reverses long-term Arctic cooling', *Science*, 325(5945): 1236–1239.

Keleher, C.J. and Rahel, F.J. (1996) 'Thermal limits to salmonid distributions in the Rocky Mountain region and potential habitat loss due to global warming: a Geographic Information System (GIS) approach', *Transactions of the American Fisheries Society*, 125: 1–13.

Keller, D.R. and Golley, F.B. (eds) (2000) *The Philosophy of Ecology. From Science to Synthesis.* Athens, GA: University of Georgia Press.

Kemp, R. (2009) 'Short-haul aviation – under what conditions is it more environmentally benign than the alternatives?', *Technology Analysis & Strategic Management*, 21(1): 115–127.

Kevan, S. (1993) 'Quest for cures: a history of tourism for climate and health', *International Journal of Biometeorology*, 37: 113–124.

King County (2011) *Cool Counties: How to Become a Cool County.* Seattle, WA: King County. www.kingcounty.gov/exec/coolcounties/HowTo.aspx

Kintisch, E. (2010) 'Scientists "convinced" of climate consensus more prominent than opponents, says paper', *Science AAAS, ScienceInsider*, 21 June. http://news.sciencemag.org/scienceinsider/ 2010/06/scientists-convinced-of-climate.html

Klein, J. and Dawar, N. (2004) 'Corporate social responsibility and consumers' attributions and brand evaluations in a product-harm crisis', *International Journal of Research in Marketing*, 21(3): 203–217.

Klein, R. (2009) 'Identifying countries that are particularly vulnerable to the adverse effects of climate change: an academic or a political challenge?', *Carbon & Climate Law Review*, 3(3): 284–291.

Klein, R. (2011) 'The cruise sector and its environmental impact', in C. Schott (ed.) *Tourism and the Implications of Climate Change: Issues and Actions. Bridging Tourism Theory and Practice.* Bingley, UK: Emerald Group Publishing.

Knutson, T., McBride, J., Chan, J., Emanuel, K., Holland, G., Landsea, C., Held, I., Kossin, J., Srivastava, A.K. and Sugi, M. (2010) 'Tropical cyclones and climate change', *Nature Geoscience*, 3: 157–163.

König, U. (1998) 'Tourism in a warmer world: implications of climate change due to enhanced greenhouse effect for the ski industry in the Australian Alps', *Wirtschaftsgeographie und Raumplanung* (University of Zurich), 28.

König, U. and Abegg, B. (1997) 'Impacts of climate change on tourism in the Swiss Alps', *Journal of Sustainable Tourism*, 5(1): 46–58.

Kont, A., Endjarv, E., Jaagus, J., Lode, E., Orviku, K., Ratas, U., Rivis, R., Suursaar, U. and Tonisson, H. (2007) 'Impact of climate change on Estonian coastal and inland wetlands: a summary with new results', *Boreal Environment Research*, 12: 653–671.

Kooiman, J. (2003) *Governing as Governance*. Los Angeles, CA: Sage.

Kotlarski, S., Bosshard, T., Fischer, E., Lüthi, D. and Schär, C. (2010) *Climate Scenarios for the European Alps and Extreme Events*, presentation at Swiss Climate Scenarios Workshop, Zurich. www.c2sm.ethz.ch/news/scen_workshop/presentations/c2sm_ws10_kotlarski.pdf

Kozak, M. (2002) 'Comparative analysis of tourist motivations by nationality and destinations', *Tourism Management*, 23(3): 207–220.

KPMG (2008) *Climate Changes Your Business*. www.kpmg.com/Global/en/IssuesAndInsights/ArticlesPublications/Documents/Climate-changes-your-business.pdf

Kragt, M., Roebeling, P. and Ruijs, A. (2009) 'Effects of Great Barrier Reef degradation on recreational reef-trip demand: a contingent behaviour approach', *Australian Journal of Agricultural and Resource Economics*, 53: 213–239.

Kraus, R. (2000) *Leisure in a Changing America: Trends and Issues for the 21st Century*. Boston, MA: Allyn and Bacon.

Krause, N.K. (2010) 'Air tax rise to mean fewer tourists – PM', *Dominion Post*, 3 November. www.stuff.co.nz/business/industries/4300903/Air-tax-rise-to-mean-fewer-tourists-PM

Kullman, L. (2007) 'Tree line population monitoring of *Pinus sylvestris* in the Swedish Scandes, 1973–2005: implications for tree line theory and climate change ecology', *Journal of Ecology*, 95: 41–52.

Lagadec, P. (2004) 'Understanding the French 2003 heat wave experience: beyond the heat, a multi-layered challenge', *Journal of Contingencies and Crisis Management*, 12(4): 160–169.

Lahiri, S. (2009) 'Bio-fuel and commons in India. A story of dispossession and colonisation', in C. D'Mello, J. McKeown and S. Minninger (eds) *Disaster Prevention in Tourism*. Chiang Mai: Perspectives on Climate Justice, Ecumenical Coalition on Tourism in cooperation with EED Tourism Watch, Germany.

Lamb, P. (2002) 'The climate revolution: a perspective', *Climatic Change*, 54: 1–9.

Lamers, M. and Amelung, B. (2007) 'The environmental impacts of tourism in Antarctica: a global perspective', in P. Peeters (ed.) *Tourism and Climate Change Mitigation. Methods, Greenhouse Gas Reductions and Policies*. Breda, the Netherlands: NHTV Breda University of Applied Sciences.

Lamothe and Periard Consultants (1988) 'Implications of climate change for downhill skiing in Quebec, Ottawa, Canada', *Environment Canada, Climate Change Digest*, 1988: 88–103.

Landau, S., Legro, S., Vlasic, S. (2008) *A Climate for Change: Climate Change and its Impacts on Society and Economy in Croatia*. Zagreb: United Nations Development Programme.

Landauer, M. and Pröbstl, U. (2008) 'Klimawandel, Skilanglauf und Tourismus in Österreich', *Naturschutz und Landschaftsplanung*, 40(10): 336–342.

Larsen, J., Urry, J. and Axhausen, K.W. (2007) 'Networks and tourism. Mobile social life', *Annals of Tourism Research*, 34(1): 244–262.

Lassen, C. (2006) 'Aeromobility and work', *Environment and Planning A*, 38: 301–312.

Launius, R.D. and Jenkins, D.R. (2006) 'Is it finally time for space tourism?', *Astropolitics*, 4(3): 253–280.

Lazar, B. and Williams, M. (2008) 'Climate change in western ski areas: potential changes in the timing of wet avalanches and snow quality for the Aspen ski area in the years 2030 and 2100', *Cold Regions Science and Technology*, 51: 219–228.

Le Blanc, D. (2009) 'Climate change and sustainable development revisited: implementation challenges', *Natural Resources Forum*, 33(4): 259–261.

Lee, D.S., Fahey, D.W., Forster, P.M., Newton, P.J., Wit, R.C.N., Lim, L.L., Owen, B. and Sausen, R. (2009) 'Aviation and global climate change in the 21st century', *Atmospheric Environment*, 43: 3520–3537.

Leggett, J., Pepper, W., Swart, R., Edmonds, J., Meira Filho, L., Mintzer, I., Wang, M. and Watson, J. (1992) 'Emissions scenarios for the IPCC: an update', in *Climate Change 1992: The Supplementary Report to the IPCC Scientific Assessment*. Cambridge: Cambridge University Press.

Leiserowitz, A.A., Maibach, E.W., Roser-Renouf, C., Smith, N. and Dawson, E. (2010) *Climategate, Public Opinion, and the Loss of Trust*, working paper. Fairfax, VA: Centre for Climate Change Communication, George Mason University.

Lemelin, H., Dawson, J., Stewart, E., Lueck, M. (2010) 'Last-chance tourism: the boom, doom, and gloom of visiting vanishing destinations', *Current Issues in Tourism*, 13(5): 477–493.

Lemelin, H., Dawson, J. and Stewart, E. (2011) *Last Chance Tourism*. London: Routledge.

Lemieux, C., Scott, D., Davis, R. and Gray, P. (2008) *Changing Climate, Challenging Choices: Ontario Parks and Climate Change Adaptation*, Climate Change Impacts and Adaptation Program. Ottawa: Natural Resources Canada.

Lemieux, C., Beechey, T. and Scott, D. (2011) 'The state of climate change adaptation in Canada's protected areas sector', *Canadian Geographer*, 55(3): 301–317.

Lenton, T.M., Held, H., Kriegler, E., Hall, J.W., Lucht, W., Rahmstorf, S. and Schellnhuber, H.J. (2008) 'Tipping elements in the Earth's climate system', *Proceedings of the National Academy of Sciences, USA*, 105(6): 1786–1793.

Leopold, A. (1966) *A Sand County Almanac*. New York: Oxford University Press.

Létard, V., Flandre, H. and Lepeltier, S. (2004) *Rapport d'Information Fait au Nom de la Mission: La France et les Français Face à la Canicule: les Leçons d'une Crise*, Report No. 195 (2003–2004) to the Parliament, Government of France.

Liang, L. and James, A.D. (2009) 'The low-cost carrier model in China: the adoption of a strategic innovation', *Technology Analysis & Strategic Management*, 21(1): 129–148.

Lim, B. and Spanger-Siegfried, E. (eds) (2005) *Adaptation Policy Frameworks for Climate Change: Developing Strategies, Policies and Measures*. United Nations Development Programme–Global Environment Facility/Cambridge University Press.

Limb, M. and Spellman, G. (2001) 'Evaluating domestic tourists' attitudes to British weather. A qualitative approach', in *Proceedings of the First International Workshop on Climate, Tourism and Recreation*, Greece, 5–10 October 2001. www.mif.uni-freiburg.de/isb/ws/papers/02_spellman.pdf

Lin, T.P. and Matzarakis, A. (2008) 'Tourism climate and thermal comfort in Sun Moon Lake, Taiwan', *International Journal of Biometeorology*, 52(4): 281–290.

Lindenmayer, D. and Burbidge, A. (2009) 'South-western Australia: a global biodiversity hotspot under stress', in W. Steffen, A. Burbidge, L. Hughes, R. Kitching, D. Lindenmayer, W. Musgrave, M. Stafford Smith and P. Werner (eds) *Australia's Biodiversity and Climate Change: A Strategic Assessment of the Vulnerability of Australia's Biodiversity to Climate Change*, a report to the Natural Resource Management Ministerial Council, commissioned by the Australian Government. Canberra: CSIRO.

Lisagor, K. and Hansen, H. (2008) *Disappearing Destinations: 37 Places in Peril and what can be done to Help Save them*. New York: Vintage Departures

Lise, W. and Tol, R. (2002) 'Impact of climate on tourist demand', *Climatic Change*, 55: 429–449.

Lohmann, M. and Kaim, E. (1999) 'Weather and holiday preference – image, attitude and experience', *Revue de Tourisme*, 2: 54–64.

Lorenzo, P., González, L. and Reigosa, M.J. (2010) 'The genus *Acacia* as invader: the characteristic case of *Acacia dealbata* link in Europe', *Annals of Forest Science*, 67(1) (doi: 10.1051/forest/2009082).

Lorenzoni, I. and Pidgeon, N.F. (2006) 'Public views on climate change: European and USA perspectives', *Climatic Change*, 77(1/2): 73–95.

Lorenzoni, I., Nicholson-Cole, S. and Whitmarsh, L. (2007) 'Barriers perceived to engaging with climate change among the UK public and their policy implications', *Global Environmental Change*, 17: 445–459.

Lynch, K. (1972) *What Time Is This Place?* Cambridge, MA: MIT Press.

Lyons, S., Mayor, K. and Tol, R.S.J. (2009) 'Holiday destinations: understanding the travel choices of Irish tourists', *Tourism Management*, 30(5): 683–692.

Machado-Filho, H. (2009) 'Brazilian low-carbon transportation policies: opportunities for international support', *Climate Policy*, 9(5): 495–507.

Maddison, D. (2001) 'In search of warmer climates? The impact of climate change on flows of British tourists', *Climatic Change*, 49(1/2): 193–208.

Magrin, G., Gay García, C., Cruz Choque, D., Giménez, J.C., Moneno, A.R., Nagy, G.J., Nobre, C. and Villamizar, A. (2007) 'Latin America', in M.L. Parry, O.F. Canziani, J.P. Palutikof, P.J. van der Linden and C.E. Hanson (eds) *Climate Change 2007: Impacts, Adaptation and Vulnerability. Contribution of Working Group II to the Fourth Assessment Report of the Intergovernmental Panel on Climate Change*. Cambridge: Cambridge University Press.

Main, M. and Dearden, P. (2007) 'Tsunami impacts on Phuket's diving industry: geographical implications for marine conservation', *Coastal Management*, 35(4): 1–15.

Mansfeld, Y., Freundlish, A. and Kutiel, H. (2004) 'The relationship between weather conditions and tourists' perception of comfort: the case of the winter sun resort of Eilat', in B. Amelung and D. Viner (eds) *Proceedings NATO Advanced Research Workshop on Climate Change and Tourism*, Warsaw, Poland. www.igipz.pan.pl/geoekoklimat/blaz/blaz28.pdf#page=116

Maplecroft (2010) *Water Security Risk Index 2010*. Bath, UK: Maplecroft. www.maplecroft.com/about/news/water-security.html

Markusen, A. (1999) 'Fuzzy concepts, scanty evidence, policy distance: the case for rigour and policy relevance in critical regional studies', *Regional Studies*, 33(9): 869–884.

Marsh, G.P. (1864) *Man and Nature; or, Physical Geography as Modified by Human Action*, ed. D. Lowenthal. Cambridge, MA: Belknap Press of Harvard University Press, 1965 edn.

Martens, P., Rotmans, J. and de Groot, D. (2003) 'Biodiversity: luxury or necessity?', *Global Environmental Change*, 13: 75–81.

Martens, W.J.M., Slooff, R. and Jackson, E.K. (1997) 'Climate change, human health, and sustainable development', *Bulletin of the World Health Organization*, 75(6): 583–588.

Martinez Ibarra, E. (2011) 'The use of webcam images to determine tourist–climate aptitude: favourable weather types for sun and beach tourism on the Alicante coast (Spain)', *International Journal of Biometeorology*, 55: 373–385.

Marttila, V., Granholm, H., Laanikari, J., Yrjölä, T., Aalto, A., Heikinheimo, P., Honkatukia, J., Jarvinen, H., Liski, J., Merivirta, R. and Paunio, M. (2005) *Finland's National Strategy for Adaptation to Climate Change*, publication 1/2005 (in Finnish). Helsinki: Ministry of Agriculture and Forestry.

Maslow, A. (1970) *Motivation and Personality*. New York: Harper and Row.

Masterton, J. (1980) 'Applications of climatology to the ski industry', in W. Wyllie and L. Maguire (eds) *Proceedings of a Workshop on the Application of Meteorology to Recreation and Tourism.* Toronto: Atmospheric Environment Service.

Mathieson, A. and Wall, G. (1982) *Tourism: Economic, Physical and Social Impacts.* New York: Longman.

Matthews, W., Kellogg, W. and Robinson, G. (eds) (1971) *Man's Impact on the Climate. Study of Critical Environmental Problems (SCEP) Report.* Cambridge, MA: MIT Press.

Maunder, I. (1970) *The Value of the Weather.* London: Methuen.

Mayor, K. and Tol, R.S.J. (2007) 'The impact of the UK aviation tax on carbon dioxide emissions and visitor numbers', *Transport Policy*, 14: 507–513.

Mayor, K. and Tol, R.S.J. (2008) 'The impact of the EU–US open skies agreement on international travel and carbon dioxide emissions', *Journal of Air Transport Management*, 14: 1–7.

Mayor, K. and Tol, R.S.J. (2009) 'Aviation and the environment in the context of the EU–US Open Skies agreement', *Journal of Air Transport Management*, 15: 90–95.

Mayor, K. and Tol, R.S.J. (2010a) 'Scenarios of carbon dioxide emissions from aviation', *Global Environmental Change*, 20: 65–73.

Mayor, K. and Tol, R.S.J. (2010b) 'The impact of European climate change regulations on international tourist markets', *Transportation Research Part D*, 15: 26–36.

Mayors Climate Protection Center (2008) *The Impact of Gas Prices, Economic Conditions, and Resource Constraints on Climate Protection Strategies in U.S. Cities: Results of a 132-City Survey, June 2008.* Washington, DC: US Conference of Mayors.

Mazanec, J. (1994) 'Image measurement with self-organizing maps: a tentative application of Austrian tour operators', *Revue de Tourisme*, 49: 9–18.

McBean, G., Alekseev, G., Chen, D., Førland, E., Fyfe, J., Groisman, P.Y., King, R., Melling, H., Vose, R. and Whitfield, P.H. (2005) 'Arctic climate: past and present', in C. Symon, L. Arris and B. Heal (eds) *Arctic Climate Impact Assessment, ACIA.* Cambridge: Cambridge University Press.

McBoyle, G. and Wall, G. (1992) 'Great Lakes skiing and climate change', in A. Gill and R. Hartman (eds) *Mountain Resort Development: Proceedings of the Vail Conference*, 18–21 April 1991. Burnaby, BC: Centre for Tourism Policy and Research, Simon Fraser University.

McBoyle, G., Wall, G., Harrison, K. and Quinlan, C. (1986) 'Recreation and climate change: a Canadian case study', *Ontario Geography*, 23: 51–68.

McBoyle, G., Scott, D. and Jones, B. (2007) 'Climate change and the future of snowmobiling in non-mountainous regions of Canada', *Managing Leisure*, 12(4): 237–250.

McCarthy, J.E. (2009) *Air Pollution and Greenhouse Gas Emissions from Ships.* Washington, DC: Congressional Research Service.

McCarthy, J.J., Canziani, O., Leary, N., Dokken, D. and White, K. (eds) (2001) *Climate Change 2001: Impacts, Adaptation and Vulnerability. Contribution of Working Group II to the Third Assessment Report of the Integovernmental Panel on Climate Change.* Cambridge/New York: Cambridge University Press.

McConnell, K.E. (1977) 'Congestion and willingness to pay: a study of beach use', *Land Economics*, 53: 185–195.

McEvoy, D., Cavan, G., Handley, J. and Lindley, S. (2008) 'Changes to climate and visitor behaviour: implications for vulnerable landscapes in the Northwest region of England', *Journal of Sustainable Tourism*, 16(1): 101–121.

McIntyre-Tamwoy, S. (2008) 'The impact of global climate change and cultural heritage: grasping the issues and defining the problem', *Historic Environment*, 21(1): 2–9.

McKercher, B., Prideaux, B., Cheung, C. and Law, R. (2010) 'Achieving voluntary reductions in the carbon footprint of tourism and climate change', *Journal of Sustainable Tourism*, 18(3): 297–318.

McKie, R. (2011) 'Ocean acidification is latest manifestation of global warming', *The Observer*, 29 May. www.guardian.co.uk/environment/2011/may/29/global-warming-threat-to-oceans? INTCMP=SRCH

McMenamin, S.K., Hadly, E.A. and Wright, C.K. (2008) 'Climatic change and wetland desiccation cause amphibian decline in Yellowstone National Park', *Proceedings of the National Academy of Sciences, USA*, 105: 16988–16993.

McMullen, C.P. and Jabbour, J. (2009) *Climate Change Science Compendium 2009*. Nairobi: United Nations Environment Programme.

Meinshausen, M., Meinshausen, N., Hare, W., Raper, S.C.B., Frieler, K., Knutti, R., Frame, D. and Allen, M. (2009) 'Greenhouse-gas emission targets for limiting global warming to 2°C', *Nature*, 458(7242): 1158–1162.

MEPC (2010) *Proposal to Designate an Emission Control Area for the Commonwealth of Puerto Rico and the United States Virgin Islands for Nitrogen Oxides, Sulphur Oxides and Particulate Matter*, submitted by the United States, MEPC 61/7/3. London: Marine Environment Protection Committee, International Maritime Organization.

MEPC (2011) *Mandatory Energy Efficiency Measures for International Shipping Adopted at IMO Environment Meeting*, MEPC 62nd session: 11–15 July. London: Marine Environment Protection Committee, International Maritime Organization. www.imo.org/MediaCentre/PressBriefings/Pages/42-mepc-ghg.aspx

Metro (2006) 'Solcharter till Skåne. Klimatförändringar ger växande badturism I Nordeuropa enligt ny forskarrapport' ['Sun-charter to Scania. Climate change leads to growing sun, sand and sea tourism in Northern Europe according to a new research report'], 8 September. www.metro.se/se/article/2006/08/28/12/2359-22/

Metz, B., Davidson, O., Bosch, P., Dave, R. and Meyer, L. (2007) *Contribution of Working Group III to the Fourth Assessment Report of the Intergovernmental Panel on Climate Change*. Cambridge: Cambridge University Press.

Meyer, D. and Dewar, K. (1999) 'A new tool for investigating the effect of weather on visitor numbers', *Tourism Analysis*, 4: 145–155.

Meyer, T. (2008) 'Soft law as delegation', *Fordham International Law Journal*, 32(3): 888–942.

Meyer, W.B. and Turner II, B.L. (1995) 'The Earth transformed: trends, trajectories, and patterns', in R.J. Johnston, P.J. Taylor and M. Watts (eds) *Geographies of Global Change: Remapping the World in the Late Twentieth Century*. Oxford: Blackwell.

Micheels, A. and Montenari, M. (2008) 'A snowball Earth versus a slushball Earth: results from Neoproterozoic climate modeling sensitivity experiments', *Geosphere*, 4(2): 401–410.

Mieczkowski, Z. (1985) 'The tourism climatic index: a method for evaluating world climates for tourism', *The Canadian Geographer*, 29: 220–233.

Millennium Ecosystem Assessment (2005) *Ecosystems and Human Well-Being: Synthesis*. Washington, DC: Island Press.

Mills, E. (2005) 'Insurance in a climate of change', *Science*, 309: 1040–1043.

Milmo, D. (2011) 'EU could ground short-haul flights in favour of high-speed rail', *The Guardian*, 18 April. www.guardian.co.uk/world/2011/apr/18/eu-transport-plan-short-haul-flights? INTCMP= SRCH

Mimura, N., Nurse, L., McLean, R.F., Agard, J., Briguglio, L., Lefale, P., Payet, R. and Sem, G. (2007) 'Small islands', in M.L. Parry, O.F. Canziani, J.P. Palutikof, P.J. van der Linden and C.E. Hanson (eds) *Climate Change 2007: Impacts, Adaptation and Vulnerability. Contribution of Working Group II to the Fourth Assessment Report of the Intergovernmental Panel on Climate Change*. Cambridge: Cambridge University Press.

Min, S.-K., Zhang, X., Zwiers, F.W. and Hegerl, G.C. (2011) 'Human contribution to more-intense precipitation extremes', *Nature*, 470: 378–381.

Ministère de L'Écologie, du Développement Durable, des Transports, et du Logement (2011) *Consommation durable: experimentation nationale de l'affichage des caractéristiques environ-nementales des produits*. www.developpement-durable.gouv.fr/-Consommation-durable,4303-.html

Minns, C. and Moore, J. (1992) 'Predicting the impact of climate change on the spatial pattern of freshwater fish yield capability in eastern Canadian lakes', *Climatic Change*, 22: 327–346.

Mintel (1991) *Special Report – Holidays*. Leisure Intelligence. London: Mintel International Group.

Mitchell, A. (2001) *Right Side Up*. London: HarperCollins Business.

Mitchell, P. (2008) 'Practicing archaeology at a time of climatic catastrophe', *Antiquity*, 82: 1093–1103.

Mitchell, R. and Hall, C.M. (2006) 'Wine tourism research: the state of play', *Tourism Review International*, 9(4): 307–332.

Moen, J. and Fredman, P. (2007) 'Effects of climate change on alpine skiing in Sweden', *Journal of Sustainable Tourism*, 15(4): 418–437.

Moore, W.R. (2010) 'The impact of climate change on Caribbean tourism demand', *Current Issues in Tourism*, Special Issue on Recession and Crisis, 13(5): 495–505 (doi: 10.1080/13683500903576045).

Moore, W.R., Harewood, L. and Grosvenor, T. (2010) 'The supply side effects of climate change on tourism', unpublished paper. mpra.ub.uni-muenchen.de/21469/

Morales-Moreno, I. (2004) 'Postsovereign governance in a globalizing and fragmenting world: the case of Mexico', *Review of Policy Research*, 21(1): 107–117.

Moreno, A. (2010) 'Mediterranean tourism and climate (change): a survey-based study', *Tourism Planning and Development*, 7: 253–265.

Moreno, A. and Amelung, B. (2009) 'Climate change and tourist comfort on Europe's beaches in summer: a reassessment', *Coastal Management*, 37: 550–568.

Moreno, A. and Becken, S. (2009) 'A climate change vulnerability assessment methodology for coastal tourism', *Journal of Sustainable Tourism*, 17(4): 473–488.

Moreno, A., Amelung, B. and Santamarta, L. (2008) 'Linking beach recreation to weather conditions. A case study in Zandvoort, Netherlands', *Tourism in Marine Environments*, 5(2/3): 111–120.

Morgan, R., Gatell, E., Junyent, R., Micallef, A., Özhan, E. and Williams, A. (2000) 'An improved user-based beach climate index', *Journal of Coastal Conservation*, 6: 41–50.

Morison, A. (2011) 'Thailand's dive industry stung by marine park site closures', *CNN GO*, 25 January. www.cnngo.com/bangkok/visit/closure-18-dive-sites-413619

Moser, S.C. and Dilling, L. (2007) *Creating a Climate for Change: Communicating Climate Change and Facilitating Social Change*. New York: Cambridge University Press.

Moss, R.H., Edmonds, J.A., Hibbard, K.A., Manning, M.R., Rose, S.K., van Vuuren, D.P., Carter, T.R., Emori, S., Kainuma, M., Kram, T., Meehl, G.A., Mitchell, J.F.B., Nakicenovic, N., Riahi, K., Smith, S.J., Stouffer, R.J., Thomson, A.M., Weyant, J.P. and Wilbanks, T.J. (2010) 'The next generation of scenarios for climate change research and assessment', *Nature*, 463: 747–756.

Müller, M. and Job, H. (2009) 'Managing natural disturbance in protected areas: tourists' attitude towards the bark beetle in a German national park', *Biological Conservation*, 142(2): 375–383.

Munich Re (2010) 'Two months to Cancún climate summit/Large number of weather extremes as strong indication of climate change', press release, 27 September. www.munichre.com/en/media_relations/press_releases/2010/2010_09_27_press_release.aspx

Murphy, R., Woods, J., Black, M. and McManus, M. (2011) 'Global developments in the competition for land from biofuels', *Food Policy*, 36: 52–61.

Nadal, J., Font, A. and Cardenas, V. (2008) 'The impact of weather variability on British outbound flows', *CRE Working Papers* (*Documents de treball del CRE*) 2008/3, Centre de Recerca Econòmica (UIB 'Sa Nostra'). www.cre.sanostra.es/internet/cre.nsf/pernomcurt/DT_adjunt/$FILE/dt2008_3.pdf

Nakicenovic, N., Davidson, O., Davis, G., Grübler, A., Kram, T., Lebre La Roverere, E., Metz, B., Morita, T., Pepper, W., Pitcher, H., Sankovski, A., Shukla, P., Swart, R., Watson, R. and Zhou, D. (2000) *Special Report on Emissions Scenarios (SRES)*. Cambridge: Cambridge University Press.

National Academy of Sciences (2002) *Abrupt Climate Change: Inevitable Surprises*, US National Academy of Sciences, National Research Council Committee on Abrupt Climate Change. Washington, DC: National Academy Press.

National Park Service (2009) *Lake Mead Proves Popular During Economic Downturn*. www.nps.gov/lake/parknews/lake-mead-proves-popular-during-economic-downturn.htm

National Research Council (1979) *Report of an Ad Hoc Study Group on Carbon Dioxide and Climate*, Woods Hole, Massachusetts, 23–27 July 1979, to the Climate Research Board, Assembly of Mathematical and Physical Sciences, National Research Council. Washington, DC: National Academies Press.

Neef, D. (2009) 'The development of a global maritime emissions inventory using electronic monitoring and reporting techniques', in *18th Annual International Emission Inventory Conference: Comprehensive Inventories – Leveraging Technology and Resources*, Baltimore, Maryland, 14–17 April 2009. Washington, DC: US Environmental Protection Authority. www.epa.gov/ttn/chief/conference/ei18/session1/neef.pdf

Network for Business Sustainability (2009) *Business Adaptation to Climate Change*. Ottawa: Network for Business Sustainability.

New, M., Liverman, D., Schroder, H. and Anderson, K. (2011) 'Four degrees and beyond: the potential for a global temperature increase of four degrees and its implications', *Philosophical Transactions of the Royal Society A*, 369(1934): 6–19.

Ngazy, Z., Jiddawi, N. and Cesar, H. (2002) *Coral Bleaching and the Demand for Coral Reefs: A Marine Recreation Case in Zanzibar*. Penang, Malaysia: WorldFish Center. www.worldfishcenter.org/pubs/coral_reef/pdf/section2-8.pdf

Nicholls, R. and Tol, R. (2006) 'Impacts and responses to sea-level rise: a global analysis of the SRES scenarios over the twenty-first century', *Philosophical Transactions of the Royal Society A*, 364: 1073–1095.

Nicholls, R.J., Wong, P.P., Burkett, V.R., Codignotto, J.O., Hay, J.E., McLean, R.F., Ragoonaden, S. and Woodroffe, C.D. (2007) 'Coastal systems and low-lying areas', in M.L. Parry, O.F. Canziani, J.P. Palutikof, P.J. van der Linden and C.E. Hanson (eds) *Climate Change 2007: Impacts, Adaptation and Vulnerability. Contribution of Working Group II to the Fourth Assessment Report of the Intergovernmental Panel on Climate Change*. Cambridge: Cambridge University Press, pp. 315–356.

Nicholls, R.J., Marinova, N., Lowe, J.A., Brown, S., Vellinga, P., de Gusmão, D., Hinkel, J. and Tol, R.S.J. (2011) 'Sea-level rise and its possible impacts given a "beyond 4°C world" in the twenty-first century', *Philosophical Transactions of the Royal Society A*, 369(1934): 161–181.

Nicholls, S. and Amelung, B. (2008) 'Climate change and tourism in northwestern Europe: impacts and adaptation', *Tourism Analysis*, 13: 21–31.

Nicholls, S., Holecek, D.F. and Noh, J. (2008) 'Impact of weather variability on golfing activity and implications of climate change', *Tourism Analysis*, 13: 117–130.

Nikolopoulou, M. and Steemers, K. (2003) 'Thermal comfort and psychological adaptation as a guide for designing urban spaces', *Energy and Buildings*, 35(1): 95–101.

Nilsson, J.H. and Gössling, S. (2012) 'Tourist responses to environmental degradation and climate change: the case of Baltic Sea algal blooms', *Tourism Planning and Development*, in press.

NOAA (2002) *Geostationary Operational Environmental Satellite System (GOES) GOES-R Sounder and Imager Cost/Benefit Analysis (CBA)*. Silver Spring, MD: National Oceanic and Atmospheric Administration, NESDIS Office of Systems Development. www.goes-r.gov/downloads/GOES-R_Sounder-Imager-Cost-Benefit-Analysis%20(CBA)/GOES-R_CBA_Final_Jan_9_2003.pdf

NOAA (2003) *Significant Climate Anomalies and Events in 2003*. Washington, DC: National Oceanic and Atmospheric Administration. www.ncdc.noaa.gov/img/climate/research/2003/ann/sig-global-extremes2003-pg.gif

NOAA (2004a) *Significant Climate Anomalies and Events in 2004*. Washington, DC: National Oceanic and Atmospheric Administration. www.ncdc.noaa.gov/img/climate/research/2004/ann/Significant_Extremes2004-pg.gif

NOAA (2004b) *Geostationary Operational Environmental Satellite System (GOES) GOES-R Sounder and Imager Cost/Benefit Analysis (CBA)*. Silver Spring, MD: National Oceanic and Atmospheric Administration, NESDIS Office of Systems Development. www.centrec.com/resources/reports/GOES%20Economic%20Value%20Report.pdf

NOAA (2005) *Significant Climate Anomalies and Events in 2005*. Washington, DC: National Oceanic and Atmospheric Administration. www.ncdc.noaa.gov/img/climate/research/2005/ann/significant-extremes2005.gif

NOAA (2006) *Significant Climate Anomalies and Events in 2006*. Washington, DC: National Oceanic and Atmospheric Administration. www.ncdc.noaa.gov/img/climate/research/2006/ann/significant-extremes2006.gif

NOAA (2007) *Significant Climate Anomalies and Events in 2007*. Washington, DC: National Oceanic and Atmospheric Administration. www.ncdc.noaa.gov/img/climate/research/2007/ann/significant-extremes2007.gif

NOAA (2008) *Significant Climate Anomalies and Events in 2008*. Washington, DC: National Oceanic and Atmospheric Administration. www.ncdc.noaa.gov/img/climate/research/2008/ann/significant-extremes2008.gif

NOAA (2009) *Significant Climate Anomalies and Events in 2009*. Washington, DC: National Oceanic and Atmospheric Administration. www.ncdc.noaa.gov/sotc/get-file.php?report=global&file=significant-extremes&year=2009&month=13&ext=gif

NOAA (2010a) *Significant Climate Anomalies and Events in 2010*. Washington, DC: National Oceanic and Atmospheric Administration. www.ncdc.noaa.gov/sotc/service/global/significant-extremes/201013.gif

NOAA (2010b) *State of the Climate in 2009 – Special Supplement to the Bulletin of the American Meteorological Society*. Washington, DC: National Oceanic and Atmospheric Administration. www.ncdc.noaa.gov/bams-state-of-the-climate

Nolan, T. (2001) 'Gold Coast businesses unhappy with weather forecasting', *PM Archive*, 4 April. www.abc.net.au/pm/stories/s271935.htm

Norgaard, R. (1985) 'Environmental economics: an evolutionary critique and a plea for pluralism', *Journal of Environmental Economics and Management*, 12: 382–394.

Norton, L., Johnson, P., Joys, A., Stuart, R., Chamberlain, D., Feber, R., Firbank, L., Manley, W., Wolfe, M., Hart, B., Mathews, F., Macdonald, D. and Fuller, R.J. (2009) 'Consequences of organic and non-organic farming practices for field, farm and landscape complexity', *Agriculture, Ecosystems and Environment*, 129(1/3): 221–227.

Nuffield Council on Bioethics (2011) *Biofuels: Ethical Issues*. London: Nuffield Council on Bioethics.

Nyaupane, G. and Chhetri, N. (2009) 'Vulnerability to climate change of nature-based tourism in the Nepalese Himalayas', *Tourism Geographies*, 11(1): 95–119.

Nygren, E., Aleklett, K. and Höök, M. (2009) 'Aviation fuel and future oil production scenarios', *Energy Policy*, 37(10): 4003–4010.

O'Brien, K. (2011) 'Responding to environmental change: a new age for human geography?', *Progress in Human Geography*, 35(4): 542–549 (doi: 10.1177/0309132510377573).

OECD (1996) *OECD Proceedings: Towards Sustainable Transportation* – The Vancouver Conference, Vancouver, British Columbia, 24–27 March 1996. Paris: Organisation for Economic Co-operation and Development. www.oecd.org/dataoecd/28/54/2396815.pdf

OECD (2009) *The Economics of Climate Change Mitigation*. Paris: Organisation for Economic Co-operation and Development.

OECD (2010a) *OECD Tourism Trends and Policies 2010*. Paris: Organisation for Economic Co-operation and Development.

OECD (2010b) *Taxation, Innovation and the Environment*. Paris: Organisation for Economic Co-operation and Development.

OECD and UNEP (2011) *Sustainable Tourism Development and Climate Change: Issues and Policies*. Paris/Washington, DC: Organisation for Economic Co-operation and Development/United Nations Environment Programme.

OECD, Morlot, J. and Agrawala, S. (eds) (2004) *The Benefits of Climate Change Policies. Analytical and Framework Issues*. Paris: Organisation for Economic Co-operation and Development.

Office for National Statistics (2011) *Transport Highlights*. www.ons.gov.uk/ons/taxonomy/index.html?nscl=Travel+and+Transport

Ogutu, J.O., Piepho, H.P., Dublin, H.T., Bhola, N. and Reid, R.S. (2008) 'Rainfall influences on ungulate population abundance in the Mara–Serengeti ecosystem', *Journal of Animal Ecology*, 77: 814–829.

O'Mahoney, B., Whitelaw, P. and Ritchie, B.W. (2006) 'The effect of fuel price on tourism behaviour: an exploratory Australian study', paper presented at Tourism After Oil, ATLAS Asia–Pacific Conference, University of Otago, Dunedin.

Ontario Ministry of Tourism and Recreation (2002) *If The Future Were Now: Impacts of Aging in the Canadian Market on Tourism in Ontario*. Toronto: Ontario Ministry of Tourism and Recreation.

Oppenheimer, M., O'Neill, B., Webster, M. and Agrawala, S. (2007) 'Climate change – the limits of consensus', *Science*, 317: 1505–1506.

Ottawa Citizen (2008) 'Ontario to sign cap-and-trade climate plan', *Ottawa Citizen*, 18 July. www.dose.ca/news/story.html?id=9ff531b5-1a6e-4864-8256-466a637b128f

Owens, B., Lee, D.S. and Lim, L. (2010) 'Flying into the future: aviation emissions scenarios to 2050', *Environmental Science Technology*, 44: 2255–2260.

Oxfam (2007) *Adapting to Climate Change – What's Needed in Poor Countries, and Who Should Pay*, Briefing Paper 104. Oxford: Oxfam. www.oxfam.org.uk/resources/policy/climate_change/downloads/bp104_adapting_to_climate_change.pdf

Oxford Economics (2009) *Valuing the Effects of Great Barrier Reef Bleaching*. Newstead, Queensland: Great Barrier Reef Foundation.

Pack, T. (2004) 'Florida tourism problem; survey shows 1 in 4 less likely to visit Florida next July to September', *Hotel Online – News for the Hospitality Executive*. www.hotel-online.com/News/PR2004_4th/Oct04_FloridaNextYear.html

Paerl, H.W., Hall, N.S. and Calandrino, E.S. (2011) 'Controlling harmful cyanobacterial blooms in a world experiencing anthropogenic and climatic-induced change', *Science of the Total Environment*, 409: 1739–1745.

Pall, P., Aina, T., Stone, D.A., Stott, P.A., Nozawa, T., Hilberts, A.G.J., Lohmann, D. and Allen, M.R. (2011) 'Anthropogenic greenhouse gas contribution to flood risk in England and Wales in autumn 2000', *Nature*, 470: 382–385.

Papanastasiou, D., Melas, D., Bartzanas, T. and Kittas, C. (2010) 'Temperature, comfort and pollution levels during heat waves and the role of sea beeze', *International Journal of Biometeorology*, 54: 307–317.

Parry, M.L., Canziani, O.F., Palutikof, J.P., van der Linden, P.J. and Hanson, C.E. (eds) (2007) *Climate Change 2007: Impacts, Adaptation and Vulnerability. Contribution of Working Group II to the Fourth Assessment Report of the Intergovernmental Panel on Climate Change*. Cambridge: Cambridge University Press, pp. 315–356.

Parry, M., Lowe, J. and Hanson, C. (2009a) 'Overshoot, adapt and recover', *Nature*, 458(7242): 1102–1103.

Parry, M., Arnell, N., Berry, P., Dodman, D., Fankhauser, S., Hope, C., Kovats, S., Nicholls, R., Satterthwaite, D., Tiffin, R. and Wheeler, T. (2009b) *Assessing the Costs of Adaptation to Climate Change: A Review of the UNFCCC and Other Recent Estimates*. London: International Institute for Environment and Development and Grantham Institute for Climate Change.

PATA (2010) 'Pacific tourism leaders slam UK "tax on travel"', Pacific Asia Travel Association, press release, 30 April. www.pata.org/press/pacific-tourism-leaders-slam-uk-tax-on-travel

Pateman, E. (2001) 'Rising energy costs cause concern in the lodging industry', *Hotel Online Special Report*, March 2001. www.hotel-online.com/News/PR2001_1st/Mar01_Pateman_Energy.html

Patterson, M. and McDonald, G. (2004) *How Clean and Green is New Zealand Tourism? Lifecycle and Future Environmental Impacts*. Landcare Research Science Series 24. Lincoln: Manaaki Whenua Press.

Paul, A. (1972) 'Weather and the daily use of outdoor recreation areas in Canada', in J. Taylor (ed.) *Weather Forecasting for Agriculture and Industry*. Newton Abbot, UK: David and Charles Publishers.

Pauli, H., Gottfried, M., Reiter, K., Klettner, C. and Grabherr, G. (2007) 'Signals of range expansions and contractions of vascular plants in the high Alps: observations (1994–2004) at the GLORIA master site Schrankogel, Tyrol, Austria', *Global Change Biology*, 13: 147–156.

Pauly, D. (1995) 'Anecdotes and the shifting baseline syndrome of fisheries', *Trends in Ecology and Evolution*, 10: 430.

Payet, R. and Obura, D. (2004) 'The negative impacts of human activities in the Eastern African region: an international waters perspective', *Ambio*, 33: 24–33.

Pearce, F. (2010) 'How the "climategate" scandal is bogus and based on climate sceptics' lies', *The Guardian*, 9 February. www.guardian.co.uk/environment/2010/feb/09/climategate-bogus-sceptics-lies

Pearce, P.L. (1993) 'Fundamentals of tourist motivation', in D. Pearce and R. Butler (eds) *Tourism Research: Critiques and Challenges*. London: Routledge, pp. 85–105.

Pearce, P.L. (2005) *Tourist Behaviour. Themes and Conceptual Schemes*. Clevedon: Channelview.

Pearce, P.L. and Caltabiano, M.L. (1983) 'Inferring travel motivation from travelers' experiences', *Journal of Travel Research*, 22(2): 16–20.

Pearce, P.L. and Lee, U.-I. (2005) 'Developing the travel career approach to tourist motivation', *Journal of Travel Research*, 43(3): 226–237.

Peel, M.C., Finlayson, B.L. and McMahon, T.A. (2007) 'Updated world map of the Köppen–Geiger climate classification', *Hydrology and Earth System Sciences*, 11: 1633–1644.

Peeters, P.M. and Middel, J. (2007) 'Historical and future development of air transport fuel efficiency', in R. Sausen, A. Blum, D.S. Lee and C. Brüning (eds) *Proceedings of an International Conference on Transport, Atmosphere and Climate (TAC)*, Oxford, 26–29 June 2006. Oberpfaffenhoven, Germany: DLR Institut für Physik der Atmosphäre, pp. 42–47.

Peeters, P.M., van Egmond, T. and Visser, N. (2004) *European Tourism, Transport and Environment*. Breda, the Netherlands: NHTV Breda University of Applied Sciences, Centre for Sustainable Tourism and Transport.

Peeters, P.M., Szimba, E. and Duijnisveld, M. (2007) 'Major environmental impacts of European tourist transport', *Journal of Transport Geography*, 15: 83–93.

Peeters, P., Gössling, S. and Lane, B. (2009) 'Moving towards low-carbon tourism: opportunities for destinations and tour operators', in S. Gössling, C.M. Hall and D. Weaver (eds) *Sustainable Tourism Futures*. London: Routledge.

Pégy, P. (1961) *Précis de Climatologie*. Paris: Masson.

Pelling, M. (2011) *Adaptation to Climate Change: From Resilience to Transformation*. London: Routledge.

Pelto, M.S. (2010) Forecasting temperate alpine glacier survival from accumulation zone observations, *The Cryosphere*, 4: 67–75.

Penner, J., Lister, D., Griggs, D., Dokken, D. and McFarland, M. (eds) (1999) 'Aviation and the global atmosphere', in *A Special Report of IPCC Working Groups I and III*. London: Cambridge University Press.

Pentelow, L. and Scott, D. (2010) 'The implications of climate change mitigation policy: volatility for tourism arrivals to the Caribbean', *Tourism and Hospitality Planning and Development*, 7(3): 301–315.

Pentelow, L. and Scott, D. (2011) 'Aviation's inclusion in international climate policy regimes: implications for the Caribbean tourism industry', *Journal of Air Transport Management*, 17: 199–205.

Perch-Nielsen, S. (2010) 'The vulnerability of beach tourism to climate change – an index approach', *Climatic Change*, 100(3): 579–606.

Perch-Nielsen, S., Sesartic, A. and Stucki, M. (2010a) 'The greenhouse gas intensity of the tourism sector: the case of Switzerland', *Environmental Science & Policy*, 13: 131–140.

Perch-Nielsen, S., Amelung, B. and Knutt, R. (2010b) 'Future climate resources for tourism in Europe based on the daily Tourism Climate Index', *Climatic Change*, 103(3/4): 363–381.

Perry, A. (1972) 'Weather, climate and tourism', *Weather*, 27: 199–203.

Perry, A. (1993) 'Climate and weather information for the package holiday-maker', *Weather*, 48: 410–414.

Perry, A. (1997) 'Recreation and tourism', in R.D. Thompson and A.H. Perry (eds) *Applied Climatology*. London: Routledge,

Perry, A.H. (2006) 'Will predicted climate change compromise the sustainability of Mediterranean tourism?', *Journal of Sustainable Tourism*, 14(4): 367–375.

Peterson, E.B., Schleich, J. and Duscha, V. (2011). 'Environmental and economic effects of the Copenhagen pledges and more ambitious emission reduction targets', *Energy Policy*, 39(6): 3697–3708 (doi: 10.1016/j.enpol.2011.03.079).

Pew Center on Global Climate Change (2010) 'Targets and actions under the Copenhagen Accord'. www.pewclimate.org/copenhagen-accord. The Pew Center is now known as the Center for Climate and Energy Solutions (C2ES).

Pham, T., Simmons, D. and Spurr, R. (2010) 'Climate change-induced economic impacts on tourism destinations: the case of Australia', *Journal of Sustainable Tourism*, 18(3): 449–473.

Phillips, M. and House, C. (2009) 'An evaluation of priorities for beach tourism: case studies from South Wales, UK', *Tourism Management*, 30(2): 176–183.

Phillips, M. and Jones, A. (2006) 'Erosion and tourism infrastructure in the coastal zone: problems, consequences and management', *Tourism Management*, 27: 517–524.

Pickering, C.M. (2011) 'Changes in demand for tourism with climate change: a case study of visitation patterns to six ski resorts in Australia', *Journal of Sustainable Tourism*, 19:767–781.

Pickering, C.M. and Buckley, R.C. (2010) 'Climate response by the ski industry: the shortcomings of snowmaking for Australian Resorts', *AMBIO: A Journal of the Human Environment*, 39(5/6): 430–438.

Pickering, C.M., Castley, J.G. and Burtt, M. (2010) 'Skiing less often in a warmer world: attitudes of tourists to climate change in an Australian ski resort', *Geographical Research*, 48(2):137–147.

Pierre, J. and Peters, G.B. (2000) *Governance, Politics and the State*. London: Palgrave Macmillan.

Pike, S. (2002) 'Destination image analysis – a review of 142 papers from 1973 to 2000', *Tourism Management*, 23: 541–549.

Pincus, R. (2003) 'Wine, place and identity in a changing climate', *Gastronomica*, 3: 87–93.

Plass, G.N. (1956) 'Effect of carbon dioxide variations on climate', *American Journal of Physics*, 24: 376–387.

Ploner, A. and Brandenburg, C. (2003) 'Modeling visitor attendance levels subject to day of the week and weather: a comparison between linear regression models and regression tress', *Journal of Nature Conservation*, 11: 297–308.

PricewaterhouseCoopers (2010) *Appetite for Change. Global Business Perspectives on Tax and Regulation for a Low Carbon Economy*. London: PricewaterhouseCoopers. www.pwc.com/appetiteforchange

Prideaux, B., Coghlan, A. and McKercher, B. (2009) 'Identifying indicators to measure tourists' views on climate change', paper presented at CAUTHE 2009 Conference: See Change: Tourism and Hospitality in a Dynamic World, Curtin University of Technology, Perth, Australia. Melbourne: Council for Australian University Tourism and Hospitality Education.

Prideaux, B., Coghlan, A. and McNamara, K. (2010) 'Assessing tourists' perceptions of climate change on mountain landscapes', *Tourism Recreation Research*, 35(2): 187–199.

Prince, B.W. (2010) *Climate Change Adaptation and Mitigation in New Zealand Snow Tourism*. Industry Report, Ministry of Economic Development, Tourism Strategy Group. Wellington: School of Geography, Environment and Earth Sciences, Victoria University of Wellington.

Raleigh, C. and Urdal, H. (2007) 'Climate change, environmental degradation and armed conflict', *Political Geography*, 26: 674–694.

Randle, H.W. (1997) 'Suntanning: differences in perceptions throughout history', *Mayo Clinic Proceedings*, 72: 461–466.

Randles, S. and Mander, S. (2009) 'Aviation, consumption and the climate change debate: "Are you going to tell me off for flying?"', *Technology Analysis & Strategic Management*, 21(1): 93–113.

Rathke, L. (2008) 'Colorful study probes climate change, fall foliage', *USA Today*, 24 September. www.usatoday.com/weather/research/2008-09-24-fall-foliage-climate-change_N.htm

Raupach, M.R., Marland, G., Ciais, P., Le Quéré, C., Canadell, J.G., Klepper, G. and Field, C.B. (2007) 'Global and regional drivers of accelerating CO_2 emissions', *Proceedings of the National Academy of Sciences, USA*, 104(24): 10288–10293.

RCCL (2010) 'Royal Caribbean and the Environment'. www.royalcaribbean.com/ourcompany/environment/rcAndEnvironment.do

Regeringskansliet (2007) Sverige inför klimatförändringarna – hot och möjligheter. Stockholm: Swedish Government Offices. www.regeringen.se/sb/d/8704/a/89334

Regeringskansliet (2011) *Klimat* (Climate). Stockholm: Swedish Government Offices. www.regeringen.se/sb/d/3188

Reisinger, Y. (2009) *International Tourism: Cultures and Behavior*. Oxford: Butterworth-Heinemann.

Republic of South Africa (2010) *Effects of Climate Change*. Pretoria: State of the Environment, Department of Environmental Affairs. soer.deat.gov.za/174.html#3561

Respect (2010) *Compulsory Carbon Footprint Labelling for Tourism and Travel Services?* Discussion Paper. Vienna: Respect.

Reuters (2009) 'Weather risk market value plunges by 53 pct-survey', *Reuters*, 3 June. www.reuters.com/article/2009/06/03/weather-risk-idUSN038150820090603

Revelle, R. and Suess, H. (1957) 'Carbon dioxide exchange between atmosphere and ocean and the question of an increase of atmospheric CO_2 during the past decades', *Tellus*, 9: 18–27.

Richardson, R.B. and Loomis, J.B. (2004) 'Adaptive recreation planning and climate change: a contingent visitation approach', *Ecological Economics*, 50(1): 83–99.

Riegl, B., Bruckner, A., Coles, S.L., Renaud, P. and Dodge, R.E. (2009) 'Coral reefs. threats and conservation in an era of global change', *Annals of the New York Academy of Sciences*, 1162: 136–186.

Rifai, T. (2010) 'Sectoral approach for international aviation emissions hits headwind in Cancun as spectre of climate financing looms', *Responsible Travel Report*, 13 December. www.responsibletravelreport.com/trade-news/spotlight/destinations/2407--sectoral-approach-for-international-aviation-emissions-hits-headwind-in-cancun-as-spectre-of-climate-financing-looms

Rignot, E., Koppes, M. and Velicogna, I. (2010) 'Rapid submarine melting of the calving faces of West Greenland glaciers', *Nature Geoscience*, 3: 187–191.

Rodríguez-Díaz, J.A., Knox, J.W. and Weatherhead, E.K. (2007) 'Competing demands for irrigation water: golf and agriculture in Spain', *Irrigation and Drainage*, 56: 541–549.

Rodríguez-Díaz, J.A., Weatherhead, E.K., García Morillo, J. and Knox, J.W. (2010) 'Benchmarking irrigation water use in golf courses – a case study in Spain', *Irrigation and Drainage*, 60(3): 381–392 (doi: 10.1002/ird.578).

Rogers, D. (2004) 'Going to extremes', *Condé Nast Traveller*, May: 93–105.

Roman, C., Lynch, A. and Dominey-Howes, D. (2010) 'Uncovering the essence of the climate change adaptation problem – a case study of the tourism sector at Alpine Shire, Victoria, Australia', *Tourism and Hospitality Planning & Development*, 7(3): 237–252.

Roman, G., Dearden, P. and Rollins, R. (2007) 'Application of zoning and "limits of acceptable change" to managing snorkelling tourism', *Environmental Management*, 39(6): 819–830.

Rosenberg, S., Vedlitz, A., Cowman, D. and Zahran, S. (2010) 'Climate change: a profile of US climate scientists' perspectives', *Climatic Change*, 101(3): 311–329.

Rosenzweig, C., Casassa, G., Karoly, D.J., Imeson, A., Liu, C., Menzel, A., Rawlins, S., Root, T.L., Seguin, B. and Tryjanowski, P. (2007) 'Assessment of observed changes and responses in natural and managed systems', in M.L. Parry, O.F. Canziani, J.P. Palutikof, P.J. van der Linden and C.E. Hanson (eds) *Climate Change 2007: Impacts, Adaptation and Vulnerability. Contribution of Working Group II to the Fourth Assessment Report of the Intergovernmental Panel on Climate Change.* Cambridge: Cambridge University Press.

Rosenzweig, C., Karoly, D., Vicarelli, M., Neofotis, P., Wu, Q., Casassa, G., Menzel, A., Root, T.L., Estrella, N., Seguin, B., Tryjanowski, P., Liu, C., Rawlins, S. and Imeson, A. (2008) 'Attributing physical and biological impacts to anthropogenic climate change', *Nature*, 453(7193): 296–297.

Ross, M., Toohey, D., Peinemann, M. and Ross, P. (2009) 'Limits on the space launch market related to stratospheric ozone depletion', *Astropolitics*, 7(1): 50–82.

Ross, M., Mills, M. and Toohey, D. (2010) 'Potential climate impact of black carbon emitted by rockets', *Geophysical Research Letters*, 37(24): L24810 (doi: 10.1029/2010GL044548).

Round, A. (2008) 'Paradise lost', *Destinations of the World News*, 21: 44–51.

Rowley, S. (2011) 'Two-thirds of UK biofuel fails green standard, figures show', *guardian.co.uk*, 27 January. www.guardian.co.uk/environment/2011/jan/27/biofuel-fails-green-standard

Royal Society (2009) *Geoengineering the Climate: Science, Governance and Uncertainty.* London: The Royal Society. royalsociety.org/Geoengineering-the-climate

Rutty, M. and Scott, D. (2010) 'Will the Mediterranean become "too hot" for tourism? A reassessment', *Tourism Planning and Development*, 7: 267–281.

Saarinen, J. and Tervo, K. (2006) 'Perceptions and adaptation strategies of the tourism industry to climate change: the case of Finnish nature-based tourism entrepreneurs', *International Journal of Innovation and Sustainable Development*, 1: 214–228.

Saarinen, J. and Tervo, K. (2010) 'Sustainability and emerging awareness of a changing climate: the tourism industry's knowledge and perceptions of the future of nature-based winter tourism in

Finland', in C.M. Hall and J. Saarinen (eds) *Tourism and Change in Polar Regions: Climate, Environments and Experiences*. London: Routledge.

Sachs, J.D. and McArthur, J.W. (2005) 'The millennium project: a plan for meeting the Millennium Development Goals', *Lancet*, 365: 347–353.

Sagoff, M. (1988) *The Economy of the Earth: Philosophy, Law, and the Environment*. Cambridge: Cambridge University Press.

Salazar, L.F., Nobre, C.A. and Oyama, M.D. (2007) 'Climate change consequences on the biome distribution in tropical South America', *Geophysical Research Letters*, 34(9): L09708 (doi: L09708).

Salkin, A. (2007) 'Tourism of doom on the rise', *New York Times*, 16 December. www.nytimes.com/2007/12/16/world/americas/16iht-tourism.1.8762449.html

San Francisco Department of the Environment (2004) *Climate Action Plan for San Francisco: Local Actions to Reduce Greenhouse Gas Emissions*. San Francisco: San Francisco Department of the Environment, San Francisco Public Utilities Commission.

Sanders, D., Laing, J. and Houghton, M. (2008) *Impact of Bushfires on Tourism and Visitation in Alpine National Parks*. Gold Coast, Australia: Cooperative Research Centre for Sustainable Tourism.

Sandvik, H. (2008) 'Public concern over global warming correlates negatively with national wealth', *Climatic Change*, 90(3): 333–341.

Sarni, W. (2011) *Corporate Water Strategies*. New York: Earthscan.

Sarnsamak, P. (2011) 'Closing marine parks "won't save coral"', *The Nation* (Thailand), 20 January. www.nationmultimedia.com/2011/01/20/national/Closing-marine-parks-wont-save-coral-30146771.html

Saunders, C., Barber, A. and Taylor, G. (2006) *Food Miles – Comparative Energy/Emissions Performance of New Zealand's Agriculture Industry*. Research report no. 285. Lincoln, NZ: Agribusiness and Economics Research Unit, Lincoln University. www.lincoln.ac.nz/documents/2328_rr285_s13389.pdf

Saunders, S., Easley, T., Logan, J.A. and Spencer, T. (2007) 'Losing ground: western national parks endangered by climate disruption', *George Wright Forum*, 24(1): 41–81.

Scarlat, N. and Dallemand, J.-F. (2011) 'Recent developments of biofuels/bioenergy sustainability certification: a global overview', *Energy Policy*, 39(3): 1630–1646.

SCBD (2009a) *Scientific Synthesis of the Impacts of Ocean Acidification on Marine Biodiversity*, Technical Series No. 46. Montreal: Secretariat of the Convention on Biological Diversity.

SCBD (2009b) *Biodiversity, Development and Poverty Alleviation: Recognizing the Role of Biodiversity for Human Well-being*. Montreal: Secretariat of the Convention on Biological Diversity.

SCBD (2009c) *Biodiversity and Climate Change Action*. Montreal: Secretariat of the Convention on Biological Diversity.

SCBD (2009d) *Connecting Biodiversity and Climate Change Mitigation and Adaptation: Report of the Second Ad Hoc Technical Expert Group on Biodiversity and Climate Change*, Technical Series No. 41. Montreal: Secretariat of the Convention on Biological Diversity.

Schafer, A. and Victor, D.G. (2000) 'The future mobility of the world population', *Transportation Research A*, 34: 171–205.

Scheidleder, A., Winkler, G., Stark, U., Koreimann, C., Gmeiner, C., Gravesen, P., Leonard, J., Elvira, M., Nixon, S. and Casillas, J. (1999) *Groundwater Quality and Quantity in Europe: Data and Basic Information*, Technical Report No. 22. Copenhagen: European Environment Agency.

Schernewski, G. and Schiewer, U. (eds) (2002) *Baltic Coastal Ecosystems. Structure, Function and Coastal Zone Management*. Berlin: Springer Verlag.

Schiermeier, Q. (2009) 'Arctic sea ice levels third-lowest on record', *Nature News*, 18 September. www.nature.com/news/2009/090918/full/news.2009.930.html

Schiff, A. and Becken, S. (2011) 'Demand elasticity estimates for New Zealand tourism', *Tourism Management*, 32(3): 564–575.

Schindler, D. (1999) 'The mysterious missing sink', *Nature*, 398: 105–106.

Schipper, E. and Burton, I. (eds) (2008) *The Earthscan Reader on Adaptation to Climate Change*. London: Earthscan.

Schleupner, C. (2008) 'Evaluation of coastal squeeze and its consequences for the Caribbean island Martinique', *Ocean and Coastal Management*, 51: 383–390.

Schmidhuber, J. and Tubiello, F.N. (2007) 'Global food security under climate change', *Proceedings of the National Academy of Science, USA*, 104(50): 19703–19708.

Schmied, M., Götz, K., Kreilkamp, E., Buchert, M., Hellwig, T. and Otten, S. (2009) *Traumziel Nachhaltigkeit. Innovative Vermarktungskonzepte nachhaltiger Tourismusangebote für den Massenmarkt*. Physica-Verlag, Heidelberg.

Schneider, S. (2001) 'A constructive deconstruction of deconstructionists: a response to Demeritt', *Annals of the Association of American Geographers*, 91: 338–344.

Schneider, S.H., Semenov, S., Patwardhan, A., Burton, I., Magadza, C.H.D., Oppenheimer, M., Pittock, A.B., Rahman, A., Smith, J.B., Suarez, A. and Yamin, F. (2007) 'Assessing key vulnerabilities and the risk from climate change', in M.L. Parry, O.F. Canziani, J.P. Palutikof, P.J. van der Linden and C.E. Hanson (eds) *Climate Change 2007: Impacts, Adaptation and Vulnerability. Contribution of Working Group II to the Fourth Assessment Report of the Intergovernmental Panel on Climate Change*. Cambridge: Cambridge University Press.

Schwartz, M.D., Ahas, R. and Aasa, A. (2006) 'Onset of spring starting earlier across the Northern Hemisphere', *Global Change Biology*, 12(2): 343–351.

Science (2001) 'The science of climate change', 292(5520): 1261 (doi: 10.1126/science.292.5520.1261).

Scott, D. (2003) 'Climate change and tourism and the mountain regions of North America', in *Proceedings of the First International Conference on Climate Change and Tourism*, Djerba 9–11 April. Madrid: World Tourism Organization.

Scott, D. (2006a) 'Climate change and sustainable tourism in the 21st century', in J. Cukier (ed.) *Tourism Research: Policy, Planning, and Prospects*. Waterloo, Ontario: Department of Geography Publication Series, University of Waterloo.

Scott, D. (2006b) 'Global environmental change and mountain tourism', in S. Gössling and M. Hall (eds) *Tourism and Global Environmental Change*. London: Routledge.

Scott, D. (2008) 'Climate change and tourism: time for critical reflections', *Tourism Recreation Research*, 33(3): 356–360.

Scott, D. (2011) 'Why sustainable tourism must address climate change', *Journal of Sustainable Tourism*, 19: 17–34.

Scott, D. and Becken, S. (2010) 'Adapting to climate change and climate policy: progress, problems and potentials', *Journal of Sustainable Tourism*, 18(3): 283–295.

Scott, D. and Jones, B. (2005) *Climate Change and Banff: Implications for Tourism and Recreation – Executive Summary*, report prepared for the Town of Banff. Waterloo, Ontario: University of Waterloo.

Scott, D. and Jones, B. (2006a) 'The impact of climate change on golf participation in the Greater Toronto Area (GTA): a case study', *Journal of Leisure Research*, 38(3): 363–380.

Scott, D. and Jones, B. (2006b) *Climate Change and Seasonality in Canadian Outdoor Recreation and Tourism – Executive Summary*, report prepared for the Government of Canada Climate Change Action Fund. Waterloo, Ontario: University of Waterloo.

Scott, D. and Jones, B. (2006c) *Climate Change and Nature-Based Tourism in Canada – Executive Summary*, report prepared for the Government of Canada Climate Change Action Fund. Waterloo, Ontario: University of Waterloo.

Scott, D. and Jones, B. (2007) 'A regional comparison of the implications of climate change on the golf industry in Canada', *Canadian Geographer*, 51: 219–232.

Scott, D. and Lemieux, C. (2009) *Weather and Climate Information for Tourism*, White Paper commissioned for World Climate Conference 3. Geneva and Madrid: World Meteorological Organization and United Nations World Tourism Organization.

Scott, D. and Lemieux, C. (2010) 'Weather and climate information for tourism', *Procedia Environmental Sciences*, 1: 146–183.

Scott, D. and Matthews, L. (2011) *Climate, Tourism & Recreation: A Bibliography*, 2010 edn. Waterloo: Department of Geography and Environmental Management, University of Waterloo. www.environment.uwaterloo.ca/geography/faculty/danielscott/publications.htm

Scott, D. and McBoyle, G. (2001) 'Using a "tourism climate index" to examine the implications of climate change for climate as a natural resource for tourism', in A. Matzarakis and C. de Frietas (eds) *Proceedings 1st International Workshop on Climate, Tourism and Recreation*, International Society of Biometeorology, Commission on Climate, Tourism and Recreation, Halkidiki, Greece, 5–10 October 2001.

Scott, D. and McBoyle, G. (2007) 'Climate change adaptation in the ski industry', *Mitigation and Adaptation Strategies to Global Change*, 12(8): 1411–1431.

Scott, D. and Suffling, R. (2000) *Climate Change and Canada's National Parks*. Toronto: Environment Canada.

Scott, D. and Vivian, K. (2012) 'Skier response to climate variability and change in New England', Proceedings of the Canadian Association of Geographers Annual Conference, 28 May–2 June 2012. Waterloo, Canada: Department of Geography, University of Waterloo.

Scott, D., Jones, B., Lemieux, C., McBoyle, G., Mills, B., Svenson, S. and Wall, G. (2002) *The Vulnerability of Winter Recreation to Climate Change in Ontario's Lakelands Tourism Region*. Waterloo: Department of Geography Publication Series, University of Waterloo.

Scott, D., McBoyle, G. and Mills, B. (2003) 'Climate change and the skiing industry in Southern Ontario (Canada): exploring the importance of snowmaking as a technical adaptation', *Climate Research*, 23: 171–181.

Scott, D., McBoyle, G. and Schwartzentruber, M. (2004) 'Climate change and the distribution of climatic resources for tourism in North America', *Climate Research*, 27(2): 105–117.

Scott, D., Wall, G. and McBoyle, G. (2005a) 'The evolution of the climate change issue in the tourism sector', in C.M. Hall and J. Higham (eds) *Tourism, Recreation and Climate Change*. Clevedon: Channelview Press.

Scott, D., Jones, B. and Abi Khaled, H. (2005b) *Climate Change: A Long-Term Strategic Issue for the National Capital Commission (Tourism and Recreation Business Lines) – Executive Summary*, report prepared for the National Capital Commission. Waterloo, Ontario: University of Waterloo.

Scott, D., McBoyle, G., Mills, B. and Minogue, A. (2006) 'Climate change and the sustainability of ski-based tourism in eastern North American', *Journal of Sustainable Tourism*, 14(4): 376–398.

Scott, D., McBoyle, G., Minogue, A. (2007a) 'The implications of climate change for the Québec ski industry', *Global Environmental Change*, 17: 181–190.

Scott, D., Jones, B. and Konopek, J. (2007b) 'Implications of climate and environmental change for nature-based tourism in the Canadian Rocky Mountains: a case study of Waterton Lakes National Park', *Tourism Management*, 28(2): 570–579.

Scott, D., Amelung, B., Becken, S., Ceron, J.-P., Dubois, G., Gössling, S., Peeters, P. and Simpson, M. (2008a) *Climate Change and Tourism: Responding to Global Challenges*. Madrid/Paris/

Geneva: United Nations World Tourism Organization, United Nations Environment Program and World Meteorological Organization.

Scott, D., Gössling, S. and de Freitas, C. (2008b) 'Preferred climate for tourism: case studies from Canada, New Zealand and Sweden', *Climate Research*, 38: 61–73.

Scott, D., Dawson, J. and Jones, B. (2008c) 'Climate change vulnerability of the Northeast US winter tourism sector', *Mitigation and Adaptation Strategies to Global Change*, 13 (5/6): 577–596.

Scott, D., Jones, B. and Konopek, J. (2008d) 'Exploring the impact of climate-induced environmental changes on future visitation to Canada's Rocky Mountain National Parks', *Tourism Review International*, 12(1): 43–56.

Scott, D., de Freitas, C. and Matzarakis, A. (2009) 'Adaptation in the tourism and recreation sector', in K.L. Ebi, I. Burton and G.R. McGregor (eds) *Biometeorology for Adaptation to Climate Variability and Change*. Dordrecht, the Netherlands: Springer.

Scott, D., Peeters, P. and Gössling, S. (2010) 'Can tourism deliver on its aspirational greenhouse gas emission reduction targets?', *Journal of Sustainable Tourism*, 18(3): 393–408.

Scott, D., Lemieux, C., Kirchhoff, D. and Milnik, M. (2011) *Analysis of Socio-economic Impacts and Adaptation to Climate Change by Québec's Tourism Industry. Technical Report 1: Climate Change Impact Assessment: Risks and Opportunities*. Waterloo, Ontario/Montréal, Québec, Canada: Interdisciplinary Centre on Climate Change (IC3), University of Waterloo and Consortium on Regional Climatology and Adaptation to Climate Change (OURANOS).

Scott, D., Sim, R. and Simpson, M. (2012) 'Impact of sea level rise on coastal tourism in the Caribbean', *Journal of Sustainable Tourism*, in press.

Scottish Government (2007) *Travel by Scottish Residents: Some National Travel Survey Results for 2004/2005 and Earlier Years*. Edinburgh: Scottish Executive, National Statistics.

Sealy-Baker, M. (2011) 'The impacts of coral reef bleaching of dive tourism in the Caribbean', Master's thesis, Department of Recreation and Leisure Studies, University of Waterloo, Ontario, Canada.

Selby, M. (2004) 'Consuming the city: conceptualizing and researching urban tourist knowledge', *Tourism Geographies*, 6: 186–207.

Serquet, G. and Rebetez, M. (2011) 'Relationship between tourism demand in the Swiss Alps and hot summer air temperatures associated with climate change', *Climatic Change*, 108: 291–300 (doi: 10.1007/s10584-010-0012-6).

Sharp, A. (2008) 'Petrol prices and drought hit Victorian tourism', *The Sydney Morning Herald*, 13 March. www.smh.com.au/executive-style/management/petrol-prices-and-drought-hit-victorian-tourism-20090518-b9mp.html

Shaw, S. and Thomas, C. (2006) 'Social and cultural dimensions of air travel demand: hyper-mobility in the UK?', *Journal of Sustainable Tourism*, 14: 209–215.

Shaw, W. and Loomis, R. (2008) 'Frameworks for analyzing the economic effects of climate change on outdoor recreation', *Climate Research*, 36: 259–269.

Shearing, S. (2007) *Here Today, Gone Tomorrow? Climate Change and World Heritage*, Macquarie Law Working Paper No 2007-11. Sydney: Macquarie University.

Shelton, D. (2006) 'Normative hierarchy in international law', *American Journal of International Law*, 100(2): 291–323.

Shih, C., Nicholls, S. and Holecek, D. (2009) 'Impact of weather on downhill ski lift ticket sales', *Journal of Travel Research*, 47(3): 359–372.

Sievänen, T., Tervo, K., Neuvonen, M., Pouta, E., Saarinen, J. and Peltonen, A. (2005) *Nature-based Tourism, Outdoor Recreation and Adaptation to Climate Change*, FINADAPT Working Paper 11. Helsinki: Finnish Environment Institute.

Silberman, J. and Klock, M. (1988) 'The recreational benefits of beach recreation', *Ocean and Shoreline Management*, 11: 73–90.

Simpson, M., Gössling, S., Scott, D. and Hall, M. (2008) *Climate Change Adaptation and Mitigation in the Tourism Sector: Frameworks, Tools and Practices.* Paris: UNEP, University of Oxford, UNWTO and WMO. www.unep.fr/pc/tourism/library/DTIx1047xPA-ClimateChange. pdf

Sivin, N. (1995) *Science in Ancient China: Researches and Reflections.* Brookfield: VARIORUM, Ashgate Publishing.

Slocum, R. (2010) 'The sociology of climate change: research priorities', in J. Hagel, T. Dietz and J. Broadbent (eds) *Workshop on Sociological Perspectives on Global Climate Change.* Arlington, VA: National Science Foundation and American Sociological Association.

Smit, B., Burton, I., Klein, R. and Wandel, J. (2000) 'An anatomy of adaptation to climate change and variability', *Climatic Change*, 45(1): 233–251.

Smith, I.J. and Rodger, C.J. (2009) 'Carbon emission offsets for aviation-generated emissions due to international travel to and from New Zealand', *Energy Policy*, 37(9): 3438–3447.

Smith, K. (1981) 'The effect of weather conditions on the public demand for meteorological information', *Journal of Climatology*, 1(4): 381–393.

Smith, K. (1990) 'Tourism and climate change', *Land Use Policy*, 7(2): 176–180.

Smith, K. (1993) 'The influence of weather and climate on recreation and tourism', *Weather*, 48: 398–404.

Smith, S. (1983) *Recreation Geography.* London: Longman.

Smith, S. (2004) 'The measurement of global tourism: old debates, new consensus, and continuing challenges', in A.A. Lew, C.M. Hall and A.M. Williams (eds) *A Companion to Tourism.* Oxford: Blackwell.

Soliva, R., Bolliger, J. and Hunziker, M. (2010) 'Differences in preferences towards potential future landscapes in the Swiss Alps', *Landscape Research*, 35(6): 671–696.

Solomon, S., Qin, D., Manning, M., Chen, Z., Marquis, M., Averyt, K.B., Tignor, M. and Miller, H.L. (eds) (2007) *Climate Change 2007: The Physical Science Basis. Contribution of Working Group I to the Fourth Assessment Report of the Intergovernmental Panel on Climate Change.* Cambridge and New York: Cambridge University Press.

Spiller, K., Green, A. and Osburn, H. (1988) 'Increasing the efficiency of angler surveys by cancelling sampling during inclement weather', *North American Journal of Fisheries Management*, 8: 132–140.

SSP Sweden (2009) *Vårt miljöarbete.* www.foodtravelexperts.com/sweden/page/about-international/ swedish

Stanton, E.A. and Ackerman, F. (2007) *Florida and Climate Change: The Costs of Inaction.* Medford, MA: Global Development and Environment Institute, Tufts University and Stockholm Environment Institute–US Center, Tufts University.

Stanton, E., Ackerman, F. and Kartha, S. (2009) 'Inside the integrated assessment models: four issues in climate economics', *Climate and Development*, 1(2): 166–184.

Stebbins, R.A. (1982) 'Serious leisure: a conceptual statement', *Pacific Sociological Review*, 25: 251–272.

Steffen, W., Burbidge, A.A., Hughes, L., Kitching, R., Lindenmayer, D., Musgrave, W., Stafford Smith, M. and Werner, P.A. (2009) *Australia's Biodiversity and Climate Change: A Strategic Assessment of the Vulnerability of Australia's Biodiversity to Climate Change*, report to the Natural Resource Management Ministerial Council commissioned by the Australian Government. Canberra: CSIRO Publishing.

Steg, L. and Vlek, C. (2009) 'Encouraging pro-environmental behaviour: an integrative review and research agenda', *Journal of Environmental Psychology*, 29: 309–317.

Stehr, N. and von Storch, H. (2010) *Climate and Society: Climate as Resource, Climate as Risk.* Singapore: World Scientific Publishing.

Steiger, R. (2010) 'The impact of climate change on ski season length and snowmaking requirements in Tyrol, Austria', *Climate Research*, 43(3): 251–262.

Steiger, R. (2011a) 'Climate change impact on skiing tourism in Tyrol (Austria–Italy)', *Tourism Geographies*, in press.

Steiger, R. (2011b) 'The impact of snow scarcity on ski tourism. An analysis of the record warm season 2006/07 in Tyrol (Austria)', *Tourism Review*, 66(3): 4–13.

Steiger, R. and Abegg, B. (2012) *Climate change impacts on Austrian ski areas*. Proceedings of the Alpine Futures Conference, Innsbruck, Austria, October 2011.

Stern, N. (2007) *The Economics of Climate Change: The Stern Review*. Cambridge: Cambridge University Press.

Sterner, T. (2007) 'Fuel taxes: an important instrument for climate policy', *Energy Policy*, 35: 3194–3202.

Stewart, E.J., Draper, D. and Johnston, M.E. (2005) 'A review of tourism research in the polar regions', *Arctic*, 58(4): 383–394.

Stoll-Kleemann, S., O'Riordan, T. and Jaeger, C.C. (2001) 'The psychology of denial concerning climate mitigation measures: evidence from Swiss focus groups', *Global Environmental Change*, 11: 107–117.

Stott, P.A., Stone, D.A. and Allen, M.R. (2004) 'Human contribution to the European heatwave of 2003', *Nature*, 432: 610–614.

Strasdas, W., Gössling, S. and Dickhut, H. (2010) *Treibhausgas-Kompensationsanbieter in Deutschland*. Abschlussbericht. Berlin, Germany: Verbraucherzentrale Bundesverband.

Stratton, A. (2011) 'Environmental campaigners angry as green laws labelled as red tape', *The Guardian*, 17 April. www.guardian.co.uk/politics/2011/apr/17/environment-green-laws-red-tape?INTCMP=SRCH

Su, Y.-P., Hall, C.M. and Ozanne, L. (2011) 'Hospitality industry responses to climate change: a benchmark study of Taiwanese tourist hotels', *Asia Pacific Journal of Tourism Research*, in press.

The Sunday Times (2006) 'It's a sin to fly, says church', 23 July.

Sussman, F. and Freed, J. (2008) *Adapting to Climate Change: A Business Approach*. Arlington, VA: Pew Center on Global Climate Change.

Suthey Holler Associates (2003) *Community-based ATV Tourism Product Model Pilot Project*. Toronto: Suthey Holler Associates.

Svenska Järnvägen (Swedish Railways) (2010) *Carbon Calculator*. www.sj.se/sj/jsp/polopoly.jsp?d=280&l=en&intcmp=13196

Szalai, K. and Ratz, T. (2007) 'Tourist perceptions of uncertainty and risk associated with extreme weather events', paper presented at *3rd International Conference on Tourism Future Trends*, Faculty of Tourism and Hotel Management, October 6 University, Egypt and Kodolányi János University College, Hungary, Sharm el Sheikh, Egypt, 26–29 October 2007.

Tasci, A.D. and Gartner, W.C. (2007) 'Destination image and its functional relationships', *Journal of Travel Research*, 45: 413–425.

Tasci, A.D., Gartner, W.C. and Cavusgil, S.T. (2007) 'Conceptualization and operationalization of destination image', *Journal of Hospitality and Tourism Research*, 31: 194–223.

Tepfenhart, M., Mauser, W. and Siebel, F. (2007) 'The impacts of climate change on ski resorts and tourist traffic', in A. Matzarakis, C. de Freitas and D. Scott (eds) *Developments in Tourism Climatology, Proceedings of International Society of Biometeorology Conference on Climate and Tourism*, Greece, May 2006.

Terrill, G. (2008) 'Climate change: how should the World Heritage Convention respond?', *International Journal of Heritage Studies*, 14(5): 388–404.

Tervo, K. (2008) 'The operational and regional vulnerability of winter tourism to climate variability and change: the case of the Finnish nature-based tourism entrepreneurs', *Scandinavian Journal of Hospitality and Tourism*, 8(4): 317–332.

Tervo, K. (2009) 'Christmas tourism in the light of changing climate', in *Human–Environment Relations in the North*, FiDiPro Seminar, Thule Institute, University of Oulu, Oulu, Finland, 23–24 February 2009.

Tesco (2009) *Our carbon label findings*. Dundee: Tesco. www.tesco.com/assets/greenerliving/content/documents/pdfs/carbon_label_findings.pdf

Thibault, K.M. and Brown, J.H. (2008) 'Impact of an extreme climatic event on community assembly', *Proceedings of the National Academy of Sciences, USA*, 105: 3410–3415.

Thomas, C.D., Williams, S.E., Cameron, A., Green, R.E., Bakkenes, M., Beaumont, L.J., Collingham, Y.C., Erasmus, B.F.N., de Siqueira, M.F., Grainger, A., Hannah, L., Hughes, L., Huntley, B., van Jaarsveld, A.S., Midgley, G.F., Miles, L., Ortega-Huerta, M.A., Peterson, A.T. and Philipps, O.L. (2004) 'Biodiversity conservation: uncertainty in predictions of extinction risk/ Effects of changes in climate and land use/Climate change and extinction risk (reply)', *Nature*, 430: 34.

Thompson, L.G., Brecher, H.H., Mosley-Thompson, E., Hardy, D.R. and Mark, B.G. (2009) 'Glacier loss on Kilimanjaro continues unabated', *Proceedings of the National Academy of Sciences, USA*, 106(47): 19770–19775.

Thornes, J. and McGregor, G. (2003) 'Cultural climatology', in S. Trudgill and A. Roy (eds) *Contemporary Meanings in Physical Geography: From What to Why?* London: Arnold.

Time Magazine (1953) 'Science: invisible blanket', 25 May. www.time.com/time/magazine/article/0,9171,890597,00.html#ixzz1Kgbj83Pc

Toba, N. (2009) 'Potential economic impacts of climate change in the Caribbean Community', in W. Vergara (ed.) *Assessing the Potential Consequences of Climate Destabilization in Latin America*, Latin America and Caribbean Region Sustainable Development Working Paper 32. Washington, DC: World Bank.

Tol, R.S.J. (2007) 'The impact of a carbon tax on international tourism', *Transportation Research Part D*, 12: 129–142.

Torres, R. (2002) 'Toward a better understanding of tourism and agriculture linkages in the Yucatan: tourist food consumption and preferences', *Tourism Geographies*, 4: 282–306.

Trafikverket (2011) 'Ökade utsläpp från vägtrafiken trots rekordartad energieffektivisering av nya bilar' (Growing emissions from road transport despite record-like efficiency gains in new cars), *Trafikverket*, Borlänge, Sweden, 18 February.

Transport & Environment (2010) *Grounded: How ICAO Failed to Tackle Aviation and Climate Change and What Should Happen Now*, published to coincide with the ICAO Triennial Assembly, Montreal, September–October 2010. Brussels: European Federation for Transport and Environment. www.transportenvironment.org/Publications/prep_hand_out/lid/606

TravelMole (2008) 'New UK departure taxes slammed', 25 November. www.travelmole.com/stories/1133123.php

Trawöger, L. (2010) 'Tourism stakeholder perceptions of climate change', paper presented at *Climate Change and Mountain Tourism Workshop*, Innsbruck, Austria, 2–4 June. University of Innsbruck.

Tucker, P. and Gilliland, J. (2007) 'The effect of season and weather on physical activity: a systematic review', *Public Health*, 121: 909–922.

Turley, C.M., Roberts, J.M. and Guinotte, J.M. (2007) 'Corals in deepwater: will the unseen hand of ocean acidification destroy cold-water ecosystems?', *Coral Reefs*, 26: 445–448.

Turner, B.L., Clark, W.C., Kates, R.W., Richards, J.F., Mathews, J.Y. and Meyer, W.B. (eds) (1990) *The Earth as Transformed by Human Action*. Cambridge: Cambridge University Press.

Turnpenny, J., Jones, M. and Lorenzoni, I. (2011) 'Where now for post-normal science? A critical review of its development, definitions, and uses', *Science, Technology, & Human Values*, 36(3): 287–306 (doi: 10.1177/0162243910385789).

Turton, S., Dickson, T., Hadwen, W., Jorgensen, B., Pham, T., Simmons, D., Tremblay, P. and Wilson, R. (2010) 'Developing an approach for tourism climate change assessment: evidence from four contrasting Australian case studies', *Journal of Sustainable Tourism*, 18(3): 429–448.

TVNZ (2010) 'UK departure tax hike could hit NZ tourism', *TVNZ News*, 2 November. tvnz.co.nz/travel-news/uk-departure-tax-hike-could-hit-nz-tourism-3875281

Tyler, N.J.C. (2010) 'Climate, snow, ice, crashes, and declines in populations of reindeer and caribou (*Rangifer tarandus* L.)', *Ecological Monographs*, 80: 197–219.

Uhlmann, B., Goyette, A. and Beniston, M. (2009) 'Sensitivity analysis of snow patterns in Swiss ski resorts to shift in temperture, precipitation and humidity under conditions of climate change', *International Journal of Climatology*, 29: 1048–1055.

UIC (2007) *Database for Energy Efficiency Technologies and related Projects for Railways*. Paris: Union Internationale des Chemins de Fer/International Union of Railways. www.railway-energy.org/tfee/index.php?ID=200

UKCIP (2004) *Costing the Impacts of Climate Change in the UK*. Oxford: UKCIP, Environmental Change Institute, University of Oxford.

UKCIP, Willows, R. and Cornell, R. (eds) (2003) *Climate Adaptation: Risk, Uncertainty and Decision-making*. Oxford: UKCIP, Environmental Change Institute, University of Oxford.

UKERC (2009) *An Assessment of the Evidence for a Near-term Peak in Global Oil Production*. Report by the Technology and Policy Assessment function of the UK Energy Research Centre. London: UK Energy Research Centre.

Umweltbundesamt (2010) *National Inventory Report for the German Greenhouse Gas Inventory 1990–2008*. Dessau-Roßlau, Germany: Federal Environment Agency (Umweltbundesamt). www.umweltbundesamt.de/uba-info-medien/mysql_medien.php?anfrage=Kennummer&Suchwort=3958

Unbehaun, W., Probstl, U. and Haider, W. (2008) 'Trends in winter sport tourism: challenges for the future', *Tourism Review*, 63(1): 36–47.

UNCLOS (1982) 'United Nations Convention on the Law of the Sea', 10 December 1982. www.un.org/depts/los/convention_agreements/texts/unclos/UNCLOS-TOC.htm

UNDP (2007) *Human Development Report 2007/2008: Fighting Climate Change. Human Solidarity in a Divided World*. New York: United Nations Development Programme. hdr.undp.org/en/reports/global/hdr2007-2008

UNDP (2010) *Human Development Report 2010. The Real Wealth of Nations: Pathways to Human Development*. New York: United Nations Development Programme.

UNECLAC (2010) *Regional Report on the Impact of Climate Change on the Tourism Sector*. Report for Subregional Headquarters for the Caribbean, UNECLAC. Santiago de Chile: United Nations Economic Commission for Latin America and the Caribbean. www.eclac.cl/publicaciones/xml/4/39774/LCARL.263.pdf

UNEP (2005a) *Marketing Sustainable Tourism Products*. Paris: United Nations Environment Programme and Regione Toscana.

UNEP (2005b) *Integrating Sustainability into Business: An Implementation Guide for Responsible Tourism Coordinators*. Paris: United Nations Environment Programme.

UNEP (2009a) *Towards Sustainable Production and Use of Resources: Assessing Biofuels*. Paris: United Nations Environment Programme. www.unep.fr/shared/publications/pdf/WEBx0149xPA-AssessingBiofuelsSummary.pdf

UNEP (2009b) *Climate Change Science Compendium 2009*. Paris: United Nations Environment Programme.

UNEP (2009c) *Sustainable Coastal Tourism. An Integrated Planning and Management Approach*. Paris: United Nations Environment Programme.

UNEP (2011a) *Global Partnership for Sustainable Tourism*. Paris: United Nations Environment Programme. www.uneptie.org/scp/tourism/activities/partnership/index.htm

UNEP (2011b) *Advancing Adaptation through Climate Information Services*. Paris: United Nations Environment Programme, Finance Initiative Climate Change Working Group. www.unepfi.org/fileadmin/documents/advancing_adaptation.pdf

UNEP and UNWTO (2005) *Making Tourism More Sustainable: A Guide for Policy Makers*. Paris: United Nations Environment Programme.

UNEP, University of Oxford, UNWTO and WMO (2008) *Climate Change Adaptation and Mitigation in the Tourism Sector: Frameworks, Tools and Practice*, United Nations Environment Programme, Oxford University, United Nations World Tourism Organization and World Meteorological Organization. Paris: UNEP.

UNESCO (2002) *Mountain Summit Opens in Kyrgyzstan*. Paris: United Nations Educational, Scientific and Cultural Organization. www.unesco.org/bpi/eng/unescopress/2002/02-88e.shtml

UNESCO (2006) *Climate Change and World Heritage*. http://whc.unesco.org/en/climatechange

UNESCO (2007) *The Impacts of Climate Change on World Heritage Properties*. Paris: United Nations Educational, Scientific and Cultural Organization. whc.unesco.org/en/climatechange

UNFCCC (1999) Compendium of Decision Tools to Evaluate Strategies for Adaptation to Climate Change: Final Report. Bonn: United Nations Framework Convention on Climate Change. www.aiaccproject.org/resources/ele_lib_docs/adaptation_decision_tools.pdf

UNFCCC (2007a) *Investment and Financial Flows To Address Climate Change – Background Paper*. Bonn, Germany: United Nations Framework Convention on Climate Change. unfccc.int/files/cooperation_and_support/financial_mechanism/application/pdf/background_paper.pdf

UNFCCC (2007b) *Climate Change: Impacts, Vulnerabilities and Adaptation in Developing Countries*. Final Report of the AIACC [Assessments of Impacts and Adaptations to Climate Change] Project. Bonn, Germany: United Nations Framework Convention on Climate Change. http://unfccc.int/files/essential_background/background_publications_htmlpdf/application/txt/pub_07_impacts.pdf

UNFCCC (2011) *Submissions from Parties in 2011*. Bonn, Germany: United Nations Framework Convention on Climate Change. unfccc.int/documentation/submissions_from_parties_in_2011/items/5900.php

United Nations (1982) *United Nations Convention on the Law of the Sea of 10 December 1982*. New York: United Nations. www.un.org/Depts/los/convention_agreements/texts/unclos/closindx.htm

United Nations (2011) *Statistics: Water Use*. New York: United Nations. www.unwater.org/statistics_use.html

United Nations and ISDR (2004) *Living with Risk – A Global Review of Disaster Reduction Initiatives*. New York/Geneva: United Nations and Inter-Agency Secretariat of the International Strategy for Disaster Reduction.

United Nations Conference on Trade and Development (UNCTAD) (2008) *UNCTAD Handbook of Statistics 2008*. New York and Geneva: UN.

United Nations Statistics Division (2010) 'Carbon dioxide emissions (CO_2), thousand metric tons of CO_2 (CDIAC)', *Millennium Development Goals Indicators*. mdgs.un.org/unsd/mdg/SeriesDetail. aspx?srid=749

Unwin, T. (1992) *The Place of Geography*. Harlow: Longman.

UNWTO (1994) *Recommendations on Tourism Statistics*. Madrid: World Tourism Organization.

UNWTO (2001) *Tourism 2020 Vision. Volume 7. Global Forecasts and Profiles of Market Segments*. Madrid: United Nations World Tourism Organization.

UNWTO (2003) *Climate Change and Tourism, Proceedings of the First International Conference on Climate Change and Tourism*, Djerba 9–11 April. Madrid: World Tourism Organization.

UNWTO (2007a) *From Davos to Bali – A Tourism Contribution to the Challenge of Climate Change*, policy document. Madrid: United Nations World Tourism Organization.

UNWTO (2007b) 'Tourism will contribute to solutions for global climate change and poverty challenges', press release, 8 March. Berlin/Madrid: Press and Communications Department, United Nations World Tourism Organization.

UNWTO (2008) *Emerging Tourism Markets – The Coming Economic Boom*. Madrid: United Nations World Tourism Organization.

UNWTO (2010a) *Historical Perspective of World Tourism*. Madrid: United Nations World Tourism Organization. www.unwto.org/facts/eng/historical.htm

UNWTO (2010b) *UNWTO World Tourism Barometer*. Madrid: United Nations World Tourism Organization. unwto.org/facts/eng/pdf/barometer/UNWTO_Barom09_2_en_excerpt.pdf

UNWTO (2010c) *Statement Regarding Mitigation of Greenhouse Gas Emissions from Air Passenger Transport*. Madrid: United Nations World Tourism Organization. www.unwto.org/climate/support/en/support.php

UNWTO (2011) *International Tourism 2010: Multi-speed Recovery*, press release, 17 January. Madrid: United Nations World Tourism Organization. 85.62.13.114/media/news/en/press_det. php?id=7331&idioma=E

UNWTO, UNEP and WMO (2008) *Climate Change and Tourism: Responding to Global Challenges*. United Nations World Tourism Organization, United Nations Environment Programme and World Meteorological Organization. Madrid: UNWTO. www.unwto.org/climate/index.php; www.uneptie.org/shared/publications/pdf/WEBx0142xPA-ClimateChangeandTourismGlobal Challenges.pdf

Upham, P., Thornley, P., Tomei, J. and Boucher, P. (2009a) 'Substitutable biodiesel feedstocks for the UK: a review of sustainability issues with reference to the UK RTFO', *Journal of Cleaner Production*, 17: 537–545.

Upham, P., Tomei, J. and Boucher, P. (2009b) 'Biofuels, aviation and sustainability', in S. Gössling and P. Upham (eds) *Climate Change and Aviation: Issues, Challenges and Solutions*. London: Earthscan.

US AID (2007) *Dominica Sustainable Tourism Policy and Marketing Strategy*, prepared for Discover Dominica Authority by Chenmonics International. Washington, DC: US Agency for International Development.

US AID (2010) *The Global Water Crisis*. www.usaid.gov/our_work/environment/water/water_ crisis.html

USA Today (2005) 'Florida's hurricanes boost Arizona tourism'. www.usatoday.com/travel/ news/2004-10-11-az-hurrican_x.htm

US CM (2008) *U.S. Conference of Mayors Climate Protection Agreement*. United States Conference of Mayors. usmayors.org/climateprotection/agreement.htm

US Department of the Interior (2003) 'Potential water supply crises by 2025', in *Water 2025: Preventing Crises and Conflict in the West*. U.S. Department of the Interior, Bureau of Reclamation. www.doi.gov/water2025/index.html (accessed 13 April 2005).

US Department of the Interior (2010) *Lake Mead at Hoover Dam, Elevation*. www.usbr.gov/lc/region/g4000/hourly/mead-elv.html

US EPA (1995) *Ecological Impacts from Climate Change: An Economic Analysis of Freshwater Recreational Fishing*. Washington, DC: United States Environmental Protection Agency.

US EPA (1999) *Global Climate Change: What Does it Mean for South Florida and the Florida Keys?*, report on the EPA public consultations in coastal cities, 24–28 May. Washington, DC: United States Environmental Protection Agency.

US EPA (2010) *Designation of North American Emission Control Area to Reduce Emissions from Ships: Regulatory Announcement*, EPA-420-F-10-015. Washington, DC: United States Environmental Protection Agency.

US EPA (2011) *Heat Island Effect*. Washington, DC: United States Environmental Protection Agency. www.epa.gov/heatisld

US Geological Survey (2003) *Glacier Retreat in Glacier National Park, Montana*. Washington, DC: US Department of the Interior.

US Government Accountability Office (2010) *Climate Change: A Coordinated Strategy Could Focus Federal Geoengineering Research and Inform Governance Efforts*. Washington, DC: US Government Accountability Office. www.gao.gov/products/GAO-10-903

Uyarra, M.C., Côté, I.M., Gill, J.A., Tinch, R.R.T., Viner, D. and Watkinson, A.R. (2005) 'Island-specific preferences of tourists for environmental features: implications of climate change for tourism-dependent states', *Environmental Conservation*, 32(1): 11–19.

Valls, J.F. and Sarda, R. (2009) 'Tourism expert perceptions for evaluating climate change impacts on the Euro–Mediterranean tourism industry', *Tourism Review*, 64(2): 41–51.

Van Lier, H.N. (1973) *Determination of Planning Capacity and Layout Criteria of Outdoor Recreation Projects*, Agriculture Research Reports 705. Wageningen, the Netherlands: Centre for Agricultural Publishing and Documentation.

van Vuuren, D., Lowe, J., Stehfest, E., Gohar, L., Hof, A., Hope, C., Warren, R., Meinshausen, M. and Plattner, G.-K. (2011). 'How well do integrated assessment models simulate climate change?', *Climatic Change*, 104(2): 255–285.

van Vuuren, D.P. and Kram, T. (2011) 'Comment on using Integrated Assessment Models for Assessing R&D Portfolios', *Energy Economics*, 33: 644–647 (doi: 10.1016/j.eneco.2010.11.016).

van Vuuren, D.P., Lowe, J., Stehfest, E., Gohar, L., Hof, A., Hope, C., Warren, R., Meinshausen, M. and Plattner, G.-K. (2009) 'How well do Integrated Assessment Models simulate climate change?', *Climatic Change*, 10 December. www.springerlink.com/content/l841558141481552

Vanham, D., Toffol, S.D., Fleischhacker, E. and Rauch, W. (2009) 'Water demand for snowmaking under climate change conditions in an alpine environment', in *Proceedings of Facing Climate Change and the Global Economic Crisis: Challenges for the Future of Tourism*, 20–21 November 2009. Bolzano, Italy: EURAC – Institute for Regional Development and Location Management. www.climalptour.eu/content/sites/default/files/Wasserbedarf_fuer_Beschneiung%5B1%5D_0.pdf

Varum, C., Melo, C., Alvarenga, A. and de Carvalho, P. (2011) 'Scenarios and possible futures for hospitality and tourism', *Foresight*, 13(1): 19–35.

Vass, B., Trevett, C. and Richards, M. (2008) 'Key battles UK flight tax in meeting with Brown', *New Zealand Herald*, 26 November. www.nzherald.co.nz/nz/news/article.cfm?c_id=1&objectid=10545097

Vera-Morales, M. and Schäfer, A. (2009) *Final Report: Fuel-Cycle Assessment of Alternative Aviation Fuels*. Cambridge: University of Cambridge, Institute for Aviation and the Environment.

Verhallen, T.M.M. and van Raaij, W.F. (1986) 'How consumers trade off behavioural costs and benefits', *European Journal of Marketing*, 20(3/4): 19–34.

Vermeer, M. and Rahmstorf, S. (2009) 'Global sea level linked to global temperature', *Proceedings of the National Academy of Sciences, USA*, 106(51): 21527–21532.

Vidal, J. (2010) 'EU biofuels significantly harming food production in developing countries', *The Guardian*, 15 February. www.guardian.co.uk/environment/2010/feb/15/biofuels-food-production-developing-countries

Vidal, J. (2011) 'Maritime countries agree first ever shipping emissions regulation', *The Guardian*, 18 July. www.guardian.co.uk/environment/2011/jul/18/maritime-countries-shipping-emissions-regulation

Vilà, M. and Pujadas, J. (2001a) 'Socio-economic parameters influencing plant invasions in Europe and North Africa', in J.A. McNeely (ed.) *The Great Reshuffling: Human Dimensions of Invasive Alien Species*. Gland, Switzerland: International Union for Conservation of Nature, Biodiversity Policy Coordination Division.

Vilà, M. and Pujadas, J. (2001b) 'Land-use and socio-economic correlates of plant invasions in European and North African countries', *Biological Conservation*, 100: 397–401.

Virgin Atlantic (2008) *Biofuel demonstration*. www.virgin-atlantic.com:80/en/gb/allaboutus/environment/biofuel.jsp

Vivekananda, D.S.J. (2007) *A Climate of Conflict: The Links between Climate Change, Peace and War*. London: International Alert. www.international-alert.org/sites/default/files/publications/A_climate_of_conflict.pdf

Vörösmarty, C.J., Green, P., Salisbury, J. and Lammers, R.B. (2000) 'Global water resources: vulnerability from climate change and population growth', *Science*, 289: 284–288.

Wainwright, J. (2010) 'Climate change, capitalism, and the challenge of transdisciplinarity', *Annals of the Association of American Geographers*, 100: 983–991.

Wall, G. (1992) 'Tourism alternatives in an era of global climate change', in V. Smith and W. Eadington (eds) *Tourism Alternatives*, Philadelphia, PA: University of Pennsylvania Press.

Wall, G. (1998) 'Climate change, tourism and the IPCC', *Tourism Recreation Review*, 23(2): 65–68.

Wall, G. and Badke, C. (1994) 'Tourism and climate change: an international perspective', *Journal of Sustainable Tourism*, 2: 193–203.

Wall, G. and Mathieson, A. (2005) *Tourism: Change, Impacts, Opportunities*. Harlow: Pearson.

Wall, G., Harrison, R., Kinnaird, V., McBoyle, G. and Quinlan, C. (1986) 'The implications of climatic change for camping in Ontario', *Recreation Research Review*, 13(1): 50–60.

Walsh, J.E. (2009) 'A comparison of Arctic and Antarctic climate change, present and future', *Antarctic Science*, 21(3): 179–188.

Walther, G.R., Beißner, S. and Potts, R. (2005) 'Climate change and high mountain vegetation shifts', in G. Broll and B. Keplin (eds) *Mountain Ecosystems*. New York: Springer Publishing, pp. 77–96.

Ward, E.R. (2003) *Border Oasis: Water and the Political Ecology of the Colorado River Delta, 1940–1975*. Phoenix, AZ: University of Arizona Press.

Ward, R. and Ranger, N. (2010) 'Trends in economic and insured losses from weather-related events: a new analysis', *Insurance Industry Brief: The Munich Re Programme of the Centre for Climate Change Economics and Policy*, November. www.cccep.ac.uk/Publications/insuranceBriefs/economic-trends-insured-losses.pdf

Wardell, R. (2010) 'Influences on transport policy makers and their attitudes towards peak oil', Master's thesis, Engineering in Transportation, University of Canterbury, New Zealand.

WBCSD (undated) *10 Key Messages*. Geneva: World Business Council for Sustainable Development. www.wbcsd.org/newsroom/key-messages.aspx

WCED (1987) *Our Common Future* (the Brundtland Report). Oxford: Oxford University Press/World Commission on Environment and Development.

Weart, S. (2003) *The Discovery of Global Warming*. Cambridge, MA: Harvard University Press.

Weather Bill, Inc. (2008) *Sunshine Guaranteed or Your Money Back: Breakthrough Weather Refund Promotions*, WeatherBill White Paper. San Francisco, CA Weather Bill, Inc. (now The Climate Corporation).

Weaver, A.J., Clark, P.U., Brook, E., Cook, E.R., Delworth, T.L. and Steffen, K. (2008) *Abrupt Climate Change – A Report by the US Climate Change Science Program*. Reston, VA: US Geological Survey.

Weaver, D. (2011) 'Can sustainable tourism survive climate change?', *Journal of Sustainable Tourism*, 19(1): 5–15.

Webb, L., Whetton, P.H. and Barlow, E.W.R. (2008) 'Climate change and winegrape quality in Australia', *Climate Research*, 36: 99–111.

Weber, C.L. and Matthews, H.S. (2008) 'Food-miles and the relative climate impacts of food choices in the United States', *Environment, Science, and Technology*, 42: 3508–3513.

WEF (2009a) *Towards a Low Carbon Travel & Tourism Sector*. Davos: World Economic Forum.

WEF (2009b) *The Travel & Tourism Competitiveness Report 2009: Managing in a Time of Turbulence*. Davos: World Economic Forum.

Weiss, M.A. (2007) *Sustainable Urban Development in the US*, report prepared for the Government of Sweden's Mistra Foundation for Strategic Environmental Research by Dr Marc A. Weiss, Chairman and CEO, Global Urban Development, 21 December.

Weitzman, M. (2009) 'Some basic economics of extreme climate change', in J.-P. Touffut (ed.) *Changing Climate, Changing Economy*. Edward Elgar. www.economics.harvard.edu/faculty/ weitzman/files/Cournot%2528Weitzman%2529.pdf

Weitzman, M.L. (2007) 'A review of the Stern Review on the Economics of Climate Change', *Journal of Economic Literature*, 45(3): 703–724 (doi: 10.1257/jel.45.3.703).

Welch, D. (2005) 'What should protected areas managers do in the face of climate change?', *George Wright Forum*, 22(1): 75–93.

Wells, V.K., Ponting, C.A. and Peattie, K. (2011) 'Behaviour and climate change: consumer perceptions of responsibility', *Journal of Marketing Management*, 27(7–8): 808–833.

Westmacott, S., Cesar, H., Pet-Soede, L. (2000) *Socioeconomic Assessment of the Impacts of the 1998 Coral Reef Bleaching in the Indian Ocean*. Resource Analysis and Institute for Environmental Science (IVM) Report to the World Bank, African Environmental Division for the CORDIO programme. www.oceandocs.net/bitstream/1834/481/1/CORDIO17.pdf

Weyl, H. (2009) *Philosophy of Mathematics and Natural Science*, revised edn. Princeton, NJ: Princeton University Press.

WGMS (2008) 'Fluctuations of glaciers 2000–2005', in W. Haeberli, M. Zemp, A. Kääb, F. Paul and M. Hoelzle (eds), *World Glacier Monitoring Service*, ICSU(FAGS)/IUGG(IACS)/UNEP/ UNESCO/WMO. Zurich, Switzerland: World Glacier Monitoring Service. www.geo.uzh.ch/ microsite/wgms/fog.html

White, M.A., Diffenbaugh, N.S., Jones, G.V., Pal, J.S. and Giorgi, F. (2006) 'Extreme heat reduces and shifts United States premium wine production in the 21st century', *Proceedings of the National Academy of Sciences, USA*, 103(30): 11217–11222.

Whittlesea, E. and Amelung, B. (2010) *Cost-a South West: What Could Tomorrow's Weather and Climate Look Like for Tourism in the South West of England?*, National Case Study. Exeter, UK: South West Tourism. ukclimateprojections.defra.gov.uk/images/stories/Case_studies/CS_ SWTourism.pdf

Wilbanks, T. and Kates, R. (1999) 'Global change in local places: how scale matters', *Climatic Change*, 43: 601–628.

Wilbanks, T.J., Romero Lankao, P., Bao, M., Berkhout, F., Cairncross, S., Ceron, J.-P., Kapshe, M., Muir-Wood, R. and Zapata-Marti, R. (2007) 'Industry, settlement and society', in M.L. Parry, O.F. Canziani, J.P. Palutikof, P.J. van der Linden and C.E. Hanson (eds) *Climate Change 2007: Impacts, Adaptation and Vulnerability.* Contribution of Working Group II to the Fourth Assessment Report of the Intergovernmental Panel on Climate Change. Cambridge: Cambridge University Press.

Wilkinson, C. (1998) 'The 1997–1998 mass bleaching event around the world', unpublished document. www.oceandocs.net/bitstream/1834/545/1/BleachWilkin1998.pdf

Williams, A.M. and Hall, C.M. (2002) 'Tourism, migration, circulation and mobility: the contingencies of time and place', in C.M. Hall and A.M. William (eds) *Tourism and Migration: New Relationships Between Production and Consumption.* Dordrecht, the Netherlands: Kluwer.

Williams, P., Dousa, K. and Hunt, J. (1997) 'The influence of weather context on winter resort valuations by visitors', *Journal of Travel Research*, 36(1): 29–36.

Williams, S.E. (2009) 'Climate change in the rainforests of the North Queensland Wet Tropics', in W. Steffen, A. Burbidge, L. Hughes, R. Kitching, D. Lindenmayer, W. Musgrave, M. Stafford Smith and P. Werner (eds) *Australia's Biodiversity and Climate Change: A Strategic Assessment of the Vulnerability of Australia's Biodiversity to Climate Change*, report to the Natural Resource Management Ministerial Council, commissioned by the Australian Government. Canberra: CSIRO Publishing.

Williams, S.E., Bolitho, E.E. and Fox, S. (2003) 'Climate change in Australian tropical rainforests: an impending environmental catastrophe', *Proceedings of the Royal Society of London: Biological Sciences*, 270: 1887–1892.

Wilton, D. and Wirjanto, T. (1998) *An Analysis of the Seasonal Variation in the National Tourism Indicators.* Ottawa, Canada: Canadian Tourism Commission.

Winston, M., Lewis-Bynoe, D. and Lewis-Bynoe, H. (2010) 'Climate change and tourism features in the Caribbean', unpublished paper. http://mpra.ub.uni-muenchen.de/21470/

Wirth, K. (2010) 'The weather preferences of German tourists', thesis, Department of Geography, Ludwig-Maximilians-Universität, Munich.

WMO (1979) *Declaration of the World Climate Conference.* Geneva: World Meteorological Organization. www.dgvn.de/fileadmin/user_upload/DOKUMENTE/WCC-3/Declaration_WCC1. pdf

WMO (1986) *Report of the International Conference on the Assessment of the Role of Carbon Dioxide and of Other Greenhouse Gases in Climate Variations and Associated Impacts*, Villach, Austria, 9–15 October 1985, WMO No. 661. Geneva: World Meteorological Organization. www. wmo.int/pages/themes/climate/international_background.php

WMO (2009) *Climate Knowledge For Action: A Global Framework For Climate Services – Empowering The Most Vulnerable*, WMO No. 1065. Geneva: World Meteorological Organization. www.wmo.int/pages/prog/lsp/congress/documents/1065_en.pdf

WMO (2010) *Weather Extremes in a Changing Climate: Hindsight on Foresight.* WMO No. 1075. Geneva: World Meteorological Organization.

Wolf, A., Callaghan, T.V. and Larson, K. (2008) 'Future changes in vegetation and ecosystem function of the Barents Region', *Climatic Change*, 87: 51–73.

Wolfsegger, C., Gössling, S. and Scott, D. (2008) 'Climate change risk appraisal in the Austrian ski industry', *Tourism Review International*, 12(1): 13–25.

Woodroffe, C.D., Thom, B.G., Chappell, J., Wallensky, E., Grindrod, J. and Head, J. (1987) 'Relative sea level in the South Alligator River Region, North Australia, during the Holocene', *Search*, 18: 198–200.

World Bank (2006) *Clean Energy and Development: Towards an Investment Framework.* Washington, DC: International Bank for Reconstruction and Development/World Bank.

World Bank (2008a) *Development and Climate Change: A Strategic Framework for the World Bank Group: Technical Report.* Washington, DC: International Bank for Reconstruction and Development/World Bank.

World Bank (2008b) *Implications of Climate Change for Armed Conflict.* Washington, DC: International Bank for Reconstruction and Development/World Bank.

World Bank (2010) *World Development Report 2010: Development and Climate Change*, advance press edition. Washington, DC: International Bank for Reconstruction and Development/World Bank.

World Bank (2011a) *Climate Change: Adaptation Guidance Notes – Key Words and Definitions.* climatechange.worldbank.org/climatechange/content/adaptation-guidance-notes-key-words-and-definitions

World Bank (2011b) *Climate Impacts on Energy Systems. Key Issues for Energy Sector Adaptation.* Washington, DC: International Bank for Reconstruction and Development/World Bank.

World Business Council for Sustainable Development (2008) *Adaptation: An Issue Brief for Business.* Washington, DC: World Business Council for Sustainable Development.

Worldwatch Institute (2006) *Vital Signs 2006–2007.* Washington, DC: Worldwatch Institute.

WTTC (2009) *Leading the Challenge.* London: World Travel and Tourism Council. www.wttc.org/bin/pdf/original_pdf_file/climate_change_final.pdf

WTTC (2010) *Climate Change – A Joint Approach to Addressing the Challenge.* London: World Travel and Tourism Council. www.wttc.org/bin/pdf/original_pdf_file/climate_change_-_a_joint_appro.pdf

Wynne, B. (1997) 'Methodology and institutions: value as seen from the risk field', in J. Foster (ed.) *Valuing Nature: Economics, Ethics and Environment.* London: Routledge.

Yates, C. (2009) 'Vulnerability of biodiversity in Western Australia to climate change', in W. Steffen, A. Burbidge, L. Hughes, R. Kitching, D. Lindenmayer, W. Musgrave, M. Stafford Smith and P. Werner, *Australia's Biodiversity and Climate Change: A Strategic Assessment of the Vulnerability of Australia's Biodiversity to Climate Change*, report to the Natural Resource Management Ministerial Council, commissioned by the Australian Government. Canberra: CSIRO Publishing.

Yates, C.J., Ladd, P.G., Coates, D.J. and McArthur, S. (2007) 'Hierarchies of cause: understanding rarity in an endemic shrub *Verticordia staminosa* (Myrtaceae) with a highly restricted distribution', *Australian Journal of Botany*, 55: 194–205.

Yeoman, I. and McMahon-Beattie, U. (2005) 'Developing a scenario planning process using a blank piece of paper', *Tourism and Hospitality Research*, 5(3): 273–285.

Yohe, G. and Tol, R.S.J. (2001) 'Indicators for social and economic coping capacity: moving toward a working definition of adaptive capacity', *Global Environmental Change*, 12: 25–40.

Yohe, G.W., Lasco, R.D., Ahmad, Q.K., Arnell, N.W., Cohen, S.J., Hope, C., Janetos, A.C. and Perez, R.T. (2007) 'Perspectives on climate change and sustainability', in M.L. Parry, O.F. Canziani, J.P. Palutikof, P.J. van der Linden and C.E. Hanson (eds) *Climate Change 2007: Impacts, Adaptation and Vulnerability. Contribution of Working Group II to the Fourth Assessment Report of the Intergovernmental Panel on Climate Change.* Cambridge: Cambridge University Press, pp. 811–841.

Yuan, L., Lu, A., Ning, B. and He, Y. (2006) 'Impacts of Yulong Mountain Glacier on tourism in Lijiang', *Journal of Mountain Science*, 3(1): 71–80.

Zahavi, Y. and Talvitie, A. (1980) 'Regularities in travel time and money expenditures', *Transportation Research Record*, 750: 13–19.

Zaninovic, K. and Matzarakis, A. (2009) 'The biometeorological leaflet as a means of conveying climatological information to tourists and the tourism industry', *International Journal of Biometeorology*, 53: 369–374.

Zemp, M., Hoelzle, M. and Haeberli, W. (2009) 'Six decades of glacier mass balance observations – a review of the worldwide monitoring network', *Annals of Glaciology*, 50: 101–111.

Zeppel, H. (2011) 'Climate change and tourism in the Great Barrier Reef Marine Park', *Current Issues in Tourism* (doi: 10.1080/13683500.2011.556247).

Zhang, D.D., Brecke, P., Lee, H.F., He, Y.-Q. and Zhang, J. (2007) 'Global climate change, war, and population decline in recent human history', *PNAS*, 104(49): 19214–19219.

Zimmerman, G., O'Brady, C. and Hurlbutt, B. (2006) 'Regional challenges of future climate change: endless summer or business as usual?', in *The 2006 Colorado College State of the Rockies Report Card*. Colorado Springs: Colorado College.

Zimmermann, E.W. (1951) *World Resources and Industries*. New York: Harper and Row.

Index